AF251575

EYE SYMPTOMS IN BRAIN TUMORS

Eye symptoms in brain tumors

SECOND EDITION

ALFRED HUBER, M.D.

*Professor of Ophthalmology and Consultant Ophthalmologist,
Neurosurgical Clinic, University of Zurich, Switzerland*

Second edition edited and newly written material translated by

FREDERICK C. BLODI, M.D.

*Professor and Head, Department of Ophthalmology, University of
Iowa College of Medicine, Iowa City, Iowa*

Foreword to the first German edition by

PROF. H. KRAYENBÜHL

Director of the Neurosurgical Clinic, University of Zurich, Switzerland

English translation of the first edition by

STEFAN VAN WIEN, M.D.

*Formerly Associate, Department of Ophthalmology, Northwestern
University Medical School, Chicago, Illinois*

Foreword to the first English translation by

DERRICK VAIL, B.A., M.D., D.Oph. (Oxon.), F.A.C.S., F.R.C.S. (Hon.)

*Formerly Professor and Director, Department of Ophthalmology,
Northwestern University Medical School, Chicago, Illinois*

With 233 illustrations

THE C. V. MOSBY COMPANY

SAINT LOUIS 1971

German edition by Dr. Alfred Huber entitled *Augensymptome bei Hirntumoren*—vol. VII of a series entitled *Innere Medizin und ihre Grenzgebiete* edited by Prof. Dr. P. H. Rossier and Prof. Dr. O. Spühler and published by Medizinischer Verlag Hans Huber, Bern and Stuttgart

DEDICATED TO THE SICK

*Whose suffering has contributed to the discovery of
new methods for better treatment of others*

Foreword TO FIRST GERMAN EDITION

Since 1918, the year Dandy introduced ventriculography as a diagnostic method of examination, the use of contrast media for roentgenologic studies in the search of brain tumors has assumed an unexpected importance. An initial opposition to pneumoencephalography and angiography used as aids in the diagnosis of gross organic involvement of the brain has given place to their acceptance to the extent that frequently the old and proved clinical neurologic methods of examination are not only neglected but their value as part of our clinical armamentarium is even questioned. The experienced clinician is aware that it would mean carrying coals to Newcastle if one would place an overemphasis or an underestimation on the importance of the various factors which produce a neurologic or neuro-ophthalmologic syndrome. Only if all these factors are taken into consideration will it be possible to gain a satisfactory and unmistakable diagnostic understanding of the morbid process in a brief period of time. Only thus will it be possible to decide what other ancillary methods should be employed for a more detailed differential diagnosis.

From this point of view, Dr. Huber's work is welcome. It had its origin in the author's need to follow an explicit method for his neuro-ophthalmologic technic of examination. Based on his neuro-ophthalmologic findings, and thanks to his profound training in neurology, his observation of individual signs merely serves as a foundation to present a fascinating and comprehensive picture of all essential factors involved. Just the same, certain details which are of equal interest to the ophthalmologist and neurosurgeon (as, for instance, the differential diagnosis of the choked disc or the significance of the blood pressure in the retinal vessels) are treated with minute consideration. This volume is particularly stimulating and important because our neurosurgical diagnostics are enriched by the wealth of experience of an ophthalmologist.

H. Krayenbühl
Zurich, Switzerland

The author of this book, Alfred Huber, a neurologist and ophthalmologist, is a distinguished younger member of the remarkable group of Swiss ophthalmologists whose names are known throughout the world for their contributions (to name but a few of these in our time, Vogt, Gonin, Amsler, Brückner, Streiff, Goldmann, and Franceschetti).

Prof. H. Krayenbühl, Director of the Neurosurgical Clinic at the University of Zurich, is as well known in his field of activity as the aforementioned are in ophthalmology. It is in this renowned clinic that Dr. Huber has obtained the necessary experience to write this extraordinarily good book on *Eye Symptoms in Brain Tumors.*

There are a number of good books on brain tumors. The majority of these have been written by neurologists and neurosurgeons. The others have been written by the very few ophthalmologists who have had the required neurologic interest, knowledge, training, and experience to interpret properly the multiple signs and symptoms that are so frequently encountered.

Because early symptoms of a patient with a brain tumor are primarily ocular, he is almost always first seen by the ophthalmologist. As a result of the modern training of ophthalmologists in this country, more and more of them are becoming excellent neurologists, competent to cope with the involved diagnostic problems. However, there is still much room for improvement, it seems to me, in this regard. It is for this reason that Dr. Huber's beautifully written and adequately illustrated work should receive a warm welcome not only by ophthalmologists but also by neurologists and neurosurgeons; in fact, by physicians in every branch of medicine.

Stefan Van Wien, of the Northwestern University Medical School, has performed an important service for us by translating Dr. Huber's text carefully, scientifically, and in a lucid style. He shows an unusual talent for clearly giving us what the author means to say, at the same time avoiding the involved and the cumbersome that tend to creep into translated manuscripts.

I consider it a great honor to have been asked by Dr. Huber to write this brief foreword and to help in the launching of his work. I am grateful to him for this, but particularly for giving to us all the fruits of his many years of intelligent and diligent studies on the eye symptoms of brain tumors.

Derrick Vail
Chicago, Illinois

Preface

Since the appearance of the first edition of this book, the disciplines of ophthalmology, neurosurgery, and especially neuro-ophthalmology have made great and unexpected progress. Thus there was no doubt that the book would have to be carefully and extensively revised in the second edition.

Several portions of the chapter on methods of neuro-ophthalmologic examination have been completely rewritten, especially the section on nystagmus. The introduction of static perimetry to the study of visual fields has added new information, and electromyography has contributed many new concepts and diagnostic possibilities to the subject of extraocular muscle palsies. A review of the new techniques of dynamometry and dynamography has been added in the section on the ocular fundus. One of the most important additions, however, is the presentation of the new method of fluorescein angiography of the fundus and the demonstration of angiographic patterns of the optic disc under normal and pathologic conditions.

In the chapter on general symptoms of increased intracranial pressure, fluorescein angiographic pictures have been added, especially in the discussion of papilledema and its differential diagnosis. Special care was taken to renew some photographs of early and chronic papilledemas and also to add some new illustrations of edema of the disc resulting from conditions other than increased intracranial pressure, such as ischemic papillitis, leukemia, spinal cord tumor, or metastatic tumor within the optic nerve.

In the chapter on local symptoms of brain tumors, new schematic drawings have been introduced, especially on sellar and suprasellar tumors. The chapter on the relationship between the type of tumor and ocular symptoms has been enriched by some new statistics. The chapter on subdural hematomas, brain abscesses, and aneurysms has been expanded to include information on cerebral pseudotumors, with corresponding illustrations of papilledemas not caused by intracranial neoplasms. In the section on intracranial aneurysms, some new concepts and newer classifications have been adopted.

The literature has been brought up to date to include the most recent textbooks, monographs, and individual articles. As in the first edition, only a selected choice of the vast literature can be given. The third edition of *Clinical Neuro-Ophthalmology* by Walsh and Hoyt is referred to frequently for information on subjects not discussed in this book.

As a consultant at the Neurosurgical Clinic of the University of Zurich, I would like to thank the chairman, Professor H. Krayenbühl, for his valuable interest in this book and especially for having given me the opportunity to use all the clinical data necessary for preparing the new edition. Special thanks also to Professor Yasargil for his good advice and especially for allowing me to publish again his outstanding human brain section, which faces page 1. I am also indebted to Professor G. Weber, as well as to the many former and present residents at the Neurosurgical Clinic, for their valuable suggestions and advice. Special credit is also due Mr. H. P. Weber, artist at the Neurosurgical Clinic, for surgical drawings and numerous new sketches.

The University Eye Clinic and its chairman, Professor R. Witmer, also made my task easier. I especially wish to thank Professor Witmer for giving me the opportunity to publish his wonderful fluorescein angiographic photographs of the normal and pathologic disc. My special thanks are also due Mr. A. Würth, the photographer of the University Eye Clinic at Zurich, who prepared the excellent fundus photographs with the able assistance of Mrs. Kleinhäny. The visual field sketches, as well as a number of other drawings, are the creation of Mr. Glitsch, artist of the Eye Clinic at Zurich; my special thanks for the care he took in this time-consuming labor. My secretary, Miss A. Brunner, typed the entire manuscript and I am greatly indebted to her for her untiring efforts.

Regrettably, Dr. Stefan Van Wien, who wrote the superb English translation, is deceased, but Dr. Frederick C. Blodi, of the University of Iowa, has kindly taken over the important task of translating the newly written chapters and revising the old text. My sincere appreciation to Dr. Blodi who by his time-consuming effort has made a great contribution to this second edition.

Alfred Huber
Zurich, Switzerland
February, 1971

Contents

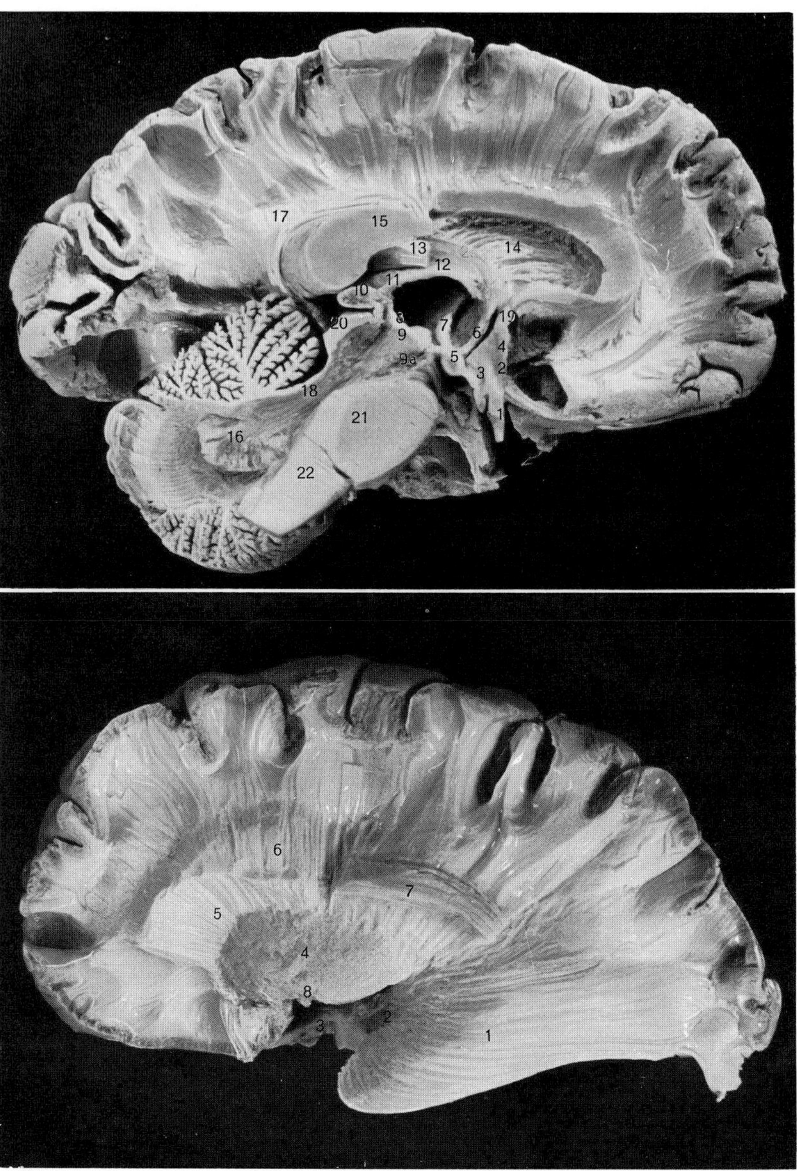

Mesial longitudinal section through the human brain demonstrating the *fiber* systems of the visual radiation, external and internal capsules. *Top:* 1, Optic chiasm; 2, chiasmal recess; 3, infundibular recess; 4, lamina terminalis; 5, mamillary body; 6, anterior column of fornix; 7, mamillothalamic tract; 8, habenular-intercrural tract; 9, red nucleus; 9a, substantia nigra; 10, pineal body; 11, habenular commissure; 12, stria medullaris and thalamus; 13, fornix; 14, internal capsule; 15, splenium of corpus callosum; 16, dentate nucleus; 17, cingulum; 18, cerebellocerebral tract; 19, anterior commissure; 20, quadrigeminate bodies; 21, pons; 22, medulla oblongata. *Bottom:* 1, Optic radiation; 2, lateral geniculate body; 3, optic tract; 4, putamen; 5, external capsule; 6, internal capsule; 7, superior longitudinal bundle; 8, anterior commissure. (Fiber specimen prepared by Prof. G. Yasargil, M.D., Neurosurgical Clinic, University of Zurich.)

Introduction

Eye symptoms appear with significant frequency in patients with brain tumors. As a general statement based on other publications and our own series of 2500 cases, it can be said that *some form* of ocular signs can be observed in *somewhat more than 50% of patients with brain tumors.* These eye symptoms are not merely insignificant accompanying signs. The following statistical data show that this actually is not the case. In about 60% of our series, *ocular symptoms were among the initial complaints in the early stages.* In about 60% of the patients, papilledema was found, generally a sign permitting a diagnosis of increased intracranial pressure or a space-consuming lesion. Furthermore, due to the close anatomic relationship between the eye and the brain, the ocular signs alone allow a fairly accurate localization of a tumor in about 25% of the patients if it is situated near the visual pathways.

On the basis of these facts there can be no doubt that *the majority of eye signs in brain tumors actually are of diagnostic and even localizing significance.* Although the ocular signs of brain tumors are the main point of interest in this treatise, we should not lose sight of other signs and symptoms. To avoid a distortion of the actual clinical picture, the significant manifestations will be considered in their totality in the discussion of the various sites of tumors.

This book is primarily intended for the ophthalmologist who frequently finds himself in a position of being the first to make a decision as to whether or not there is an increase in intracranial pressure or to make a diagnosis as to whether an impairment of the visual pathways or of the extraocular muscles is caused by a tumor. The tremendous *responsibility of the ophthalmologist who has to make an early diagnosis of a brain tumor needs no further emphasis* if one considers that subjective ocular symptoms are among the early signs of brain tumors in 60% of such patients. This also demonstrates the *importance of close cooperation between the various specialties.* The discovery of minute and early ocular signs by means of suitable technical equipment should remain within the realm of the ophthalmologist—in spite of a more generalized awareness by others of the significance of some of these signs in brain tumors. At the same time, the ophthalmologist should never evaluate his findings as isolated facts but should always coordinate them with those of the neurologist, the internist, the radiologist, the otologist, and the neurosurgeon.

Eye signs in brain tumors are discussed in a number of publications on neuro-ophthalmology (Rea; Walsh and Hoyt; Lyle; Kestenbaum; Bing and Brückner; Kyrieleis; Smith; Brégeat; Dubois-Poulsen; Guillaumat, Morax, and Offret) as well as in the standard texts on neurosurgery (Cushing; Bailey; Dandy; Olivecrona; Tönnis; Krayenbühl). Most of these publications are part of the English literature. Still we do not believe that this book is superfluous. Perhaps our method of presentation, limited to the field of brain tumors with its ever-increasing importance, is as novel as is the fact that it is *based on the large material of an up-to-date neurosurgical department,* which was made available to me through the courtesy of Prof. Krayenbühl. The great privilege of being associated with such an institution as consulting ophthalmologist also places one under a certain obligation. The large number of patients examined and followed up by me personally may justify this addition to an already ample literature on the subject. *Personal experience in the evaluation of this material* seems to us to be of the utmost importance.

Regarding terminology, we include among ocular symptoms not only changes involving the eye itself but also anomalies of motility and sensory anomalies and impairments of the optic nerve, the visual pathways, and the higher visual centers. Thus the term "ocular symptoms" is used in the broadest sense of the word; it would perhaps be more appropriate to speak of "symptoms of the visual organ." In this book we have limited ourselves to the neoplasms of the brain tissue proper and the meninges. However, a concluding chapter with a brief separate presentation on ocular symptoms typical of aneurysms and other tumorlike lesions (pseudotumors, abscesses, hematomas, etc.) has been included.

In view of the tremendous wealth of literature dealing with brain tumors and their neuro-ophthalmologic aspects, the bibliography should be considered only as a random selection, especially of the more recent publications. Whenever an author is quoted, the appropriate reference is listed. In addition, the bibliography includes a number of references and publications not expressly cited in the book. The bibliography has been divided according to the individual chapters and divisions to facilitate its use.

 *Methods of neuro-ophthalmologic
examination in patients with
suspected brain tumors*

The neuro-ophthalmologic methods of examination have been dealt with in
a general manner in numerous older as well as some more recent publications
(Wilbrand and Saenger; Walsh and Hoyt; Lyle; Kestenbaum; Rea; Bing and
Brückner; Brégeat; Dubois-Poulsen; Guillaumat, Morax, and Offret). The excellent book by Kestenbaum is arranged according to the principles of the technique of examination; the pathologic changes are discussed in individual chapters
according to important ocular signs. Many valuable references to neuro-
ophthalmologic diagnostics in brain tumors can be found in numerous articles
(see bibliography) as well as in textbooks on neurosurgery (Cushing; Bailey;
Dandy; Sachs; Krayenbühl; Olivecrona; Tönnis) and ophthalmology (Schieck
and Brückner; Bailliart, Coutela, Redslob, and Velter; Duke-Elder; Amsler,
Brückner, Franceschetti, Goldmann, and Streiff; and others).

In view of the large amount of material available, it would almost seem
superfluous to present a separate discussion on the methods of examination
at this point. However, we are anxious to give our *own experience with and our
evaluation* of various methods and their usefulness for the neuro-ophthalmologic
diagnosis of brain tumors. In general, the neuro-ophthalmologic method of
examination in patients with brain tumors does not differ from that in patients
with other morbid changes of the central nervous system. However, we aim
to elaborate on some of the peculiarities of these methods and the interpretation of their results. These aims may justify the following rather condensed
presentation of methods of examination.

HISTORY AND SUBJECTIVE SYMPTOMS

Even the most detailed neuro-ophthalmologic examination—carried out with
the most refined ancillary instruments—has to be preceded by a carefully
taken history, an unavoidable procedure, as in any other clinical diagnostic
problem. This may not always be an easy task in patients with a brain tumor
if increased intracranial pressure has dimmed or completely abolished consciousness or if the tumor has caused a motor or sensory aphasia. In such patients

it is frequently imperative to obtain additional data concerning the course of the disease from relatives of the patient. In a large number of patients the spontaneous history has to be supplemented with leading questions by the examining physician. The severity and importance of the ocular symptoms may easily be outweighed by other neurologic symptoms. It should be stated emphatically that *the patient may be quite unaware of certain visual sensations and defects.*

Visual field defects

In regard to the previous statement, we want to recall particularly homonymous hemianopic field defects which, in one fourth of all patients, either are not appreciated at all or are not disturbing, especially in symmetrical defects and in defects that spare the central fixation. Critchley, in particular, concerned himself with the interesting phenomenon of *"awareness and nonawareness" of hemianopic field defects.* He concluded that the defectively perceived object is frequently "filled in"—an interpolation, so to speak, causes the patient not to be aware of the field defect. We know of numerous patients who had a resection of the occipital lobe because of a tumor who hardly noticed a complete homonymous hemianopic defect (without sparing of the macula) except (in right-sided cases) when reading. Such a patient, for instance, sees not just one half of his face in a mirror (or at least one half covered by a veil). On the contrary, he sees his entire face. In part this may be due to a constant lateral wandering of the eyes toward the side of the hemianopia. However, the "visual completion" realized by the higher visual centers is a necessary complementary factor (Poppelreuter; Goldstein and Gelb; Fuchs). Other conditions that frequently are not realized by the patient are bitemporal defects or concentric constrictions of the field. In some patients this phenomenon can be explained by the fact that the *hemianopia* caused by the tumor *is only relative in the sense of a so-called unilateral visual inattention.* Only qualitative partial functions of the visual perception (dyschromatopsia, spatial disorientation, retarded form recognition, indistinct outlines, or changes in the perception of movements) are disturbed in the involved halves of the visual field. The patients are not aware of these defects in everyday life. However, even a superficial simultaneous testing of both fields (for instance, with the hand) will demonstrate hemianopia. This diagnostically important phenomenon of "visual inattention" will be discussed further on p. 164.

Pseudohemianopia, which occurs in unilateral gaze palsies, and the absence of optic attention in the missing "field of fixation" will be discussed in detail in connection with tumors of the frontal lobe (p. 164).

The patient may be unaware of the visual field defect as such but not of its consequences in his everyday experiences. In addition to patients who are not at all aware of such defects, there are others *who complain, either spontaneously or when questioned, that they frequently collide to one (homonymous hemianopia) or both sides (bitemporal hemianopia) with persons or objects without any apparent reason.* Such statements strongly suggest hemianopia. The motorist hits the wall of the garage. The cyclist overlooks a pedestrian on one side of the street. A patient hurriedly writes past the margin of the paper. These are some examples selected from case histories to demonstrate the consequences of hemianopia in everyday life. Some patients do not even mention such details in their histories. Their hemianopia is masked by the complaint of cloudy or poor vision (they blame insufficient lighting at their place of work!) or the statement that one or both eyes are visually disturbed (a homonymous hemianopia is projected by the patient to the eye corresponding to the side of the hemianopia; bitemporal hemianopia is projected to both eyes).

Disturbance of simple and higher visual functions

A history of *loss of vision* is an important symptom in tumors near the optic nerve, the chiasm, or the optic tract. The patient, however, does not always make such a specific statement. If he is not alert, he does not appreciate a deterioration of his visual acuity for quite a while. We can cite from our material numerous cases of patients with pituitary tumors who suspected at first that their glasses were not adequate and consulted the optician or ophthalmologist for stronger lenses! Loss of vision is not necessarily the result of direct pressure of a tumor on the distal visual pathways. It could be the precursory sign of an imminent optic atrophy resulting from a chronic papilledema—prognostically, a particularly unfavorable situation. The *early stages of progressive loss of vision* are not always easy to detect. However, just these symptoms should be of enormous importance to the neurosurgeon. In taking a careful history, one may learn of *loss of vision only under poor lighting* (impaired dark adaptation as an early sign of damage to the optic nerve!) or of certain changes or *disturbances of the color sensation* (likewise, symptoms of damaged fibers of the optic nerve). Occasionally patients report a veil, fog, or smoke before the eyes in early stages of optic atrophy due to a tumor. Not infrequently loss of vision in one eye is not appreciated because the other eye functions well and vicariously substitutes for the other. Thus amaurosis in one eye may develop without the patient being aware of it. Only a deterioration of vision in the second eye causes alarm and unexpectedly brings forth this catastrophic situation. One should always interpret somewhat skeptically subjective complaints of a sudden visual disturbance and particularly a sudden decrease in vision. As a rule, defects of the conductivity of the visual pathways due to a tumor show a progressive development. It is often only the patient who suddenly discovers the loss! We shall discuss the so-called *amblyopic attacks* in patients with brain tumor in greater detail later; these brief attacks of blackouts or fogginess (not infrequently associated with violent headaches) are well-known accompanying symptoms of papilledema (especially in its chronic form) and should not be missed (p. 102).

Photopsia (that is, the sensation of sparks, lightning, luminous rings, etc.—mostly in both visual fields) is not likely to be missed in taking the history because it is quite an impressive and actually frightening phenomenon. According to Ethelberg and Jensen, it occurs mostly in association with the attacks of amblyopia or obscurations already mentioned—either as a prodromal symptom, simultaneously with these attacks, rarely after the attack, or as an isolated symptom.

Various forms of photopsia show an imperceptible transition to *optic hallucinations* that should always be searched for if a brain tumor is suspected. In principle, optic hallucinations may occur as consequence of a lesion anywhere along the visual pathway—namely, the retina, the optic nerve, or the higher visual pathways (Weinberger and Grant). We are not in agreement with Weinberger and Grant's opinion that the type of hallucination is completely useless for purposes of localization. Rather, we agree with Walsh and Hoyt and with Parkinson, Rucker, and McCraig in the belief that the customary classification of optic hallucinations is grossly correct: *unformed crude light sensations* (fiery globes, luminous rings, stripes, circles, discs, blinking lights) are typical *for occipital lesions,* whereas *formed, differentiated, and complex sensations* (objects, animals, persons, entire scenes), not infrequently associated with dreamy states and olfactory or gustatory hallucinations, are more typical for *tumors of the temporal lobe.* If one can learn about the occurrence of such hallucinations while taking the history, their nature may be of some value in localizing the site of the tumor. This is the more welcome because tumors in the temporal and occipital lobes develop in relatively "silent" zones. In the majority of patients the hallucinations occur in the region of the hemianopic field defects (Wilbrand; Henschen; Cushing; Horrax). Depending on the nature of the hallucinations, the site of the tumor can be suspected with some degree of probability. Only the proof of hemianopia will provide certainty, because time and again cases are reported of hallucinations either in the intact half of the visual field or on both sides (Weinberger and Grant; Sanford and Blair). Unfortunately, optic hallucinations do not occur very frequently. According to Horrax and Putnam and to Parkinson, Rucker, and McCraig, their frequency in patients with occipital tumors varies between 15 and 20%. In our own series their occurrence in patients with tumors of the temporal lobes as well as the occipital lobes was even rarer (only a few percent).

Micropsia and *macropsia* are occasionally mentioned by patients with brain tumors. Such symptoms should be further analyzed (peripheral causes such as affections of the macula or disturbances of accommodation and convergence have to be ruled out!) before they can be interpreted and evaluated as phenomena of central origin. Bender and Savitsky are of the opinion that, in addition to affections of the parietal and occipital regions, other lesions of the visual pathways (for instance, tumors near the chiasm) may be responsible for micropsia and macropsia; in other words, these symptoms are not of localizing value. We have observed micropsia in a patient with glioblastoma multiforme of the occipital lobe. This patient saw objects in her surroundings (especially furniture in her home) strikingly smaller than usual. She reported this sensation simultaneously with a homonymous loss of the visual field. Hemimacropsia in the right halves of the visual field (without hemianopia) has been observed by Brégeat, Klein, Thiébaut, and Bouniol in a patient with a tumor of the left occipital lobe.

The visual hallucinations and the phenomena of micropsia and macropsia actually bring us to the symptomatology of *disturbances of the higher visual functions* (cortical optosensorics and optognostics) that may also have to be considered in taking the history (Fig. 1-1).

Tumors in the occipital cortex (area 17 = "visual cortex"), especially in the calcarine fissure, result in optosensory disturbances of the true *cortical blindness* type if the involvement is bilateral. Characteristic findings are loss of all visual sensations, including the differentiation between light and dark, preservation of the pupillary reaction to light, and loss of the lid reflex to bright light or finger movements before the eyes. Occasionally cortical blindness is accompanied by unformed visual hallucinations. A particularly interesting phenomenon in cortical blindness is the fact that the *patients frequently are not aware of their loss of visual sensations,* as if they were "blind for their blindness" (Lhermitte). The

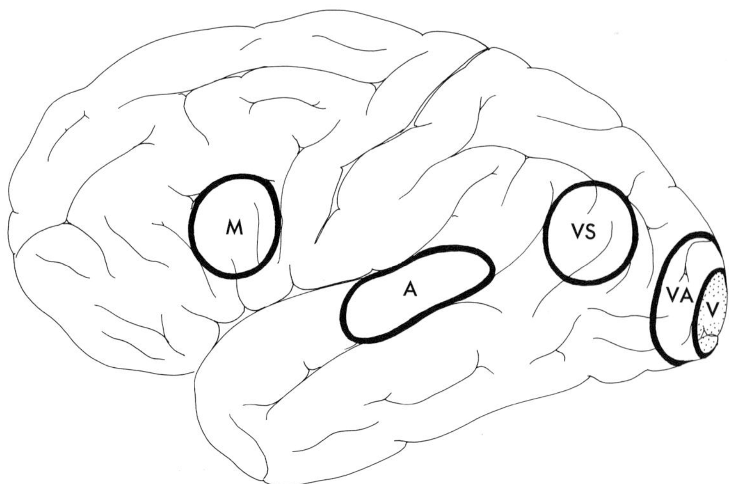

Fig. 1-1. Sketch of dominant hemisphere demonstrating the position of the visual centers and the centers for reading, speech, and understanding of symbolic expression. Lesions of these areas may lead to the different forms of central disorders of visual integration. **V,** Visual cortex (cortical blindness); **VA,** visual associational area (object agnosia, soul or mind blindness); **VS,** visual speech area (alexia, word blindness); **A,** auditory speech area (sensory aphasia); **M,** motor speech area (motor or verbal aphasia).

patients behave as if they actually could see and describe (as in hallucinations) their surroundings, which they believe they see in the most minute and fanciful details. We have followed up two such patients with complete cortical blindness. In one the cause was an x-ray necrosis of both occipital lobes following irradiation after extirpation of a temporal tumor; it remained irreversible until death. The other patient showed cortical blindness that appeared suddenly following a bilateral ventriculography performed through the occipital lobes. Obviously it was caused by an acute local edema. Upon recovery this patient had a complete restitution of the visual functions. As a rule, this recovery is not complete at once but progresses from the stage of mere light-dark perception to the recognition of form and finally to color perception and all the other attributes of normal vision. Generally, tumors involve the optic radiation rather than the visual cortex and, at that, more often only one side. Thus the phenomenon of cortical blindness in patients with neoplasms of the brain is rarely encountered.

Lesions at a higher level of the optognostic pathways (the border area between the occipital, temporal, and parietal lobes) may cause central disorders of visual integration of the type of so-called *mind blindness* (visual object agnosia), which is the inability to appreciate the meaning of a visual perception. The patients are able to describe details of objects but are unable to classify the objects in their totality in the storeroom of their memories and to identify them accordingly. However, the moment they are aided by another sense (for instance, when they touch the object they could not identify), they recognize it instantly. Since most objects in everyday life are identified by means of the visual sense, mind blindness means an unpleasant and even disastrous condition and one which patients are hardly able to dissimilate. There is no unanimity of opinion as to the exact localization of this higher cortical visual disturbance. Probably it is in the parieto-occipital region near the angular gyrus (Wernicke; Wilbrand). Other schools do not believe that there are precise centers for higher cortical functions: they explain such disorders by interruption of association pathways at certain sites related with certain functions. The lesion can be unilateral. It does not seem to make any difference whether the dominant (as a rule, the left) or the other side of the brain is involved. This type of visual disturbance is also found infrequently in patients with cerebral tumors.

Tumors in the area of the angular gyrus and in the border area between the parietal and occipital lobes of the dominant hemisphere (mostly on the left side) are more likely to result in *sensory-aphasic disturbances* of the type of *word and letter blindness* or *alexia*. Although the vision is intact, the patient is unable to read. In other words, he is unable to comprehend the significance of written or printed words, musical notes, or digits (this special form of "visual agnosia" is limited to the partial function of the process of reading). Since the motor function depends on the sensory function and its evaluation, alexia is almost perforce combined with a motor-aphasic phenomenon, namely, *agraphia* (the

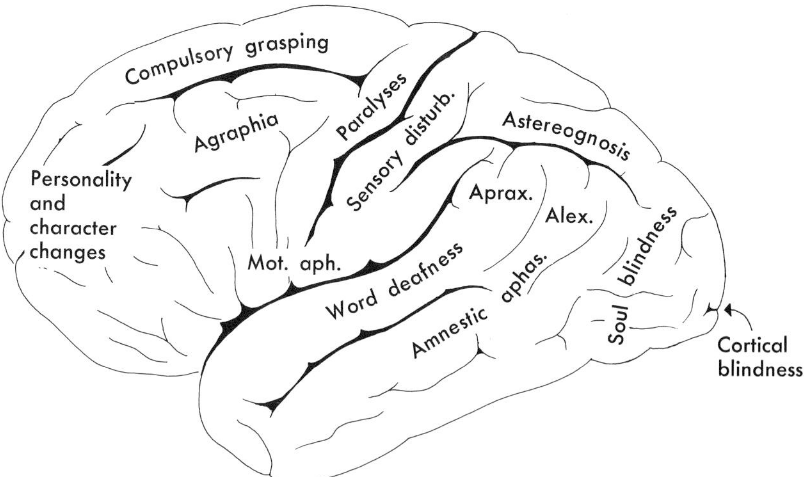

Fig. 1-2. Sketch of left lateral surface of the brain hemisphere indicating functional disturbances that occur if certain areas or connections to them are damaged. Of special interest are cortical blindness as the result of lesions of both visual occipital cortex areas and soul (or mind) blindness and alexia as the result of damage to the parieto-occipital region.

inability to write, either spontaneously or in dictation). The combination of alexia and agraphia occurs particularly in patients with so-called cortical alexia. In lesions of the left occipital lobe and of the splenium (disconnection syndrome) one finds pure alexia without marked disturbance of the ability to write—the patient is unable to read what he has written. Usually this condition is found in combination with a disturbance of conductivity in the optic radiation, that is, with right-sided homonymous quadrantanopia or hemianopia. We have frequently observed alexia with agraphia either as isolated disorders or in combination with apractic as well as other sensory-aphasic disturbances in patients with tumors of the parietal and occipital lobes (Fig. 1-2).

When the history is taken, alexia and agraphia must be carefully differentiated from peripheral disturbances of the function of the visual organ. As a rule, such a differentiation should cause no difficulties. The situation will only become complicated and difficult if, for instance, because of a chronic papilledema with optic atrophy, the vision has already become so impaired that reading and writing are no longer possible. If one takes into consideration, however, that *alexia and agraphia rarely occur alone* but are *frequently combined with other sensory-aphasic* disturbances such as *word deafness* (the inability to understand speech in the presence of intact hearing), *the inability to understand gestures and mimic actions, and particularly the inability to express thoughts in words, gestures, or mimic actions,* one can hardly fail to evaluate such conditions properly.

Disturbances of extraocular muscles

Paralysis of extraocular muscles usually causes a very disturbing *double vision*, especially if the vertical muscles are involved. Thus, as a rule, the patients mention diplopia spontaneously. *In patients with a less severe paresis, especially of the horizontal rotators (medial rectus and lateral rectus),* there usually is no double vision at all in part of the field of fixation. Even if the patient looks toward the side of the paretic muscle, he may perhaps have no true double vision but merely complains of an *uncertain blurring of objects* or a glimmer that he has difficulty in describing. This is, so to speak, an expression of a minimal disparity of the two images, a minimal displacement of contours without complete lateral or vertical separation of the objects. In patients with brain tumor such slight pareses quite frequently are the sequelae of the general increase of the intracranial pressure (involving especially the sixth nerve). In taking a careful history, one should suspect such pareses that may be masked by vague statements on the part of the patient.

In cases of manifest diplopia one should inquire whether the images are separated in the vertical, in the horizontal, or in both planes. Furthermore, one should determine in *which direction of gaze is the maximal horizontal or vertical separation of the two images.* On the basis of these data ascertained from the history, one is frequently in the position to diagnose the paretic muscle with the aid of a simple diagram (Fig. 1-13), provided only one muscle is responsible for the diplopia.

As a rule, a patient is not aware of a supranuclear paresis or *gaze palsy* unless, in the course of the examination, his attention is called to his inability to shift his gaze sideways or up. This fact can be understood if one considers that such palsies can be easily compensated for in everyday life by certain movements or positions of the head. A certain exception is the vertical palsy present in patients with *Parinaud's syndrome.* In eight such cases, usually caused by tumors in the region of the roof of the midbrain, we were able to obtain a history of *vertical diplopia* in six—obviously an expression of the disparity in the vertical impairment of the individual eye. This type of diplopia, in combination with pupillary disturbances (of the Argyll Robertson type) and the unmistakable vertical gaze palsy, permits the definite diagnosis of Parinaud's syndrome, pathognomonic for lesions in the region of the roof of the midbrain or the quadrigeminate plate. A few times we have seen Parinaud's syndrome combined with so-called *paralysis of convergence.* Distant objects were seen single; at *close range* there was *crossed diplopia* with only a slight horizontal disparity of images that showed little change during lateral conjugate movements. Paralysis of convergence, like Parinaud's syndrome, suggests a mesencephalic lesion (Perlia's nucleus is the hypothetical center of convergence). We find *uncrossed diplopia* with a corresponding convergent position of the eyes *for distance* but not for near in the so-called *paralysis of divergence.* The horizontal separation of the images in all directions of gaze remains unchanged.

The ability to converge is intact. There is an impairment of the divergence, although each lateral rectus muscle functions properly. A number of autopsy findings (Savitsky and Madonick; Lippmann) suggest lesions caused by a tumor in the area of the brain stem near the sixth nerve nuclei.

Gaze nystagmus or *central vestibular nystagmus,* usually occurring in patients with brain tumors, practically never causes subjective symptoms. Only the peripheral vestibular nystagmus (either due to a lesion in the inner ear or the region of the vestibular nucleus) may cause ocular disturbances such as vertigo with whirling of the surroundings around the patient, associated headaches, nausea, and a general sensation of insecurity (the well-known symptoms of Ménière's disease).

In the history given by patients with neuro-ophthalmologic problems, symptoms of a disturbance of vision or ocular motility are quite preeminent because they are appreciated relatively early and distinctly. Other ocular symptoms in patients with brain tumors may give rise to subjective sensations which, in the overall picture, are less important for diagnosis and localization but will be discussed briefly.

Pain

Headache, as a sign of increased intracranial pressure, occasionally radiates into the eyes and is interpreted as an associated eye ache. However, it is minor compared with the general cephalalgia. *If the eye ache is the outstanding symptom, a localized disease of the eye has to be considered,* which in turn may cause radiating pain into various regions of the head. Acute or subacute glaucoma, in addition to a deep-seated eye ache, may cause diffuse headaches, nausea, or even vomiting, symptoms that could easily be mistaken for those of a general increase in the intracranial pressure. The ocular signs are increased intraocular pressure, ground-glass appearance of the cornea, a relative dilated pupil, ciliary congestion of the globe, and possibly glaucomatous excavation of the disc, signs that usually pose no diagnostic problems.

The eye is within the area of distribution of the ophthalmic branch of the trigeminal nerve; thus tumors affecting this cranial nerve may cause ocular pain. However, an isolated pain in the eye is rare. Rather, there are neuralgias within the entire area of distribution of the first branch and possibly also in those of the other two branches—that is, the maxillary and mandibular nerves. In the *superior orbital fissure syndrome* and in the *cavernous sinus syndrome* affecting the third, fourth, fifth, and sixth cranial nerves, there may be referred pain, especially in the area of the first branch of the trigeminal nerve involving the face and eye. Unlike idiopathic trigeminal neuralgia, there is no paroxysmal onset of pain in the case of a tumor or an aneurysm of the internal carotid artery within the cavernous sinus. The pain is rather persistent and sharply localized within the area of distribution of one of the trigeminal branches—occasionally with sensory disturbances. A similar type of pain, also radiating into the eye, may be caused by a tumor within the gasserian ganglion itself (a neurinoma of the trigeminal nerve), by a tumor in the middle fossa (cave of Meckel), or by a tumor in the posterior fossa (meningioma, acoustic neuroma, and others) if there is a corresponding involvement of the trigeminal nerve. One should not forget that ocular pain in the area of distribution of the first branch of the trigeminal nerve is frequently accompanied by increased *lacrimation* and painful *photophobia*.

Photophobia

Excessive sensitivity to light in the form of photophobia is often painful and so annoying that the patient will mention it spontaneously. The *peripheral* form referred via the trigeminal nerve has already been discussed in connection with pain. We have seen abnormal photophobia in a few patients with tumor of the midbrain, especially combined with Parinaud's syndrome. One of these patients would cover his eyes with a dark cloth, even indoors, so the photophobia would not annoy him. It disappeared completely after a course of x-ray treatments. Milder forms of photophobia, according to our own observations, occur quite commonly in any type of increased intracranial pressure— even in cases of subdural hematomas. This type of photophobia should be regarded as *central* in origin and be differentiated from the peripheral trigemi-

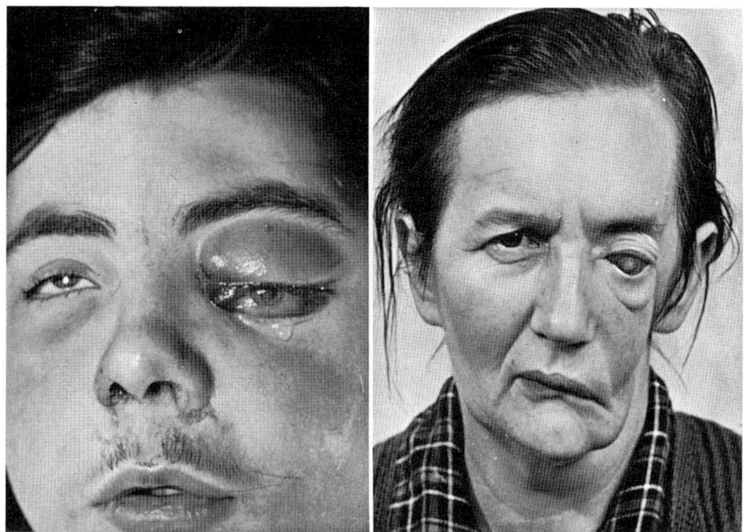

Fig. 1-3. Lagophthalmic corneal ulcer. In the patient on the left the condition resulted from incomplete lid closure due to a poor general condition and stupor. Thrombosis of the cavernous sinus, exophthalmos, lid edema, chemosis of the conjunctiva, and hypopyon. In the patient on the right the condition resulted from incomplete lid closure due to peripheral facial palsy. Loss of physiologic folds of face, drooping of the angle of the mouth and of the lower lid, and widening of the lid fissure.

nal form. It is possible that the state of adaptation of the retinal elements or perhaps the response of the synapses in the external geniculate body is regulated in a retrograde fashion by the mesencephalon and diencephalon (perhaps the hypothalamic roots of the optic nerve). Neoplasms in this region of the brain (in the case of tumors located elsewhere, a remote effect must be assumed) can cause a defective regulation resulting in faulty adaptation or increased excitability of the visual pathways and consequently a pathologic photophobia. This hypothesis for a central form of photophobia is supported by the observation of a type that may occasionally be quite severe, occurring after minor but more frequently after severe trauma to the skull, in meningitis, in migraine, and in subarachnoidal hemorrhage.

Observations made by the patient when looking into a mirror

Patients will often mention observations they have made while looking into a mirror, observations that will assume considerable importance during the course of the examination.

A unilateral *widening of the lid fissure* will be quite conspicuous to the patient. It is the result of central or peripheral facial palsy, a frequent phenomenon in brain tumor—not only due to localized pressure but also to a generalized increase in the intracranial pressure. A complete lagophthalmos (that is, the inability to close the eye) occurs much more rarely except as a surgical complication; for instance, during extirpation of an acoustic neuroma, damage to the facial nerve occasionally cannot be avoided because of technical reasons (Fig. 1-3).

Widening of the lid fissure may also be an early sign of *exophthalmos*. In the initial stage of exophthalmos only the change in the width of the lid fissure, not the proptosis, will be obvious to the patient. An exophthalmos does not necessarily have an orbital cause. We have observed numerous cases of unilateral exophthalmos caused by intracranial tumors. It is most common in meningiomas of the sphenoid wing (Fig. 1-5). Less frequent causes are sellar, parasellar, and suprasellar tumors and occasionally even a tumor of the frontal or temporal lobe. In the advanced stages, exophthalmos may cause symptoms such as pressure sensation or even dull pain in the orbit in addition to eventual diplopia.

Narrowing of the lid fissure may be noticed by the patient even earlier. He will always consider a *ptosis* (even of the slightest degree) as a blemish and will certainly mention it in his history. Ptosis may be part of *Horner's syndrome* (miosis, ptosis, enophthalmos). Damage to the central sympathetic system caused by tumors may occur in the cervical region, the medulla oblongata, the pons, and occasionally the thalamus. Much more frequent and pronounced, even with a complete inability to raise the upper lid, is the ptosis in patients with *paresis or paralysis of the oculomotor nerve*. If combined with a paralytic mydriasis, a slight ptosis may be part of the so-called *"clivus ridge syndrome"* (Fischer-

Brügge), a syndrome that may be characteristic for a generalized increase in the intracranial pressure caused by tumors. According to Fischer-Brügge, displaced brain substance presses the oculomotor nerve at the site of its emergence from the brain stem, either against the clivus ridge of the tentorial notch or against the superior cerebellar artery, thus causing its damage (p. 152). However, the clivus ridge syndrome may be of localizing value in tumors of the temporal lobe with direct pressure on the oculomotor nerve due to a herniation of the lobe through the tentorial notch. Pareses or paralyses of the oculomotor nerve causing a ptosis may be caused by pressure due to tumors anywhere along the entire course of the nerve from the brain stem to the orbit. Their localizing value will be discussed in greater detail on p. 22. In the subjective symptomatology it is important to remember that even an incomplete ptosis *(as long as the upper lid covers the pupil)* will eliminate the eye from the act of binocular vision, *thus masking diplopia due to paresis of the oculomotor nerve.* In such an instance one is tempted to talk about a "rational" protective mechanism on the part of nature that makes it unnecessary for the physician to prescribe a frosted lens. In such a severe ptosis it is imperative to raise the upper lid if one wishes to analyze the diplopia (Fig. 1-18).

Differences in the size of the pupil also will be mentioned by an observant patient. Of course, in most instances he will be unable to distinguish between an abnormal dilation on one side or an abnormal constriction on the other side. *Unilateral miosis* as part of Horner's syndrome causes no subjective sensations. In the history of patients with brain tumors a *unilateral mydriasis* is of greater importance. If the pupil is fixed, *dazzling* (an abnormally large amount of light will enter the eye) and even photophobia and *blurring of vision* (the spherical and chromatic aberration of the crystalline lens is greater with a large than with a small pupil) will almost always be in evidence. If the unilateral mydriasis is caused by an impairment of the oculomotor nerve, it will usually be associated with a more or less distinct *paresis of accommodation* (Fig. 1-8). The patient will have fairly good distance vision if the dazzling is not too annoying. However, his near vision at 20 to 30 cm. will be blurred to the extent that he will be unable to read without the aid of a 3- or 4-diopter convex lens.

The simultaneous involvement of the pupillary reaction and the accommodation is due to the fact that the iris sphincter and the ciliary muscle are both innervated by the efferent parasympathetic fibers of the oculomotor nerve. Unilateral impairment of accommodation may also cause micropsia or macropsia of the involved eye. Unilateral mydriasis as a result of peripheral damage to the third nerve has already been discussed in connection with the clivus ridge syndrome. In the *differential diagnosis of unilateral mydriasis* we have to consider, in addition to lesions of the oculomotor nerve, *the effect of certain drugs* (homatropine, atropine, epinephrine), a *congenital anisocoria,* and in particular Adie's syndrome (abolished light reaction of the mydriatic pupil, tonic reaction and relaxation in convergence, and tonic accommodation).

The patient may be able to discover a *paralytic strabismus* when he looks into a mirror if the third, fourth, or sixth cranial nerve is involved, as happens quite frequently with cerebral neoplasms. However, in the majority of patients the subjective sensation of double vision will be the dominating factor and will consequently be stressed by the patient. A patient would be unable to notice nystagmus himself; someone else would have to call his attention to such a disturbance of the ocular motility. Likewise, it would be impossible for him to observe gaze palsies.

From our dissertation so far it should seem amazing *how many important diagnostic clues can be gained merely from the history of patients with brain tumors.* It pays to "lose time" and take a detailed history! We therefore elaborated on the material in this opening chapter which, in our opinion, is so very important.

OBJECTIVE EXAMINATION
External aspect of the eyes

Some aspects included in this section have been treated in the preceding discussion of the subjective symptoms because external changes of the eyes may be conspicuous to the patient. One should listen to the patient's side of the story but should not make the mistake of depending on it too much. Merely a careful glance at the eyes and their surroundings may yield diagnostic clues in the course of the neuro-ophthalmologic examination.

Unilateral widening of the lid fissure, as mentioned before, *is primarily the result of a facial paresis* or more rarely of an irritation of the ocular sympathetic fibers. The supranuclear and peripheral forms of Bell's palsy are frequent signs in brain tumors. The frontal branch is not involved in the supranuclear but only in the peripheral form. The orbicularis oculi may or may not be involved in a supranuclear paresis. The loss of tone may cause raising of the upper lid and sagging of the lower lid; the sclera above and below the limbus is not covered by the lids and thus is visible. In testing the strength of the active lid closure, the patient is asked to close his eyes as forcefully as possible. The examiner attempts to open the lids with his fingers. He should be able to diagnose a paresis from the weak or missing resistance and by a comparison with the opposite side. A total paralysis of the orbicularis muscle leads to *lagophthalmos,* the inability to close the eyes completely. It is seen mostly in peripheral Bell's palsy. If the patient attempts to close the eye, it characteristically rotates upward (Bell's phenomenon). A deficiency in the moistening of the cornea by the conjunctiva of the upper lid in patients with lagophthalmos may lead to a lagophthalmic corneal keratitis manifested by a perilimbal ciliary congestion as well as by cloudiness and infiltration of the lower part of the cornea. The early damage to the corneal epithelium caused by the exposure can be demonstrated by a positive stain after instillation of 2% fluorescein. It may progress to ulcer formation of the cornea and after a secondary infection produce the severe picture of a

serpiginous ulcer with accumulation of pus in the anterior chamber (hypopyon).
We have observed lagophthalmic keratitis in stuporous or unconscious patients
with a brain tumor when—even in the absence of a facial paralysis or perhaps
one of a minor degree—no proper lid closure was possible as a result of the
poor general condition (lagophthalmos in sopore), increasing the danger of
corneal exposure. A particularly serious situation exists in patients who have been
operated on for a tumor of the cerebellopontine angle. In addition to a facial
palsy, there is usually a lesion of the trigeminal nerve. The exposure of the cornea
is complicated by neurotrophic disturbances—a combination of lagophthalmic
keratitis and neuroparalytic keratitis (Figs. 1-3 and 1-7).

If there is a difference in the width of the two lid fissures, it is not always
easy to decide whether there is a pathologic widening on one side or narrowing
on the other side. Testing the active tone of the orbicularis muscle will usually
facilitate such a decision. An examination of the pupils (p. 22) should supply
additional data. A widening of the lid fissure in patients with Bell's palsy will
not be associated with pupillary disturbances. A narrowing of the lid fissure, on
the other hand, will often be combined with a pathologic miosis (impairment
of the sympathetic fibers) or mydriasis (paresis of the oculomotor nerve).

A *unilateral narrowing of the lid fissure* is usually caused by an abnormal
drooping of the upper lid, that is, a *ptosis*. This may be due either to a paresis of
the oculomotor nerve or to a lesion of the sympathetic fibers. In the first instance, the ptosis results from a paresis of the levator muscle. In the latter, the
ptosis is due to involvement of the superior and inferior tarsal muscles (Müller's
muscle). Associated signs may help in the differential diagnosis. According to a
classic rule, *a wider pupil on the side of the ptosis indicates a paresis of the oculomotor nerve, and a narrower pupil on the side of the ptosis* (Horner's syndrome)
means a sympathetic lesion. Relatively often a ptosis will occur without pupillary
changes. In that event the evaluation of the position of the lower lid will be of
importance (Kestenbaum). It will be normal in "oculomotor ptosis." In "sympathetic ptosis," however, the lower lid on the involved side is somewhat higher than
on the normal side (the downward pull of the inferior tarsal muscle is missing).
In other words, the lower lid margin covers a larger segment of the limbus on
the involved eye. Thus the "sympathetic ptosis" can be diagnosed on the basis of
the difference in the limbus segment covered by the lower lid, which will be
larger on the involved side. The "oculomotor ptosis" plays a more important role
in brain tumors. It has already been mentioned as a manifestation of a general
increase in the intracranial pressure if associated with mydriasis (clivus ridge
syndrome) or of a localized pressure caused by tumors of the temporal lobe (Fig.
2-45).

In the differential diagnosis of ptosis a *unilateral blepharospasm* must be ruled
out. It may be a sign of a local irritative process in the eye, or it may be an
idiopathic phenomenon, occasionally associated with a facial spasm. Charcot's
"eyebrow sign" occasionally is of some value. In actual ptosis the eyebrow is at

the level of the upper orbital margin, whereas it is distinctly lower in blepharo-spasm.

Under certain conditions, *exophthalmos* may simulate a widening of the lid fissure. A unilateral proptosis of the globe can be recognized only by comparison with the other side. The following method is suitable for a preliminary examination. The examiner stands behind the patient and studies the prominence of the globes by looking down from above; he raises the upper lids and directs the patient to look straight ahead. Quantitative measurements can be obtained with a *ruler,* using a simple test suggested by Kestenbaum. A ruler is held vertically so it touches both the superior and inferior orbital rims. With the eye closed and the globe in a normal position, the ruler will just touch the skin of the upper lid over the corneal apex. In the patient with exophthalmos the ruler will be in contact with the inferior orbital rim but will miss the superior orbital rim by a distance dependent on the prominence of the globe. The distance of the ruler from the superior orbital rim is measured and the amount divided by two. The value obtained represents the extent of the exophthalmos. More reliable figures are obtained with the *Hertel* or the *Krahn exophthalmometer.* The principle of these instruments is based on the determination of the relative distance between the apex of the cornea and the lateral orbital rim. This is accomplished by an ingenious arrangement of mirrors. If the readings are repeated on the same patient, it is important to always employ the same base line between the mirrors. Since there are considerable individual variations in the anatomic configuration of the lateral orbital rims, the figures obtained with Hertel's instrument are not absolute values. In normal eyes they vary between 12 and 17 mm. The method is, however, suitable for comparative exophthalmometry (that is, for a comparison between the two eyes and also for repeated measurements on the same patient on different occasions). Variations or differences of less than 1 mm. are not significant.

Exophthalmos is the cardinal symptom of an orbital tumor (Fig. 3-87). The exophthalmos is unilateral. Characteristically, a distinct resistance is felt if one attempts to push the globe into the orbit. Tumors of the anterior and middle fossa may cause exophthalmos by direct invasion of the orbit or by compression of the cavernous sinus. Exophthalmos is an almost constant sign in patients with *meningiomas of the sphenoid wing* (especially of the medial and middle thirds) (Fig. 3-74). It is seen less frequently with large olfactory meningiomas or pituitary tumors. Actually, brain tumors in almost any location can cause an exophthalmos that is usually bilateral. This is not just a simple proptosis of the globe but rather an "extrusion" of the orbital contents—obviously the result of an extension of the increased intracranial pressure through the orbital fissures. The frequency of exophthalmos in brain tumors is reported to be between 2 and 8% (Cushing; Uhthoff; Elsberg, Hare, and Dyke), usually due to meningiomas (which, according to Cushing, may frequently cause a unilateral exophthalmos) and tumors at the base of the brain. Exophthalmos of a lesser degree may occur in patients with tumors of the posterior fossa, a sign of increased pressure (p. 155). This complication may be due in part to the relatively frequent and early development of an internal hydrocephalus in patients with brain tumors. An exophthalmos is of little value for purposes of localization. If it is unilateral, it may point out the side of the tumor.

Pareses of the extraocular muscles, a frequent event with brain tumors, may create some degree of proptosis because of the missing retraction of the globe. In such instances it may be difficult to rule out "pressure exophthalmos."

One should not forget *Graves' disease* in the differential diagnosis of exophthalmos. Although, as a rule, there is a bilateral proptosis, an asymmetrical or even unilateral exophthalmos is not unusual. Numerous characteristic signs (Graefe, Moebius, Stellwag) as well as the determination of the basal metabolic rate, protein-bound iodine, radioactive iodine uptake, and the Werner triiodothyronine suppression test should help make the correct diagnosis in most cases.

An *intermittent exophthalmos* (proptosis of the globe on bending or on extreme turning of the head) is caused by varicose veins or a varix within the orbit. *Pulsating exophthalmos* is an important sign of an aneurysm of the internal carotid artery within the cavernous sinus (Fig. 5-15). The pulsation can be demonstrated by the rhythmic movements of the reflex to a source of light focused on the cornea. In addition to the pulsation, a marked stasis of the episcleral veins (which should not be confused with a conjunctivitis) is frequently conspicuous in the involved eye. The grave picture of an inflammatory *thrombosis of the cavernous sinus* can easily be differentiated from other forms of exophthalmos. Usually there is a bilateral proptosis with lid edema and severe chemosis, immobilized globes, corneal anesthesia, and general signs of septicemia (Fig. 1-3).

Swelling of the lids with or without chemosis of the conjunctiva is not necessarily caused by an inflammatory process in the orbit or the cavernous sinus. We have observed patients with meningiomas of the sphenoid ridge in whom this sign was quite prominent in addition to a pronounced exophthalmos. Obviously there is an impediment of the venous circulation of the ophthalmic veins or the cavernous sinus caused by tumor compression (Fig. 3-75).

By definition, exophthalmos is merely a proptosis of the globe in the anteroposterior direction. There may also be a *vertical or lateral displacement of the eye*. Intraorbital tumors or tumors invading the orbit from the middle fossa frequently cause a grotesque displacement of the globe downward (Fig. 1-4) or downward and outward ("chameleon eye") (Fig. 1-5). The motility of such eyes is, of course, severely impaired. As a result of such a displacement, pseudolagophthalmos (possibly with signs of keratitis), chemosis, and lid edema are frequent accompanying signs.

The sclera and its vessels may be of some importance in the evaluation of the external aspect of the eye. As mentioned, pressure caused by a tumor in the region of the cavernous sinus or of the ophthalmic vein may cause a *stasis of the*

episcleral and scleral or even the conjunctival veins. The veins in the "white" of
the eye are dilated, tortuous, more conspicuous (thus appearing more numerous),
and occasionally distinctly prominent. Such a condition can easily be mistaken
for a conjunctivitis or episcleritis. We have observed a patient with a post-
traumatic carotid-cavernous fistula that was not recognized at first and was
treated as chronic conjunctivitis in the presence of obvious signs of stasis (Fig.
1-6).

Corneal changes have been listed in the discussion of lagophthalmos and the

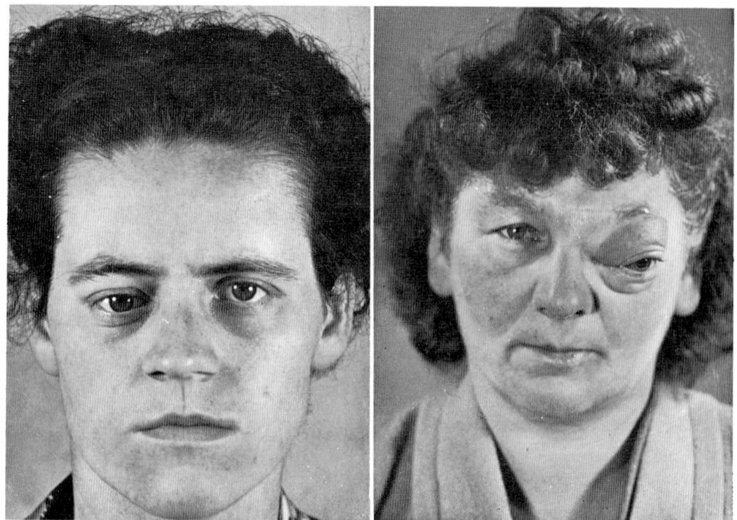

Fig. 1-4

Fig. 1-5

Fig. 1-4. Exophthalmos of the right eye with downward displacement. Extradural dermoid
cyst of the right frontal region with pressure atrophy of the corresponding orbital roof.
Fig. 1-5. Exophthalmos of the left eye with downward and outward displacement of the
globe ("chameleon eye") due to recurrence of a meningioma of the left sphenoid ridge
(mesial and middle thirds) with invasion into the orbit. Enormous lid swelling, extensive
impairment of motility of globe, and complete optic atrophy with amaurosis and amaurotic
mydriasis. Recurrence 12 years after first surgical intervention.

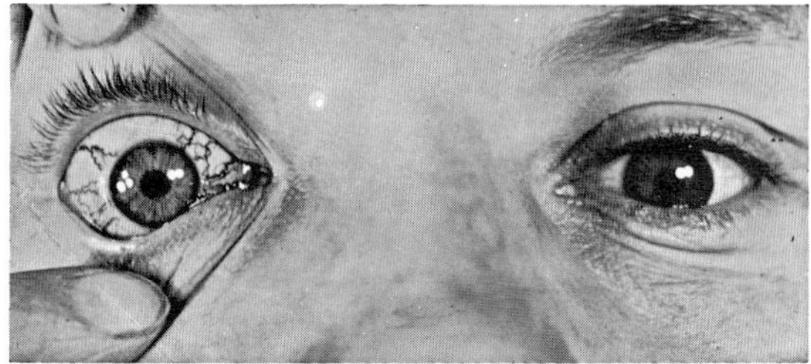

Fig. 1-6. Venous stasis of conjunctival and episcleral vessels of the right eye in a patient
with aneurysm of the internal carotid artery in the cavernous sinus (carotid-caverous fistula).

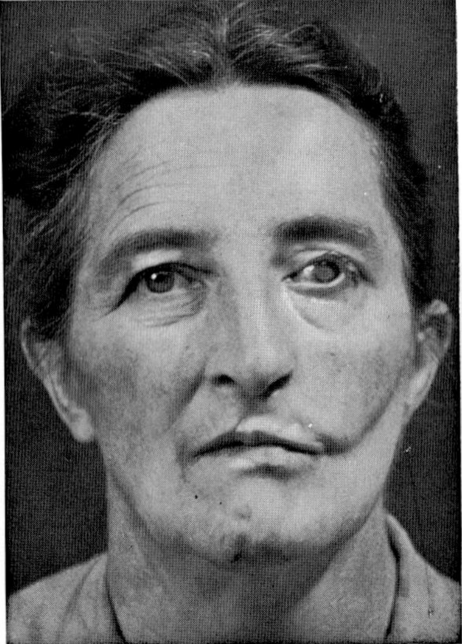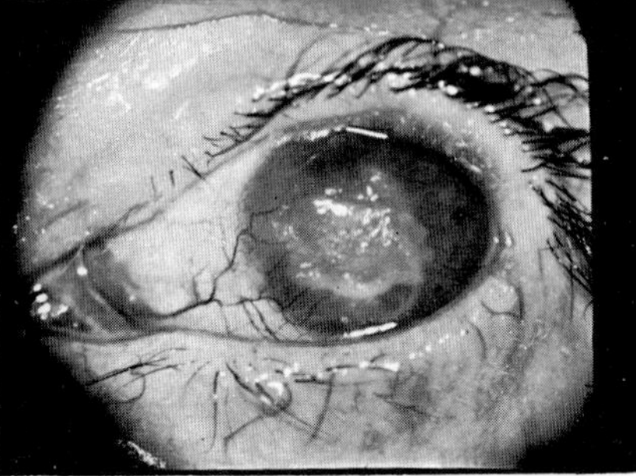

Fig. 1-7. Neuroparalytic keratitis of the left eye due to a trigeminal lesion after surgical removal of an acoustic neuroma. The illustration on the right is an enlargement of the ulcerated, dense, central corneal opacity with superficial vascularization originating from the limbus. (A coexisting facial palsy has been greatly compensated by plastic surgery.)

resulting *keratitis.* The ulcer seen in this type of keratitis has a typical shape. The upper border is a more or less horizontal line that corresponds approximately to the position of the upper lid margin during sleep. The lower border is a curved line parallel to the limbus. It is difficult to differentiate this form of ulcer from *neuroparalytic ulcer* (Fig. 1-7). Here we also have a perilimbal ciliary or mixed injection, a superficial cloudy zone (consisting mostly of delicate vesicles) in the center of the cornea or slightly below it, positive staining of the damaged epithelium with 2% fluorescein, and eventually outright ulceration. In contrast to lagophthalmic keratitis, the annoying symptoms of foreign body sensation, photophobia, and blepharospasm are missing in patients with neuroparalytic keratitis because the impairment of the trigeminal nerve has caused a loss of sensitivity. The testing of the *corneal sensitivity* (p. 21) should be particularly helpful in the differential diagnosis. On the other hand, in cases of lagophthalmic keratitis a paresis or paralysis of the orbicularis muscle should be taken into consideration. *A combination of a lesion of the facial nerve with one of the first branch of the trigeminal nerve,* as seen frequently in patients with tumors of the cerebellopontine angle, especially postoperatively, will cause a *particularly severe* involvement of the cornea. If one does not perform an early tarsorrhaphy in such patients, the corneal ulceration may spread, the stroma may undergo a dense, snow-white infiltration, and the cornea may show superficial and deep vascularization—complications that may reduce the central vision to mere light perception. A superimposed infection or perforation of such ulcers may lead to the complete functional loss of the eye and frequently necessitate enucleation. We were fortunate to have observed such cases only on rare occasions. However, the gravity of such a complication and the failure of our therapeutic measures always made a deep impression on us. Corneal transplants usually are out of the question because of the severe neurotrophic disturbances and the extensive vascularization.

Disturbances of motility as well as *pupillary changes,* some of the most important changes in the external aspects of the eyes and of great diagnostic significance, will be discussed separately.

Corneal sensitivity

Corneal sensitivity, one of the attributes of the first branch of the trigeminal nerve in its area of distribution, plays an extremely *important part* in the neuro-ophthalmologic examination for a group of brain tumors, mainly because diagnostically valuable disturbances occur before the appearance of hyperesthesia or hypesthesia of the skin.

Corneal sensitivity is tested with a small paper strip, a small bead on the end of a pencil (Kearns), or preferably a small, moist, pointed wisp of cotton. The examiner carefully touches the upper and lower halves of the corneas of both eyes with these objects and compares the force of the orbicularis reflex in the various areas of contact. The direct as well as the consensual corneal reflex on the opposite side should not be neglected. In testing the upper quadrants of the cornea, one cannot avoid, as a rule, lifting the upper lids a bit with the finger and asking the patient to look down. This is desirable because the force of the closure reflex of the lids can be felt with the finger. It is also important to ask the patient on which side he feels the touch of the test object less distinctly. If the difference between the two sides is slight, the test may be equivocal. In brain stem lesions it is important to differentiate fifth (sensory) from seventh (motor) nerve impairment of the corneal reflex. Examining the sensitivity of the conjunctiva, the skin of the lids, or even the remainder of the skin in the area of distribution of the first branch of the trigeminal nerve (particularly the frontal and temporal area) or even of the second trigeminal division may possibly supply additional information. A more exact method employs the use of Frey's hairs. For clinical purposes only the modification suggested by Boberg-Ans might be suitable.

In a unilateral hypesthesia or anesthesia of the cornea it may be necessary to rule out local conditions (status after trauma, zoster, or herpes simplex of the cornea). *The unilateral absence or diminution of the corneal reflex* may represent an *important sign of a tumor in the cerebellopontine angle.* We have noted this condition in practically every case of acoustic neuroma in our series of patients. We have emphasized that the branch of the ophthalmic division of the trigeminal nerve supplying the skin need not necessarily be involved. Disturbances of corneal sensitivity, though combined with lesions of other branches of the trigeminal nerve, are found in the *cavernous sinus,* the *superior orbital fissure,* and the *apex syndromes* (various combinations of involvement of the third, fourth, fifth, and sixth cranial nerves; in the case of the apex syndrome, also involvement of the optic nerve, occasionally associated with exophthalmos).

Pupils and pupillary reactions

The true significance of pupillary disturbances due to brain tumors is based on the protracted course of the reflex arc and the central position of the pupillary center in the midbrain. Even more important, the afferent part of this reflex arc follows the optic nerve and optic tract and the efferent part follows the oculomotor nerve, both structures quite frequently involved by cerebral neoplasms (Fig. 1-10).

During the examination the *size* of the pupils should be observed first. Even under normal conditions it varies with age (wider pupils in children), the color of the iris (wider pupils in blue eyes), the refractive state (wider pupils in myopic persons), and the tone of the autonomous system (wider pupils in sympatheticotonic persons). A relatively symmetrical dilatation of both pupils varying from moderately wide to maximum wide occurs in the terminal stage of increased intracranial pressure. This phenomenon has been interpreted by Fischer-Brügge as a bilateral clivus ridge syndrome due to damage of both oculomotor nerves immediately after their emergence from the brain stem; others (Scharfetter) have interpreted it to be a consequence of irritation of the sympathetic nervous system. An abnormal bilateral miosis occasionally is a sign of a pontine tumor.

More important for diagnostic and localizing purposes are *differences in size of the pupils* (anisocoria). It is advisable in such cases to measure the size of the pupils to have some data on record for the sake of comparison at a later date. It is best to use a *pupillometer,* a small ruler with graduated circles indicating pupillary sizes from 2 to 8 mm. The actual size of the pupil examined can be determined by comparing it with a circle of the corresponding size on the ruler. For repeated examinations on subsequent occasions a constant illumination (possibly with the aid of a light meter) is, of course, essential. Obviously it is not always easy to decide whether a difference in size of the pupils indicates a pathologic mydriasis on one side or an abnormal miosis on the other side. Usually there are other signs that should make a differential diagnosis easier, and these will be stressed in the discussion to follow.

A unilateral mydriasis is caused either by *stimulation* of the *sympathetic fibers* or by *paralysis of the sphincter of the pupil* secondary to an *involvement of the oculomotor nerve.* A unilateral irritation of the sympathetic branches is rarely caused by tumors. It is evidenced by a widening of the lid fissure and exophthalmos, signs that are occasionally observed only unilaterally in Graves' disease. The paralytic mydriasis plays an important part in brain tumors. In describing subjective symptoms, we have already mentioned that it is frequently associated with a more or less complete paralysis of accommodation. The *impair-*

ment of accommodation can be determined by testing the near vision. The patient is asked to read the smallest print possible at a distance of 20 to 25 cm.— first with the good eye, next with the eye involved. If he is unable to read small print with the latter, perhaps even at an increased distance, a paresis of accommodation can be assumed. This can be compensated by a convex lens (up to 4 diopters), which allows the patient to read at a normal near point. *Mydriasis and paresis of accommodation* are typical *signs of a nuclear lesion of the oculomotor nerve* in the midbrain. Tumors may cause such a lesion in this area, usually in combination with a vertical gaze palsy (Parinaud's syndrome). A unilateral mydriasis (with or without paresis of accommodation) may also occur in patients with a peripheral oculomotor lesion. It has been mentioned repeatedly as one of the signs of the clivus ridge syndrome (Fischer-Brügge) (Fig. 2-46). This may be either the *manifestation* of a *generalized increase in the intracranial* pressure or a *localizing sign of a temporal lobe tumor* (here may be an early slight homolateral ptosis) due to pressure on the peripheral oculomotor nerve against the clivus ridge by displaced brain substance. In the early stage of the clivus ridge syndrome the mydriatic pupil will still react slightly to light. During the course of the disease the reaction will become increasingly more sluggish. Once the patient has lapsed into unconsciousness, the pupil will be fixed. In this connection we should like to mention the importance of unilateral mydriasis for

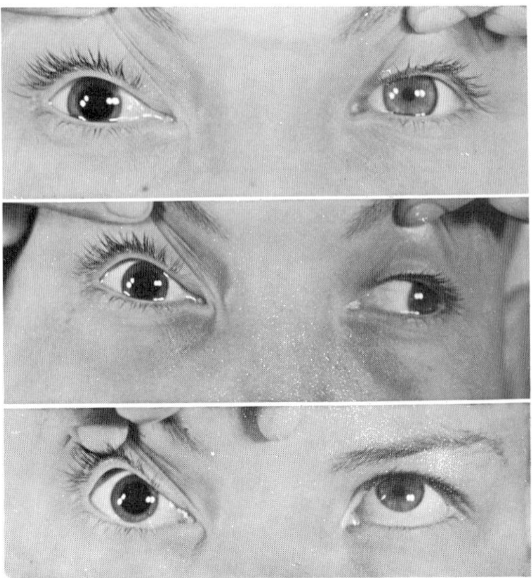

Fig. 1-8. Unilateral mydriasis with no pupillary reaction to light and on convergence; paresis of accommodation. This is an absolute pupil rigidity in a patient with right basal oculomotor paralysis caused by an intracavernous aneurysm of the internal carotid artery. The lower photographs illustrate the inability of the right eye to adduct (medial rectus muscle) and to elevate (superior rectus muscle).

the localization of epidural or subdural hematomas (Krayenbühl and Noto). A unilateral mydriasis in *combination with paresis of the extraocular muscles supplied by the oculomotor nerve* plus ptosis is seen in even more distal lesions of the oculomotor nerve, as for instance in the syndromes of the cavernous sinus, the superior orbital fissure, or the orbital apex—all of which could be caused by tumors in addition to other factors (Fig. 1-8). There are also numerous reports of pituitary tumors that may cause peripheral oculomotor pareses with mydriasis and ptosis during the course of an extrasellar lateral expansion.

In the differential diagnosis of a unilateral mydriasis the following conditions should be ruled out: *amaurosis* of the eye in question (absence of the pupillomotor stimulus), *congenital anisocoria, drug mydriasis* (homatropine, atropine, cocaine), and in particular *Adie's syndrome* (a wide pupil that does not react to light and that will dilate even more in subdued light, a pupil with a tonic convergence reaction and prolonged, slow redilation when looking at distance, and a tonic accommodation; this syndrome is mostly unilateral and occurs more frequently in females). Local conditions of the eyes (such as iritis, glaucoma, and traumatic sphincter tears) can, as a rule, be easily excluded by a closer inspection of the eye.

A *unilateral miosis* is caused either by a stimulation of the iris sphincter or by a *paralysis* of the dilator. The former occurs rarely with brain tumors. Fischer-Brügge describes the unilateral miosis as quasi a first stage of the clivus ridge syndrome. There is at first a stimulation of the oculomotor nerve, which later gives way to a partial and, in the end, a complete palsy. We have not observed a unilateral miosis as an early sign of a general increase in the intracranial pressure among our patients. The other possibility, a *paralysis of the sympathetic fibers,* results in a *Horner syndrome* characterized by the triad of miosis, ptosis, and enophthalmos. Occasionally the skin of the forehead on the same side is paler and drier than on the opposite side due to a dyshidrosis. In young individuals with brown irides, a unilateral lesion of the cervical sympathetic fibers may be followed by a depigmentation of the iris, causing the involved eye to appear lighter or bluish (heterochromia). It may become necessary to differentiate an "oculomotor" from a "sympathetic" miosis, especially if there are no distinctive accompanying symptoms. The *cocaine test* (Hughlings Jackson) is particularly useful for this purpose. One instills a few drops of a 4% cocaine solution into the eye to be examined and observes the behavior of the pupil. If the pupil fails to dilate with cocaine, a sympathetic stimulator, there must be a lesion in the second or third neuron (ciliospinal center, superior cervical ganglion, or eye). If the pupil dilates, there is either an irritation of the oculomotor nerve or a sympathetic lesion in the first neuron (between the hypothalamus and the ciliospinal center). For a precise localization of a sympathetic lesion with a Horner syndrome the pharmacodiagnostic scheme of Foerster and Gagel is recommended (Table 1).

The first neuron (hypothalamus to ciliospinal center) is the one most likely to be affected by brain tumors. We have reviewed our series of patients with

Table 1. Scheme for localization of a break in the chain of the three neurons of the sympathetic system

Testing substance	Mydriatic effects		
	First neuron	*Second neuron*	*Third neuron*
Atropine, 1%	+	+	+
Cocaine, 4%	++	–	–
Adrenaline, 1%	–	–	+++

First neuron = Hypothalamus to ciliospinal center.
Second neuron = Ciliospinal center to superior cervical ganglion.
Third neuron = Superior cervical ganglion to eye.

tumors or with pressure symptoms in the hypothalamic region (craniopharyngiomas). Despite the fairly large number of patients, there was only one with a slight miosis but none with a fully developed Horner syndrome. More distal would be the region of the pons. A unilateral or bilateral "pinpoint-sized" miosis has been mentioned as typical for pontine and intrapontine tumors. There was no case of a pontine tumor with such miosis in our series. Primary or metastatic tumors located in the medulla oblongata, in the upper cervical spine, or directly in the ciliospinal center (lowest part of the cervical or highest part of the dorsal spine) are also possible sites for a Horner syndrome. An intracranial lesion of the sympathetic system supplying the eye could, however, also occur in the most distal part, that is, in the third neuron; sympathetic fibers accompany the internal carotid and ophthalmic arteries in their course through the cavernous sinus and the superior orbital fissure. Such neoplasms usually cause a simultaneous oculomotor paresis or paralysis, resulting in additional extraocular muscle pareses, a ptosis, and a homolateral mydriasis; in other words, the sympathetic miosis would be superseded by the parasympathetic mydriasis. It is not easy in such cases to demonstrate the sympathetic lesion because the miosis, the outstanding sign, is missing. Instillation of atropine will cause a less extensive mydriasis of the homolateral pupil because of the obscuring damage to the sympathetic fibers.

An *Argyll Robertson* pupil is the primary consideration in the differential diagnosis of a unilateral miosis. It is also known as *reflex pupillary paralysis* and is characterized by its inactivity to light and the so-called *"spinal miotic triad"* (miosis, anisocoria, irregularities of pupils), the latter a sign of a spinal sympathetic lesion.

After one has noted a possible difference in the size of the two pupils, the *reaction to light* is tested. A small but bright pencil flashlight serves quite well for clinical purposes. It is rapidly moved from one eye to the other, with the light beam in the visual axes. It is best if the examining room is not too brightly illuminated so the light source of the flashlight furnishes the differential threshold for the testing of the pupillary reaction. The room should not be completely dark since this makes the evaluation of the consensual light reaction difficult or impossible. In testing the direct reaction to light, the opposite eye should be covered by hand or, even better, by a black cloth, lest a consensual reaction interfere with the test, giving a false result. Note the rate of speed of the pupillary contraction, the extent of the contraction, whether the contraction is uniform or limited only to parts of the pupillary margin, and the length of time the contraction is sustained. Electronic pupillography may be used to measure accurately pupillary size, detect differences in size, and evaluate abnormal pupillary reactions, but this technique is still largely a research method. Even under normal conditions the light reaction shows considerable variations. It is less active in older than in younger people, less active in more darkly pigmented irides, and less active in small pupils. The consensual reaction normally is exactly equal to the direct reaction. In a completely amaurotic eye the pupil is wider than that of the other eye.

Disturbances of the pupillary reaction to light are of considerable importance in brain tumors. In patients with *amaurotic pupil rigidity* the pupil on the blind side is wider than that on the other side. If the amaurotic eye is exposed to light, neither pupil will react. Both pupils will react if the healthy side is exposed to light. Amaurotic pupil rigidity occurs in patients with tumors of the optic nerve and its sheaths, with pituitary tumors, with tumors of the tuberculum sellae, and with meningiomas of the sphenoid ridge—provided the function of one optic nerve has been destroyed completely by one of these tumors. If the interruption of the optic nerve is partial rather than complete, the direct reaction to light on the involved side and the consensual reaction on the other side are distinctly diminished. Generally, there is a correlation between the impairment of the visual function and the degree of the pupillary reaction to light. Occasionally, however, there is still a trace of light reaction in a blind eye with a secondary optic atrophy after papilledema. It is considered evidence in favor of the hypothetical separate pupillomotor fibers of the optic nerve, which are supposed to be more resistant. If only one of the optic nerves has been damaged anywhere between the globe and the chiasm (for instance, by the type of tumors just mentioned), Kestenbaum's modification of the *Marcus Gunn pupillary sign* may be significant—under uniform illumination of the face the pupil of the affected eye will be larger with the normal eye covered than the pupil of the normal eye with the affected eye covered. The pupillomotor strength of the eye involved, with identical illumination, is weaker than that of the healthy eye. If one is able to rule out a purely local cause (such as lens or vitreous opacities or some retinal involvement) in a patient with unilateral decrease of vision, a positive Marcus Gunn sign is rather definite proof of a retrobulbar interference in the conductivity of the homolateral optic nerve between the disc and chiasm (Fig. 1-9).

An interruption within the optic tract is known to cause a homonymous hemianopia. If the defect is located between the chiasm and the point where the pupillary reflex fibers branch off just distal to the lateral geniculate body, there is necessarily an impairment of the pupillary reaction of the Wernicke type

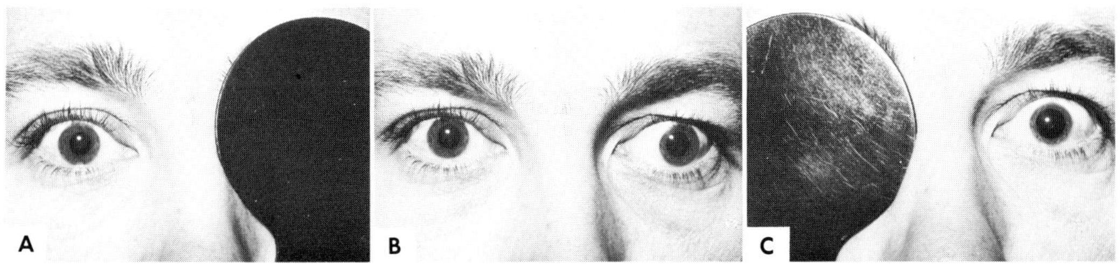

Fig. 1-9. Positive left Marcus Gunn pupillary sign after a traumatic lesion of the left optic nerve (white pallor of whole disc, amaurosis). The pupil of the affected eye with the normal eye covered is larger, **C**, than the pupil of the normal eye with the affected eye covered, **A**. **B**, Both eyes under uniform illumination of the face. Note practically equal size of both pupils and left affected eye in slight divergent deviation (concomitant squint).

(homonymous hemianopic pupil rigidity). Neither pupil will react if the light falls on the blind retinal halves, but both pupils will react promptly if the light falls on the intact halves. The Wernicke sign is of some importance in cases of homonymous hemianopia caused by a tumor if one wishes to decide whether to place the lesion in the optic tract or the optic radiation. Actually, it is enormously difficult to demonstrate a hemianopic pupil rigidity because of the problem of projecting the light strictly on one half of the retina; there is interference both by scattering of the refractive media and the transparency of the sclera. Furthermore, a physiologic factor has to be taken into consideration as a possible source of error in testing for a hemianopic pupil rigidity: the nasal half of the retina always shows a more distinct pupillomotor effect than the temporal half. The significance of Wernicke's sign can be summarized as follows: if it is definitely positive, the lesion is situated in the optic tract; if it is equivocal, the lesion may be either in the tract or the radiation. Certain observations by Harms tend to make this question even more confusing. He was able to demonstrate a hemianopic pupil rigidity even in certain hemianopias due to lesions in the optic radiation, an observation that enabled him to actually delineate the field defect. This should be further proof that Wernicke's phenomenon, or rather the influence of the higher visual centers on the pupillary reflexes, should be carefully reevaluated. One would expect a bitemporal hemianopic pupil rigidity in patients with bitemporal hemianopia caused by pituitary tumors. However, we have never observed it in a fairly large number of patients. (See Fig. 1-10.)

In addition to the reaction to light, the pupillary *reaction on convergence* should always be tested. Its presence or absence may be of importance not only from the point of etiology but also for the purpose of localization. The illumination of the room should be even and not too bright. The patient, after having gazed into distance, is asked to fix on an object (such as a finger or a pencil) held close to his nose. The examiner should note the contraction of the pupils that occurs normally and observe any deviation from this normal reaction,

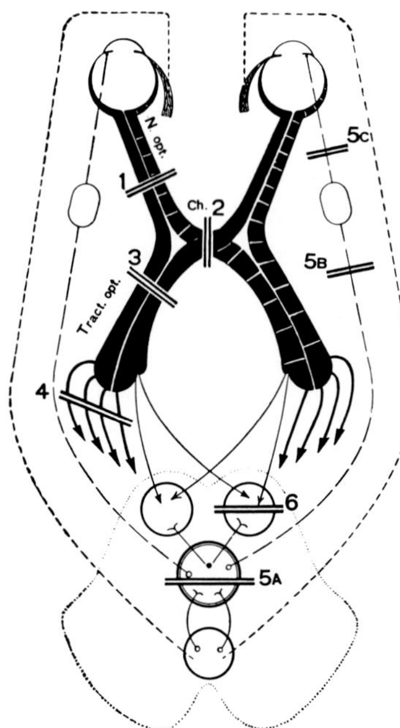

1, Lesion in optic nerve = Amaurotic pupil paralysis

2, Lesion in chiasm = Bitemporal hemianopic pupil rigidity

3, Lesion in optic tract = Homonymous pupil rigidity (right)

4, Lesion in optic radiation = Homonymous hemianopia (right) without hemianopic pupil rigidity

5a, Lesion in sphincter nucleus (Edinger-Westphal nucleus) = Absolute pupil rigidity

5b and 5c, Lesion of sphincter fibers in oculomotor nerve = Absolute pupil rigidity

6, Lesion in synapsis between sensory and motor part of reflex arc = Light rigidity of pupil (Argyll Robertson)

Fig. 1-10. Pathways for the pupillary reflexes. The circle below the sphincter nucleus, **5a,** indicates the center of convergence (Perlia's nucleus?). It is the point of origin for impulses for the convergence of the eyes to a near point, for accommodation, and for the near reflex of the pupils.

such as diminution or complete absence or a difference between the two sides as well as the dilatation of the pupils on decrease in convergence.

The so-called *light rigidity of the pupil* (Argyll Robertson) is characterized by a usually bilateral absence of any reaction to light in the presence of a well-preserved or even exaggerated reaction on convergence (Fig. 1-10). A unilateral or bilateral miosis due to a superimposed sympathetic lesion may or may not be an additional sign. It is essential to demonstrate (if necessary, with the aid of a loupe or slit lamp) the complete absence of the reaction to light. In its typical form (possibly associated with anisocoria, miosis, and the failure to dilate after the instillation of atropine) the Argyll Robertson pupil is pathognomonic for syphilis of the central nervous system, tabes as well as general paresis. The exact seat of the lesion is still quite controversial. It is very likely the gray matter surrounding the Sylvian aqueduct in the midbrain, causing an interruption of the afferent arc between the primary visual centers and the Edinger-Westphal nucleus. We have purposely described the typical picture of the Argyll Robertson pupil in such detail because *atypical forms* are observed in *tumors of the midbrain,* especially those of the anterior quadrigeminate plate, the third ventricle, and the Sylvian aqueduct. Among ten patients with tumors in this location, we saw eight with bilateral pupillary disturbances of the Argyll Robertson type and one unilateral case. Half of these patients had pinealomas. The disturbances were atypical either because the pupils were not completely refractory to light or because there was an incomplete or even complete absence of reaction on convergence. If the pupils, in addition to being totally or partially fixed to light, show a complete or partial loss of convergence reaction, it would be more correct to speak of a *complete or incomplete general pupil rigidity*. It is more appropriate to use these terms for the pupillary reactions of some of the tumors in the midbrain than to talk about an atypical Argyll Robertson pupil. For practical clinical purposes these differences are of a more academic interest because the nuclei for the light and near reaction in the midbrain are very close together. Neoplasms may damage both nuclei or involve just one. Furthermore, such tumors only rarely cause an isolated involvement of the pupillary nuclei in the midbrain but rather additional (and not less characteristic) signs such as isolated palsies of extraocular muscles innervated by the oculomotor nerve, vertical gaze palsies, and paralysis of convergence—all signs which, in their entirety, form *Parinaud's syndrome*.

A *general pupil rigidity* (that is, absence of light and near reaction, together with mydriasis and perhaps even paralysis of accommodation) suggests most of all a focus involving the sphincter nucleus (Edinger-Westphal). We have seen such a case of an isolated unilateral pupillary disturbance following encephalography. We made a tentative diagnosis of encephalitis or multiple sclerosis activated by the diagnostic procedure. The complete or incomplete general pupil rigidity may, however, be caused by lesions localized in other areas, *lesions involving the efferent pupillary reflex arc (that is, the oculomotor nerve)*. The various forms of insults to the oculomotor nerve in its distal portion caused by tumors were discussed in the section on unilateral mydriasis (clivus ridge, cavernous sinus, superior orbital fissure, and orbital apex syndromes). The degree of the individual factors that form the general pupil rigidity (light reaction, near reaction, mydriasis) may vary. In the more severe and in the complete lesions there will be a greater tendency toward a complete paralysis to light and on convergence with a maximal mydriasis (for instance, in the clivus ridge syndrome in the advanced stage of unconsciousness). For the sake of completeness it should be stated that damage to the peripheral oculomotor nerve may cause any combination of pareses or palsies of extraocular muscles in addition to the pupillary impairment (Fig. 1-8).

The differentiation of a light or general pupil rigidity from *Adie's syndrome* usually presents no difficulty. The observation of a tonic convergence reaction and a slow redilation on divergence will help to establish the diagnosis. There are, however, atypical forms of the Adie pupil (for instance, if it is bilateral) that are not so easy to differentiate from the pupillary anomalies just discussed. Certain pharmacologic tests may then be helpful. Topical instillation of 2.5% methacholine (Mecholyl), a cholinergic substance, usually causes contraction of the Adie pupil but not of the normal pupil (Scheie). Atropine will cause dilation of the Adie pupil but not the Argyll Robertson pupil.

Testing the pupillary reaction to light may be of importance in yet another respect in the case of brain tumors. It may allow *differentiation of cortical blindness and peripheral amaurosis* (for instance, one caused by a pituitary lesion). In cortical blindness the pupils are of normal width and show a perfectly normal direct and consensual reaction to light. In bilateral (peripheral) amaurosis the pupils are rather large and show a complete absence of reaction to light in both eyes (or at the most a minimal reaction). The presence of a pupillary reaction to light enabled us to make a rather definite diagnosis of cortical blindness in the case of a bilateral x-ray necrosis of the occipital lobes cited on p. 8.

Motility of the eyes

Even though certain motor anomalies are merely general symptoms of increased intracranial pressure (Chapter 2) and scarcely of diagnostic value, there are other anomalies (accompanying especially tumors located at the base of the brain) that actually point to a definite area, thus making a specific diagnosis possible. The more or less sudden appearance of *diplopia,* the subjective equivalent of a motor anomaly, is quite frequently one of the first symptoms of a brain tumor in its initial stage and, as such, important for a topical diagnosis. Pareses of the extraocular muscles in the later stages should be regarded rather as manifestations of a general increase in the intracranial pressure and should be evaluated with caution. In the discussion of history and subjective symptoms it was emphasized that *motor anomalies do not always cause diplopia,* especially if they are slight or if they are of the supranuclear type. One of the objects of the clinical examination is to discover motility disturbances, even though they might be latent, and to analyze them in detail. Such an analysis includes not only the determination of the muscle involved but also the evaluation of secondary disturbances of the muscle equilibrium such as overaction and contractures of the antagonists. Such secondary changes are known to modify greatly and mask the original picture in pareses of long standing. The localization of certain disturbances of motility is facilitated if we classify them as extraocular muscle palsies, gaze palsies (that is, palsies of the conjugate eye movements), and nystagmus.

Extraocular muscle palsies

Extraocular muscle palsies indicate a *disturbed function of a single or several extraocular muscles caused by a process in the muscle itself, the neuromuscular junction, or its nerve or by a nuclear lesion in the brain stem* (Fig. 1-11).

One begins the test for a possible muscle paresis or paralysis by observing the *position of the eyes* in the primary position. The more obvious palsies result in a change in position of the eyes already in the primary position—less an

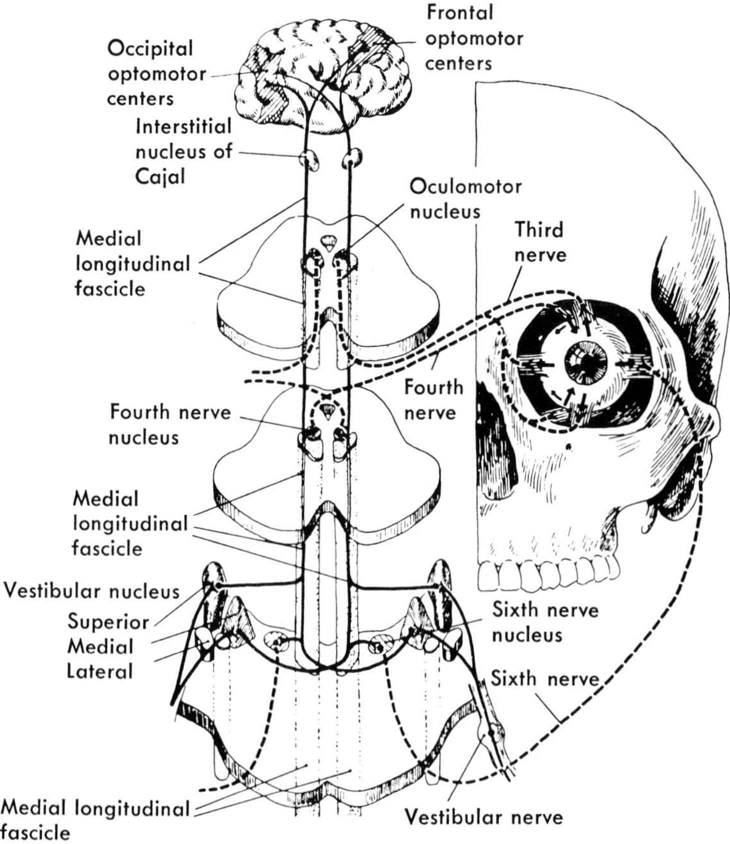

Fig. 1-11. Pattern of innervation of the extraocular muscles. Cortical optomotor centers, supranuclear centers for conjugate movements in the brain stem, posterior longitudinal bundle, nuclei of vestibular nerves, Cajal's nuclei, oculomotor nuclei, oculomotor nerves, and effective organs (six extraocular muscles). (After Netter.)

effect of the muscle paresis as secondary to an overaction of one or more antagonists. In patients with unilateral sixth nerve paresis, for instance, the involved eye frequently turns in because of the overaction of the homolateral medial rectus muscle. In a paresis of the medial rectus the eye turns out due to overaction of the lateral rectus muscle. In the first instance there seems to be a convergent strabismus; in the second instance, a divergent strabismus. In pareses of the vertical rotators there is also a compensatory torsion, depression, or inclination of the head—signs that are generally neglected. It should be stated quite emphatically that these are cases of *paralytic strabismus* that should be strictly differentiated from concomitant squint. The mere inspection of the position of the eyes permits no such differentiation, especially if performed only when the eyes are in the primary position. Only a careful study of the excursion of the eyes in the various directions of gaze will help to clarify the situation.

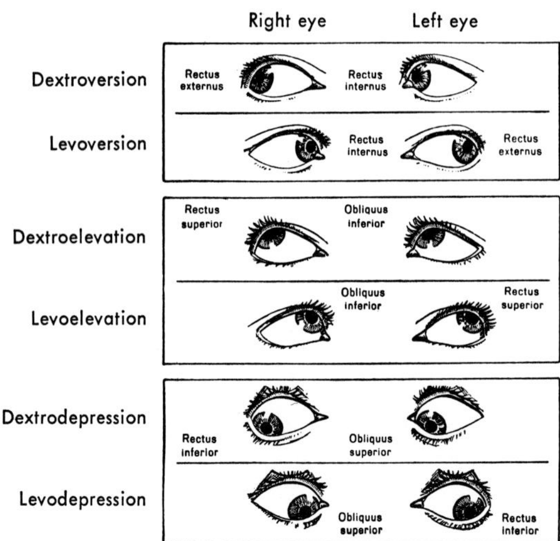

Fig. 1-12. The diagnostically important six cardinal directions of gaze with an indication of which individual muscles of each eye are involved in each instance. (After Lyle.)

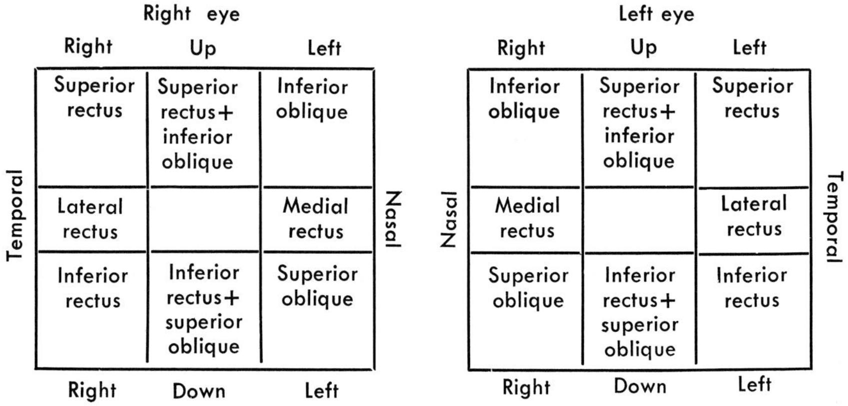

	Right eye					Left eye			
	Right	Up	Left			Right	Up	Left	
Temporal	Superior rectus	Superior rectus + inferior oblique	Inferior oblique	**Nasal**	**Nasal**	Inferior oblique	Superior rectus + inferior oblique	Superior rectus	**Temporal**
	Lateral rectus		Medial rectus			Medial rectus		Lateral rectus	
	Inferior rectus	Inferior rectus + superior oblique	Superior oblique			Superior oblique	Inferior rectus + superior oblique	Inferior rectus	
	Right	Down	Left			Right	Down	Left	

Fig. 1-13. Diagram of the principal action of the individual muscles of both eyes, disregarding torsion. Analysis of the impairment of motility of the eye according to this diagram should permit the determination of the paretic or paralytic extraocular muscle in simple cases. (After Franceschetti.)

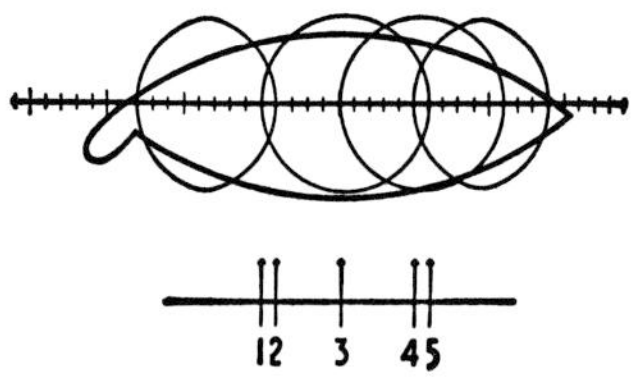

Fig. 1-14. Limbus test, according to Kestenbaum, for measurement of the excursion capacity of the eye: 5-2, normal internal rotation (9 to 10 mm.); 1-4, normal external rotation (9 to 10 mm.); 1-3, diminished external rotation (5 mm.). The horizontal graduated line represents the millimeter graduation of the transparent ruler.

Table 2. Franceschetti's diagram for examination and interpretation of diplopia with red-green goggles or with the Maddox test

Differential diagnosis of double images in the horizontal plane		
Type of separation (first criterion)	*Maximum of separation (second criterion)*	*Diagnosis*
Uncrossed (homonymous) diplopia	To the right	Paralysis of right lateral rectus
	To the left	Paralysis of left lateral rectus
	To the right as well as to the left	Paralysis of both lateral rectus muscles
	In the primary position	Paralysis of divergence or convergence spasm
	No maximum Diplopia rare	Convergent concomitant strabismus
Crossed (heteronymous) diplopia	To the right	Paralysis of left medial rectus
	To the left	Paralysis of right medial rectus
	To the right as well as to the left	Paralysis of both medial rectus muscles
	In the primary position	Convergence paralysis
	No maximum Diplopia rare	Concomitant divergent strabismus

Differential diagnosis of double images in the vertical plane			
Type of separation (first criterion)	*Increase in vertical separation (second criterion)*	*Maximum of vertical separation (third criterion)*	*Diagnosis (paralyzed muscle)*
Positive vertical separation (position of right eye higher; image of left eye seen higher)	Up	To the right and up	Left inferior oblique
		To the left and up	Left superior rectus
	Down	To the right and down	Right inferior rectus
		To the left and down	Right inferior oblique
Negative vertical separation (position of right eye lower; image of left eye seen lower)	Up	To the right and up	Right superior rectus
		To the left and up	Right inferior oblique
	Down	To the right and down	Left superior oblique
		To the left and down	Left inferior rectus

tinction between the recti and oblique muscles: *if the vertical separation of the two images reaches its maximum with the involved eye in abduction, the muscle in question is a rectus (superior or inferior); if the maximum occurs with adduction of the involved eye, the muscle in question must be an oblique.*

The more or less distinct inclination of the images as well as a slight horizontal disparity that occurs in pareses of the vertical rotators should be disregarded in these tests.

In our opinion these methods are perfectly adequate to arrive at a fairly accurate qualitative decision during a bedside examination. For quantitative purposes and for a repetition of the examination on different occasions, more exact methods have to be employed. The same is true for palsies of long standing and for those with pronounced secondary changes (overaction of the homolateral antagonist or the contralateral yoke muscle; secondary paresis of the contralateral antagonist). Such changes may be difficult to elicit with the simple clinical tests.

The *Hess screen* is ideally suited for this purpose. Also based on subjective diplopia, the separation of the two images is accomplished with the use of red-green goggles. The great advantage of this method is that, at the end of the test, the result can be read immediately from a diagram. It gives information on the paretic muscles as well as the overaction of the ipsilateral antagonists and contralateral synergists. If *several ocular pareses overlap,* as is likely to occur in patients with brain tumors, we have used the Hess screen as the *method of choice.* Lately we also use a modified Hess screen with a green and red dot projected by an electric flashlight on a tangent screen divided into squares. The examiner points one dot successively to the various squares on the screen. He directs the patient, whose head is in a fixed position and who wears the red-green goggles, to move his flashlight in such a manner that the other dot seems to cover the first one (Lancaster test). In the case of a muscle paresis the patient will project his dot in certain directions incorrectly, thus modifying the Hess diagram in such a manner that it will allow the determination of the type of muscle disturbance. The smaller "field of fixation" always

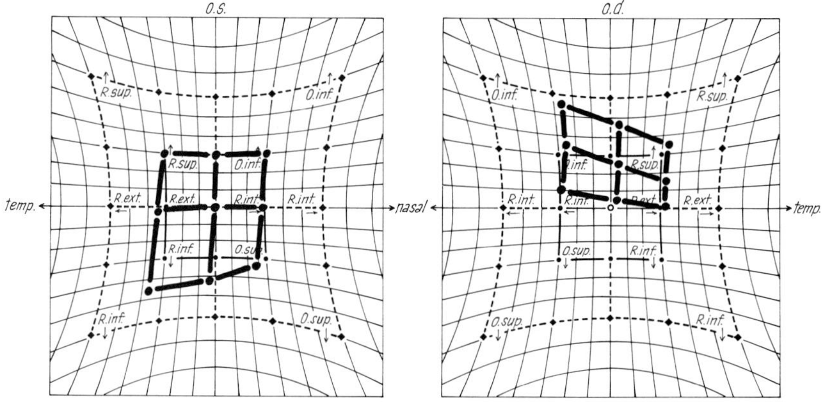

Fig. 1-15. Graphic account of disturbances of ocular motility obtained with the Hess coordinometer. The example demonstrates a paresis of the right superior oblique with an overaction of the left inferior rectus and a contracture of the right inferior oblique.

belongs to the eye with the paretic muscle (Fig. 1-15). For other details of the Hess-Lancaster method refer to the various textbooks of ophthalmology as well as to Lyle and Jackson's *Practical Orthoptics in the Treatment of Squint*.

For a quantitative and even more accurate analysis of motility disturbances we have used the *Maddox rod test* as an additional method. It is performed at a distance of 5 meters and is based on the method and interpretation elaborated by Franceschetti. We have examined patients with minute pareses that would not manifest themselves on the Hess screen but could be demonstrated without difficulty on the Maddox cross.

The Maddox test utilizes a tangent scale arranged in the form of a cross with a small source of light in its center. The patient fixes on this light, at a distance of 5 meters, with the red Maddox rod in front of one eye. This changes the point source of light to a streak that runs at a right angle to the direction of the rod. The patient should indicate whether the streak of light passes directly through the light seen by the other eye or whether there is a lateral or a vertical displacement. According to the position of the light streak, the lateral or vertical deviation can be read directly in angle degrees on a properly calibrated tangent scale. With paralyzed horizontal rotators, the displacement has to be determined only in the primary position and in dextroversion and levoversion. With paralyzed vertical rotators, it is advisable to measure the vertical displacement in the nine cardinal directions, with the head turned in the appropriate position. The diagram thus obtained is analyzed according to Franceschetti's suggestion (Table 2).

According to the literature as well as our own material, muscle pareses occur in 10 to 15% of patients with brain tumors. Sixth nerve pareses are by far the most frequent, with those of the oculomotor nerve in second place. Pareses of the superior oblique are rare.

As part of the test for an ocular paresis, one should attempt to get an exact *localization of the basic lesion,* an important factor in brain tumors. It is almost impossible to pinpoint a lesion from the type of ocular paresis. It is rather the accompanying signs that might supply such important clues. The following criteria may serve as guides in the examination of the various innervational disturbances of the nerve supplying the extraocular muscles (Fig. 1-16).

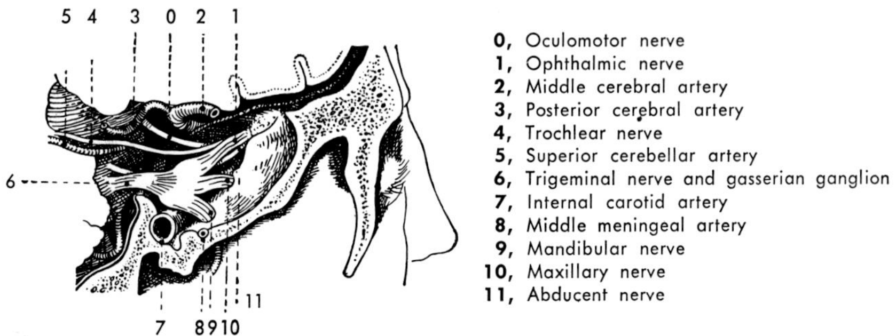

Fig. 1-16. Vertical section through the base of the brain and the base of the skull illustrating the intracranial course of the oculomotor, trochlear, abducent, and trigeminal nerves.

Abducent nerve (VI). The characteristic sign of the motility disturbance is a horizontal diplopia with maximal separation of the images in lateral rotation to the involved side. The eye will assume a convergent position if the paresis is severe. Because of the long and exposed intracranial course of the sixth nerve, it may suffer from distant effects of brain tumors in a variety of locations, perhaps due to displacement of the brain stem. For that reason an isolated sixth nerve paresis is of little interest as far as localization is concerned! Only in combination with other neurologic defects will it permit certain conclusions. (See Fig. 1-17.)

Nuclear types

Millard-Gubler syndrome. Paresis of the lateral rectus muscle, homolateral peripheral facial palsy, crossed hemiplegia (including crossed hypoglossal paresis).
Foville's syndrome. Paresis of the lateral rectus muscle, homolateral peripheral facial palsy, homolateral horizontal gaze palsy (due to additional damage to the homolateral posterior longitudinal bundle), possibly combined with a Horner syndrome.

Root types

Paresis of the lateral rectus muscle usually combined with a homolateral peripheral facial palsy.

Basal types

Paresis of the lateral rectus muscle combined with a varying combination of the third, fourth, fifth, sixth, seventh, and eighth homolateral cranial nerves.
    Clinical picture in cerebellopontine tumors (acoustic neuromas!)
    Gradenigo's syndrome. Homolateral paresis of the lateral rectus muscle and trigeminal lesion (pain, hypesthesia) as sign of an affection of the apex of the petrous bone (otitis, tumor).
    Syndromes of the cavernous sinus, superior orbital fissure, and orbital apex. Paresis of the lateral rectus muscle combined with pareses of the third, fourth, and fifth cranial nerves; perhaps also the optic nerve in the apex syndrome.
    (Tumors, especially metastases and aneurysms, in the cavernous sinus; pituitary adenomas with lateral extension; meningiomas of the middle fossa; trigeminal neurinomas; meningiomas of the sphenoid ridge.)

Oculomotor nerve (III). Characteristic signs of the motility disturbance include, in the fully developed picture, ptosis, paralysis of the superior rectus, medial rectus, inferior rectus, and inferior oblique muscles, and mydriasis as well as a more or less sluggish pupillary reaction to light and on convergence. In the primary position the involved eye turns out and down. There is horizontal and vertical diplopia in the primary position. The vertical diplopia increases in elevation and depression. The horizontal diplopia increases in rotation to the side of the uninvolved eye. (See Figs. 1-8 and 1-18.)

Nuclear types

Pareses of a single or of a few extraocular muscles supplied by the oculomotor nerve in one or both eyes. There may or may not be pupillary disturbances (mydriasis, sluggish pupillary reaction) and paresis of accommodation. In tumors within or near the midbrain (pinealomas) there is a combination of isolated muscle pareses with *vertical gaze palsy,* possibly a disturbance of convergence, and nystagmus retractorius (Parinaud's syndrome, Sylvian aqueduct syndrome, pineal syndrome).

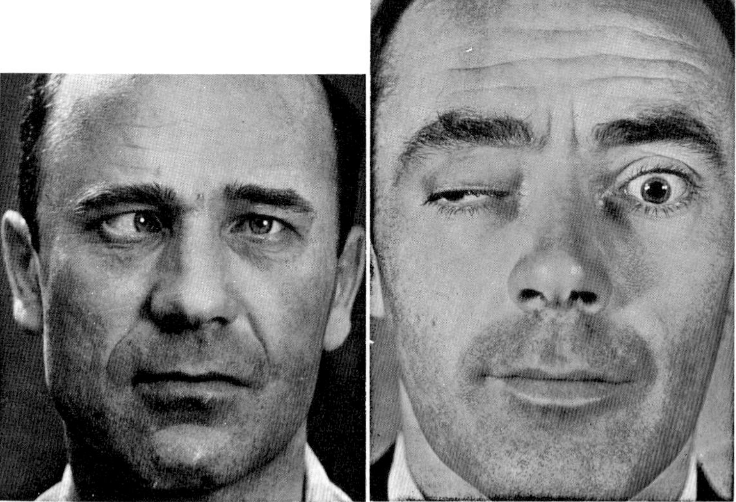

Fig. 1-17

Fig. 1-18

Fig. 1-17. Paresis of the right lateral rectus muscle. Convergent position of the right eye is due to overaction of the homolateral medial rectus muscle.
Fig. 1-18. Complete right oculomotor paresis. Ptosis, globe turns out and down, and mydriatic absolute pupil rigidity (not visible because of the ptosis).

Fascicular types

Dorsal form. Unilateral oculomotor paresis with crossed hemitremor (Benedict's syndrome), possibly also with crossed hemianesthesia.
Ventral form. Unilateral, mostly complete oculomotor palsy with crossed hemiplegia (Weber's syndrome), possibly with crossed central facial and hypoglossal palsies.

Root types

Unilateral oculomotor paresis with crossed hemiplegia (Weber's syndrome).

Basal types

A localized lesion shows a monosymptomatic oculomotor paresis with varying affections of the sphincter of the pupil and the ciliary muscle (aneurysms of the posterior communicating artery or rather at the site of its branching off from the internal carotid artery) (Figs. 1-8 and 5-11). A large diffuse lesion at the base of the skull (tumors, basal meningitis) is characterized by an additional involvement of the fourth, fifth, sixth, and possibly the seventh and eighth cranial nerves.

    Cavernous sinus syndrome. Unilateral oculomotor paresis (mostly with mydriasis and general rigidity of the pupil) combined with pareses of the fourth, fifth, and sixth cranial nerves (Fig. 3-91) Sinus sign! (Tumors, especially metastases and aneurysms in the cavernous sinus; laterally extending pituitary adenomas; meningiomas of the middle fossa; trigeminal neurinomas.) Posterior form: first and second (possibly also third) branches of trigeminal nerve involved. Anterior form: first branch of trigeminal nerve involved.
    Superior orbital fissure syndrome. Unilateral oculomotor paresis combined with a paresis of the fifth (first branch) and sixth cranial nerves (Fig. 1-19); absence of sinus signs!
    Apex syndrome. Pareses of the third, fourth, fifth (first branch), and sixth cranial nerves combined with lesions of the optic nerve (central scotoma; peripheral visual field defects; optic atrophy; or possibly papilledema). Exophthalmos in the case of tumors.

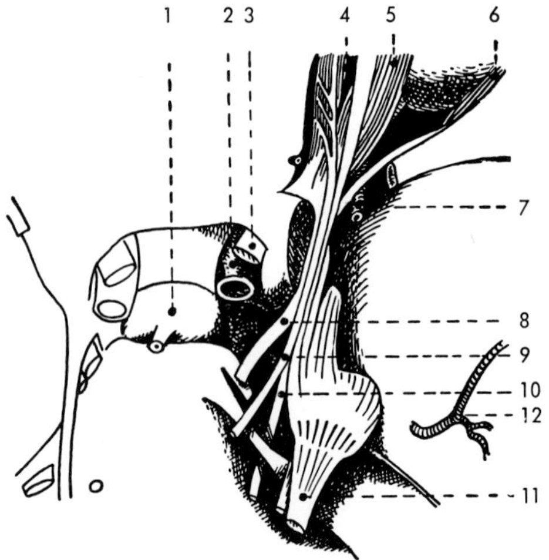

Fig. 1-19. Schematic illustration of the area of the chiasm and the cavernous sinus. Course of the nerves supplying the extrinsic ocular muscles as they enter the orbit from the middle cranial fossa.

Trochlear nerve (IV). Characteristic signs of the motility disturbance include slight upward deviation of the involved eye that increases with inward rotation and depression (there is a predominantly vertical deviation, with the maximum in the in and down position), a compensatory torsion, and a tilting of the head to the opposite side.

An isolated trochlear paresis is rare. It is of no localizing significance because the lesion could be nuclear, somewhere in the brain stem, or could affect the peripheral nerve.

Nuclear types

A trochlear paresis combined with a homolateral oculomotor paresis, occasionally in association with vertical gaze palsies, convergence spasm or convergence palsy, and pupillary disturbances is seen in tumors of the *roof of the midbrain* or *pinealomas* (pineal syndrome).

Peripheral types

See cavernous sinus, superior orbital fissure, and apex syndromes discussed with oculomotor lesions.

• • •

It is obvious that in the differential diagnosis of the extraocular muscle pareses many other causes in addition to tumors have to be considered that were not discussed in detail (encephalitis, meningitis, neuritis, syphilis, tuberculosis, exogenous poisons, diabetes, vascular accidents, traumas, multiple sclerosis, syringobulbia, and others). Although a large number of etiologic factors have to be ruled out by whatever special type of investigation is indicated, there is one in particular that should always be considered and which we have en-

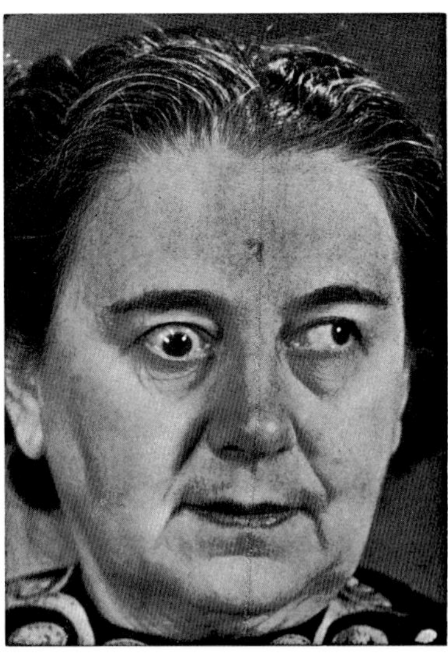

Fig. 1-20. Multiple pareses of the extraocular muscles (right superior and medial rectus muscles, left medial and inferior rectus muscles, intermittent ptosis of the left eye) in myasthenia gravis. In this patient the involvement of the ocular muscles was the only manifestation of the disease.

countered time and again in connection with patients suspected of brain tumors—namely, *myasthenia gravis.* Even though there is an actual involvement of the neuromuscular junction, the resulting picture can easily be mistaken for muscle pareses of neurogenic origin. As a rule, the muscle disturbances in patients with myasthenia show conspicuous fluctuations in degree. In the morning, after the patient has had a good night's rest, they may be missing completely, only to return (together with a ptosis of the upper lid) during the course of the day. Signs of fatigue in other muscle groups (muscles of speech and mastication, muscles of the extremities) may be completely missing initially. The disease may even remain limited to the ocular muscles. The edrophonium (Tensilon) test (intravenous injection of 5 to 10 mg. edrophonium), which is almost specific for myasthenia gravis, will, in many cases, cause the paresis and, with it, the subjective diplopia to disappear for a short time. This is not always the case. We have observed one patient with myasthenia gravis who had several bilateral multiple ocular pareses that did not respond to edrophonium (Fig. 1-20). They became permanent and required surgical correction. This patient had been referred to the department of neurosurgery because a tumor in the brain stem had been suspected!

The electromyographic examination of an extraocular muscle during the edrophonium test is indispensable in patients suspected of having myasthenia but in whom neither the ptosis nor the extraocular paresis improved clinically after administration of edrophonium. The electromyogram (EMG) can give

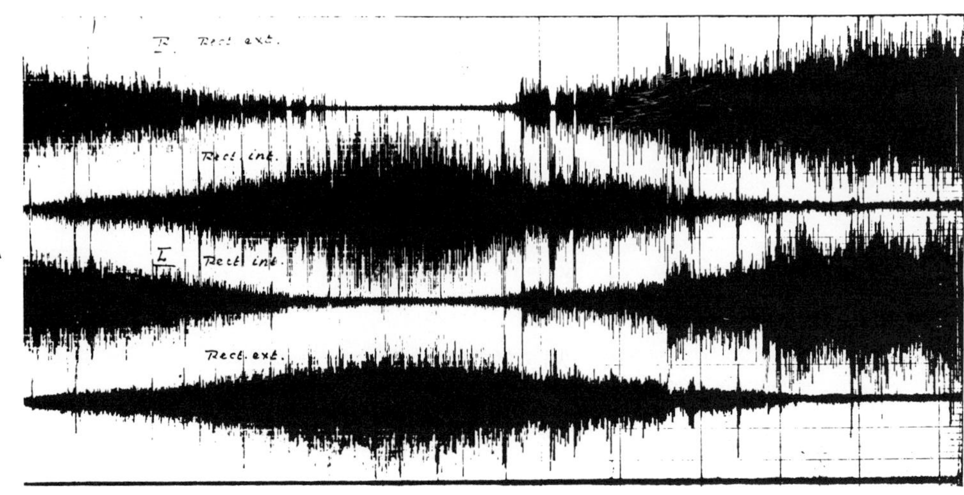

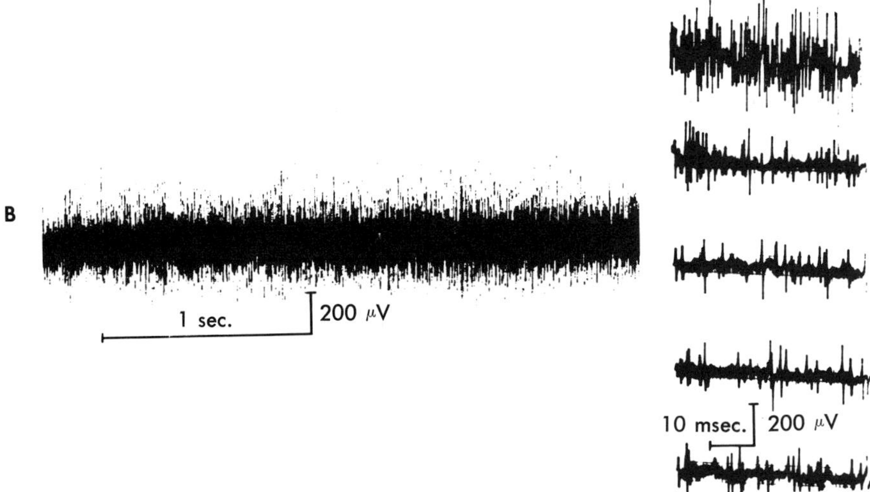

Fig. 1-21. A, EMG's of the four horizontal recti of both eyes. Augmentation of activity in the agonist is accompanied by a concomitant parallel decrement of activity in the antagonist. On rotation out of the field of action, reduction of electric activity to the firing of a few motor units; on rotation into the field of action, increase of frequency of discharge and appearance of new motor units (interference pattern). **B,** EMG of ocular myopathy. Intensive interference pattern upon effort despite little or no movement of the affected muscle. Increased number of polyphasic potentials (see curves on the right).

a positive edrophonium effect (transient improvement of the neuromuscular junction disturbances and activation of motor units) even when no visible effect on the extraocular muscles is observed clinically (Fig. 1-21).

Electromyography plays an important role today in the diagnosis of paresis of extraocular muscles (Breinin; Huber; Esslen and Papst; Blodi; Tamler and Jampolski). This examination method makes it possible to differentiate between myopathies (myositis, muscular dystrophy, myotonia), myasthenia (dis-

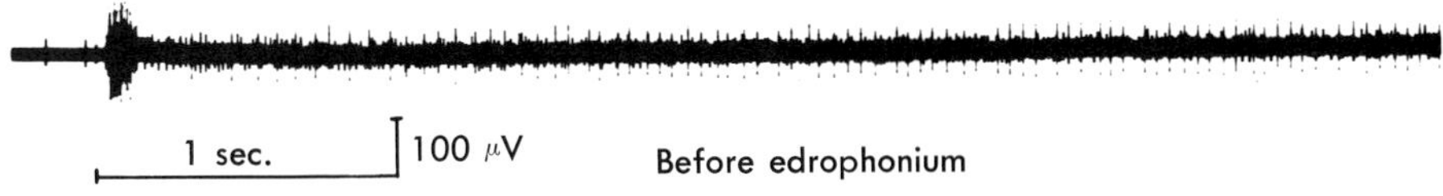

C

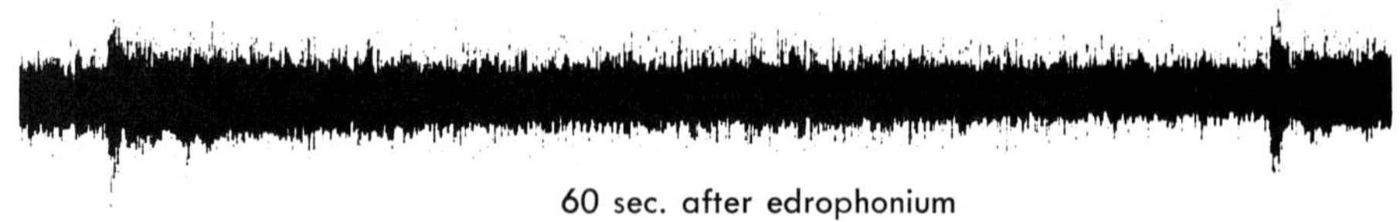

D

Fig. 1-21, cont'd. **C,** EMG of myasthenia gravis. Progressive decrease in electric activity of the muscle during sustained effort (fatigue pattern) (top curve). Prompt building up of electric activity up to the interference pattern after intravenous injection of edrophonium (Tensilon) (bottom curve). Reaction pathognomonic for myasthenia! **D,** EMG of neurogenic palsy. Irregular or sparse recruitment and falling out of motor units, poorly sustained discharge, and loss of interference pattern (top curve). Appearance of fibrillations (discharge of single muscle fiber potentials without relation to volition) as cardinal sign of denervation after 2 to 3 weeks (bottom curve).

turbances of the neuromuscular junction), peripheral neurogenic paresis, and supranuclear disturbances (gaze palsies, internuclear ophthalmoplegia, convergence palsies) (Fig. 1-27). This differentiation is of the utmost importance when dealing with ocular muscle paresis produced by cerebral tumors. In these cases we are dealing either with a peripheral neurogenic or with a supranuclear form. These have to be differentiated from myopathies or myasthenia. Depending upon the case, the electric response has to be obtained from one or several of the extraocular muscles, occasionally even at different time intervals.

In patients with an isolated paresis of the sixth nerve (less frequently in those with pareses of the oculomotor nerve), *multiple sclerosis* always has to be considered. Occasionally it causes ocular symptoms long before the occurrence of other typical neurologic symptoms. If such a paresis is accompanied by vague chronic headaches (we have seen such cases), a brain tumor may be rightly suspected. With the modern diagnostic methods that the neurosurgeon has at his disposal, brain tumor usually can be ruled out with a great degree of probability. Examination of the spinal fluid and perhaps the demonstration of other very minute early signs of multiple sclerosis are helpful in establishing the diagnosis.

Next to the possibility of a brain tumor, an isolated peripheral oculomotor paresis with involvement of all the extrinsic muscles, mydriasis, and sluggishness of the pupillary reactions should arouse suspicion of an existing *aneurysm at the base of the brain.* Aneurysms, in the broadest sense of the word, are tumors and therefore will be briefly discussed in Chapter 5 (see p. 295 and Figs. 1-8 and 5-11).

Palsies of conjugate eye movements

In contrast to the isolated extraocular palsies, with a disturbance of one or several individual muscles, *gaze palsies are disturbances of the higher centers involving movements that combine the two eyes, so to speak, into a single visual organ that makes uniform movements in space.* Gaze palsy is the inability to look in a given direction (that is, to the side, up, or down) or to converge the eyes and fix on a near object. As an example, in a right horizontal gaze palsy the right lateral and left medial rectus muscles do not function in an attempted movement to the right. If these muscles are tested individually or in connection with a different type of movement (for instance, convergence or divergence), they appear intact; they fail only if called upon to act within the higher function of the conjugate lateral movement.

Although gaze palsies are caused by lesions in various rather centrally located areas of the brain (midbrain, pons, hemispheres), they rarely occur in patients with brain tumors. Their frequency varies between 3 and 6%.

From a clinical and in particular from a diagnostic point of view it is expedient to subdivide the gaze palsies into horizontal, vertical, convergence, and divergence types.

A *horizontal gaze palsy,* by definition, is the inability to look to one side with both eyes. There are different forms of horizontal conjugate movements, such as an optically induced movement, a voluntary or command movement (or schematic movement), a pursuit movement, or a vestibular movement. All of these or one particular component movement might be disturbed. An analysis of these individual components is important for the topical diagnosis of a horizontal gaze palsy. The optically induced movement is tested by attempting to attract the patient's attention by approaching him from one side with some object (for instance, a flashlight). If the reflex of regard is positive, the eyes move in the direction of the object so that its image moves from the retinal periphery to fall upon the macula. Command movements are easy to test. The patient is asked to look to the left, to the right, up, and down, and the way his eyes react to these instructions is observed. Pursuit movements are induced by holding a moving object (such as a pencil or a finger) before the patient's eyes and asking him to follow the object. Vestibular movements are examined by moving the patient's head quickly in one direction and then observing whether or not there is a compensatory reflex movement of the eyes in the opposite direction. It should be stated quite emphatically that a pure horizontal gaze palsy does not cause diplopia because the impairment is equal in both eyes, thus causing no change in the relative position of the two eyes.

On the basis of such a detailed analysis of the lateral conjugate movements, it is quite possible to decide whether, in a given patient, the lesion is located in the hemispheres, the posterior longitudinal bundle, or the supranuclear centers of the pons (Fig. 1-11).

An isolated palsy of the voluntary and command movements with preservation of the other types of movement is suggestive of a localization in the gaze center of the *frontal lobe* (second frontal gyrus, Brodmann's area 8) or in the corresponding *internal capsule.* In a unilateral process the gaze palsy is toward the side opposite the lesion. Usually there is simultaneously a conjugate deviation toward the side of the lesion (the patient looks at the lesion!). As a rule, a horizontal gaze palsy is associated with facial palsy as well as hemiparesis or hemiplegia toward the side of the gaze palsy. Even if caused by a neoplasm, the gaze palsy is of a relatively short duration because obviously the opposite hemisphere vicariously compensates for the lost functions. Numerous cases of frontal lobe tumors or conditions after lobectomy because of such a tumor have been reported in which the patients showed such a gaze palsy. We were unable to observe a lateral gaze palsy in the form just described among a fairly large number of patients with frontal lobe tumors. However, seizures with conjugate deviation of the eyes can be observed in frontal lobe tumors. The so-called *adversive seizure* is actually a focal motor epilepsy involving, in part, the frontal cortical center for lateral gaze and is observed as tonic and clonic contractions of the extraocular muscles that move the eyes away from the irritative focus. It is commonly accompanied by a facial twitching on the affected side and turning of the head in that direction. Often a jacksonian progression of motor epilepsy over the extremities of the side contralateral to the focus follows or precedes this event. Termination of the seizure is characteristically followed by paralysis of the muscles involved in the ictus, including a transient loss of lateral gaze in that direction, with the result that the eyes are often deviated to the side opposite the irritative lesion.

According to modern research on the neurophysiology of cortical control mechanisms, there appears to be a broad representation in oculomotor function throughout the cortex. It has been emphasized time and again that, in addition to the anterior hemispheric oculomotor centers, there are also posterior centers in the parietal and occipital lobes (parastriate and peristriate areas). Experiments with electric stimuli suggest that similar horizontal conjugate disturbances may occur in patients with lesions of the parietal and occipital lobes. The distinguishing feature of lesions in the *occipital oculomotor centers* (Brodmann's areas 18 and 19) supposedly is a pronounced involvement of the fixation mechanism (fixation with convergence and accommodation, optically induced movements, pursuit movements, optokinetic nystagmus) (Walsh; Holmes; Holmes and Horrax). Except for the optokinetic nystagmus, we have been unable to find examples of such disturbances in the patients with hemispherical tumors just mentioned. This may perhaps be due to the fact that the frequently present hemianopias make a proper evaluation of the lateral conjugate movements difficult, if not impossible (for instance, in the case of the so-called reflex of regard).

The horizontal gaze palsies caused by lesions in the supranuclear horizontal gaze centers of the *pons* and in the adjacent *posterior longitudinal bundle* (connecting the various nuclei of the extraocular muscles) are much more frequent and definitely of importance in the localization of tumors. If in a lateral gaze palsy all the individual forms of gaze movement are involved (that is, command and pursuit movements, optically induced movements, and vestibular movements), the focus in question can be assumed to be located in the pons or the posterior longitudinal bundle (Fig. 1-11). The resulting gaze palsy is toward the side of the lesion. If there is a conjugate deviation, it is to the side opposite the lesion. Even though the gaze palsies in pontine lesions are of a lesser degree than those of cortical or subcortical foci, as a rule they appear to be more enduring and are not compensated for in the course of time. The supranuclear gaze centers in the pons are relatively close together; thus bilateral horizontal gaze palsies are much more likely to occur than in cortical lesions. In contrast to the latter, bilateral lesions in the pons usually cause only horizontal but no vertical palsies because the seat of the vertical gaze center is higher. Likewise, the convergence center is situated in a higher part of the brain stem (between the oculomotor nuclei). For that reason many *pontine lesions,* particularly tumors, are notable because of a *horizontal gaze palsy with an intact convergence mechanism.* We have observed a patient with a glioma of the pons who had a bilateral horizontal gaze palsy. A command to look to the side or up caused a distinct convergence spasm (Fig. 3-110). The close proximity of the nuclei of the sixth and facial nerves as well as of the pyramidal tracts to pontine lesions results frequently in important accompanying symptoms in addition to the horizontal gaze palsies. A horizontal gaze palsy with a homolateral peripheral facial palsy, a homolateral nuclear sixth nerve palsy, and a hemiplegia of the opposite side (the patient always looks to the paralyzed side!) form the so-called *Foville-Millard-Gubler* syndrome (p. 280). It is caused by a paramedian lesion in the lower pons. Among the etiologic factors involved in supranuclear horizontal gaze palsies, tumors of the pons play an important part, occurring especially frequently in children. In judging the localizing value of horizontal gaze palsies, one should not lose sight of the fact that a compression of the pons and its structures, especially the posterior longitudinal bundle, may be caused by neighboring processes. Tumors of the cerebellopontine angle (we have observed a patient with an acoustic neuroma with horizontal gaze palsy to the side of the tumor), the cerebellum, or the medulla oblongata may occasionally cause horizontal gaze palsies which, however, are usually less distinct and can easily be overlooked.

Vertical gaze palsies are distinguished by the patient's inability to look up or down. As in the horizontal type, similar individual component movements can be differentiated. Their analysis, however, is of no significance for the localization of a lesion. There is usually an inability to look up (Fig. 3-99), less frequently an inability to look up and down. The rarest form is an inability to look down only. The supranuclear nature of a vertical gaze impairment can be demonstrated by observing a positive Bell's phenomenon (elevation of the eyes on closing the lids or during sleep). Vertical gaze palsies point to lesions in the vertical supranuclear gaze centers, that is, in the upper end of the midbrain in the region of the superior colliculus and thus near the pineal gland and the third ventricle.* Isolated vertical disturbances practically never occur in association with cortical lesions and do not play an important clinical role. The immediate proximity to the pupillary centers, to the convergence centers, and to the oculomotor nuclei is responsible for lesions in this area frequently giving rise to pupillary disturbances (for instance, Argyll Robertson pupils or general pupil rigidity), convergence palsies, convergence spasms, as well as nuclear oculomotor pareses, in addition to the vertical gaze palsies. Diplopia, however, may also be caused merely by a dissimilar involvement of the two eyes in their vertical conjugate movements. A vertical gaze palsy impairing depression is particularly frequently associated with disturbances of convergence and accommodation. The symptoms just described, occasionally combined with nystagmus retractorius and ectopic pupils (Wilson), form, in a variety

---

*More recently, in view of stimulation experiments, the concept of centers in the brain stem for control of gaze movements in various directions has been criticized and found to be misleading.

of combinations, *Parinaud's syndrome*. It is caused by tumors near the *area of the Sylvian aqueduct*, the *third ventricle*, the thalamus, the splenium of the corpus callosum, the quadrigeminate plate, and in particular the *pineal gland*. This may be a direct effect of the tumor on the center, or it may be a remote effect. Among our patients, *pinealomas* accounted for 50% of cases, thus being the major cause of Parinaud's syndrome (therefore also called pineal syndrome). Of ten patients with tumors manifesting a vertical gaze palsy (six with pinealomas), nine showed impairment of elevation, one showed impairment of elevation and depression, and none showed an isolated impairment of depression. We observed convergence palsy three times, convergence spasm twice, and nystagmus retractorius once. Pupillary disturbances of the Argyll Robertson type were seen eight times in the bilateral form and once in the unilateral form.

Convergence palsy is the inability of the eyes to converge and to fix on a near object. There is crossed diplopia at close range, with the separation of the two images practically remaining constant for lateral and vertical movements of the eyes. Distant objects are seen single. In its pure form a convergence palsy shows no other limitations of ocular motility. A pure convergence palsy is quite rare but occurs usually in combination with other symptoms—for instance, a vertical gaze palsy (see discussion of Parinaud's syndrome in the section on vertical gaze palsies)—a fact that is related to the location of the convergence center in the midbrain (that Perlia's nucleus is the midbrain center for convergence is still a controversial issue). A convergence palsy, together with a vertical gaze palsy, occurs in patients with tumors of the *midbrain* and pinealomas, and it may be combined with a horizontal gaze palsy in patients with expansive tumors of the *pons*. It has been mentioned that smaller pontine lesions leave the convergence intact. Convergence palsy must be differentiated from convergence insufficiency, a functional disturbance. In convergence palsy a supranuclear disturbance, an adduction of the visual axes, is all but impossible: a prism base out before one eye already causes diplopia for distance. In convergence insufficiency some degree of adduction can always be demonstrated: up to a certain power a base-out prism can be overcome. Only if the impaired adduction is taxed too much will diplopia result.

Divergence palsy is the inability to bring the eyes to a parallel position in looking at a distant object. There is homonymous diplopia with a corresponding convergent position of the eyes for far but not for near. There is no impairment of the function of the individual muscles, in particular the lateral recti. The separation of the double images remains constant in all directions of gaze, but the greater the distance the object is, the greater the separation. It is possible to confuse a divergence palsy with a convergence spasm. Perhaps the two terms refer to an identical phenomenon. Autopsy findings revealed tumors in the brain stem between the sixth nerve nuclei, the hypothetical divergence center (Savitsky and Madonick; Robbins; Lippmann).

Internuclear ophthalmoplegia is characterized by a failure of the medial rectus on the side of the lesion to act in horizontal gaze to the opposite side and nystagmus of the abducting eye. If, for instance, the patient is asked to look to the right, the right external rectus works normally and brings the right eye into abduction; on the contrary, the left internal rectus does not contract and the left eye remains immobile in a median position. There is no palsy of this left medial rectus: it is able to function normally in convergence. The abducted right eye manifests a horizontal nystagmus. Internuclear ophthalmoplegia can also occur bilaterally: both medial recti fail to work when an attempt is made to look to the side; both medial recti work normally in convergence. For detecting milder cases of internuclear ophthalmoplegia, two subtle signs, the "optokinetic phenomenon" and the "ocular dysmetria sign," have recently been described (Cogan). Since unilateral internuclear ophthalmoplegia is vascular in origin in three fourths of the cases and bilateral internuclear ophthalmoplegia is pathognomonic for multiple sclerosis, cerebral tumors rarely have to be considered. Internuclear ophthalmoplegia is supposed to result from a unilateral or bilateral lesion of the posterior longitudinal bundle in the upper part of the pons or the caudal region of the midbrain.

In the *Hertwig-Magendie squint position* (skew deviation), one eye deviates downward and inward, the other eye upward and outward. The pathogenesis of this rare phenomenon of disturbance of conjugate movement is still controversial: no doubt one has to deal with lesions (also tumors) within the brain stem, especially in the lateral pons and rarely in the cerebellum.

It has been mentioned previously that conjugate movements are governed from frontal as well as from occipital gaze centers. The spontaneous and command movements originate in the frontal gaze centers (Fig. 1-11). The centers for the more primitive reflexlike ocular movements (for instance, the pursuit movements, the reflex of regard, etc.) lie in the occipital oculomotor centers in the convex side of the occipital poles (areas parastriata and peristriata). *Lesions of these occipital centers and their efferent tracts* (which supposedly run toward the anterior quadrigeminate bodies in the optic radiation in a direction opposite to the optic fibers) must result in *impairment of the type of conjugate movements just mentioned*—among others, the *pursuit movement.* In patients with homonymous hemianopia caused by lesions in the middle or posterior parts of the optic radiation, Kestenbaum found a disturbance of the pursuit movement toward the side opposite the hemianopia. Instead of an ordinary pursuit movement, there is (with the eyes following a slowly moving object) a series of jerky intervals, a *saccadic movement* (so-called cogwheel movement). The test for the conjugate pursuit movement in the case of a homonymous hemianopia may therefore give some indications concerning the site of the lesion causing the hemianopia. If the normal pursuit movement toward the side opposite the hemianopia is changed into a saccadic movement, the lesion is most probably in the middle or posterior part of the optic radiation (that is, in the parietal or occipital lobes). A normal pursuit movement is of no value in localizing the lesion. Still,

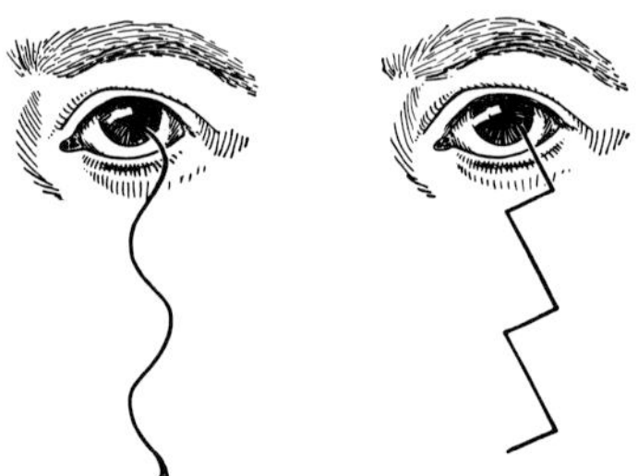

Fig. 1-22. The two general types of nystagmus. Left, Pendular nystagmus with undulating movement of the eyes. Right, Jerk nystagmus with a biphasic rhythm; the quick phase is to the nasal side and the slow phase is to the temporal side.

proof of the "cogwheel" phenomenon may in certain cases facilitate the differential diagnosis of a tract and a radiation hemianopia.

Optokinetic nystagmus is a type of movement with a certain similarity to the pursuit movement. It also is governed by occipital optomotor centers and their efferent pathways. These pathways run centrifugally from the visual cortex in the so-called stratum sagittale internum of the visual radiation and reach the posterior longitudinal bundle via the brain peduncles and the midbrain. It is also of some significance in the evaluation of homonymous hemianopia from the standpoint of localization. It will be discussed in detail in the next section.

Nystagmus

Nystagmus is an involuntary rhythmic oscillating type of movement of one or both eyes in all or some fields of gaze. There are two principal forms of nystagmus (Fig. 1-22):

1. *Pendular* nystagmus, in which both phases of the movement are of equal speed
2. *Jerk* nystagmus, which shows a biphasic rhythm with a slow phase to one side and a quick phase to the other side

Pendular nystagmus rarely remains pendular in all fields of gaze: outside the "neutral" zone, on lateral gaze, there is often a change from pendular to jerky movements.

According to its *plane,* horizontal, vertical, and rotary nystagmus (pendular or jerk) can be distinguished; there may also be combinations of these types (horizontal-rotary, oblique). *In jerk nystagmus it is customary, by convention, to designate its direction according to the direction of the quick phase.* At the same time it is important to know that the direction may change in various fields of gaze. For clinical purposes it is convenient to use the following classification based on *intensity* (Fig. 1-23):

First degree
No nystagmus with the eyes straight; nystagmus only when the eyes turn toward the side of the quick phase

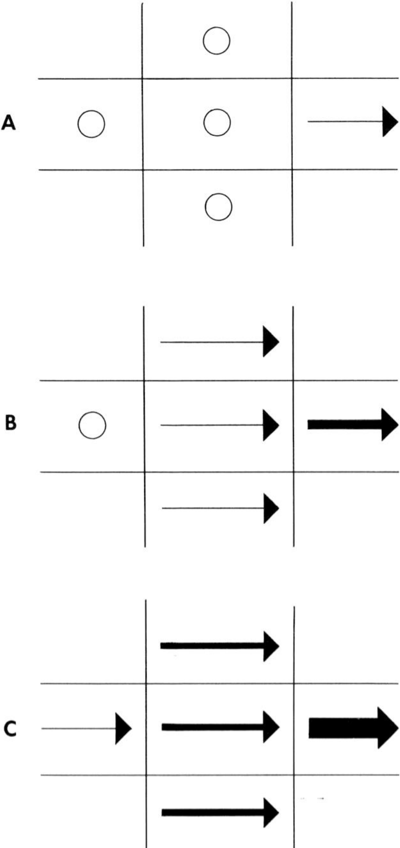

Fig. 1-23. Schematic representation of horizontal jerk nystagmus. **A,** First-degree horizontal jerk nystagmus. **B,** Second-degree horizontal jerk nystagmus. **C,** Third-degree horizontal jerk nystagmus.

Second degree
Slight nystagmus with the eyes straight; very active nystagmus if the eyes turn toward the side of the quick phase; no nystagmus if they turn to the opposite side.

Third degree
Very active nystagmus with the eyes straight; extremely active nystagmus if the eyes turn toward the side of the quick phase; slight nystagmus if the eyes turn to the opposite side.

Nystagmus can be also divided according to its *amplitude* into a fine type (excursions of less than 3 degrees), medium type (excursions between 5 and 15 degrees), and coarse type (excursions of more than 15 degrees). As a general rule, the amplitude of jerk nystagmus increases when the patient looks toward the side of the quick phase. The *frequency* of the nystagmus should also be considered. In general, the faster the rate, the smaller the amplitude and vice versa. In some cases the amplitude may be so small that the existence of nystagmus is only noted by observing the optic nerve head with the direct ophthalmoscope.

Jerk nystagmus usually shows the same amplitude in both eyes; unilateral or incongruent jerk nystagmus, however, may occur.

The *technique of examining* patients with nystagmus is simple and can be performed for clinical purposes without complicated instruments. First, the examiner notes whether or not there is nystagmus in the primary position of the eyes. Next, he directs the patient to look to the sides, up, and down; these movements can be performed on command or by following a finger or a pencil. The examiner then notes whether nystagmus occurs which was not present in the primary position or in which direction an existing nystagmus increases. The type of nystagmus (pendular or jerk), its plane (horizontal, vertical, rotary), its degree (first, second, third), its amplitude (fine, medium, coarse), and its frequency are observed, if necessary with the aid of a loupe to magnify the ocular movements and at the same time to eliminate fixation. *Bartels'* spectacles with +20-diopter lenses or, even better, *Frenzel's* goggles are suitable for this purpose. The latter have in addition a built-in illumination that dazzles the patient to the extent that fixation is impossible if the examination is performed in an almost or completely darkened room. Visual fixation will increase the tonus of the extrinsic muscles to the extent that a slight nystagmus (for instance, first degree) may be suppressed, and thus not infrequently the eye may appear at complete rest although a nystagmus is actually present. Therefore *for an absolutely reliable observation, particularly if the nystagmus is of weak intensity* (either the spontaneous or induced type), *it is imperative to eliminate fixation.* Only the elimination of fixation will make it possible to determine the degree of the nystagmus accurately. Generally, nystagmus will increase by one class after fixation is eliminated. In other words, if there seems to be a first-degree nystagmus with the ordinary finger method, it will prove to be a second-degree nystagmus with the Frenzel goggles. The determination of the true class of the nystagmus after fixation has been eliminated is quite essential for its proper diagnosis. A first-degree nystagmus is in many cases of no pathologic significance. On the other hand, a *second-degree nystagmus always indicates a pathologic process.*

A correct analysis of nystagmus includes, in addition to examination in the erect position, testing with the head tilted to the left and to the right, forward and backward, if necessary with the patient in a supine position, and after shaking his head. It is frequently useful to repeat the examination several times with the Frenzel goggles in the position in which the patient reports vertigo. Prerotatory and postrotatory nystagmus, caloric nystagmus, galvanic nystagmus, and compression nystagmus as methods of examination should, for technical reasons, be left to the otologist.

As already mentioned, the direction of the jerk nystagmus is determined by the direction of its quick phase. The direction of the slow phase as a rule corresponds to the side to which the patient has a tendency to fall if there are disturbances of gait or to the side of passpointing in the finger-nose test.

As a rule, there is some relationship between the direction of gaze and the direction of nystagmus: a nystagmus to the right occurs more readily when looking to the right, and a nystagmus to the left occurs more readily in gaze to the left. The most frequent combination is right nystagmus in gaze to the right, together with left nystagmus in gaze to the left.

In recent years, *nystagmography* (Ohm; Coppez), especially *electronystagmography* (ENG), has come to play an increasingly important role in the examination of the eyes for spontaneous or induced nystagmus. It permits not only more exact observations of frequency, amplitude, and direction of the nystagmus but also permanent objective recording of these ocular movements. In addition to its use in spontaneous nystagmus, ENG is gaining increasing practical value in the examination of induced nystagmus (caloric, rotational, optokinetic).

The principle of ENG consists in picking up the corneoretinal potentials of both eyes by two electrodes placed as near as conveniently possible to the left and right outer canthi. D.C. recording and amplification are definite needs in ENG. Special devices permit recording of the concomitant gaze positions of the eyes at the same time: in any ENG recording this information is as essential as the information about character and magnitude of the nystagmus. Apart from the facility to control the direction of gaze, there must also be a means to measure it accurately when fixation is removed. Fixation can be eliminated either by closing the eyes or by testing the subject in total darkness. It is a well-known fact that removing optic fixation brings about a marked increase of vestibular nystagmus (either spontaneous or induced) of peripheral origin. In a spontaneous nystagmus due to a lesion above the level of the vestibular nuclei (for example cerebellar tumor), eye closure and darkness cause complete abolition of the nystagmus.

Refer to special textbooks and papers for further information on the technical and clinical problems of ENG (Franceschetti; Cawthorne, Dix, and Hood; Hallpike, Hood, and Trinder; Jung and Kornhuber; Hart).

All spontaneous pathologic nystagmus can be related to four basic biologic mechanisms: the fixation mechanism, the mechanism of the vestibular apparatus, the mechanism of conjugate eye movements or gaze mechanism, and the convergence mechanism. Therefore it is customary to divide the numerous forms of pathologic nystagmus into four main groups:

1. *Fixation nystagmus*
2. *Vestibular nystagmus*
3. *Gaze nystagmus*
4. *Convergence nystagmus*

Fixation nystagmus. This group includes above all the so-called *ocular* nystagmus arising from disturbances of the retina and its connections with the visual cortex: *pendular nystagmus* (beginning during the first few months and persisting throughout life; often associated with severe refractive errors, media opacities, aniridia, macular defects, albinism, or chromatopsia; sometimes hereditary; occuring only in males), *spasmus nutans* (head tilting, head nodding, and horizontal pendular nystagmus beginning during the first 18 months and invariably disappearing by the third year of life), *latent nystagmus* (elicited by covering one eye; jerk nystagmus of both eyes away from the covered eye, often associated with strabismus, especially esotropia and unilateral amblyopia), *miner's nystag-*

mus (pendular, horizontal, or oblique form generally restricted to the upper field of gaze), and *congenital jerk nystagmus* (typical head turning of the patient so that he is looking toward the point of reversal, thus giving himself maximum visual acuity).

Apart from the ocular forms, this group of fixation nystagmus also contains *neurologic* types due to lesions in the posterior fossa, and these have neuro-ophthalmologic importance: *acquired pendular nystagmus* (usually horizontal; becoming vertical on upward gaze; acquired later in life; sensation of moving of the surroundings; most frequently seen in multiple sclerosis), *ocular flutter* and *ocular dysmetria* (the former a periodic series of pendular movements while fixating an object; the latter a series of dampening movements when the object of regard is shifted; both signs of cerebellar disease), *see-saw nystagmus* of Maddox (a disjuncted pendular nystagmus characterized by alternating elevation and depression of the eyes; frequently associated with bitemporal defects due to a chiasmal lesion), and *postural nystagmus* (occurring with a shift of the head; presumably caused by traction or pressure exerted on the brain stem by a neoplastic or vascular lesion in the posterior fossa).

Vestibular nystagmus. *Experimental* vestibular nystagmus such as prerotatory, postrotatory, caloric, galvanic, and compression nystagmus will not be discussed here (see textbooks on physiology and otolaryngology).

Spontaneous vestibular nystagmus may appear as a *peripheral* nystagmus due to lesions of the labyrinth or its nervous connection to the brain stem or as a *central* nystagmus due to a lesion of the vestibular nucleus or due to a lesion of the central connections of the vestibular nuclei (such as the connections with the cerebellum or with the eye muscles via the reticular formation). *Vestibular nystagmus, the peripheral as well as the central type, is always jerky, usually horizontal-rotary,* and may be first degree, second degree, or third degree.

Peripheral vestibular nystagmus remains in the same direction in all fields of gaze and is independent from the position of the body. An irritation of the labyrinth (for example, labyrinthitis) usually causes a horizontal-rotatory nystagmus in both directions (that is, in gaze to the right, nystagmus to the right; in gaze to the left, nystagmus to the left). Destruction of the labyrinth or an interruption of the vestibular nerve results in nystagmus to the other side, away from the lesion. A prominent symptom is *vertigo,* usually associated with acoustic disturbances such as transient or permanent *deafness* and *tinnitus.* Peripheral vestibular nystagmus and the associated vertigo show a gradual decrease in intensity and disappear after a few weeks even if the underlying cause continues. ENG reveals the peripheral vestibular nystagmus to be enhanced (if present) or made manifest (if not) by eliminating fixation by closing the eyes or testing in darkness.

Central vestibular nystagmus, while it is also horizontal-rotary, may change its direction on right and left lateral gaze, and it may even become vertical on upward or downward gaze, a phenomenon never observed with a peripheral lesion. Therefore the determination of the direction of the nystagmus is not conclusive evidence of the side of the lesion. Changes in the position of the body frequently cause central nystagmus to change its direction. Vertigo may be present, but is not nearly so severe as in peripheral vestibular nystagmus. Tinnitus and deafness are not present. In contrast to peripheral vestibular nystagmus, the central type may persist for months and years if the underlying cause does not disappear. ENG findings show central nystagmus (due to a lesion above the level of the vestibular nuclei) to be abolished by eye closure or by testing in darkness.

From the clinical point of view, central vestibular nystagmus may be due to a process in the brain stem or in the cerebellum.

In *tumors of the pons,* central nystagmus is an almost constant symptom. We found such a nystagmus in twelve out of thirteen patients with pontine tumors. Seven showed a horizontal type, four a combined horizontal-vertical type, and only one a vertical type. Most authors are of the opinion that a vertical nystagmus is due to a long-distance effect on the quadrigeminate plate. The direction of the nystagmus in patients with lesions of the brain stem does not permit localization to a particular side unless there is an asymmetry (p. 274). Long-distance effects of certain tumors on the brain stem are well known. Thus tumors of the occipital lobe may cause nystagmus of the central vestibular type (p. 184). A similar effect has been observed with tumors of the mesencephalon or the pineal body. Whether the nystagmus occasionally seen in frontal lobe tumors is caused by such long-distance effects is a matter of dispute.

Nystagmus occurs with extraordinary frequency in patients with *cerebellar tumors* (50% or more). This *cerebellar nystagmus* usually has the character of a central vestibular nystagmus and sometimes also of symmetrical gaze or vertical gaze nystagmus (p. 267). Cerebellar nystagmus is predominantly horizontal in nature. It may be spontaneous or only appear during lateral movements. It is distinguished by the fact that the quick phase occurs in the direction of the patient's gaze. In unilateral cerebellar lesions the amplitude varies with the direction of gaze: despite some exceptions, it can be generally stated that the *direction of gaze associated with the coarse form of nystagmus corresponds to the side of the lesion.* In case of a median location of the tumor (for instance, in the vermis) there may be a purely vertical nystagmus (pressure effect on the collicular plate?). According to Spiller, this also indicates a considerable anterior extension of the tumor from the posterior fossa anteriorly. It is still a controversial subject whether the nystagmus associated with cerebellar tumors represents truly a cerebellar sign or whether it is the result of a long-distance effect on the brain stem—on the pons or the posterior longitudinal bundle, respectively. In patients with tumors of the *cerebellopontine angle,* especially acoustic neuromas, a horizontal nystagmus can be observed as an almost constant sign. This horizontal jerk nystagmus is usually coarser in its excursions toward the side of the tumor than toward the opposite side. In some instances we have seen in addition to the horizontal nystagmus a rotary or, with eyes up, even a vertical component. The fact that nystagmus in association with cerebellopontine angle tumors is persistent and does not disappear suggests that it is less the result of a lesion of the vestibular nerve than a phenomenon of compression of the brain stem or the cerebellum. In other words, it is a central type of nystagmus similar to the one seen in association with cerebellar tumors. This is equally true for the very frequent occurrence of a vertical component (probably due to a remote effect on the quadrigeminate plate), especially if it occurs as an isolated phenomenon and without a horizontal component.

Gaze nystagmus. One of the most common causes of nystagmus is a failure of the gaze mechanism: such a nystagmus is called gaze nystagmus. It is characterized by the *absence of nystagmus when looking straight ahead and the* *appearance of nystagmus in one or more fields of gaze, with the fast component always in the direction of gaze.*

The so-called physiologic *end-position nystagmus* represents a form of gaze nystagmus occurring in many healthy individuals (either with or without a latency period) as an expression of weakness or fatigue of gaze; instead of a steady position of the eyes in lateral gaze, repeated quick impulses to the side of gaze and slow deviations toward the midline alternate. This form of nystagmus has no clinical significance and is usually exhausted after a short time (maximally after ten or fifteen jerks). Tired and nervous individuals are especially subject to end-position nystagmus (fatigue nystagmus, neurasthenic nystagmus).

Muscle paretic nystagmus is caused by a paresis of a single extraocular muscle. A progressive exhaustion (slow phase) alternates with a contraction (quick phase) induced by a renewed impulse. Since the repeated impulses are automatically sent equally to both eyes, the nystagmus occurs in both eyes. The movements in the normal eye are even greater than in the paretic eye. Its direction corresponds to the direction of action of the paretic muscle. This type of muscle paretic nystagmus does not occur with a completely paralytic muscle. The paresis may be of myopathic, synaptic, or peripheral neurogenic origin. Here it may be mentioned that myasthenic patients sometimes manifest peculiar jerky movements of the eyes following the administration of a cholinergic drug.

Related to muscle paretic nystagmus is gaze nystagmus, also called gaze paretic nystagmus. In such an instance not a single muscle but a group of muscles executing a conjugate movement is involved. End-position nystagmus and fatigue nystagmus actually belong to this group. However, they differ from gaze nystagmus because they are exhaustible. Nevertheless, it is possible to mistake an end-position nystagmus, which after extended observation changes to a fatigue nystagmus, for an inexhaustible gaze nystagmus. This mistake can be avoided if the test for gaze nystagmus is never performed in the extreme lateral position. In other words, pathologic gaze nystagmus always appears in lateral gaze before the end position is reached.

There are two types of pathologic gaze nystagmus: *horizontal symmetrical gaze nystagmus and horizontal asymmetrical gaze nystagmus* (Fig. 1-24). "Symmetrical" means a nystagmus appears at the same distance from the primary position to the right and to the left, with an identical increase in its intensity in both directions. "Asymmetrical" nystagmus appears earlier to one side, with a more pronounced increase in intensity to this side (a sort of gaze weakness to the same side). These forms of horizontal gaze nystagmus are almost purely horizontal, without a rotary component. They differ from end-position and fatigue nystagmus because in lateral movements they appear before the terminal position is reached. The quick phase is toward the lateral side and the slow phase toward the midline. In severe cases of symmetrical gaze nystagmus there may also be a vertical component, which is of no importance in itself. On the other hand, an isolated *vertical gaze nystagmus,* when looking up or down, may be the prodromal stage of a vertical gaze palsy, suggesting a lesion in the roof of the

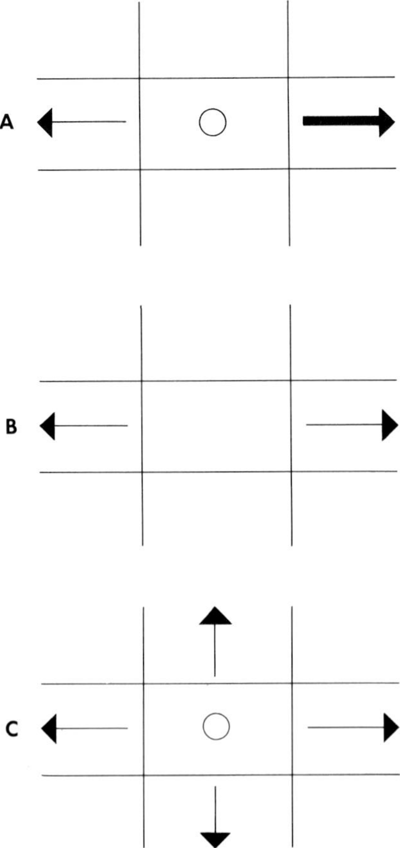

Fig. 1-24. Schematic representation of gaze nystagmus. **A,** Horizontal asymmetrical gaze nystagmus. **B,** Horizontal symmetrical gaze nystagmus. **C,** Horizontal and vertical gaze nystagmus.

midbrain (p. 264). In symmetrical or asymmetrical gaze nystagmus the underlying lesion is located in the brain stem. According to Kestenbaum, asymmetrical gaze nystagmus is a localizing sign indicating the site of the lesion to be within the pons. Also, cerebellar tumors may present with asymmetrical gaze nystagmus with the head turned to the side of the lesion. The causes of symmetrical or asymmetrical gaze nystagmus include various possibilities: vascular disease, tumor, multiple sclerosis, degenerative disease, intoxication, etc.

Convergence nystagmus. Convergence nystagmus is characterized by *intermittent convergence movements produced on horizontal gaze and often exagerated on attempts at near vision or upward gaze.* In severe cases the convergence movements may be associated with more or less rhythmic jerking retraction movements of the eye: *retraction nystagmus,* or *nystagmus retractorius.* This convergence nystagmus is diagnostic for a lesion of the Sylvian aqueduct. In its full form the so-called *Sylvian aqueduct syndrome* consists of vertical gaze

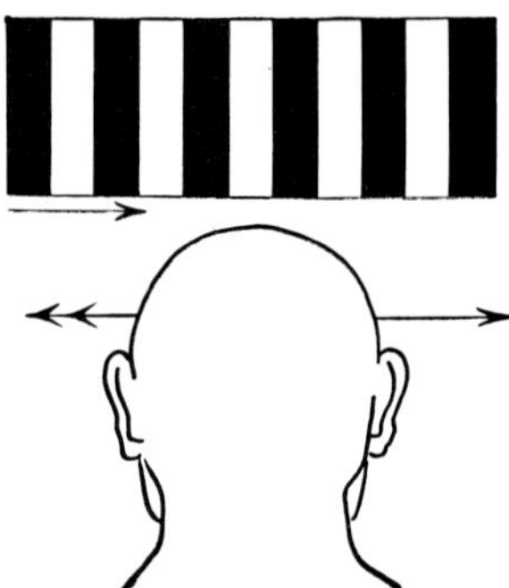

Fig. 1-25. Optokinetic nystagmus, or "train nystagmus." The patient fixes on a drum that moves in the direction of the arrow. The drum is painted with broad black stripes. The resulting jerk nystagmus has a slow phase in the direction of the movement of the drum and a quick phase in the opposite direction. (After Duke-Elder.)

palsy, pupillary disturbances, convergence and retraction nystagmus, and paresis of one or more extrinsic eye muscles (p. 264).

In addition to the various pathologic types, one form of experimental nystagmus based on the normal fixation mechanism should be considered here: *optokinetic nystagmus,* which may play a significant role in the topical diagnosis in patients with homonymous hemianopia. This nystagmus was known for a long time as "railroad nystagmus." If the patient observes a revolving drum with a broad black stripe (a so-called optokinetic drum), a characteristic horizontal nystagmus is induced with its slow phase in the direction of the movement of the drum and its quick phase to the opposite side (Fig. 1-25). The drum should be about 13 cm. high with a diameter of 25 cm.; its surface should be painted with vertical black and white stripes of 1.5 cm. width. If an optokinetic drum is not available, a large newspaper can be used. It is slowly moved before the patient's eyes in a horizontal direction with the printed lines in a vertical position. If the patient is inattentive (for instance, in a slight stupor or in aphasia), the optomotor strength of the black stripes on the drum is insufficient. A better result may be obtained in such patients if pictures of persons or objects are substituted for the stripes, a method that is also more suitable for children. *Optokinetic nystagmus may be elicited horizontally or vertically.* Under normal conditions it is equal in both horizontal and vertical directions, but it may be greater on downward than upward movement.

The *slow phase* of optokinetic nystagmus is identical to the pursuit reflex and is mediated through the optomotor field of the occipital lobe. The efferent impulses pass directly from the occipital lobe to the brain stem. The pathway for the *fast phase* of optokinetic nystagmus requires frontal lobe activity and is probably identical with the pathway for voluntary gaze (as for the fast phase of vestibular nystagmus).

Horizontal optokinetic nystagmus disturbances can only be interpreted when there is a definite asymmetrical response between the two sides. If the optokinetic nystagmus is identical or approximately identical for both directions of the drum (tested according to the method just described), this is called a *negative opto-kinetic nystagmus sign.* In a patient with homonyous hemianopia this indicates, with a certain degree of probability, that the lesion is in the optic tract, the external geniculate body, the anterior part of the optic radiation, or the cortex itself near the calcarine fissure (Kestenbaum; Stenvers; Cords). A *positive nystagmus sign* is present if no optokinetic nystagmus or only an attenuated nystagmus can be elicited toward the side of the hemianopia. In other words, the nystagmus is absent or diminished when the drum moves out of the blind side. In patients with asymmetrical optokinetic nystagmus the most significant abnormality is the loss of the fast phase of the optokinetic reflex to the opposite side (disconnection of the frontal lobe from the occipital lobes). According to different authors (Kestenbaum; Ling and Gay), *the positive nystagmus sign indicates quite reliably a lesion in the white matter of the parietotemporal or the parieto-occipital region, in other words a lesion in the middle or posterior part of the optic radiation.* Abnormal optokinetic nystagmus is therefore observed most frequently in lesions of the *parietal lobe.* Stadlin has described a patient with a parieto-occipital astrocytoma with homonymous hemianopia. Following surgical removal of the tumor, the optokinetic nystagmus, which was equal on both sides before the operation, disappeared toward the side of the hemianopia, only to return after an interval of almost 2 months, at first in an irregular and saccadic manner. This and many similar observations indicate a mechanism that compensates for such disturbances of the optokinetic nystagmus. Moreover, one has to know that occasionally the optokinetic nystagmus may be missing or may be easily exhaustible in normal individuals.

In summary, the practical importance of the optokinetic nystagmus can be stated in the following manner. *In cases of homonymous hemianopia a missing or attenuated optokinetic nystagmus toward one side points toward a lesion in the medial or posterior part of the optic radiation, most frequently in the parietal lobe.* If, in the presence of hemianopia, the optokinetic nystagmus is more or less identical on both sides, no localizing value can be established.

In diffuse frontal lobe lesions there may also be an impairment of the optokinetic nystagmus. The fast phase of the optokinetic nystagmus is impaired, but fixation and pursuit responses are intact. In addition, voluntary conjugate eye movement to the opposite side is defective or absent; also, the caloric response is impaired (especially the fast phase). Unilateral occipital lesions produce no abnormality of optokinetic nystagmus. Bilateral occipital lesions lead to complete loss of response to optokinetic stimulation only if there is complete cortical blindness.

Asymmetry of optokinetic response in the vertical plane may suggest a *disturbance in the brain stem,* but one must always be aware that a slight asymmetry may be normal. Symmetrical disturbances in the horizontal and vertical plane have no localizing value. Only asymmetrical disturbances between the horizontal and the vertical responses suggest a lesion in the brain stem.

Refer to other publications for details of the theoretical and, to some extent, hypothetical principles of these empiric observations (Bárány; van Bogaert; Cords; Holmes; Ohm; Stenvers; Kestenbaum; Cawthorne; Dix and Hood; Ling and Gay).

It should be of interest to examine briefly the relationship between disturbances of optokinetic nystagmus and those of pursuit movement (cogwheel movement, p. 49). Kestenbaum's studies on this subject are particularly penetrating. A disturbed pursuit movement to one side is always accompanied by an impaired optokinetic nystagmus to the other side. On the other hand, impairment of the optokinetic nystagmus is not always associated with a cogwheel movement. This may justify the assumption that the pathways of the occipitomesencephalic tracts for the conjugate movements, producing changes in either the optokinetic nystagmus or the pursuit movements, do not coincide completely. *Anomalies of either one of these phenomena* (optokinetic nystagmus or pursuit movement) *are of localizing value only if differences between the two sides can be demonstrated.* In such an event the lesion responsible for the hemianopia lies either in the middle or the posterior part of the optic radiation (especially the parietal lobe). The absence of optokinetic nystagmus and the pursuit movement or symmetrical changes of these phenomena to both sides are of no diagnostic importance. If the lesion involves the anterior part of the optic radiation or the area of the calcarine fissure, the optic nystagmus and the pursuit movement are, as a rule, normal.

Fundus

The external aspects of the eyes, the corneal sensitivity, the pupils, and the motility can easily be observed directly and examined without major accessories. The pupils impose a formidable barrier to the examination of the eyegrounds. This barrier is created by the usually small size of the pupils or by some opacities of the refracting media (lens and the vitreous). A special instrument, the ophthalmoscope, is necessary for the visualization of the eyeground. The examiner needs some dexterity and considerable personal experience in the use of this instrument to obtain reliable results.

We cannot always count on the pupils to remain in satisfactory mydriasis or perhaps even be rigid to light. In older people there is frequently a very annoying miosis which, in combination with a physiologic nuclear sclerosis of the lens, can render ophthalmoscopy all but impossible. In spite of these obstacles, one should *attempt at all cost to examine the eyeground of a patient with a suspected brain tumor without instilling a mydriatic.* Any drug-induced mydriasis will mask the original state of the pupils for a varying length of time. During this time complications might occur that would make evaluation of the pupils (for instance, a unilateral dilation in increased intracranial pressure) highly desirable. Ophthalmoscopy without artificial mydriasis is facilitated by *darkening the examining room,* which will tend to dilate the pupils. If that is impossible, the patient should turn his face from a bright window so that his eyes are shaded by his own head. Occlusion of the other eye eliminates the consensual miotic impulse of the eye to be examined. To minimize the pupillomotor effect of the ophthalmoscope light, it is best to begin the examination with the disc which, as a "blind spot," is insensitive to light. The light bundle of the ophthalmoscope entering the eye through a small pupil should be as narrow as possible to avoid scattering and reflection of the light at the pupillary area of the iris. The modern direct ophthalmoscopes are equipped with apertures of different sizes that permit the selection of a light bundle in accordance with the size of the pupil.

Only after these measures have failed is one justified in *dilating the pupil artificially* if a satisfactory view of the eyeground cannot be obtained because of a small pupil or because of opacities of the refracting media. The pupillary reaction to light and in convergence as well as the power of accommodation should be tested before drops are instilled. The mydriatic selected should be short-lasting and should interfere as little as possible with accommodation.

In many cases, tropicamide (Mydriacyl) is suitable for this purpose. From 2 to 4 drops, a few minutes apart, are instilled into the conjunctival sac. The maximal mydriatic effect is reached after 10 to 15 minutes. It disappears rapidly and is scarcely noticeable after an hour. Disturbances of accommodation are practically nonexistent or at the most quite insignificant and transitory. Cyclopentolate (Cyclogyl) should be used only if tropicamide does not cause a satisfactory mydriasis. It takes effect a little more slowly and lasts much longer (8 to 15 hours). It also immobilizes the accommodation almost completely for a considerable length of time. As a rule, the use of cyclopentolate cannot be avoided in patients with diabetes, in whom it is difficult to dilate the pupils. Atropine should never be used because its cycloplegia lasts for days (up to 1 week) and because of the danger of precipitating a glaucoma in older people.

The artificial mydriasis not only eliminates the difficulty of performing ophthalmoscopy with small pupils but also increases the visibility of the eyeground in the presence of slight *opacities of the media.* At the same time, such opacities of the lens or vitreous become visible, especially if a +10 lens is used, allowing some magnification. Any type of opacity manifests itself as a black moving (vitreous) or fixed (lens) shadow on a red background. If indicated (for instance, in pseudopapilledema seen in hyperopia), retinoscopy can be done with the pupil dilated. (The use of homatropine is essential for this purpose.)

It would seem almost superfluous to discuss the *ophthalmoscope* within the frame of this dissertation. Nevertheless, personal experience prompts us to make a few pertinent remarks. In our opinion *direct ophthalmoscopy* (upright image) is best suited for neuro-ophthalmologic examinations, for it frequently has to be performed by nonophthalmologists. The modern ophthalmoscopes have a brilliant light source. They are provided with a number of apertures to adjust the width of the light bundle, with rheostats to regulate the light intensity, with a Recoss disc that will permit a sharply focused image of the fundus and compensation for refractive errors of the eyes of the patient or the examiner, and perhaps even with filters for red-free light. These qualities should make the direct ophthalmoscope an excellent tool for examination. Compared with indirect ophthalmoscopy, it has the advantage of greater simplicity and of higher magnification (16×) of the fundus picture. Its disadvantage is that the total retinal area visible at one time is smaller than that seen with the indirect method. This is, however, of no great importance because, for the neuro-ophthalmologic examination, the central part of the retina (namely, the disc and the macula) is of primary interest.

With adequate mydriasis, however, the peripheral parts of the fundus can be evaluated fairly completely with direct ophthalmoscopy. The instrument, supplied with a battery handle, makes the examiner completely independent of an external light source. Those who like the indirect method in addition to direct ophthalmoscopy can use the electric instrument, with a +12- or 13-diopter lens held in front of the patient's eye and with a distance of 40 to 50 cm. between this lens and the ophthalmoscope.

As the first landmark in examining the fundus, the disc should be visualized by asking the patient to turn the eye to be examined slightly nasally. The examiner should attempt to relax his accommodation as much as possible; this can be facilitated if he imagines himself looking into a spacious room through the ophthalmoscope. A refractive error of the patient's eye can be compensated by the lenses in the Recoss disc. After examining the papilla and the retinal vessels, one continues with the macula and its surroundings and finally with the periphery of the fundus in all meridians. Details of eyeground changes in tumors and the differential diagnosis of such conditions will be discussed in more detail in subsequent chapters. Essentially, these changes are papilledema and optic atrophy. The ophthalmoscopic aspects of these phenomena will be described and differentiated from similar affections (p. 94).

The ophthalmoscopic evaluation of the retinal vessels also affords the opportunity of measuring *arterial and venous blood pressures*. Under certain conditions such data can be of importance in the diagnosis of increased intracranial pressure and may help rule out complications such as carotid artery occlusion, vertebrobasilar artery occlusion, or carotid-venous communications. Bailliart's or H. K. Müller's ophthalmodynamometer (clock-dial type or straight model) is suitable for this purpose, especially for measuring the arterial pressure. There are similar instruments with a finer calibration and a lighter spring action for testing the venous pressure. For accuracy of results the pupils are dilated and the eyes anesthetized topically. The intraocular pressure is measured by Schiøtz or applanation tonometry. Direct or indirect ophthalmoscopy may be used. The advantages of indirect ophthalmoscopy, especially with a binocular indirect ophthalmoscope, are stressed by many ophthalmologists familiar with the difficulties of ophthalmodynamometry. Some authors (Rintelen; Weigelin and Lobstein; Smith; Gay) suggest that two observers perform the dynamometric examination: one places the instrument on the globe in the lateral canthal angle, taking care that the footplate is always tangential and that the plunger points at all times toward the center of the globe, gradually increasing the pressure inside the eye; the other looks through the ophthalmoscope and observes the behavior of the retinal vessels as they emerge from the disc. If the force is gradually increased, first there will be a venous pulsation. As soon as this occurs, the pressure of the spring is noted; it expresses the venous pressure in grams. The pressure on the plunger is continued until pulsation (more precisely an intermittent collapse) of the central artery at the site of its branching becomes visible:

the spring tension measured is the equivalent of the *diastolic arterial pressure* in grams. The *systolic pressure* is reached with a spring pressure that causes the arterial pulsation to cease completely. To eliminate the errors of only one reading, it is advisable to perform several measurements of both the systolic and the diastolic pressure and to take an average of the spread of pressure values found. Since this blood pressure is naturally a function of the systemic blood pressure, immediately after measuring systolic and diastolic pressures in both eyes the brachial blood pressure is taken and recorded. Comparative measurements of the pressure in the ocular vessels should be performed with the patient either in a sitting or a supine position (but always in the same position).

The actual ophthalmodynamometric values read on the instrument (Bailliart or Müller) are grams. They have to be converted to millimeters of mercury (mm. Hg) and brought into relation with the intraocular pressure (Schiøtz, applanation). For this purpose the conversion tables of *Bedavanija* and *Niesel* are preferred to convert the pressure values in grams measured at a certain intraocular pressure level directly into the real pressure values of the ophthalmic artery in millimeters of mercury.

The ophthalmic artery pressures are of significance for the analysis of disorders of the cerebral circulation only when brought into relation with the brachial pressure. In this connection one has to consider the *equation of Weigelin and Lobstein,* which is based on vast statistical material:

$$P_{m\ ophth} = 0.73\ (P_{m\ brach} - T) + T$$

where

$P_{m\ ophth}$ = Median ophthalmic artery pressure (diastolic pressure + 42% of the difference between systolic and diastolic pressure)

$P_{m\ brach}$ = Median brachial pressure (diastolic pressure + 42% of the difference between systolic and diastolic pressure)

T = Intraocular pressure (Schiøtz, applanation)

There are special tables (Bedavanija) that replace the calculations according to this equation. They give the normal values of median ophthalmic artery pressures expected for a given intraocular and brachial pressure. To make a decision as to whether there exists a disorder of the intracranial circulation or not, these normal values have to be compared with those found in the patient.

An abnormal increase of the ophthalmic arterial pressure can be a sign of dilatation of the carotid artery or of obliteration or contraction of the cerebral arteries, whereas an abnormal decrease can point to a stenosis of the carotid or to a dilatation of the peripheral cerebral vessels. The behavior of the vascular pressure in cases of brain tumors will be discussed in the section on the choked disc (p. 109). For a more specific consideration of the pathophysiologic principles involved in the pressure of the ophthalmic artery and its measurement, consult the monographs by Bailliart; Streiff and Monnier; and Weigelin and Lobstein as well as the contributions by Hollenhorst; Smith; Gay; Barut; and others.

A relatively new method of "objective" ophthalmodynamometry is represented by the *ophthalmodynamography* based on the instrument and technique devised by Hager. The ophthalmodynamograph is a tambourlike apparatus which, applied to the surface of the eye, transmits and records pulsations from the orbital vascular structures. The tracings obtained give information on the systolic and diastolic ophthalmic artery pressure, the pulse volume, and the pulse-wave speed within the internal carotid. Abnormal ophthalmodynamographic patterns are especially characteristic in carotid artery occlusive disease, low-tension glaucoma, ischemic papillitis, migraine disease, and temporal arteritis (Hager; Bettelheim; Salmon and Gay). Special ophthalmodynamographic data in cases of increased intracranial pressure have been collected by Finke. These findings will be discussed in the section on papilledema (p. 110).

Fundus photographs, black and white or color, easily performed today with a hand camera (Kowa, Olympus, etc.) as a bedside procedure, furnish important records which, by comparison at different time intervals, give valuable information on the development and morphologic changes of any fundus process (papilledema, optic atrophy, vascular affections, etc.). A new method to record function and structure in the living eye became available with the introduction of serial photography of fluorescein in the circulating blood of the human ocular fundus (Novotny and Alvis).

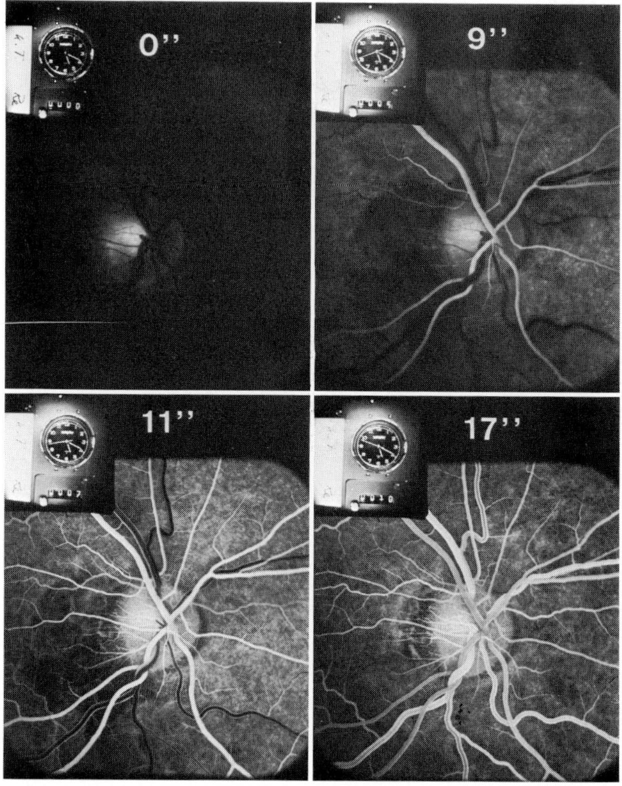

Fig. 1-26. Fluorescein angiogram of the normal fundus and disc. At time mark 0 a faint glow (incomplete filtering) of the disc is visible. Nine seconds after intravenous injection of fluorescein the retinal arteries become stained with the dye, and the background of the fundus shows mottling as a result of staining of the big choroidal vessels. At time mark 11 seconds the retinal arteries are fully stained, the disc manifests distinct fluorescence with numerous fine capillary vessels, and the veins begin to show laminar inflow of fluorescein. At time mark 17 seconds the retinal veins are fully stained (laminar flow in some places still visible), the arteries become less fluorescent, the disc still reveals the fine capillary vessels, and the background fluorescence reaches its maximum.

Fluorescein angiography of the ocular fundus (Fig. 1-26) has not only become a popular method for studying the circulation and vascular changes of the eye but also for visualizing some aspects of the appearance and function of the vasculature of the optic nerve head not revealed by ordinary ophthalmoscopy.

For fluorescein fundus photography, 10 ml. of 5% fluorescein is injected into the antecubital vein. Any retinal camera can be modified for fluorescein fundus photography. All that is required is the insertion of two light filters; the first a blue filter (Kodak Wratten No. 47) over the aperture of the illumination diaphragm, the second a green-yellow filter (Kodak Wratten No. 15) in front of the film. This filter cuts out the blue exciting light and allows as much as possible of the fluorescent light to reach the film. The power supply for the electronic flash has to be strong enough to allow pictures to be taken at a time interval of 1 to 2 seconds. Twenty-four serial pictures taken at an interval of 1.5 seconds are sufficient to demonstrate the characteristic arteriolar, arteriolar-venous, and venous filling phases, whereas demonstration of the arteriolar-venous recirculation phases requires further pictures for 5 to 20 minutes.

If a fundus camera is not available, the examination of the fundus and the observation of the passage of the fluorescein into the vessels and the tissues can also be observed with the ordinary direct or indirect ophthalmoscope using an additional blue filter (Kodak Wratten No. 47 or No. 47a). The personal observation of the patient's fundus during fluorescein passage may sometimes supply even better information than photography; it becomes equivalent in a certain way to *fluorescence cinematography* of the ocular fundus, which has the advantage of allowing visualization of all the circulatory phases without interruption.

Normal and pathologic patterns of fluorescein angiograms will be discussed in connection with papilledema (p. 102) and optic atrophy. For further information concerning the method and results of fluorescein fundus photography, refer to the monographs by Wessing and by Jütte and Lemke as well as to the numerous contributions by Amalric, Bessou, and Aubry; Charamis, Katsourakis, and Mandras; Ferrer; Matsui, Koh, and Tashiro; Norton; Smith; and Witmer.

Visual acuity

In the course of a neuro-ophthalmologic examination it is absolutely necessary to determine the visual acuity as one of the functions of the retina and the visual pathways. The visual acuity and the visual fields, possibly with the addition of the color sense, are the principal criteria used in the evaluation of the functional state of the optic nerve and the higher visual pathways under normal and pathologic conditions. The examination of every patient with proved or suspected brain tumor must include the measurement of visual acuity.

The *distance* vision is tested with the aid of a chart to be used at 5 or 6 meters (Snellen or Birkhäuser type). These charts are arranged in rows of letters or numbers of certain sizes, the smallest of which should just be readable by the normal eye with a visual acuity of 1.0 (subtending an angle of 1 arc minute).*

---

*Actually, these characters subtend an angle of 25 arc minutes, with five squares in the horizontal and five squares in the vertical line. Each of these individual squares subtends an angle of 1 arc minute. It should stimulate an isolated retinal cone. If one cone lies between two cones thus stimulated by two black squares, these two squares are appreciated separately; otherwise they are seen as just one black area; "minimum separabile."

The actual visual acuity of the patient depends on the size of the line he is just able to read. It can be expressed in the form of a decimal (1.0; 0.1) or common fraction (6/6; 6/60). For the sake of uniformity it would be desirable if the value of the visual acuity were always expressed in the form of a decimal fraction. If the patient is unable to read the largest characters on the chart (corresponding to a visual acuity of 0.1), he should be tested at half the distance (3 or 2.5 meters). The resulting values, accordingly, should be divided by 2. If the vision is still poorer, the reading distance can be further reduced. For a distance of 1 meter or less it is better to have the patient count fingers and then express the visual acuity as "finger counting" (for instance, "f.c. at 50 cm."). If the vision is impaired very severely (for instance, secondary optic atrophy after papilledema), fingers may not be recognized—only the gross movement of the entire hand. The visual acuity is then recorded as recognition of hand movements (for instance, "H.M. at 50 cm."). If even the recognition of form is impossible, only the *light localization* can still give some idea of the visual faculty. A flashlight or the beam of an ophthalmoscope is brought in from the periphery in various meridians, and the patient is asked to indicate when he perceives the light. This type of examination also gives a rough idea of the visual field in patients with severely impaired visual acuity (p. 73).

The *near* vision is tested with suitable reading charts (Jaeger; Birkhäuser), which consist of text matter printed in paragraphs of varying size corresponding to a visual acuity that ranges from 0.1 to 1.5. For this examination one should maintain a normal reading distance of 30 cm. The near vision test is unreliable if there is a paresis of accommodation.

Whenever possible, the *binocular visual acuity* should be determined after each eye has been tested individually; usually it is somewhat higher than the monocular vision. Proper illumination of the charts is essential for a reliable vision test. It should always be of the same intensity so that the results of the tests, if repeated, can readily be compared. Refractive errors are a factor that always has to be considered in the determination of the visual acuity. As a rule, the physician who performs the neurologic examination has neither a trial case at his disposal nor the time to concern himself with the problem of refraction. Patients with high refractive errors (myopia, astigmatism) usually already have a correction. The glasses should be worn, of course, for the testing of the visual acuity. The hyperopia of younger patients is of little importance for the testing of distance vision because it can be compensated without difficulty by accommodation. Like presbyopia, hyperopia definitely has to be taken into consideration for the determination of the near vision; if necessary, it must be corrected with convex lenses. If no trial case is available, one may dispense with the determination of near vision in older people (older than 45 years) and in persons with hyperopia because the results would not be of much value.

During the performance of a neuro-ophthalmologic examination in patients with suspected brain tumors, *malingering* probably will have to be considered only rarely in the determination of the visual acuity. This problem is of considerably greater importance for medicolegal evidence with regard to persons having suffered from brain injuries. A large number of "tricks" for the discovery of malingerers feigning poor vision has been suggested. They are described in detail in textbooks of ophthalmology. We want to mention here only that it is *possible to use optokinetic nystagmus for an objective determination of visual acuity* (Ohm; Goldmann; Günther).

A pattern of minute black dots or squares is moved evenly either horizontally or vertically before the eyes of the patient to be examined. He is asked to observe this pattern without too much concentration. If his vision is keen enough to recognize this pattern and its movements, he will show an involuntary optokinetic nystagmus. The patient then is moved away from the instrument until the nystagmus is just barely noticeable or shows a tendency to disappear. The ratio between this distance and the actual visual acuity has been determined empirically for subjects with normal vision. This method permits a very reliable objective determination of the visual acuity. Since this form of examination is based on special instruments, it can be performed only in well-equipped eye departments. Those satisfied with less accurate results can easily improvise a similar arrangement with an optokinetic drum (p. 58).

In the section on symptomatology it has been stressed that it is necessary for the patient to recognize and understand letters and words if his visual acuity is to be tested. This may cause a great deal of difficulty for patients with *sensory-aphasic disturbances* of the *alexia* type (for instance, those with tumors near the angular gyrus and the adjacent areas of the temporal or occipital lobes). In most instances such patients also will not understand gestures or pantomimes; hence attempts to test the vision in such a manner will also be unsuccessful. If the descending optomotor fibers in the middle and posterior part of the optic radiation are intact, an attempt can be made to determine the visual acuity roughly with the aid of the optokinetic nystagmus (p. 58). The situation is by far less complicated in patients with *motor aphasia* (tumors of the inferior posterior part of the left frontal lobe) because the patient understands quite well what he reads. Even though he may not be able to express what he has read, he can indicate distinctly to the examiner by signs or gestures (for instance, nodding) whether or not he is able to read a certain line on the visual acuity chart. (See Fig. 1-1.)

Under certain conditions, discriminative and minute testing of the visual acuity may uncover other value details among the symptoms found in patients with brain tumors. The near vision test may disclose a *homonymous hemianopia*—especially if it is on the right side—because of the difficulty the patient has in reading a whole sentence fluently or because of his complete inability to form a correct sentence from the words he has read. *Concentric constrictions of the visual field* may become evident if it is difficult for the patient to locate the chart or if he sees only a few letters instead of a whole line. *Central scotomas* are particularly annoying to the patient while reading. If they are quite large, they prevent him from reading altogether. Smaller scotomas cause the fading of letters in a word or of complete words in a sentence. *Metamor-*

phopsias (micropsia, macropsia, irregular distortion) become quite obvious in reading. As a matter of fact, letters are an ideal test object for pathologic distortion, magnification, or diminution of an image. However, it should be emphasized that central metamorphopsias are rare (any lesion along the entire visual pathway may cause them!) (p. 6). Before such a diagnosis is made, retinal changes always have to be ruled out. A *paresis of accommodation* should be suspected if the patient's distance vision is satisfactory but he is unable to read small print at a distance of 25 to 30 cm. Presbyopia and high hyperopia have to be ruled out in such a case. Crossed diplopia on near vision without double vision for distance may be the result of a central *convergence palsy* (due to tumors in the midbrain) (p. 263). It must be differentiated from the purely functional convergence insufficiency.

In special cases in which the determination of visual acuity is difficult or impossible (cataract, vitreous opacities, cortical blindness, malingering) *electroretinograms* (ERG's), eventually combined with *electroencephalograms* (EEG's) (evoked potentials, determination of retinocortical time), can be extremely helpful and provide important additional information on the functional state of the retina and the optical pathways, including the occipital cortex.

Color sensation

Any damage to the optic nerve or optic tract by an intracranial neoplasm results in a more or less severe disturbance of conductivity; this can also modify or completely destroy the color sensation. According to experience, the *color sensation for red is most vulnerable and will be modified or destroyed before alterations of the sensation for white appear.* Sensations for green and blue will be involved only later. This rule applies to inflammatory and toxic disturbances of the conductivity of the nerve and tract as well as to those caused by a tumor. Traquair is of the opinion that this disturbance of conductivity for red, which is so important for an early diagnosis, is not due to a selective vulnerability of the fibers conducting the red sensation. Red is generally considered a particularly suitable and refined clinical test for disturbances of conductivity, which actually could be demonstrated in the early stages just as well for white and blue if one would take the trouble to search for them with suitable test objects. Impairments of the color sensation in disturbances of conductivity caused by a tumor are usually concomitant with the appearance of a more or less advanced optic atrophy. This may take the form of a primary descending atrophy as the result of direct pressure of a tumor on the nerve, chiasm, or optic tract (pituitary tumors, meningiomas of the tuberculum sellae or the sphenoid ridge, craniopharyngiomas, tumors of the optic nerve or its sheaths), or it may be an atrophy secondary to a chronic papilledema associated with tumors in almost any location. *Disturbance of the color sensation frequently precedes the visible manifestation of the atrophy, thus assuming important diagnostic significance as an early sign of a retrobulbar disturbance of conductivity.* As a rule, this disturbance is not a general depression. In the early stages it usually manifests itself as a more or less extensive *central scotoma for color,* especially for red; it will expand from the point of fixation toward the periphery. Such a scotoma for red may at first be only *relative* (that is, the red target will appear less saturated in a certain central area than in the remainder of the field). In a later stage the relative scotoma for red becomes *absolute.* In other words, the form of the red target of a given size used for the test will be recognized but its color cannot be perceived any longer. In even later stages, scotomas will develop for green, blue, and white (in that order) until the disturbance of conductivity and the resulting atrophy of the nerve fibers terminate in the complete extinction of all color sensations. This experience is encountered time and again in patients with severe chronic atrophic papilledema.

In every case of an acquired disturbance of the color sensation a congenital form (red-green blindness, yellow-blue blindness) must be excluded. This is customarily done with the standard *pseudoisochromatic plates by Ishihara or Hertel.* These plates may also be used for color disturbances in nerve and tract lesions, but only in the relatively severe stages of general depression of the color sensation over wide areas of the visual field. In the early stages with only scotomatous defects the results are unsatisfactory. To demonstrate a color scotoma, one uses either a *perimeter* or *campimeter* (pp. 74 and 76). This is the best and most reliable method. For a quick survey to determine the presence of a scotoma for color the so-called *scotometer* is suitable. A black wand is provided with colored dots of varying size and color. For the examination the patient is instructed to fix on a certain colored dot with one eye (it is important to make sure that the patient fixes properly!) and to indicate whether he perceives color and, if so, what color. Dots of the same color but of a different size are shown to the patient. In this simple manner, scotomas for red and other colors can be rapidly demonstrated and, to a certain extent, their size can be determined. Such a color scotometer can be improvised whenever the need arises, and without great technical resources. Even colored crayons can be used for this purpose; they are covered by black paper with holes of varying sizes.

For the more exact methods of testing the color vision, especially for a description of the use of the *anomaloscope,* textbooks of ophthalmology should be consulted.

Although the peripheral disturbances of conductivity involving the visual pathways (namely, the optic nerve, the chiasm, and the optic tract) are the principal causes of a defective color sensation in connection with brain tumors, it should be remembered that there are also *central forms* associated with certain lesions of the visual cortex or the higher visual centers (angular gyrus) (Fig. 1-2).

In describing cortical blindness (p. 7), it has been stated that color sensation is last to return during the phase of recovery (that is, after light and form sensation). In partial lesions of the visual cortex the color sensation seems to be the most vulnerable part of the visual function and the one that is most easily impaired. In lesions of the calcarine area, a homonymous hemianopia in a unilateral process or an extensive concentric constriction of the visual field in a bilateral process with sparing of the macula may be associated with a complete or partial disturbance of color sensation (achromatopsia or dyschromatopsia). A disturbance of space and form perception, which usually occur simultaneously, should make it possible to differentiate it from the peripheral forms.

An entirely different type of disturbance of the color sensation is *amnestic color blindness* (Wilbrand) in patients with lesions in or near the angular gyrus. In this type of disorder the visual pathway ending in the calcarine cortex is intact. Only the associative interpretation of the color impression via the higher visual centers is impaired. The patient is unable to recognize the color he sees or, in less severe forms, confuses colors and calls them by their wrong names. Again, associated symptoms such as simultaneous object agnosia, alexia, and apractic and sensory-aphasic symptoms (word deafness, Fig. 1-2) will usually suggest central origin.

Dark adaptation

Dark adaptation is the faculty of the eyes to adjust to varying intensities of light by an increased or decreased response of the light-sensitive retinal elements. This adaptation is primarily the function of the rods of the retina.

It should be recalled in this connection that the very *first sign of a disturbed conductivity of the optic nerves and tracts is an impairment (that is, a reduction or complete loss)* of the dark adaptation (Behr; Rutgers; Gasteiger). It can be determined before the appearance of changes in the visual acuity, the visual fields, the color sensation, or the eyegrounds. A dark adaptation test therefore may be of considerable importance for the early diagnosis of a tumor causing disturbance of the conductivity of the optic nerve, the chiasm, or the optic tract. Unfortunately, the value of this test is limited because it requires relatively complicated instruments that are not always available. The determination of the visual field in the dark-adapted eye and its comparison with that of the light-adapted eye may, to a certain extent, serve as a suitable alternative (Goldmann).

The adaptometers of Nagel and Della Casa are useful for dark adaptation tests. Goldmann has constructed a self-registering adaptometer that is ideal for research purposes (manufactured by Haag-Streit in Bern).

Visual field

Next to ophthalmoscopy, the examination of the visual field is one of the most important procedures in the neuro-ophthalmologic evaluation of brain tumors. It is so important because *visual field changes occur so frequently in association with brain tumors.* They occur in about 50% of the total number of patients with brain tumors and, according to Cushing and to Sanford and Bair, in 70 to 80% of patients with temporal lobe tumors. Furthermore, certain types of defects have a definite *localizing value* (in about 25% of the patients). This fact is so important because the determination of the visual field is a procedure that can be performed relatively simply as part of a neurologic examination. A reliable field can be plotted by every physician without special training. We said intentionally "by every physician" because we are of the opinion that testing the visual field is too important a task to be entrusted to a technician. Such a person is not familiar with the physiologic principles involved and does not possess the necessary basic training. The modern neurosurgical methods for an exact localization (such as ventriculography, arteriography, electroencephalography, and others) should in no way lessen the localizing value of the visual field. By its use the patient may be spared a disagreeable and potentially harmful diagnostic procedure. It would be wrong for the neurosurgeon to take the attitude that examination of the visual field should be performed by the referring physician or by an ophthalmologist. On the contrary, he should insist on performing the test himself according to the principles of exact perimetry and campimetry so he can form his own opinion regarding the usefulness of the method and its results in a given patient. Naturally this does not exclude cooperation with the ophthalmologist, especially in difficult and obscure cases.

During the past years the methods of perimetry and campimetry have become incredibly refined and perfected. In particular, the instruments for perimetry (for instance, the one designed by Goldmann) come equipped with all sorts of perfections that should render examination of the visual fields quite valuable and furnish, in certain cases, data of the utmost importance. However, reliable *cooperation* on the part of the patient is an absolute necessity when such instruments are used. *Unfortunately, this is not always the case, especially in patients with brain tumors.* For one thing, the patient may be drowsy because of increased intracranial pressure. Also, a tumor causing involvement of the associative centers may make it impossible for the patient to comprehend certain events in the surroundings (for instance, a visual field test). He may be unable to perform a task requested of him (sensory aphasia) or to express his thoughts in words (motor aphasia). The complicated instruments used in perimetry and campimetry frequently fail with such patients because of their lethargic or impaired cooperation. The technical perfection of the instrument is out of proportion to the mental alertness of the patient. There is no advantage in using such minute test objects with a variety of light intensities if a dull and indifferent patient is unable to follow and appreciate them! We ophthalmologists are the last to deprecate the value of the modern perimeter. *Yet, in our contact with patients with brain tumor we have come to recognize that, in certain cases, crude and seemingly simple methods of testing the visual field may be just as valuable, if not more so, as some such highly complicated instruments.* For this reason we shall discuss "crude bedside methods" before discussing the methods of perimetry and campimetry.

Qualitative testing of the visual fields without special instruments (crude methods)

For the so-called *confrontation test* the patient and physician face each other, with the light (daylight or artificial light) in back of the patient. The patient is instructed to cover one eye with his hand and to fix the eye to be examined on the examiner's opposite eye. The latter moves his entire hand or extends only the index finger in a frontal plane equidistant between the patient and himself in various (preferably eight) meridians from the periphery toward the center. As soon as the patient sees the finger in his field, he should indicate this. With this method, the examiner's own field serves as a control. Testing by confrontation can also be performed with a pencil or by asking the patient to count the examiner's fingers in the various quadrants of the visual field. Traquair has modified this test by using a white disc, 2 to 6 cm. in diameter, mounted on a stick. As a further refinement of this method he has advised reducing the illumination of the room because a reduction in the brightness of the test object may render pathologic changes of the visual field more conspicuous. The *simul-taneous binocular testing of the visual fields* is quite an important variant of the confrontation field. Both eyes of the patient are uncovered. The examiner moves both hands or extended index fingers symmetrically from the temporal sides toward the center. It is important to have even illumination. The patient has to indicate whether he sees one or two hands and whether he recognizes one hand earlier or more distinctly. If there is a difference in perception of the two sides or perhaps no perception on one side, one speaks of a so-called *relative hemian-opia* (p. 4). This is in contrast to the absolute hemianopia that becomes manifest in the separate examination of each individual eye. *The relative hemianopia must be considered an incomplete form of the absolute hemianopia.* In the former only partial functions of vision of a qualitative nature (recognition of form, perception of movement, and others) are disturbed. Ordinary perimetry performed separately for each eye will not uncover a relative hemianopia, not even with the most minute methods, unless a test object from the nasal side and one from the temporal side are moved simultaneously from the periphery toward the point of fixation (Dufour; Head; Thiébaut, Guillaumat, and Brégeat). Since, as mentioned previously, only relatively primitive partial visual functions are disturbed, only a relatively crude method such as the one outlined will uncover such a relative hemianopia, also called a *"unilateral visual inattention."* An example such as this should emphasize how valuable and at times even indispensable such simple methods can be in the testing of the visual fields. In case one does not find field changes with the exact methods of perimetry in patients with tumors of the parietal or parieto-occipital area, he should always, by means of the confrontation test, search for a relative hemianopia, this important form of "visual inattention" which is in the same class as other manifestations of the so-called *anosognosia* (for instance, the unawareness of a hemiplegia or a central deafness).

Kestenbaum describes another modification of the confrontation test which, in our opinion, is also valuable and which supposedly will furnish even more accurate results. He has observed that the *visual field in the plane of the face corresponds with a surprising exactness to the outline of the face.* Whereas in the ordinary confrontation test the field of the examiner serves as a control, the outline of the patient's face serves this purpose in Kestenbaum's *"outline perimetry."*

The patient is instructed to look straight ahead. One eye is covered. The examiner brings a pencil or his finger from the periphery into the field of vision. This is done in a plane not more than 2 or 3 cm. in front of the patient's face and repeated in eight to twelve meridians. As soon as the test object appears in the periphery of the field, the patient should indicate this. If the visual field is normal, it will coincide with the outline of the face (nasally with the border of the nose). On the basis of this normal outline, a pathologic field can be recognized without difficulty. Kestenbaum is of the opinion that the margin of error with this method is less than 10 degrees in the hands of an experienced examiner. It would be expedient to include this outline perimetry in the general screening of every patient even if no field defects are suspected at first.

As a bedside test, especially in patients who fatigue easily or who are slightly stuporous or in aphasic patients showing poor cooperation, it may, except for the confrontation tests, be the only way to determine the visual field. This also applies to patients whose visual acuity is reduced to finger counting or recognition of hand movements and in whom ordinary perimetry would fail.

Patients with a brain tumor may not even respond to outline perimetry. In such instances even cruder methods may be called for. For the so-called *reflex lid closure test* the examiner abruptly approaches the eye to be examined with the hand or fist from both sides as well as from above and below. The presence or absence of the lid closure reflex will allow some conclusions as to whether the visual field is normal or abnormal. Other *optically induced reflex movements of the eyes* may be used instead of the lid closure reflex. An article that will attract the patient's attention, such as a pencil, cigarette, key, coin, or food (for instance, a piece of fruit), is moved toward the patient. If his field is intact, he will perceive the object already in the periphery of his field and will direct his gaze toward it (optically induced reflex movement, p. 49). The eight meridians should be tested for each eye, and the response should be recorded. The persistent absence of a reflex in a particular meridian should be considered as an indication of a field defect. In aphasic patients the small test objects mentioned are inadequate. Kestenbaum is right when he mentions that the strongest stimulus to gain the attention of such patients is the human face. The examiner stands behind the patient in a position that enables him to bring his own face into the patient's visual field from various directions. If the patient turns his eyes toward the physician's face, he probably has an intact field for that particular meridian, and vice versa. This crude test is particularly useful for the determination of field defects not only in aphasic patients but also in children.

If the visual field is reduced to light perception, *testing light projection* is the only means of gaining information concerning the visual field. A bright point source of light (for instance, an ophthalmoscope) is moved toward the eye in various meridians. The patient has to indicate the direction from which the light comes and, if possible, the moment a light sensation is perceived. This test can be performed similar to Kestenbaum's "outline perimetry."

Quantitative determination of visual fields

If a field defect has been discovered with these crude methods, it should be investigated, if possible, with the *more refined methods of perimetry and campimetry.* If a superficial screening has shown no definite defect, a further and more subtle analysis is even more indispensable.

The customary *quantitative examination of the visual field* is based on the *principle of the determination of the differential threshold.* This can be accomplished in two ways.

The first way is to expose the target in a fixed position and to vary its size or its light intensity until the patient recognizes it. This determines the differential threshold for a given point of the retina (static perimetry). Point after point of the field is tested in this manner— an extremely cumbersome procedure but one that furnishes extremely reliable results, especially in neurologic patients (Harms; Ferree and Rand). The second way, and the one followed in most perimetric procedures (including those which will be described here), follows a procedure whereby a target of a certain size and luminosity is moved from the periphery toward the point of fixation (or in the opposite direction) until the patient recognizes it. This is the determination of retinal points with the same differential threshold (kinetic perimetry). All points thus determined are connected to form a curve, the so-called *isopter,* which represents a line formed of points with the same differential threshold for the chosen test. In the more common types of perimeter the illumination and the contrast between background and target are constant. The only variable is the size of the target (perimetry with one variable). Newer instruments have been designed that permit a variation of the size and luminosity of the target according to a definite ratio (perimetry with two variables; for instance, the Goldmann perimeter).

For an exact evaluation of the visual field in its entirety it is still advisable to use two instruments—a *perimeter* and a *campimeter* (tangent screen). *The former is used primarily for testing the peripheral field and the latter for the intermediate and particularly the central areas.* The need for two separate procedures is a consequence of the enormous difficulty of constructing targets sufficiently small for the exploration of the central area on the perimeter. It is not the purpose of this book to describe the exact technique used for perimetry (static and kinetic) and campimetry. We recommend the excellent and detailed monographs on this subject by Traquair; Lauber; Malbrán; Harrington; Hughes; and Dubois-Poulsen as well as the valuable articles by Goldmann; Harms; Aulhorn; Walsh and Hoyt; Schmidt; and others. Nevertheless, we want to stress a few points that are important in the neuro-ophthalmologic examination (especially in brain tumors) and that are, in part, based on personal experience.

Examination of the visual field with the perimeter

A large number of various instruments are available for this purpose. Essentially they are modified and improved types of the classic *Foerster* perimeter. It can still be recommended if it is equipped with a suitable illuminated arc. More up-to-date models have been designed by Ferree and Rand, Magitot, Maggiore, and Aimark. Next to the Goldmann perimeter, the two last-named instruments are most popular in our eye clinics. The Maggiore, Aimark, and Goldmann models are so-called projection perimeters because the test marks are replaced by projected light dots. The instruments employing test marks are for perimetry with one variable. Projection of a light target allows perimetry with two variables (that is, a variation in size and brightness of the light marks). As a rule, the radius of the perimeter arc is 33 cm. The test objects or light marks usually have a diameter of 10, 5, 3, 2, or 1 mm.

For the determination of the absolute outer limits of the visual field on the perimeter, one begins with a 10 mm. target. It is moved slowly from the periphery toward the point of fixation at a steady rate of about 2 degrees per second. Next follows the exploration of the isopters. The differential threshold varies considerably in pathologic cases; thus no set rules can be stated for the determination of the isopters. Nevertheless, one should be guided by the following directions: the isopters determined should be rather evenly distributed from the periphery toward the point of fixation and, if possible, should include most zones of the peripheral visual field. The central zones are part of the examination on the tangent screen. In addition to the absolute outer limit, which is determined with a 10 mm. target, it is desirable to have an isopter of the 60-degree zone (with a test object of 5 or 3 mm. for the normal individual) and one in the 30-degree zone (with a test object of 2 or 1 mm. for the normal individual). At least twelve meridians have to be tested in a routine examination. The use of targets smaller than 1 mm. is unreliable and should be avoided in perimetry. Instead of moving the target from the "nonseeing" to the "seeing" part of the field, this method can be reversed. This will result in a somewhat greater sensitivity of some retinal areas (desirable in cases of poor cooperation); for instance, scotomas will appear smaller, and the peripheral limits of the field will extend farther out.

The isopters obtained in this fashion will be transferred to a visual field chart and identified in the form of a fraction (10/330, 5/330, 3/330, 2/330, 1/330). It is best to *use the same chart for perimetry and campimetry and to record the isopters found by the two different methods on the same chart* (Fig. 3-12).

With projection perimeters, the number of isopters can be increased by the use of each of the different-sized targets with varying degrees of brightness.

This is perhaps the appropriate time to describe the *Goldmann projection perimeter* which, as a perimeter with two variables, can be considered one of the most modern, most complete, and most efficient instruments. The two variables are the size of the object and its brightness or, more precisely, its contrast with the background of the perimeter bowl. The illumination of this bowl is such that an extended adaptation of the patient is superfluous. By means of a telescope-like arrangement, the examiner can constantly check the patient's eye against a thread cross to assure central fixation, which is of the utmost importance for neuro-ophthalmologic field determinations. A pantograph permits the direct registration of the findings from the projection arm to the visual field chart. Six targets of different sizes and four different intensities of illumination make the determination of a large number of evenly scaled isopters possible. The relationship between the size of the objects and the intensity of illumination has been arranged according to the law of summation, so the same isopters can be determined either with a large but less luminous or with a small but bright target.

It has been observed that in patients with certain affections (p. 109) it is indispensable to determine not only the isopters but also the capacity for summation of various retinal areas (Dubois-Poulsen; Goldmann). This can be easily accomplished on the Goldmann perimeter by means of either the large, dimly illuminated or the small, brightly illuminated test marks that have been arranged according to the law of summation. A number of methods permits the determination of the sensitivity of any retinal area between the examined isopters:

Goldmann's skiascotometry, Harms' method of static perimetry, and flicker fusion perimetry as elaborated by Hylkema, Weekers and Roussel, and Miles. Since it is possible to vary the brightness of the inner surface of the bowl of the Goldmann perimeter, the use of this instrument makes it possible to examine and compare visual fields for the light-adapted (photopic perimetry) and the dark-adapted (scotopic perimetry) eye.

For a detailed description of the technique used with the Goldmann perimeter, we recommend the original publications of Goldmann as well as the monograph by Dubois-Poulsen, whose material was mostly obtained with the aid of this perimeter. One of the chief advantages of this instrument, so extremely valuable for neuro-ophthalmologic examinations, is the ability to *control fixation.* Also, unlike any other perimeter, it furnishes reliable results *for the peripheral as well as the central areas of the field.* The use of the Goldmann perimeter is simple and rapid—in experienced hands more rapid than that of other perimeters. We regard it as the instrument of choice for a detailed and exact visual field test in patients with brain tumors, provided there is proper or at least halfway satisfactory cooperation. Examples of visual fields determined with the Goldmann perimeter can be found in later chapters. (See, for instance, Figs. 3-34 and 3-39.)

Goldmann and Schmidt briefly make the following suggestions for a proper and suitable technique with this perimeter—suggestions that prove to be satisfactory in our experience. The good eye or the eye with the better vision is examined first. The size of the pupil is recorded. For the examination of the central and intermediate field the patient should wear his correction (if indicated, a presbyopic correction!). There should be a preadaptation of the patient on the illuminated perimeter for 4 to 5 minutes. The examination begins with target size 1 and with intensity 4 (the corresponding isopter is recorded as "1/4"). For healthy persons between the ages of 20 and 30 years this isopter should extend to the normal peripheral limits of the field. If a constriction of the field is found, the size of the target should be increased, and the test should be repeated for 2/4 and, if necessary, 3/4 or even 4/4 and 5/4. The target should be brought from the periphery toward the center in the meridian examined with a steady speed of about 5 degrees per second. This is performed every 30 degrees, but every 5 degrees or at even lesser intervals for the meridians of a field defect. After the peripheral limits of the field have been determined, one proceeds with the investigation of the intermediate and central isopters with the use of targets of less intense illumination or smaller size (for instance, 1/3, 1/2, 1/1, and 0/1). The outline of the blind spot should be determined with the target for the isopter that just includes the blind spot (usually 1/2). The target should be moved from the blind toward the seeing area. In the case of a scotoma the extent of the scotoma is first determined with the target corresponding to the isopters including the scotoma. The test is repeated first with a target of higher intensity and next with one of larger size. For instance, if the scotoma lies in the 1/2 isopter, the test is performed with the 1/2, 1/3, 1/4, 2/4, and 4/4 targets. Paracentral scotomas are tested by flashing the 1/1 and 0/1 targets. If malingering is suspected, the binocular field should be examined. Even the Goldmann perimeter does not furnish absolutely reliable findings in the area central to the blind spot, particularly around the point of fixation and its immediate surroundings. For the exact evaluation of this part of the field one still cannot dispense with the time-proved method of campimetry.

Visual field examination with the campimeter (tangent screen)

For this purpose it is best to use the *Bjerrum screen* at a distance of either 1 or 2 meters. We definitely prefer the large screen at a *distance of 2 meters* for neurologic patients, especially those with brain tumors. A large number of the fields reproduced in this book have been obtained by this method. For the very smallest central defects Krayenbühl recommends Vogt's method of testing at a distance of 6 meters (see Wiesli). The tangent screen permits the use of objects that are very small as measured in degrees and yet large enough in actual extent to be easily handled. The examined dimensions being larger than at the perimeter, the examination can be performed with more accuracy.

The Bjerrum screen should be provided with concentric circles, if possible, in addition to the radial meridians. It should be evenly and intensely illuminated with floodlights. The patient is seated at a distance of 2 meters exactly opposite the fixation mark, with his eyes level with the latter. His head should be supported by a chin rest. The examiner should wear a black coat, black gloves, and, if possible, a black mask of the type used in the operating room.

First the target just inside the boundary of the screen for the eye examined is determined. With this target the most peripheral isopter is plotted. It is also used for a superficial survey of the size and position of the blind spot. This enables one to check the patient's fixation at the same time. Other isopters are determined by choosing the targets in such a manner that one isopter is outside the outer border and one inside the inner border of the blind spot. The most suitable targets are 3, 2, and 1 mm., the last one for the central and paracentral zone. After the preliminary test, the blind spot should also be examined with the smaller 2 and 1 mm. targets—a test that may be of particular help when demonstrating an increase in size of the blind spot due to a papilledema.

The isopters thus obtained are recorded on the blank and labeled as 3/2000, 2/2000, and 1/2000. If possible, perimetric and campimetric results should be plotted on one and the same blank (Fig. 3-52).

We use *color test objects* for the *visual field examination* only as a test for the central color sensation (that is, to demonstrate a scotoma for red in early disturbances of conductivity of the optic nerve and tract). As a rule, relatively larger objects should be used for color perimetry (for instance, 5/330) as well as for campimetry. In agreement with Dubois-Poulsen and Goldmann, we are of the opinion that *nothing can be accomplished with color perimetry that could not be accomplished with a carefully performed "white" perimetry—* particularly because the former is rendered less valuable by a number of disadvantages in the form of a large number of inconstant factors, most of them of a technical nature. Many authors are still of the opinion that the initial disturbance in the conductivity of the optic nerve, the chiasm, or the optic tract is a disturbance in the central red field. A possible explanation for this misconcept is that a detailed "white test" has been omitted or performed in a faulty manner in such cases. An examination on the perimeter and tangent screen that has been carefully performed according to the previously described principles will eliminate the need for color perimetry and campimetry in the neuro-ophthalmologic examination.

The *Amsler-Landolt grid* is quite useful in certain patients for testing the very central part of the field (that is, the macula and its surroundings), especially for the demonstration of an absolute or relative central scotoma. Each individual square of this grid subtends a 1-degree angle if viewed at a

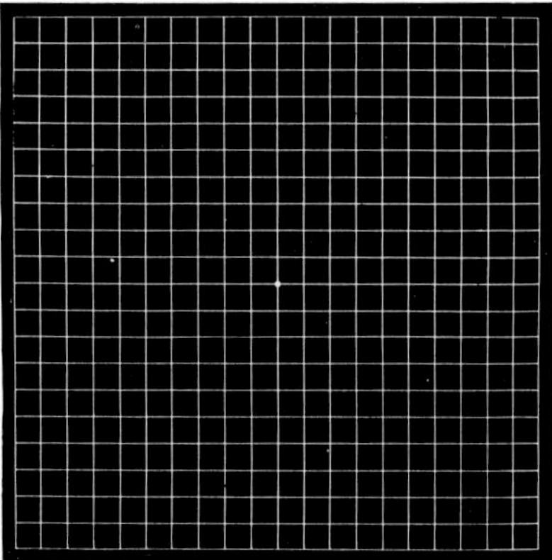

Fig. 1-27. Amsler-Landolt grid in reduced (½) reproduction. In the original the length of each small square measures 5 mm. and corresponds to a visual angle of 1 degree at a distance of 28 to 30 cm. Under such conditions the entire grid occupies the central area of a 10-degree radius around the point of fixation in the visual field.

distance of about 30 cm. (Fig. 1-27). Even the central part of a hemianopic disturbance becomes manifest on this grid provided it extends to the point of fixation and provided the patient has no "unilateral visual inattention" such as, for instance, in a relative hemianopia. In such instances the patient himself may be able to indicate the distance that separates the fixation point from the line dividing the seeing from the blind area of the visual field. This method has been quite helpful in our field studies with patients after occipital lobectomies, especially in answering the question as to whether or not there was sparing of the macula (Fig. 1-28).

Cuendet and Dufour have suggested that the area enclosed by each isopter (especially the intermediate and central isopters found on the Goldmann perimeter) be determined with the help of a planimeter, using the simplified figures thus obtained for the magnitude of the visual field as a basis of comparison in repeat examinations.

The perimetric and campimetric methods are subjective methods of examination. They require full cooperation on the part of the patient. Naturally they could be the source of numerous errors, which should always be considered judiciously. Fatigue of the patient should be taken into consideration. High refractive errors should be corrected by glasses. The response may be slow and the cooperation weary, especially with patients suffering from brain tumors. The result may be a spiral or, more rarely, a concentrically constricted field, although there is no organic defect of the visual pathways. Such an apparent contraction of the field caused by inattentiveness of the patient can be eliminated if, during the perimetric or campimetric test, the examiner moves the targets away from the fixation point rather than toward it.

As has been mentioned previously, most methods of qualitative determination of visual fields (perimeter and campimeter) are based on the procedure

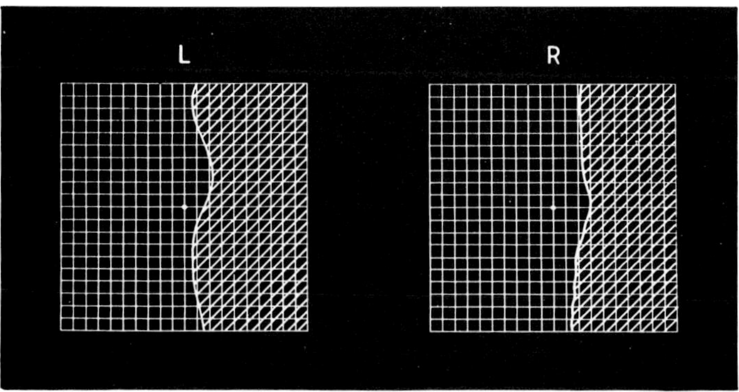

Fig. 1-28. Right homonymous hemianopia determined for the central visual field on the Amsler-Landolt grid. The shaded squares mark the field defect. The sparing of the macula amounts to 2 degrees on the left side and to 3 degrees on the right side, measured from the point of fixation. The patient had a tumor of the left occipital lobe.

of determination of retinal points with the same differential threshold and thus the principle of determination of isopters. For neuro-ophthalmologic purposes, however, it is especially desirable to also know the retinal sensitivity of any retinal area between the isopters. Two methods of perimetry making this possible deserve special consideration in this connection—flicker perimetry (Hylkema; Weekers and Roussel; Miles) and static perimetry (Harms).

Flicker perimetry consists in plotting visual fields by determining thresholds for a flickering light.

It is performed using an electronically driven glow modulator tube, presenting a target of about a 2-degree visual angle under the same conditions of illumination as in standard perimetry. Flicker thresholds are determined for twenty-six points in each eye at the 10- , 20- , and 30-degree circles.

Flicker perimetry appears to be more sensitive for central nervous system dysfunctions than standard field examination. Before it can be stated, on the basis of the standard examination, that central lesions (tumor, trauma, cerebrovascular accident) are present, extreme impairment of flicker discrimination must be demonstrable (Parsons, Chandler, Teed, and Haase).

In *static perimetry,* a stationary target is used that can be placed in any part of the field, and the differential threshold to light of the tested area measured.

For static perimetry Harms has designed a special instrument (Tübinger perimeter, manufactured by Oculus) that at the same time can also be used for kinetic perimetry under different adaptations (photopic, scotopic, mesopic), for flicker fusion perimetry, and for adaptometry. With special additional devices, the Goldmann perimeter can also be used for static perimetry. The target used can be varied in diameter, enabling the exploration of the whole visual field. In the chosen retinal area it remains absolutely immobile. Its luminosity decreases gradually, the time of presentation being less than 1 second and two

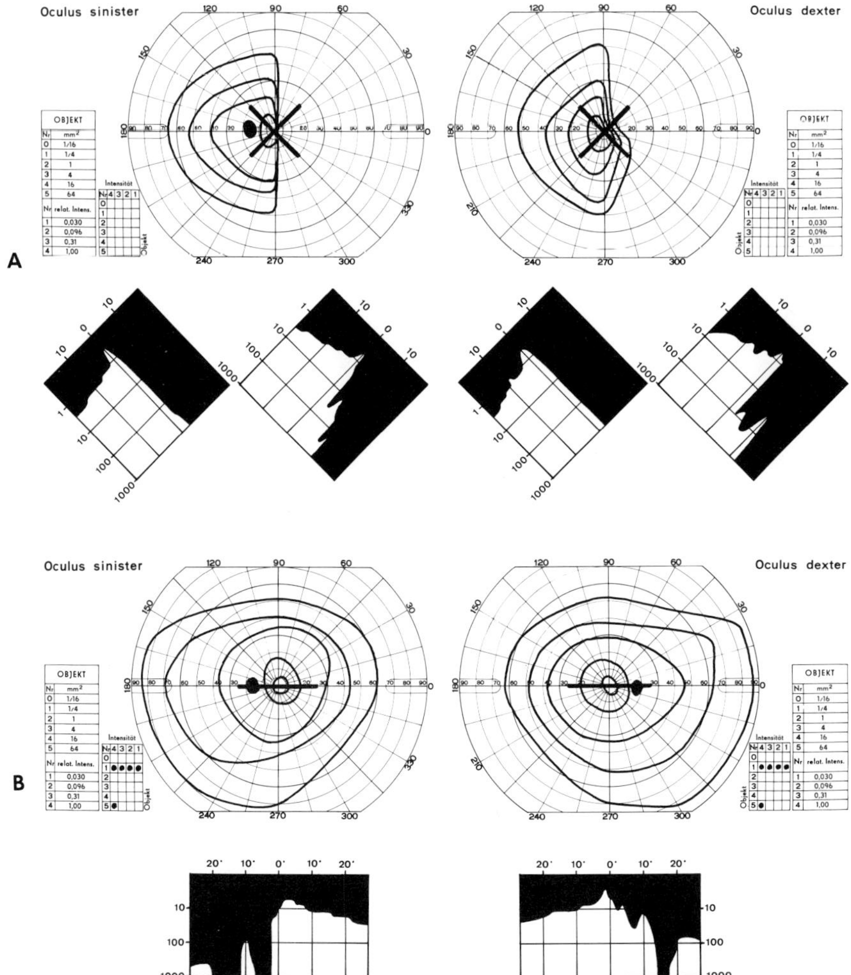

Fig. 1-29. **A,** Demonstration of kinetic and static perimetry in combination in a patient with a tumor of the left parietotemporal region. Above, Kinetic perimetry (Goldmann perimeter) shows right homonymous hemianopia with sparing of the macula. Below, Static perimetry (Harms perimeter) performed in the 45- and 135-degree meridians: the "function mountain" of the retina is cut away to the right side with distinct sparing of the macular area. Note the striking congruity of the field defects and its borderlines, which becomes manifest only in the static fields (in contrast to the kinetic fields, where by mistake not enough meridians have been examined in the left field [right inferior area], and therefore apparent incongruity appears!). **B,** Kinetic and static perimetry in a patient with suprasellar meningioma. Above, Kinetic visual fields (Goldmann perimeter), apart from a slight temporal indentation of the innermost isopters, show nothing pathologic (blind spots of normal size and site). Below, Static visual fields (Harms perimeter) manifest in the 0-degree meridian bitemporal paracentral scotomas (extending from 3 to 10 degrees), which escaped kinetic perimetry because they are situated just between two isopters. (After Harms.)

different stimulations being separated at intervals of 2 to 3 seconds. The results obtained are points on one and the same meridian, which are plotted on a chart: the distance from the fixation point in degrees on the abscissa, the luminosity of the threshold stimulus in asb on the ordinate. Different meridians of the visual field are examined according to the defect.

The results of static perimetry can be compared with meridional sections through the "function mountain" of the retina (Harms) and thus represent on the whole a three-dimensional visual field where defects, especially of neurologic origin (scotomas, hemianopias), can be identified and mapped earlier and with more accuracy (especially very small defects) than by means of the two-dimensional kinetic perimetry. In static perimetry, meridians going through the center of the visual field are examined. The results obtained inform therefore only about the distribution of visual function in certain meridians. What meridians to choose can be decided beforehand by kinetic perimetry. *Especially for neuro-ophthalmologic examinations, the combination of kinetic and static perimetry is the ideal procedure of choice* (Fig. 1-29). Kinetic perimetry, giving a rough idea of the existing field defects, will instruct the perimetrist what meridians to use in static perimetry; static perimetry, on the other hand, will furnish important data for deciding what size and luminosity of target is necessary for successful kinetic perimetry.

General pathology of the visual field in brain tumors

The pathologic visual fields will be discussed in detail in connection with the various tumor sites. Nevertheless, we believe it is important to list first in a brief but systematic manner the field changes caused by disturbances in the conductivity of the visual pathways. This *brief general pathology of the visual field* is intentionally based on the *shape of the visual field* to provide the examiner with a guide that should enable him to localize a lesion from a given form of visual field change (Fig. 1-30).

The visual field changes observed in patients with brain tumors may be divided into five "morphologic" groups: general and concentric constriction (centric or eccentric), unilateral sector-shaped defects, scotomas (absolute or relative), bitemporal field defects, and homonymous field defects.

General and concentric constriction (centric or eccentric). A *bilateral* concentric constriction should always be evaluated with caution, especially in a patient with a brain tumor, because, as has just been mentioned, it could be merely the result of a general apathy in a lackadaisical subject rather than a manifestation of an actual organic interference with the visual pathways. We have had occasions to observe such concentrically contracted fields or spiral fields in patients with frontal lobe tumors (Fig. 3-7). A contraction progressing to a typical "gun barrel" field in a *chronic atrophic choked disc* is caused by a nutritional disturbance of the retina rather than by a damaged optic nerve (Fig. 2-18). A bilateral "gun barrel" field may also result from a bilateral homonymous hemianopia, an event suggesting a bilaterally expanding tumor

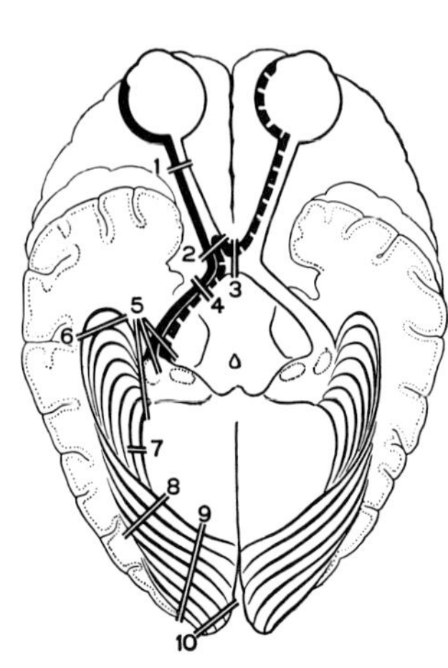

1, Lesion in optic nerve = Ipsilateral amaurosis
2, Lesion in optic nerve close to chiasm = Ipsilateral amaurosis with contralateral temporal hemianopia
3, Median lesion in chiasm = Bitemporal hemianopia
4, Lesion in optic tract = Incongruous homonymous hemianopia
5, Lesion in posterior part of tract, lateral geniculate body, or anterior part of optic radiation = Hemianopia without sparing of macula
6, Lesion in anterior (Meyer's) loop of optic radiation = Incongruous superior homonymous quadrantanopia
7, Lesion in inner part of optic radiation = Slightly incongruous inferior homonymous quadrantanopia
8, Lesion in middle of optic radiation = Slightly incongruous hemianopia without sparing of macula
9, Lesion in posterior part of optic radiation = Congruous homonymous hemianopia, frequently with sparing of macula
10, Lesion in area of occipital pole = Congruous homonymous hemianopic central scotomas

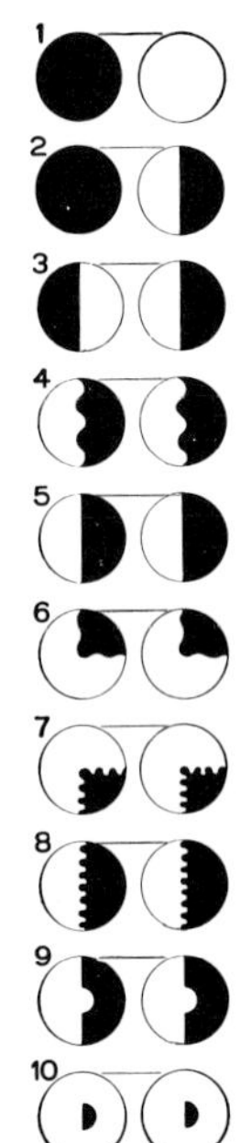

Fig. 1-30. Diagram of the most important visual field defects for lesions in various parts of the visual pathways. (After Duke-Elder.)

process in the region of the calcarine fissure. A *unilateral* concentric constriction suggests a *lesion of the optic nerve between the globe and chiasm,* provided an even more peripheral etiology such as a diseased retina or glaucoma can be ruled out; *usually there is an additional central or cecocentral scotoma* (Fig. 3-3). The etiology for such a defect may be a tumor of the optic nerve itself or of its sheaths; a meningioma of the tuberculum sellae, the sphenoid ridge, or the olfactory groove; a craniopharyngioma; a pituitary adenoma with an extrasellar extension; a tumor at the inferior surface of the brain in the anterior or middle fossa (there may be an accompanying cavernous sinus or superior orbital fissure syndrome); and occasionally a frontal lobe tumor (presenting itself in the form of a Foster Kennedy syndrome). Especially in a unilateral concentric contraction of the visual field, it is not always easy (or may even be impossible) to differentiate disturbances of conductivity caused by a tumor from those due to other etiologies—whether inflammatory, toxic, or degenerative. In the differential diagnosis one should always consider glaucoma, tapetoretinal degeneration of the retina, secondary atrophy after optic neuritis, multiple sclerosis, tabes, toxic amblyopia (tobacco, alcohol, quinine), and traumas.

Concentric contraction can also be the expression of generally diminished vision and must not be confused with real loss of the periphery of the visual field due to a local or neurogenic process. True field defects and such "spurious" concentric contractions can be differentiated as follows: spurious contraction always parallels a corresponding decrease of the central vision; the outline of the field becomes normal again if a large enough object is used; and in a true defect of the periphery the bigger target will give the same or only a moderately larger field, but not a normal outline.

Tubular fields are generally contracted fields that become smaller as the patient moves away from the tangent screen. They are always a sign of malingering or hysteria (Goldmann).

Unilateral sector-shaped defects. Unilateral sector-shaped defects may occur *in combination with central scotomas due to tumor compression of the optic nerve* (see section on concentric constriction)—an indication that only part of the optic nerve is damaged (Fig. 3-77). Sector-shaped defects in retrobulbar disturbances of the optic nerve are much rarer than concentric constrictions. Sector-shaped or wedge-shaped defects extending from the periphery of the field toward the point of fixation are more *characteristic for an involvement of the disc*—for instance, secondary optic atrophy after choked disc (with a predilection for the nasal side) papillitis, glaucoma (in the early stages also predominantly on the nasal side), and juxtapapillary chorioretinitis. In the differential diagnosis, vascular changes of the retina must be considered, especially branch embolisms of the central retinal artery. The latter usually cause a sector-shaped defect involving an entire quadrant, with the apex characteristically touching the blind spot—in contrast to nerve fiber bundle disturbances of the disc or optic nerve where the apex of the sector defect points toward the point of fixation or actually coincides with it.

Scotomas (absolute or relative). Scotomas may be classified as central, paracentral, or cecocentral (extending from the blind spot to the point of fixation). A scotoma is relative if it exists only for color or small white targets; it is absolute if it exists for color and white targets of any size. The *blind spot* is a physiologic paracentral scotoma. Characteristically, it shows a symmetrical enlargement in papilledema (Fig. 2-10). Many authors attribute some importance to this enlargement of the blind spot in the early differential diagnosis of choked disc (Chamlin and Davidoff; Davis; and others).

Central, especially cecocentral, scotomas suggest a *disturbance of conductivity of the nerve fibers of the papillomacular bundle within the homolateral optic nerve* and are sometimes associated with a concentric constriction of the visual field (see above). They are *negative* (that is, they are not appreciated by the patient) in slowly progressing lesions of the optic nerve or tract (for instance, tumors) but *positive* (causing the sensation of veils, gray spots, or shadows) in an acute embarrassment of the infrageniculate pathways (for instance, an inflammatory process or a hemorrhage). The papillomacular bundle, the anatomic structure involved in these scotomas, forms a separate entity not only in the optic nerve but also in the chiasm. Thus a tumor causing a lesion in the *chiasm* may also be responsible for scotomas, which may occur unilaterally or bilaterally and which in their typical form, may assume a characteristic *temporal hemianopic* shape. Frequently it is difficult to demonstrate them (Fig. 3-39). Color campimetry may be justified for this type of scotoma: scotomas for red may represent an early sign of an impaired conductivity of the papillomacular bundle. However, even in these early stages it should be possible to demonstrate similar defects if careful campimetry (or even perimetry) is carried out with small white targets. *Homonymous hemianopic central scotomas* in principle may be caused by a lesion anywhere along the suprachiasmal visual pathway. It is known that they are rarely caused by tumors along the visual tracts. If they occur in connection with the visual radiation, they will almost always take the form of a negative scotoma. The *lesion* is most likely cortical or subcortical (in other words, in the *region of the occipital pole or the calcarine fissure*).

We have observed a typical case of such homonymous hemianopic scotomas in a patient with jacksonian epilepsy (Fig. 3-24). There was a visual aura (flashing, sparking, and blazing lights). This was the residue of an old meningitis. Surgical exploration revealed a circumscribed internal hemorrhagic pachymeningitis over one occipital pole.

Bitemporal field defects. Bitemporal field defects generally indicate a *lesion of the central part of the chiasm* (that is, a lesion of the crossing fibers). The involvement of the fibers may originate from above, from below, from in front, or from behind the chiasm. The typical bitemporal hemianopia is usually caused by *pituitary tumors*. It may also occur in an incomplete or modified form, depending on the pressure effect on the chiasm and its surroundings from a different direction. A *bitemporal quadrantanopia* of the superior quadrants is usually produced in the initial stages of a pituitary tumor. A symmetrical progress of the defect is characteristic for a pituitary tumor. An *asymmetrical bitemporal hemianopia* or *quadrantanopia* with incongruous field defects progressing in an asymmetrical manner is more characteristic for extrasellar (parasellar or suprasellar) neoplasms such as craniopharyngiomas, meningiomas of the tuberculum sellae, or meningiomas of the olfactory groove. The asymmetrical growth of these tumors frequently causes a direct lesion of the optic nerve if the tumor is in an anterior location or a direct lesion of the optic tract if the tumor is in a posterior location, resulting in *amaurosis of one eye with temporal hemianopia of the other eye*. Occasionally such a picture is also seen in association with a pituitary tumor, but here as sign of a lesion involving the entire half of the chiasm (Figs. 3-38 and 3-60).

Homonymous field defects. Homonymous field defects indicate a *contralateral lesion in the region of the optic tract or the geniculocalcarine optic radiation*. A homonymous hemianopia is caused by a complete or almost complete interruption of the pathway, and a *homonymous quadrantanopia* is caused by an incomplete interruption of the corresponding upper or lower section of the optic tract or radiation. An incongruous onset and progress of the field defects, a limiting line through the point of fixation without sparing of the macula, and pallor of the discs (only the temporal half of the disc on the side of the lesion but of the entire disc on the contralateral side), which may develop after months, are indicative of a *lesion of the tract.* Such a lesion may be caused by a meningioma of the sphenoid ridge, a tumor of the temporal lobe, a tumor of the middle fossa, a tumor of the thalamus, or a tumor of the quadrigeminate plate (Fig. 3-27). According to Cushing, a *superior homonymous quadrantanopia* (damage to Meyer's loop) is almost pathognomonic for *tumors of the temporal lobe* (Fig. 3-12). An *inferior homonymous quadrantanopia* occurs with tumors of the parietal lobes. A *complete homonymous hemianopia* with a striking congruence of both defects is characteristic for a large number of *tumors of the occipital lobe* (Fig. 3-22). Sparing of the macula, which becomes most pronounced in tumors near the occipital pole, is seen much more frequently in such tumors than a vertical boundary line of the field through the point of fixation. In our opinion the absence of *sparing of the macula* does not necessarily rule out a suprageniculate site but may indicate a very extensive involvement of the entire optic radiation—for instance, in cases of occipital lobectomies (Huber). Of course, an involvement of the entire optic radiation is much more likely to occur in its anterior part, where the fibers are close together, than in the posterior occipital section. Thus, if there is sparing of the macula, the tumor probably is in the posterior part of the optic radiation. This is of some significance for purposes of localization. If there is no sparing of the macula, it is of no value in this respect. Other methods for the localization of a lesion causing homonymous field defects have already been discussed (optokinetic nystagmus, saccadic pursuit movement—pp. 49 and 58).

CHAPTER TWO *General symptoms of increased intracranial pressure in patients with brain tumors*

The symptomatology of a brain tumor can be divided into two different groups of symptoms: *general symptoms,* mostly signs of increased intracranial pressure which are of no localizing value and which are more or less independent of the seat of the tumor, and *focal symptoms,* which are of decided localizing importance and which are the results of irritation of or damage to a specific part of the brain and to its immediate surrounding area. It is always important to keep in mind the chronologic sequence of these groups of symptoms. As a rule, the focal signs are the first to occur and depend strictly on a lesion of a specific part of the brain (jacksonian epileptic seizures, pareses, sensory disturbances, ataxia, agnosias, anosmia, etc.). As the growth increases in size, there may be remote effects. Finally, there will be the general signs of increased intracranial pressure caused by a space-consuming lesion. This *sequence* of *focal symptoms–general symptoms* is significant for a large number of brain tumors, especially those which develop in a highly differentiated part of the brain: here even a minute circumscribed focus may lead to manifestations of neurologic disturbances. In the diagnosis of brain tumors this *steady and more or less rapid progression of focal symptoms,* which are complicated by the general symptoms of increased intracranial pressure only in the later stages, is an important and pathognomonic feature. However, the chronologic development does not always follow this pattern. It may not be observed with tumors of silent parts of the brain (for instance, in the frontal or occipital lobes) or with tumors that tend to cause an early internal hydrocephalus (for instance, tumors of the third or fourth ventricles). In such instances the general symptoms of increased intracranial pressure may be the first indications of a tumor. They may remain the only symptoms, or the *sequence may be reversed* and the general symptoms may precede the focal symptoms.

This discussion already clearly indicates a general *classification of the ocular symptoms* in brain tumors. Analogous to other neurologic signs, we distinguish between *general symptoms* that are independent of the seat of the tumor and that primarily indicate an increased intracranial pressure and *symptoms of local-*

izing significance that represent the direct result of damage to a specific part of the visual system caused by the brain tumor.

Such a classification is justified for a number of reasons. There are actually ocular signs associated with a brain tumor which are only general signs but which are so typical that they alone frequently enable one to make a general diagnosis of increased intracranial pressure caused by a space-consuming lesion. *Papilledema* is such a sign. These symptoms do not depend on the seat of the tumor and develop regardless of whether the tumor is in direct contact with the visual apparatus. On the other hand, intracranial neoplasms may produce quite specific ocular focal symptoms if there is a direct interference with the visual apparatus. Such symptoms frequently permit the localization of the tumor in an amazingly exact manner. This fact must be attributed in no small way to the highly differentiated structure of the visual apparatus and its close topographic and functional relationship to the brain. As the sensory organ trusted with visual perception, the eye is particularly well equipped to register the minutest disturbances and to make the patient conscious of functional disorders that may not always be so obvious in other sensory organs and nervous elements.

The nonocular general symptoms of increased intracranial pressure are headaches, vomiting, circulatory and respiratory disturbances, and psychic changes. We purposely want to stress these symptoms briefly in order to put the ocular general symptoms into the proper perspective and to clarify their relationship to the nonocular symptoms.

There is no question that *headache* is one of the most frequent symptoms of increased intracranial pressure and thus of a brain tumor. The headache is rather diffuse in character and its intensity typically increases with an additional rise in the ventricular pressure, which may be caused by coughing, sneezing, or defecating. If the headache is localized, it may occasionally indicate the seat of a superficial tumor, but more often not. According to present opinion, the headache caused by a brain tumor is due to the effect of traction on sensitive intracranial structures (large venous sinuses, meningeal blood vessels, arteries on the base of the brain, dura of anterior and posterior fossa). This may be either a local effect of the tumor or a remote effect due to displacement of brain substance.

Vomiting is usually associated with the headache. It affords relief from the steady or paroxysmally aggravated headache. Vomiting that occurs without nausea and without relationship to the intake of food should always arouse suspicion of increased intracranial pressure.

Circulatory disturbances associated with increased intracranial pressure are a slow pulse (vagus effect!) and a pressure pulse. Together with *respiratory disturbances* (increased rate, Cheyne-Stokes breathing), they are, from the prognostic point of view, unfavorable signs which, as a rule, occur in the late stages of acute increased intracranial pressure.

Psychic disturbances usually are a manifestation of diffuse damage to the brain. Clouding of the consciousness ranging from slight somnolence to severe coma is one such disturbance. Delirium is a form of disturbance of consciousness in which the patients exhibit a pronounced motor unrest because of hallucinations and delusions. An increase of the intracranial pressure may also produce an *organic psychic syndrome* with disturbances of apprehension, memory, and power of association, a tendency for perseverance and confabulation, and increased emotional lability, yet a general lack of impulse and dullness of emotional activity.

The *ocular general signs of increased intracranial pressure consist of papilledema, nonspecific ocular pareses (especially of the abducent nerve), the clivus ridge syndrome (pupillary disturbances), and occasionally a bilateral exophthalmos.* They will be discussed in the order of their importance. Without doubt, the choked disc has to be named first because its occurrence alone frequently permits one to make a diagnosis—not only as an ocular general sign but as the most important of all general signs of increased intracranial pressure.

The *local symptoms* will be discussed in a separate chapter. In certain tumor localizations the ocular symptoms are prominent—that is, in patients with tumors that directly involve the structures of the visual apparatus (visual pathways, nerves and nuclei of the extrinsic muscles). In patients with tumors in other sites the ocular local signs are of equal importance to the other neurologic signs. Again, there are other types of tumors that cause the ocular signs to take a somewhat subordinate place of importance in the total clinical picture. It is just for such reasons that it will be necessary to consider the ocular local signs not as isolated phenomena but in their proper relation to the entire neurologic status of a given patient. Thus we want to *appeal to the ophthalmologist* to never evaluate either the ocular general or the ocular local signs in patients with suspected brain tumors as isolated from the total clinical picture but to always *consider them with the entire neurologic symptomatology and the results of specific tests.*

PAPILLEDEMA

Krayenbühl in his contribution on neurosurgery in *Lehrbuch der Chirurgie* states: *"Among the symptoms of increased intracranial pressure, the papilledema should be considered the most important one."* This statement of a well-known neurosurgeon points out the prominent position papilledema occupies as a general sign in the diagnosis of increased intracranial pressure. The significance of this fact for the general practitioner, for the neurologist, and in particular for the ophthalmologist is obvious. The recognition of increased intracranial pressure and thus of a brain tumor may depend on the familiarity of the examiner with this fundus picture. Its recognition could be of absolutely vital importance for the patient in view of the chances of successful outcome with modern neurosurgical procedures.

First a few terms should be defined. The German term *"Stauungspapille"* indicates a *passive edema of the optic disc that is caused by increased intra-cranial pressure but is not associated with primary inflammatory changes and frequently not with functional disturbances.* Thus the term "Stauungspapille" is, to a large extent, identical to the English term "papilledema." The German term "Papillenödem," however, is in no way identical to the English term. It is, so to speak, a neutral expression for an edematous swelling of the disc that may be caused by stasis, by inflammation, or by toxic agents. In our opinion it is a mistake to reserve the term "Stauungspapille" for the more extensive forms of swelling of the disc (Uhthoff) and to call lesser forms "Papillenödem." One should limit the term "Stauungspapille" to cases with proved or—at least— probable increased intracranial pressure. Minor stages of "Stauungspapille" can be simply designated as incipient "Stauungspapille." However, if one is un- certain as to the etiology of the swelling of the disc and if, on the basis of the ophthalmoscopic findings and the clinical picture in its entirety, he has to con- sider other etiologies (for instance, a malignant hypertensive retinopathy), he is permitted to use the neutral terms "Papillenödem" or "Papillenschwellung" without being compelled to use a term implying a specific interpretation of the underlying etiologic factors. *"Papillitis"* ("optic neuritis" in the English nomen- clature) indicates an *inflammatory swelling of the nerve head with functional loss.* It must be differentiated from papilledema, which it resembles quite closely from the standpoint of appearance. The terms "pseudopapilledema" and "pseu- doneuritis" refer to ophthalmoscopic pictures, which will be treated in greater detail in the discussion of the differential diagnosis of papilledema.*

The *presence of papilledema,* especially in bilateral form, is now considered *pathognomonic for increased intracranial pressure* provided an ocular or orbital etiology can be ruled out. A number of etiologic factors can produce such an increase in the intracranial pressure (for instance, displacement of brain sub- stance by a tumor or a hemorrhage, increase in the volume of the cerebro- spinal fluid, diffuse or circumscribed edema of the brain, disproportion between the cranium and the volume of its contents). By far the greatest number of cases are due to tumors of the brain (about 75% according to Uhthoff's fig- ures)—perhaps because the etiologic factors just mentioned are frequently superimposed (volume of tumor + edema of the brain + hydrocephalus). In

---

*This entire dissertation may be of little interest to the English-speaking reader. There is no similar ambiguity in the terms used in the English literature. Nevertheless, this paragraph has been retained in the English translation to guard the English reader in his perusal of the German literature against any misunderstanding that might arise from this inconsistency in the German terminology. (Stefan Van Wien.)

In modern French literature (Brégeat; Bonamour, Brégeat, Bonnet, and Juge) the corre- sponding terms are clearly differentiated: "oedème papillaire pure" means an interstitial edema of the disc without primary affection of the nerve fibers (= papilledema, *Stauungspapille*), whereas "oedème papillaire accompagnée" refers to edema of the disc with primary affection of the nerve fibers (= papillitis, optic neuritis, ischemic papillitis, toxic and traumatic syn- dromes of the optic nerve). (Frederick C. Blodi.)

consideration of these facts one may state that papilledema indicates the *presence of a brain tumor* (in the broad sense of the word) *with a very high degree of probability.* Its absence, on the other hand, does not exclude the possibility of a brain tumor. As a matter of experience, papilledema usually is not an early general sign of increased pressure, or it may even be missing in patients with slowly growing tumors (for instance, a benign meningioma) in spite of a marked increase of the intracranial pressure. The infratentorial tumors, especially those of the cerebellum, are known to cause a relatively early papilledema quite frequently (obviously as a result of their tendency to interfere with the circulation of the cerebrospinal fluid at an early stage), whereas supratentorial tumors cause a choked disc less frequently (if a choked disc is present, it occurs in the later stages and progresses slowly).

Papilledema is much more frequently observed in children than in older people (81% in the age group between 1 and 5 years and 43% in the age group between 60 and 70 years according to Bonamour, Brégeat, Bonnet, and Juge and to Brégeat). The relative frequency of choked discs in children can be explained by the special character of infantile cerebral tumors (infratentorial seat, early blockage of cerebrospinal fluid circulation, rapid growth) and by the soft nature of the disc tissue and its vessels. In elderly men, papilledema is rarely found or even completely absent (rarity of cerebral tumors at this age, sclerosis of disc tissue).

Although papilledema frequently does not develop simultaneously and symmetrically in both eyes, a *bilateral form is the rule.* The unilateral form (for instance, a Foster Kennedy syndrome) is rare (8% according to Bonamour's statistics). A *unilateral papilledema is usually ocular or orbital in nature,* or if this is not the case, there must be a local cause (unilateral optic atrophy, unilateral disc anomaly, unilateral glaucoma, unilateral congenital anomaly of the sheaths of the optic nerve) that prevents the increased intracranial pressure from producing swelling of the disc on the corresponding side. There will be a fuller discussion of unilateral papilledema in a separate section. Formerly it was frequently maintained that the more advanced papilledema corresponded to the side of the tumor location (Gibbs). However, modern neurosurgical experience has taught us that certain tumors, especially those of the temporal and parietal lobes, may cause a unilateral internal hydrocephalus of the opposite side due to displacement of brain substance and a partial obstruction of the interventricular foramina of Monro; in other words, the more advanced papilledema in such cases corresponds to the side opposite the tumor. *Differences in the two sides regarding the rate of development as well as the appearance of the fully developed choked discs are therefore of no value for diagnosing the laterality of the tumor.*

Our statement that not every brain tumor causes papilledema needs some elaboration by examining the ratio of tumors with and without papilledema as well as the influence of the localization and type of tumor on this ratio. Our own statistical material includes 1166 patients with brain tumors: 698 (59%) showed papilledema, whereas in 468 (41%) the ophthalmoscopic findings were *negative in this respect*. In other words, *a little more than half of all patients showed a papilledema*. Compared with older statistics (Gowers; Paton; Uhthoff) in which papilledema was reported in up to 80% of patients with brain tumor, our figure of 59% may appear somewhat low. However, it is in striking agreement with a newer publication by Petrohelos and Henderson who analyzed 358 cases and reported an incidence of 59%. One may wonder what could cause such a discrepancy between the more recent and the older figures. We believe that we are justified in assuming that *these figures unmistakably reflect the progress made in the diagnosis and treatment of brain tumors during the past few decades*. In other words, thanks to our modern diagnostic methods (electroencephalography, arteriography, encephalography, ventriculography, scintigraphy, and others), we are able to make an earlier diagnosis of a brain tumor in the average case. Thus fewer patients reach the state of increased intracranial pressure showing papilledema. In a scrutiny of the total number of our own cases (Table 3) with regard to the occurrence of papilledema in pa-

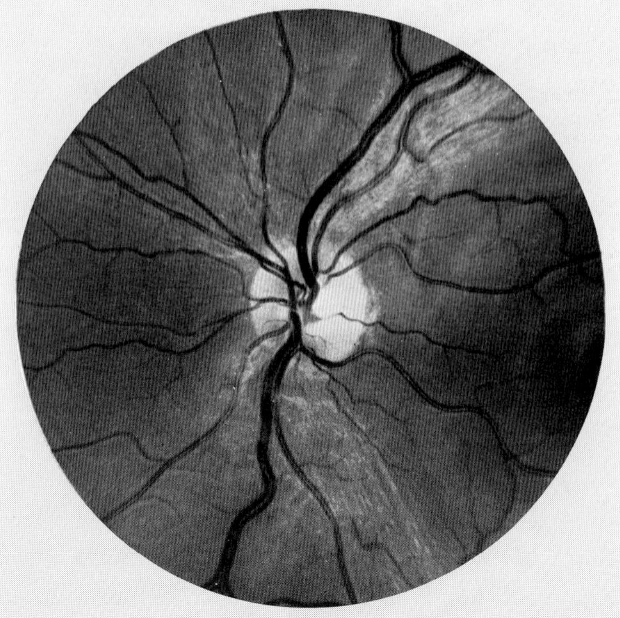

Fig. 2-1. Photograph of a normal fundus (left eye). (Enlargement ×10.) The papilla is contrasted with the surrounding eyeground as a pale red disc with sharply outlined margins except for slight blurring of the nasal border due to the larger number of nerve fibers crossing it. The level of the papilla is in plane with the surrounding area. Because of the papillomacular bundle, the temporal half of the disc normally is lighter in color than the nasal half. The bright red arteries and the dark red veins are of normal structure and caliber. The caliber ratio of veins to arteries is 3:2. The normal diameter of the disc is 1.5 to 1.7 mm.

Table 3. Manifestation and frequency of papilledema in 1166 patients with brain tumor*

Localization	Glioblas-tomas		Astrocyto-mas and astroblas-tomas		Meningio-mas		Metastases		Medul-loblas-tomas		Heman-giomas		Cranio-pharyn-giomas		Others		Total	
Papilledema	+	−	+	−	+	−	+	−	+	−	+	−	+	−	+	−	+	−
Frontal	60	50	37	25	38	16	14	11	2	0	0	1	—	—	18	16	169	119
Temporal	23	23	22	6	14	2	8	4	1	0	1	1	—	—	13	6	82	42
Parietal	4	12	4	6	7	6	2	4	1	0	0	1	—	—	5	7	23	36
Occipital	19	13	11	2	1	2	10	3	—	—	0	1	—	—	6	2	47	23
Sagittal, parasagittal	—	—	—	—	15	9	—	—	—	—	—	—	—	—	—	—	15	9
Third ventricle	3	0	2	0	—	—	3	0	2	0	—	—	11	2	7	2	28	4
Fourth ventricle	4	1	12	0	—	—	—	—	7	0	—	—	—	—	6	0	29	1
Base of brain	—	—	—	—	—	—	1	1	—	—	—	—	—	—	5	2	6	3
Lateral ventricle	0	2	—	—	1	0	—	—	—	—	—	—	—	—	0	2	1	4
Cerebellum	0	6	24	6	6	4	10	4	15	8	11	9	—	—	9	5	75	42
Cerebellopontine angle	4	2	2	0	2	3	1	0	3	0	4	2	—	—	0	3	16	10
Not localized	60	62	40	33	26	19	19	16	11	1	5	3	—	—	46	41	207	175
Total	177	171	154	78	110	61	68	43	42	9	21	18	11	2	115	86	698	468
	50%		66%		64%		61%										59%	41%

*Of the total 1166 patients with brain tumors, 698 (59%) had papilledema and 468 (41%) did not have papilledema.

tients with supratentorial and infratentorial tumors, we find confirmation of the familiar and well-established fact that there is a *distinct prevalence of choked disc in patients with infratentorial tumors (70%) as compared with those with the supratentorial type (60%)*. These figures are also in close agreement with those of other authors. Petrohelos and Henderson report an incidence of 53% in patients with supratentorial tumors and 75% in those with infratentorial tumors. *Bilateral papilledema occurred in practically all our patients.*

Apart from the localization of the tumor, the presence of a papilledema also depends on the type and particularly the rate of growth of the tumor. Papilledema is seen in 50% of our patients with rapidly growing and malignant glioblastomas. In patients with the rather slowly progressing and benign group of astrocytomas and meningiomas, there is an incidence of 65%. The figure in patients with cerebral metastases is 61%. Even though these numerical dif-

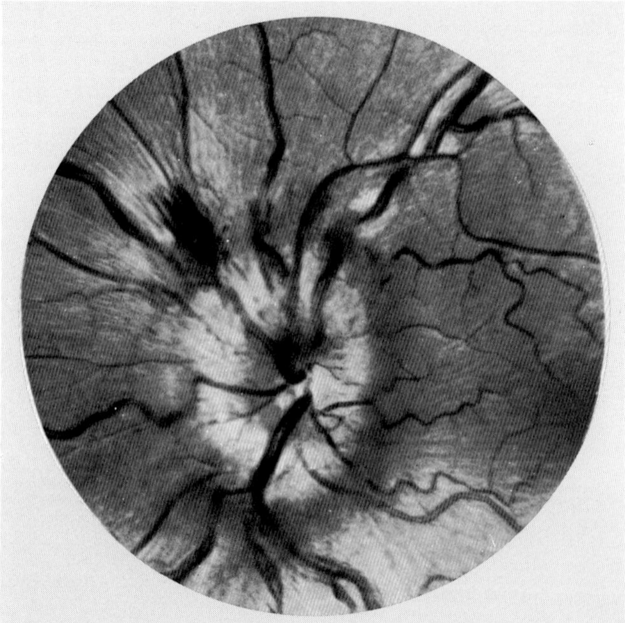

Fig. 2-2. Fully developed papilledema (left eye) in a 20-year-old patient with hemangioma of the right tonsil and hemisphere of the cerebellum. Enormous enlargement of the diameter of the disc, in areas up to twice the normal diameter. Blurring of the nasal and temporal disc margins. Mushroomlike swelling and a prominence of about 2.5 diopters of the nerve head. Loosening and veiling of the texture of the disc, with capillary congestion. Preservation of the vascular funnel. Congestion, dilation, and tortuosity of the retinal veins, with ratio of veins to arteries of 4:2 to 5:2. Deflection of the vessels at the disc margin, with partial concealment of the vessels by the edema of the papilla. Peripapillary retinal edema with markedly increased display of the nerve fiber pattern. "Flame-shaped" radial hemorrhages in the nerve fiber layer at the 11 o'clock position of the disc margin. Intraocular arterial pressure, O.U. 40 grams diastolic (Bailliart). Visual fields, except for an enlarged spot on both sides, normal. Visual acuity, O.U. 1.0, distance and near.

ferences are not very large or significant, there seems to be a slightly greater tendency for the development of papilledema in patients with slowly progressing tumors than in those with the rapidly expanding malignant forms. For further details see Table 3.

Ophthalmoscopic appearance of papilledema

The clinical picture of *fully developed papilledema* (Figs. 2-2 and 2-3) is relatively distinct and can be summarized as follows:

1. Increase of the disc diameter
2. Nasal and temporal indistinctness and blurring of the disc margins
3. Elevation of the disc and mushrooming of the nerve head into the vitreous
4. Reddish discoloration of the disc (capillary stasis)
5. Venous congestion and tortuosity of the veins, with relatively normal arteries (increase of ratio in caliber of veins and arteries)
6. Deflection of the vessels over the disc margin
7. Hemorrhages at the disc margin and within the disc
8. White exudates over the surface and at the margin of the disc
9. No primary disturbances of the sensory functions

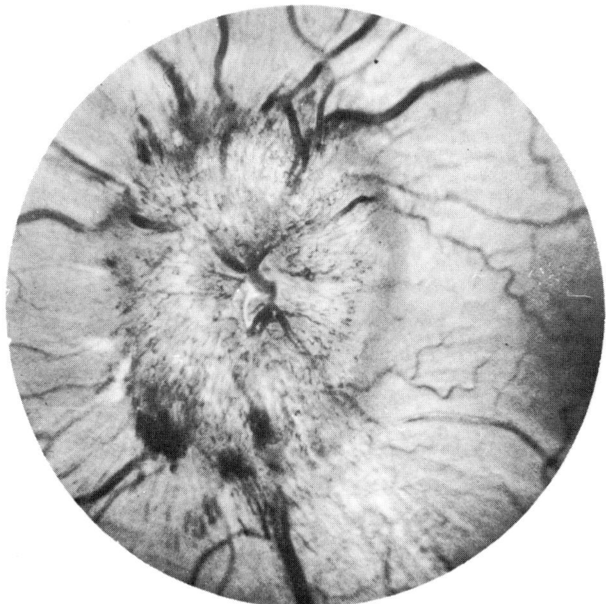

Fig. 2-3. Fully developed papilledema in a 46-year-old patient with hemangioma of the cerebellum. "Champagne cork" protrusion of the enlarged edematous papilla (prominence of about 3 diopters), intensive congestion of venules and capillaries, obscuration of the major vessels, nerve fiber layer hemorrhages at the upper and lower margin of the disc, and two white "exudates" within the retina at the nasal border. Visual acuity, O.U. 1.0, distance and near.

Anybody who has seen fully developed papilledema a few times should thereafter recognize its ophthalmoscopic picture without difficulty. The reddish discoloration of the disc that abolishes the contrast with the surrounding retina may render its recognition quite difficult, especially in ophthalmoscopy through a small pupil. However, the retinal vessels converging toward the disc will assist the examiner in finding the optic nerve head. One should also adhere to the rule mentioned in the section on ophthalmoscopy (p. 62)—the patient should rotate the eye to be examined slightly nasally, a maneuver that will make it easier to find and recognize the disc.

Fully developed papilledema is truly indicative of an increase in the intracranial pressure. The force of this pressure has caused the "mushrooming" of the

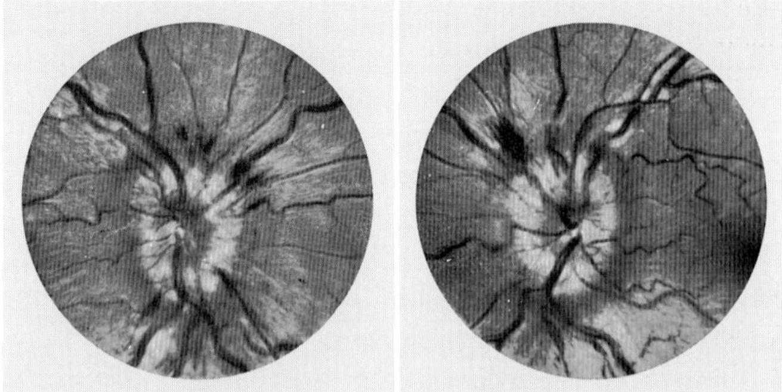

Fig. 2-4. Fully developed papilledema in a patient with hemangioma of the right tonsil and hemisphere of the cerebellum (same patient as in Fig. 2-2). (Enlargement ×6.) The left illustration corresponds to the right eye; the right illustration, to the left eye.

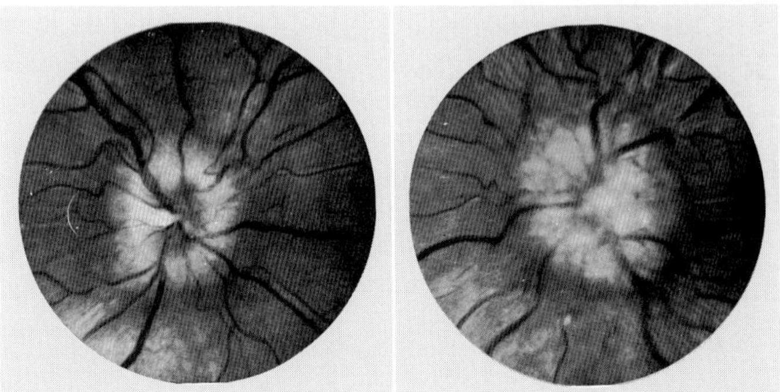

Fig. 2-5. Advanced stage of bilateral papilledema in a 27-year-old patient with neurinoma of the choroidal plexus in the area of the left lateral ventricle. The papillae show a mushroomlike prominence of approximately 3 diopters. Conspicuous edematous veiling of the tissues of the disc with complete filling of the left cup. Slight radial folding of the retina near the disc, especially toward the left fovea. Visual acuity, O.U. 1.0. Both visual fields intact. The left illustration corresponds to the right eye; the right illustration, to the left eye.

nerve head into the vitreous, drawing the tortuous and dilated veins with it—a picture that could be called the *grotesque other extreme of increased intra-ocular pressure* (that is, *a glaucoma, with its excavation of the disc into the opposite direction*), in other words, an "inverse glaucoma." In spite of the round shape of the disc, which is usually retained, the disc *diameter* is usually more or less *increased* in papilledema. This enlargement is due to the swelling of the nerve head itself and the spreading of the edema into the surrounding retina. The functional equivalent of this extension of the disc with a consecutive *lateral displacement of the adjacent retina* (which can occasionally be seen ophthalmoscopically as a *concentric peripapillary folding*) is an enlargement of the blind spot, which will be discussed below (Fig. 2-10).

A parallel development of the enlargement is an indistinctness and *blurring of the disc margin* caused by the edema which, especially in the early stages, may be less pronounced on the temporal side than on the nasal side. This diffuse indistinctness of the disc margin, however, is due not only to the edema but also to scattering of the light reflexes by the edematous tissues of the nerve head and the surrounding retina. This edema also separates the nerve fibers. Even with normal light, but more so with red-free light (which makes the nerve fibers more visible), the pattern of the *nerve fibers becomes manifest in the form of grayish white streaks.* The serous infiltration of the disc, however, is not limited to the margin but involves the disc in its entirety. The funnel is usually filled, and a previously existing physiologic cup has been completely smoothed out and can no longer be recognized. As a result of the serous infiltration, the *entire tissue of the disc is loosened,* less compact than under normal conditions, and of a *finely striated or reticular structure,* which is in distinct contrast to the rather uniform and smooth appearance of the normal optic papilla (Fig. 2-1).

The *mushroomlike prominence of the choked disc* is one of its most conspicuous signs. In the end stage of this development the prominence usually is approximately equal on the nasal and temporal sides. In the incipient stages a difference is possible, with the nasal margin progressing more rapidly and more strikingly than the temporal rim. Even with the monocular ophthalmoscope, the elevation of the papilledema is obvious at once, provided this elevation is of a fairly marked degree. Naturally the stereoscopic picture obtained with a binocular ophthalmoscope makes this situation even more conspicuous; thus this instrument is a favorite in the early diagnosis of a beginning prominence. One point that should aid in recognizing an early mushrooming is the *deflection of the vessels at the disc margin* which, so to speak, underscores and accentuates this particular contour.

The difference in the planes of the retina and of the nerve head in which the retinal vessels course leads to the phenomenon of the *parallax:* its magnitude can be used to express the degree of prominence of the papilla.

In order to determine the parallax the examiner has to move his head, together with the ophthalmoscope, a little up and down as well as to the right and to the left without, of course,

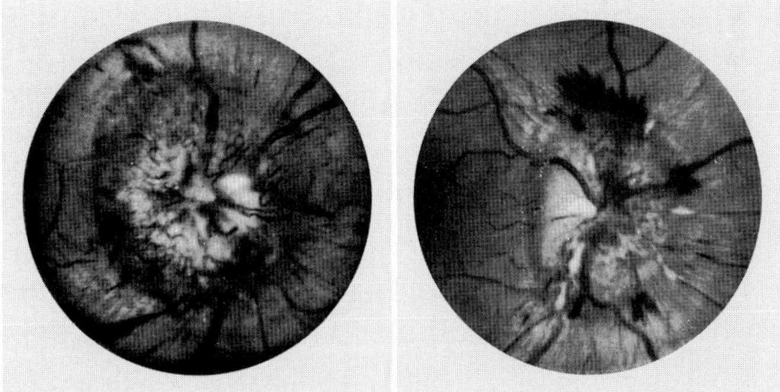

Fig. 2-6. Advanced bilateral papilledema in a 52-year-old patient with meningioma of the left sphenoid ridge. Enormous enlargement of the disc diameter, especially on the right side (up to three times the normal size). Pronounced venous congestion with intense congestive hyperemia of the fine vessels of the disc and conspicuous capillary loops at the summit of the discs (especially the right one). Numerous streaklike and punctate hemorrhages on and near the disc. Retinal arteries obscured by the edematous tissues of the disc. White and grayish white dots (cytoid bodies) on the discs and the disc margins. Distinct parallax between the vessels in the plane of the retina and those on the disc. Visual acuity, O.D. 1.0; O.S. 0.6. The left illustration corresponds to the right eye; the right illustration, to the left eye.

losing sight of the fundus. He should compare the behavior of one and the same vessel first in the plane of the nerve head and then in the plane of the retina. The section of the vessel that is closer (that is, on the disc) will seem to move in a direction opposite to the movement of the head of the examiner, whereas the more distant section will move in the same direction. The more pronounced this relative movement of the two sections of the vessel is, the greater should be the prominence of the disc (Fig. 2-6).

The prominence of the disc can be quantitatively measured with the ophthalmoscope. Although this is a relatively crude method, it is quite adequate for our clinical needs provided it is performed properly.

The prominence is measured in diopters because the lenses in the Recoss disc of the ophthalmoscope are used with this method. For a more conventional expression of this magnitude it should be mentioned that *3 diopters correspond approximately to 1 mm. of prominence.* The elevation of the disc is determined in the following manner: the ophthalmoscope is first focused on the most prominent vessel on the nerve head, and the strongest plus or weakest minus lens giving a clear picture of this vessel is noted. It should be understood that the examiner must be emmetropic or wear his correction. The next step is to select a vessel, parallel to the first observed vessel, near the disc margin in the retinal plane and again to find the strongest plus or weakest minus lens allowing a clear image of this vessel. The difference in the strength of these two lenses expresses the prominence of the disc in diopters. This can be easily converted into millimeters.

A more exact method for the determination of the prominence of the disc has been described by Heinz. He retinoscopes the fundus, beginning at a point temporal to the macula, and using successive points, he progresses over the disc into the nasal part of the horizontal meridian. In this manner he obtains a sort of profile of the papilla and the surrounding parts of the retina. He uses the Maddox cross for fixation and progresses from degree to degree in the horizontal or vertical meridian.

The *prominence of the choked disc* parallels the progress of its development. Prominence and enlargement of the papilla, however, need not always be parallel. It may reach values as high as 8 or 9 diopters. Based mostly on our material, the *average is between 2 and 4 diopters.* There are a number of references in the literature (Uhthoff; Kestenbaum; and others) to the effect that the diagnosis of a papilledema of less than 2 diopters prominence is quite difficult and unreliable. According to our own experience, we cannot agree with such an opinion but would rather lower this value to about 1 diopter of prominence before it actually becomes difficult to decide whether or not there is a choked disc. In our opinion a prominence of more than 1 diopter can be definitely interpreted as favoring a diagnosis of papilledema if other symptoms are taken into consideration. In reviewing our cases of brain tumors we were able to observe a fairly close correspondence between the increase in the intracranial pressure found on the operating table (tension of the dura, force of escape of the cerebrospinal fluid after ventricle puncture) and the degree of prominence of the disc. In the absence of increased intracranial pressure there was no choked disc. If the pressure was high, we usually found a choked disc. Regression of the papilledema and of the elevation of the disc usually occurs

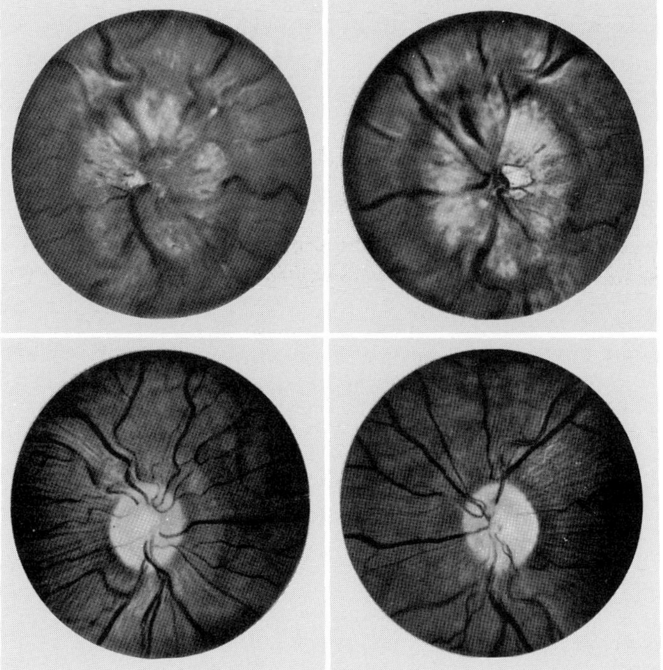

Fig. 2-7. Bilateral choked discs showing all the characteristic signs of the advanced stage in a 33-year-old patient with astrocytoma of the left temporal lobe. The lower photographs were taken 7 months after extirpation of the tumor. They demonstrate the capacity of the edema of the disc to regress almost completely. Only a slight atrophy of the discs and a circumpapillary radial folding of the retina, evidenced by the reflex lines, are visible. The left illustrations correspond to the right eye; the right illustrations, to the left eye.

either after successful neurosurgical intervention (Fig. 2-7) or together with the onset of a secondary optic atrophy (p. 115). Spontaneous regression can be observed in cases of serous meningitis or pseudotumoral encephalitis. Sometimes regression of the papilledema seems incomplete. This is due to a more or less pronounced proliferation of glial tissue. A recurrence of the elevation and of the swelling of the disc is possible if there is a recurrence of the increase of the intracranial pressure. However, this recurrence cannot occur if the nerve fibers have degenerated and glial cicatricial tissue has invaded the papilla.

The *reddish discoloration of the disc* is a result of *venous congestion* extending into the smallest capillaries and rendering them frequently visible as very small vessels in the edematous tissue of the disc (Fig. 2-3). This "plethora" of the disc is also an important early sign and one that is conspicuous by its absence in other conditions that might be mistaken for papilledema. As a result of this reddish discoloration, the color of the disc is almost identical with that of the surrounding retina. This factor, together with the blurring of the disc margins, makes it even more difficult to differentiate the disc from the remaining fundus. The direction of the large vessels converging toward the disc will always make it possible to identify the latter. Whereas the arteries maintain their normal caliber or actually become somewhat attenuated, the *retinal veins show a characteristic congestion in the case of a choked disc, which can be recognized as a broadening of their lumens* (the ratio of veins to arteries, which is normally 3:2, increases to 4:2 or even 5:2) and a distinctive *increased tortuosity.* The deflection of the vessels at the disc margin caused by a prominence of the papilla has already been mentioned. The sloping or almost vertical direction of the vessels at the neck of the papilla usually causes them to lose their wall reflex in this area. They may actually become invisible if they are buried in the edematous tissues of the disc. In such cases they seem to be interrupted in their course (Fig. 2-3).

A direct result of the venous and capillary congestion is the occurrence of *hemorrhages* (either on the disc itself or in its immediate surrounding area) near the large branches of the retinal veins. If the hemorrhages are in the nerve fiber layer, they are flame shaped and have a radial arrangement (Fig. 2-4). Occasionally they are in the outer nuclear layer, in which case they appear punctate. Sometimes these may also be subhyaloid hemorrhages extending into the vitreous. Only rarely are the hemorrhages far from the disc. This is an important point in the differential diagnosis of a central retinal vein thrombosis, in which the hemorrhages reach far out into the periphery. Hemorrhages do not occur with any great regularity in papilledema. They may appear as an important early sign (for instance, in a rapidly developing choked disc, Fig. 2-12), or they may be missing completely in some advanced stages. It should be emphasized that the *presence of hemorrages on the disc itself or in the area surrounding it is not absolutely pathognomonic for papilledema.* Similar hemorrhages are seen, for instance, in an inflammatory papillitis (Fig. 2-25).

The soft, white patches seen occasionally on the disc or at its margin, especially in the later stages, are definitely not exudates. They consist of varicosities and so-called cytoid bodies resulting from terminal nerve fiber swellings of Cajal, identical with those of "cotton-wool exudates" in the retina. Some of the white or grayish white spots in the retina surrounding the disc are remnants of absorbed hemorrhages. In reviewing our material, we found hemorrhages and white dots on the disc in patients with tumors of the occipital lobe (Fig. 3-25) and the cerebellum with striking frequency.

If the edema increases in intensity, it may extend into the adjacent retina to reach the macular area. There is a turbid milky appearance of the retina with a striate pattern due to the manifestation of the arrangement of the nerve fibers. *As a result of the edema,* droplets of fluid occasionally accumulate underneath the internal limiting membrane between the *disc* and the *macula*

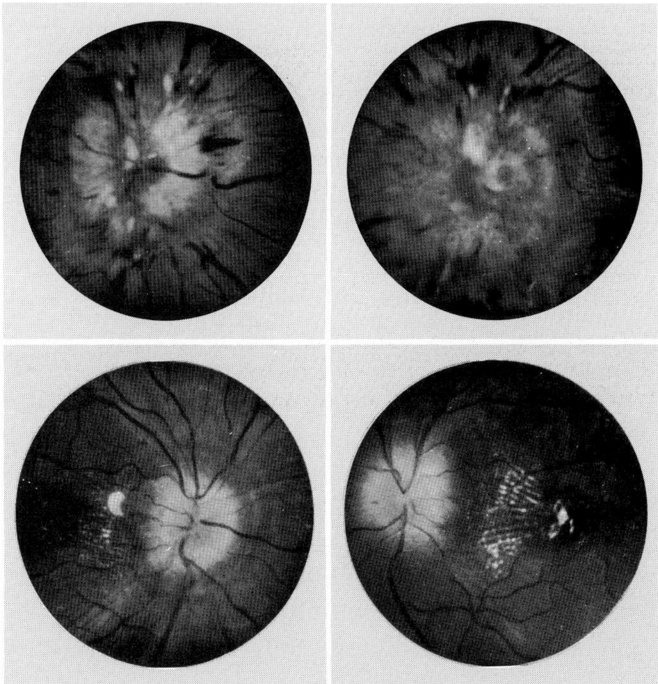

Fig. 2-8. Chronic atrophic papilledema with beginning star figure in the fovea in a 26-year-old patient with an astroblastoma of the right side of the frontal lobe. The upper photographs already show transition to atrophy of a papilledema measuring 4 diopters before surgical intervention. The lower photographs show the fundus after craniotomy (7 weeks later). The edema of the papilla is considerably less distinct. The prominence is only 1.5 diopters. There is a conspicuous pallor, a sign of atrophy. The right fundus shows an incomplete and the left fundus shows an almost complete macular star consisting of brilliant white dots in a radial arrangement. A suggestion of these changes can already be recognized in the preoperative stage. The left visual field shows a marked nasal constriction. The left illustrations correspond to the right eye; the right illustrations, to the left eye.

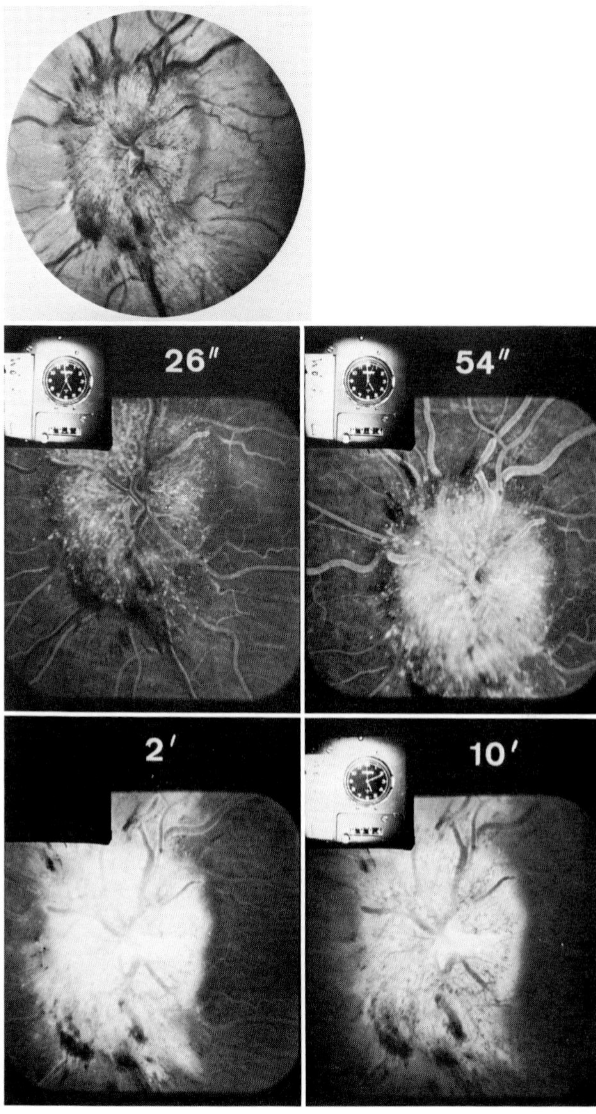

Fig. 2-9. Fluorescein angiogram of fully developed papilledema in a 46-year-old patient with hemangioma of the cerebellum (same patient as in Fig. 2-3). Typical for the arterial and early venous phase (26 seconds) in papilledema is the net of dilated congested capillary vessels with irregularities of caliber and sometimes microaneurysms. The fluorescence of the disc increases gradually to reach its maximum in the late venous phase of the angiogram (2 and 10 minutes), at which time the capillary stasis on the fluorescent background of the disc is distinctly visible. The diffuse staining of the papilla, limited to the disc and its immediate surroundings, may last for several hours and reflects the uptake of the dye by the edema itself.

and become visible as *minute brilliant white dots in a radial arrangement* and assume the *shape of a fan*. This picture contrasts with the macular fan seen in malignant hypertensive retinopathy, which is composed of lipids and lipid-laden macrophages. Such a fan-shaped macular structure may recede completely with the resolution of the papilledema (Fig. 2-8), and the related impairment of central vision disappears.

In the fluorescein angiogram of the fundus the papilledema shows characteristic signs (Fig. 2-9). The first one occurs during the arterial and the early venous phase. During this phase a net of dilated, tortuous capillaries is visible on the disc. They come from the depth of the physiologic cup and show great irregularities of caliber and occasionally aneurysmatic dilatations. The margins between the vascular net on the elevated disc and the surrounding retina seem to be remarkably sharp. The second sign is a diffuse fluorescence of the entire disc tissue. This begins in the late venous phase and may last several hours; it is also remarkably well demarcated and confined to the disc itself. In other types of elevation and edema of the disc such as papillitis, pseudopapilledema, and drusen (p. 140), the fluorescein angiogram looks entirely different.

In this connection the angiogram of a normal disc should be mentioned. The optic nerve head has a weak bluish green autofluorescence. Immediately after the dye reaches the ocular fundus, the disc begins to fluoresce. This occurs a fraction of a second before the retinal vessels are filled and is due to the circulation of the dye in the opticociliary capillary net deriving from Zinn's anastomoses. A little later the capillary branches of the central retinal artery are filled, together with the main artery. The disc is covered by a net of fine fluorescing vessels densely covering the nerve head. For this reason the expected optic illusion, which usually occurs in areas of autofluorescence, does not appear. This illusion appears a little later, approximately 5 seconds later, when the dye has disappeared from the ciliary capillary network. The background of the disc now appears dark. The fluorescing capillary branches of the central retinal artery are still yellowish green and contrast vividly with the dark, nearly black, background. This darkness remains for 30 seconds. With decreasing fluorescence of the entire fundus, in the late stages the disc regains its original autofluorescence (Fig. 1-26).

Symptomatology

The *symptoms* in fully developed papilledema usually are insignificant. In spite of a considerable prominence, even one of long standing, visual acuity and visual fields may be completely intact. *This very preservation of the visual functions is characteristic of the papilledema* and may become quite important in the differentiation of choked disc from similar ophthalmoscopic pictures. However, we should be aware that patients with distinct papilledemas occasionally have subjective symptoms that will bring them to the ophthalmologist—such as fleeting "attacks" of obscuration lasting only seconds or minutes, blurred vision, or even transient amaurosis, so-called *amblyopic attacks*.

An intelligent patient who had an acoustic neuroma with bilateral papilledema of about 2.5 diopters described his attack as follows: Especially at night under artificial light, there was a more or less dense but always transparent fog which started at times from the right side, at other times from the left side, but always from the side. There was no headache or any other discomfort. The sensation of this fog lasted from 3 to 5 seconds. No obvious cause (such as, for instance, some strain) could be observed to provoke an attack. Rather the sensation of the fog occurred during a period of quiescence. For a few days, four or five times daily, a peculiar luminous ring was seen in the upper left field and lasted 20 to 30 seconds.

If one searches for these amblyopic attacks in a large series of patients with papilledema—they occur in about one fourth of all patients with brain tumor (Ethelberg and Jensen)—he will be able to differentiate *three degrees of intensity of these obscurations:*

1. Sudden appearance of blurred vision as if a heavy fog, smoke, or a veil obscured the surroundings. There is a simultaneous impairment of color sensation.
2. Sudden appearance of a bluish or bluish green cloud; daylight changes to twilight, or the patient has the sensation of a momentary night blindness. There is a striking shift of color sensation toward blue-green.
3. Sudden appearance of complete darkness or actual blindness.

The average duration of such obscurations and amblyopic attacks is spectacularly short and lasts, as a rule, not more than 30 seconds—mostly a few seconds up to 10 seconds. Attacks lasting for minutes are quite rare. This observation is important for the differential diagnosis of migraine, with its obscurations lasting much longer (that is, minutes, a quarter of an hour, or more). Another characteristic sign is the complete restoration of the visual function after the obscuration. This occurs just as abruptly as the onset of the attack itself. In the majority of patients the obscuration involves the entire visual field uniformly. Occasionally it is limited to a central scotoma or to one half of the field, similar to a homonymous hemianopia. Attacks may occur simultaneously in both eyes or may alternate between them. At times patients report peculiar uncharacteristic and *unformed photopsias,* which they describe as sparks, stars, lightning, or luminous spheres or rings and which occur during, before, or after the amblyopic attacks (occasionally even independent of such attacks). The patient who described his attacks as quoted previously also had photopsias. If one probes the conditions that are likely to trigger these obscurations, he frequently may find that they are prone to occur when the patient arises (a change from the horizontal to the vertical position of the body) or when he turns his head abruptly. They may also occur after bodily exertion. *Occasionally the onset of the amblyopic attacks is accompanied by a fit of violent headache,* or an already existing headache may become more intense. There seems to be a direct relationship between the frequency of the attacks and the state of development of the papilledema (Ethelberg and Jensen), respectively, the severity of increased intracranial pressure.

The pathogenesis of these obscurations and amblyopic attacks is still quite controversial. A number of theories have been proposed, such as cortical origin as an equivalent to epileptic amaurosis (Jackson), sudden pressure of the dilated third ventricle on the chiasm (Leber; Paton; Holmes), transient spasms of the retinal arteries (Harms), or compression of the optic nerves in the optic canals (Behr). Ethelberg and Jensen relate the amblyopic attacks (which they have seen frequently in association with changes of the muscular tone, that is, hypertonia and hypotonia, or an impaired sensorium, ranging from numbness to coma) to a transient strangulation of the mesial part of the temporal lobe in the tentorial notch, which causes a temporary vascular disturbance due to the compression of the posterior cerebral arteries, or rather their branches, especially the calcarine artery, supplying the visual cortex.

The amblyopic attacks, just like the papilledema itself, occur relatively late in the course of a brain tumor. With tumors in silent zones of the brain, they may be the first symptom that brings the patient to the physician. Considering the typical and unmistakable ophthalmoscopic picture of papilledema, the diagnosis of amblyopic attacks, as a rule, causes no great difficulty. *Migrainelike obscurations* last from several minutes to several quarter hours and are often accompanied by typical scintillating scotomas. The fogginess of *acute and subacute angle-closure glaucoma* lasts longer. The sensation of rainbow colors (a result of the edema of the corneal epithelium!) as well as an increased intraocular pressure, a glaucomatous excavation of the disc, and visual field changes (starting with sector-shaped nasal defects) usually help in confirming such a diagnosis. The situation becomes somewhat more complicated if the *obscurations occur as the premonitory signs of impending epileptic attacks.* The latter may be the well-known accompanying signs of a growing brain tumor, but yet there may be a different etiology, such as a cicatricial or an idiopathic epilepsy. Careful ophthalmoscopic examinations in such cases should also aid in differentiating ambylopic attacks associated with increased intracranial pressure and with papilledema from other types.

In spite of the occasional occurrence of obscurations, the visual acuity and the peripheral fields in a patient with papilledema remain intact for a long time. The patient is not conscious of the *concentric enlargement of the blind spot* caused by the edematous enlargement of the disc and the resulting lateral displacement and compression of the adjacent retina (Fig. 2-10). The normal blind spot lies in an area between 13 and 18.5 degrees temporal to the point of fixation.

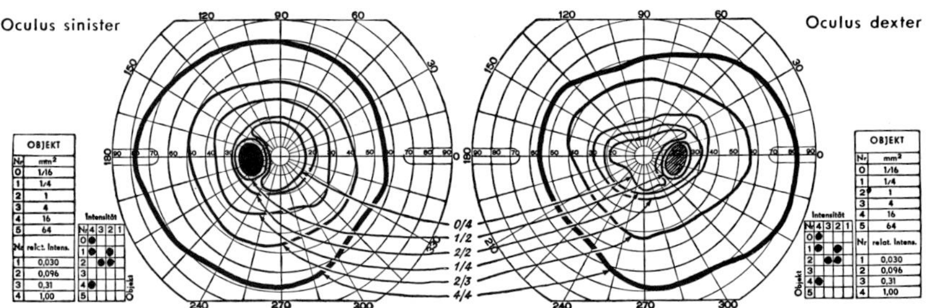

Fig. 2-10. Visual field changes (Goldmann perimeter) in bilateral papilledema: (1) marked enlargement of the blind spot (up to three times the normal size) and (2) disturbance of the law of summation of the isopters surrounding the blind spot. The 2/2 isopters with the larger-sized targets include Mariotte's spot, whereas the 0/4 isopters with the smaller targets bare it. (After Dubois-Poulsen.)

Its medium width in emmetropes is 5.5 degrees and its height 7.5 degrees, with a 5 or 10 mm. target used on the tangent screen at a distance of 2 meters. According to Chamlin and Davidoff, an additional relative blind zone of 1 degree surrounding the blind spot can be demonstrated with smaller targets (for instance, a 2 mm. object at the same distance of 2 meters). This zone corresponds to retinal elements of lesser sensitivity. Thus it can be stated that the *normal blind spot tested with a 2 mm. white target at a distance of 2000 mm. has a width of 7.5 degrees and a height of 9.5 degrees on the tangent screen.* An increase of these figures, if it amounts to 1 degree or more, has to be considered as enlargement of the blind spot. The enlargement in the horizontal direction is the more important one because the mapping of the upper and lower poles is rendered somewhat unreliable by angioscotomas. The enlargement is generally as much as three to four times, although enlargement of as much as eight times can be observed. These figures naturally apply only if there are no circumpapillary or parapapillary changes such as a scleral crescent, peripapillar choroidal atrophy, or medullated nerve fibers. We agree with numerous other authors that *in practically every case of papilledema there is an enlargement of the blind spot.* This empiric fact is easily explained by the pathologic-anatomic findings. There is such a striking correlation between the enlargement of the blind spot and the extent of the papilledema that some authors prefer the size of the blind spot as a criterion of the status of the edema to its prominence as determined using the ophthalmoscope (Davis; Chamlin and Davidoff).

A further perimetric symptom of papilledema is the relatively *smooth slopes of the borders of the blind spot,* which contrast with the abrupt limits of the normal disc. Dubois-Poulsen and Brégeat rightly call attention to a *sign typical for papilledema* that can be determined on the Goldmann perimeter. This sign indicates an *anomaly of the law of summation of the isopters surrounding the blind spot* (photometric disharmony): the shape of some isopters determined by targets with the same additive qualities do not correspond any longer. In other words, isopters that should coincide recede from each other in such a manner that the isopters for the larger targets surround the blind spot, whereas those for the smaller ones bare it and are situated centrally to it (Fig. 2-10). However, this phenomenon is only the result of the edema of the nerve head and thus not absolutely pathognomonic for papilledema due to increased intracranial pressure. For this reason we feel we cannot agree with Dubois-Poulsen and consider this sign as an early indication of a choked disc. Also we can hardly agree with the opinion expressed by some researchers (de Schweinitz; Chamlin and Davidoff; Brégeat) that an enlargement of the blind spot can be demonstrated before the choked disc manifests itself ophthalmoscopically. *On the basis of our personal experience, we believe it is neither possible nor reliable to depend solely on the size of the blind spot for the early diagnosis of a papilledema.* There is some justification in using an enlargement of the blind spot for such purpose if combined with other early signs; it should never serve as the basis for a diagnosis if it is the only sign. In our opinion the real importance of an enlarged blind spot lies in the fact that it permits us to differentiate papilledema from conditions with a similar ophthalmoscopic appearance (p. 141). If the edema of the nerve head expands toward the macula, the enlargement of the blind spot will extend toward the point of fixation. A macular edema will produce metamorphopsia and a *relative central scotoma* with a slightly reduced central visual acuity. According to Traquair, such a scotoma shows a relative blue-blindness, which he considers characteristic for a disturbance of the outer retinal layers. In the even more severe and advanced stages of papilledema the enlarged blind spot and the central scotoma may merge to form a relative cecocentral scotoma. Edema of the nerve head extending into the macular region generally causes only a minor loss of central vision—an important criterion in favor of papilledema. A more extensive impairment of the visual acuity (possibly in combination with small or even large absolute scotomas) is caused by secondary changes such as hemorrhages or the fan-shaped accumulations of white retinal dots in the macular region described previously.

The *perimetric syndrome of papilledema,* characterized by enlargement of the blind spot, smooth slopes of its borders, and an anomaly of the law of summation of the isopters surrounding it can be easily demonstrated by photopic and scotopic perimetry and campimetry. Scotopic perimetry and campimetry show proportionally more pronounced pericecal defects around the blind spot than photopic methods. In the fully developed stage (without any signs of atrophy) the papilledema manifests normal electrophysiologic signs (ERG, objective and subjective frequency of fusion, occipital response in the EEG, retinocortical time).

Incipient papilledema

Although fully developed papilledema should not offer any diagnostic problems, its early stages may present a much more delicate and complex situation. Yet it is in these early stages that we as ophthalmologists are called in consultation by the neurologist and the neurosurgeon to make such a weighty and grave decision. Occasionally it depends on this very decision whether the patient will be subjected to all the diagnostic procedures of modern neurosurgery and neuroradiology, procedures that are not only disagreeable to the patient but are potentially dangerous, even with all possible precautions.

Indeed, incipient papilledema (Fig. 2-11) is frequently mistaken for similar conditions with the same ophthalmoscopic picture but with a different etiology and significance. Which are the more or less reliable signs of incipient papilledema? In agreement with numerous authors, we believe they are as follows: *hyperemia and redness of the papilla* (due to dilation of the capillaries within the substance of the optic disc), together with *blurring of its margins* at the superior and inferior poles, later on the nasal side and finally on the temporal

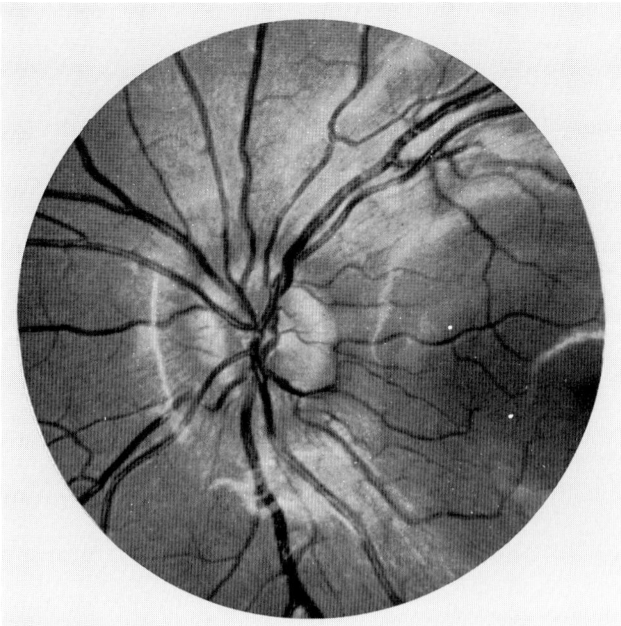

Fig. 2-11. Incipient papilledema of the left eye in a 9-year-old boy with a tumor of the posterior fossa. Slight enlargement of the diameter of the disc. Blurring of the entire disc margin. Beginning flat prominence of the papillary margin (about 0.5 diopter) with slight deflection of the retinal vessels. Preserved vessel cup. Moderate dilation and tortuosity of the retinal veins. Discrete peripapillary edema with a pronounced pattern of the nerve fibers. Visual acuity, O.U. 1.0. Diastolic pressure of arteries, 40 to 50 grams (Bailliart).

side; a slight *elevation* of the papillary margin with a corresponding deflection of the vessels at first on the nasal side and then on the temporal side; and *widening of the veins and increased visibility of the capillaries of the disc* (Fig. 2-12). Venous distention is a sign of incipient papilledema, but is significant only if accompanied by progressing hyperemia and blurring of the disc margins. The early stages of papilledema are characterized by minimal prominence of the disc margins. More pronounced or even measurable elevation of the optic disc is not an early sign of papilledema (elevation of the disc may also be congenital and absolutely unrelated to increased intracranial pressure). In the presence of disc hyperemia and disc blurring, *small hemorrhages* in the nerve fiber layer at the border of the disc represent a definite and reliable sign of incipient papilledema. In this connection even a single splinter bleeding may be significant.

Other authors also mention filling of the physiologic cup as well as grayish sheathing of the vessels in the funnel of the disc (widening of the perivascular lymph spaces) as an important early symptom. Furthermore, there is a grayish white radial striation of the retina at the papillary margin and in its immediate surrounding. In other words, we have all the characteristic signs

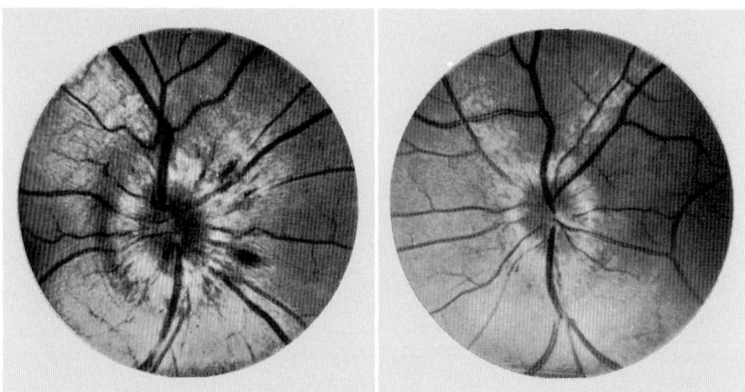

Fig. 2-12. Incipient papilledema in a 35-year-old patient with meningioma in the area of the right temporal region. There is a distinct difference between the two eyes in the development of the choking. The right papilledema is already fully developed and shows pronounced venous congestion and hemorrhages at the nasal disc margin. The left disc shows discrete bulging of about 0.5 diopter of its margin, with slight deflection of the vessels, blurring of the entire disc margin, and slight venous congestion—all signs of an incipient papilledema. Diastolic pressure of arteries, O.U. 50 grams (Bailliart). Visual acuity, O.U. 1.0. The left illustration corresponds to the right eye; the right illustration, to the left eye.

that have already been mentioned in the description of the fully developed choked disc here in an attenuated form. The correct diagnosis is rendered more difficult because these signs in the early stages do not involve the entire nerve head but are frequently confined to only part of it, especially the nasal part. Even the experienced ophthalmologist will hesitate to consider these early signs as absolute and unequivocal evidence, because they may occur in a very similar form (either as a single one or several of them) in various physiologic (for example, hyperopia with pseudoneuritis and congenital tortuosity of the retinal vessels) and pathologic (for example, incipient hypertensive malignant retinopathy, papillitis) conditions (p. 123). It is absolutely essential to caution against a rash evaluation of these signs!

For further evaluation of early papilledema the *examination of the optic disc with the slit lamp using the contact glass* may furnish important data. In this connection, Goldmann mentions two important symptoms of incipient papilledema: hyperrefringency of the papillary border and elevation of the prepapillary internal limiting membrane. A further biomicroscopic sign is the enlargement of the perivascular lymphatic spaces (Koeppe).

There is no doubt that *fluorescein angiography of the fundus* can contribute to the early diagnosis of papilledema. In favor of such a diagnosis are a rapid diffuse coloration of the disc tissue and the increased visibility of dilated capillary vessels (p. 102). Especially important in this connection is the exact timing of the observed phenomena, which is different from other pathologic conditions of the disc.

In the opinion of some authors (de Schweinitz; Chamlin and Davidoff; Dubois-Poulsen; Brégeat) the enlargement of the blind spot (p. 83) is considered a reliable early sign that may even precede the ophthalmoscopic manifestation of swelling of the papilla. We tend to follow Traquair and cannot agree with this opinion. After all, the enlargement of the blind spot is merely the nonspecific evidence of an edema of the disc which, in addition to increase of intracranial pressure, may be caused by other factors such as an inflammation or a toxic agent. For early papilledema we are tempted to accept Dubois-Poulsen's *observation on the Goldmann perimeter that indicates an anomaly of the law of summation for the isopters surrounding the blind spot.* However, again we have to make the reservation that this phenomenon is merely a nonspecific sign of damage to the retina surrounding the disc as a result of the edema (Fig. 2-10).

There is a loathsome uncertainty regarding the early signs of papilledema. If papilledema is a direct consequence of increased intracranial pressure, proof of the latter should assist in arriving at a diagnosis. A spinal or occipital puncture could establish the existence of such an increase. However, it is a well-founded fact that such a puncture in patients with increased intracranial pressure harbors the immense danger of incarceration of the brain stem into the foramen magnum, an event that would lead to a grave aggravation of the existing condition, requiring immediate neurosurgical intervention. *A lumbar or occipital puncture for the verification of an incipient or an already fully developed choked disc is strictly contraindicated and would betray extremely poor judgment on the part of the physician.*

The *pressure in the ophthalmic artery* is not only related to the systemic blood pressure but also to the intracranial pressure. This is obvious if one is mindful of the origin of these ocular vessels from the large cerebral vessels and of their partly intracranial course. Thanks to the fundamental investigations of Bailliart, it is possible to determine the arterial as well as the venous pressure of the ocular vessels with the aid of the ophthalmodynamometer (p. 62) with some accuracy. Bailliart and, with him, the majority of investigators, especially of the French school (Coppez, Rasvan, Magitot, Pereyra, Spinelli, Gallois, Bauwens, Winther, Ascher), originally were of the opinion that there is always an increased pressure in the arteries in patients with brain tumors. De Morsier, Monnier, and Streiff were able to prove that this belief is not always correct. They made the observation on twenty-one patients with brain tumors (verified either during operation or postmortem) that *tumors of the anterior or middle fossa were usually associated with normal or decreased pressure of the ophthalmic arteries,* whereas *those of the posterior fossa were, as a rule, associated with increased arterial pressure.* Streiff was able to confirm this observation in an additional series of ten patients. He concluded that the variations of the pressure in the arteries were not a direct result of the space-consuming process but an indirect effect on the central mechanism governing the circulation of the cerebral vessels. Bailliart later arrived at the conclusion, confirmed by numerous authors (Rossano; Gauddisart; Suvina; Streiff and Monnier;

Redslob), that in *brain tumors an increase in pressure in the ophthalmic arteries precedes the appearance of the papilledema* and that once the latter develops this pressure becomes normalized or even gives way to a decrease in pressure. This increase of the pressure in the arteries is an *isolated discordance* since the pressure in the brachial artery in patients with space-consuming lesions is usually normal.

Bailliart calls this drop of pressure in the arteries at the onset of papilledema "asystoly." The ophthalmic artery seems to resist the increased intracranial pressure for a while (initial increase in pressure) only to give up in the end; the drop in pressure is quasi a sign of an insufficiency of the arterial wall. Bailliart believes that the fall in pressure of the arteries, together with the resulting insufficiency of the venous return flow, is the true basis for the pathogenesis of the choked disc.

According to Weigelin, the ophthalmodynamometric results in patients with brain tumors (with or without papilledema) do not give any concordant results from which one could draw practical conclusions.

In thirty-one patients with cerebral tumors with papilledema, Weigelin registered the median ophthalmic arterial pressure in relation to the median arterial brachial pressure: in four patients the pressure was elevated, in nine it was lowered, and in eighteen it was normal. In thirty-four patients with cerebral tumors without papilledema the median ophthalmic arterial pressure in relation to the median brachial pressure was found to be elevated in eight, lowered in four, and normal in twenty-two.

Therefore Weigelin came to the conclusion that there is no characteristic alteration of the ophthalmic arterial pressure due to increased intracranial pressure. Increase of intracranial pressure no doubt causes an increase of the resistance of flow within the intracranial vessels, which at the same time, however, may be compensated by a dilatation of the terminal vessels. This is just the phenomenon that explains so many normal values of intraocular arterial pressure in patients with papilledema. As this compensation varies greatly from subject to subject, the measurement of the intraocular arterial pressure, according to Weigelin, cannot be used for evaluation of the intracranial pressure.

Important data gained with the method of *ophthalmodynamography* are reported by Finke, who examined twenty patients with intracranial tumors using the method of Hager.

Fourteen patients with no or only minimal signs of increased intracranial pressure revealed normal conditions in the ophthalmodynamogram. In seven patients with distinct signs of increased intracranial pressure, ophthalmodynamography revealed a normal general pressure with relative cranial hypertonia in two, a general hypertonia with ophthalmobrachial isotonia in one, and a general hypertonia with relative cranial hypotonia in four. The five patients with general hypertonia had never before had any signs of hypertonia of an essential nature. In these cases the increased pressure in the brachial artery is an expression of the tendency to overcome the increased intracranial pressure, as mentioned by Cushing in 1902. Of special interest in this series are the patients with general hypertonia and relative cranial hypotonia. Finke calls this the *decompensated intracranial pressure:* in spite of an increase of the brachial artery pressure, there occurs a further decrease of the ophthalmic artery pressure because the increased intracranial pressure can be overcome incompletely or not at all.

With regard to ophthalmodynamography, Finke comes to the conclusion that this method gives information as to whether and how far a compensatory increase of the brachial artery pressure is accompanied by an increase of the cranial blood pressure. It may also give information in certain cases of increased intracranial pressure as to whether, in spite of an increase of the arterial brachial pressure, there is a decrease of the ophthalmic artery pressure (decompensated intracranial pressure). Furthermore, the ophthalmodynamogram is able to control the effect of dehydration of the brain on the ophthalmic artery pressure. If a relatively low ophthalmic artery pressure does increase in spite of an unaltered brachial artery pressure, this is a sign of improvement of the so-called decompensated intracranial pressure.

In summary, the following rule is suggested for practical purposes: the determination of the pressure in the ophthalmic artery (ophthalmodynamometry) for the early diagnosis of a papilledema as well as of a brain tumor must be regarded with caution and should never be the single fact on which such a diagnosis is based. *In a patient with suspected incipient papilledema an increase in the pressure of the ophthalmic artery may be interpreted as an indication of increased intracranial pressure.* Its absence or even a decrease, however, never rules out increased intracranial pressure. One always has to keep in mind that in cases of pronounced increased intracranial pressure there might be a general arterial hypertension with relative intracranial hypotonia manifested by a relatively low ophthalmic artery pressure, a phenomenon that can be registered especially well by ophthalmodynamography. The relation between the pressures within the brachial artery and the ophthalmic artery gives valuable information as to whether the increased intracranial pressure still permits an increase of the cranial blood pressure or leads to a decompensation of the arterial cerebral blood flow.

Since a beginning venous congestion is an early sign of papilledema, the determination of the *pressure of the veins* should be of special interest. The intraocular venous pressure depends on the intracranial venous pressure, which, on the other hand, is a function of the pressure of the cerebral spinal fluid. Even in recent publications we frequently find the opinion expressed that an increased intracranial pressure can be excluded if a spontaneous venous pulsation is present or occurs after exerting slight pressure on the eye (Lauber; Sobanski; Duke-Elder; Redslob; and others). We are in agreement with Streiff, Serr, and Williamson-Noble that the disappearance of spontaneous venous pulsation is of no significance for the diagnosis of an increased intracranial pressure, especially not for the early diagnosis of choked disc. In numerous patients with brain tumors with an incipient papilledema as well as a fully developed choked disc, we have seen a positive venous pulse or at least an immediate collapse of the retinal veins upon slight pressure on the eye. Actual measurement of the retinal venous pressure with a special dynamometer likewise does not supply additional information, although various authors (Baurmann; Lauber; Sobanski; Marchesani; Redslob) regard an increased venous pressure as all but pathognomonic for increased intracranial pressure. There is a considerable division of opinion concerning the significance of the venous pressure. Other authors (Riser, Calmettes, Garipuy, and Pigassou) are convinced that they found a pronounced lowering of the venous pressure in the majority of their verified cases of brain tumor. Such discrepancies demand a cautious interpretation. Perhaps these discrepancies are related to the technical difficulty of the method. Streiff is so right when he states in his monograph on the retinal blood pressure: *"The exact determination of the pressure of the retinal veins is, however, very difficult and subject to many sources of errors; for that reason one should not attach too much significance to it, neither to the indirect evaluation of the intracranial pressure—an attempt that is made time and again."* Our own experience in the evaluation of the pressure of the retinal veins for the early diagnosis of a papilledema is in full agreement with Streiff's opinion.

This discourse should make it quite plain how difficult the diagnosis of an incipient papilledema is. Often it cannot be made merely on the basis of the ophthalmoscopic examination and some of the ophthalmologic ancillary methods. We maintain that *we, as ophthalmologists, should not evaluate the ophthalmologic findings as data detached from the complete clinical picture but that we should consider the entire neurologic picture and the results of appropriate tests in our interpretation of changes of a disc if papilledema is suspected.* If it is impossible to arrive at a definite decision as to whether or not there is an incipient papilledema, there are, in our opinion, two avenues of approach. If the general condition of the patient permits it, the ophthalmoscopic examination can be repeated at a later date (perhaps 1 or 2 weeks later); a definite progress of the signs of edema of the disc may be observed at that time. We would like to stress in this connection the enormous importance of having a complete record of the previous findings for the sake of comparison; *photographs* of the fundus (also fluorescein angiograms) taken on different occasions are particularly valuable. If the situation in a suspected papilledema is more urgent, especially in view of the general condition of the patient, one should not hesitate to refer the patient to the neurologist or the neurosurgeon. With the advanced modern diagnostic facilities, it is preferable to make such a referral once too often!

Unilateral papilledema

As mentioned previously, the swelling of the disc resulting from an increased intracranial pressure usually occurs bilaterally (in 20% of the cases there is a prominence on one side). A unilateral form is rare. This bilateral manifestation of a choked disc is sometimes forestalled by local factors. One of these factors is a *previous unilateral atrophy of the optic nerve.* An atrophic optic nerve is incapable of developing edema. This is one reason why pituitary tumors that produce a descending optic atrophy do not cause papilledema, even though there might be a fairly large space-consuming lesion. Another possible mechanism is the *blockage of the communication between the vaginal spaces of the sheaths of the optic nerve and the subarachnoidal space* due to some inflammation, compression, or congenital anomaly. The *Foster Kennedy syndrome* is cited as a typical example. Its characteristics are an optic atrophy on the side of the tumor (caused by direct pressure on the intracranial part of the optic nerve) and a choked disc on the opposite side. In addition to this fully developed picture of the Foster Kennedy syndrome, there are a number of incomplete forms that belong to the same group: there may be a preponderance of signs of atrophy on the side of the tumor, with a preponderance of swelling of the disc on the opposite side, but definitely a bilateral papilledema with a distinct difference on the two sides; there may be a normal disc with central scotoma on the tumor side and a papilledema on the opposite side; or there may be an atrophic papilledema (secondary optic atrophy) on the side of the tumor and an ordinary papilledema on the opposite side. We shall consider the Foster Kennedy syndrome in the discussion of the ocular symptoms found in patients with frontal lobe tumors and in those with meningiomas of the olfactory groove and the sphenoid ridge. Apart from tumors (68%), the Foster Kennedy syndrome may also be caused by nontumoral conditions (32%) such as arteriosclerosis, optochiasmatic arachnoiditis, and carotid aneurysms. In these cases the unilateral optic atrophy may be explained by some pressure on the optic nerve, but the contralateral papilledema is not explained by an increased intracranial pressure. Here the swelling of the disc must have some relation to circulatory disorders of the optic nerve itself (for instance, as in ischemic papillitis).

A much less known and recognized cause of unilateral papilledema associated with increased intracranial pressure is *unilateral myopia* of medium or higher degree. A medium myopia is one of 5 to 10 diopters; high myopia is one of 10 to 15 diopters or more. The anatomic peculiarities of the myopic disc and the area surrounding it (myopic crescent) either prevent the development of a papilledema or cause its development at a later stage and in an atypical form (Marchesani; Bietti; Morone). In bilateral myopia a papilledema may never become manifest in spite of an existing increased intracranial pressure.

If increased intracranial pressure can be ruled out as the cause of a unilateral choked disc, an *orbital process* should be considered next. As a rule, it will produce a more or less distinct *exophthalmos,* which is of considerable importance for diagnostic purposes. Usually the closer such a process is to the globe, the more pronounced is the edema of the disc. The etiologic factor of such a unilateral papilledema may be tumors, aneurysms, inflammatory conditions, abscesses, or an edema of the orbit. It should be mentioned that occasionally tumors of the middle and anterior fossa (for instance, meningiomas of the sphenoid ridge) may extend into the orbit and cause a unilateral papilledema in such a manner. However, such a condition will mostly cause a Foster Kennedy syndrome with papilledema on the side opposite the tumor.

In reviewing our material, we have observed a unilateral papilledema only rarely as the result of increased intracranial pressure (8%). This may occasionally be the case in cerebral abscesses or in temporal lobe tumors. The Foster Kennedy syndrome, likewise, is not so frequent as is generally assumed. Orbital and purely ocular conditions can be recognized with relative ease. Such ocular conditions are hypotension of the globe (for instance, after a fistulating operation for glaucoma), swelling of the papilla in case of a unilateral iridocyclitis,

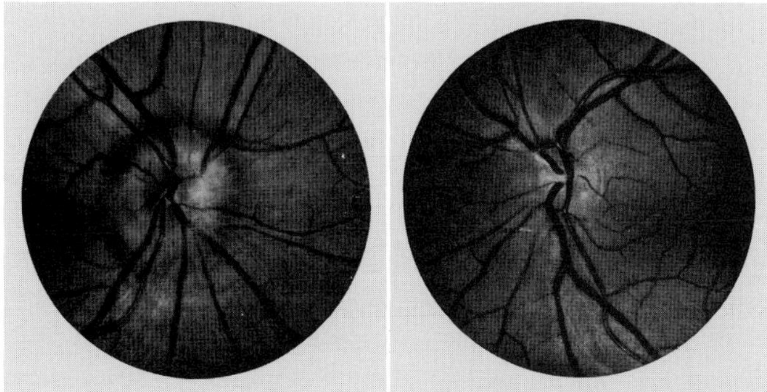

Fig. 2-13. Unilateral pseudopapilledema in a 20-year-old patient. Enlarged disc diameter. Blurred disc margin. Slight bulging of the right disc margin. Insignificant deflection of the vessels but no venous congestion of either the large veins or the capillaries of the disc. The left disc appears blurred on the nasal side but not prominent. The left illustration corresponds to the right eye; the right illustration, to the left eye.

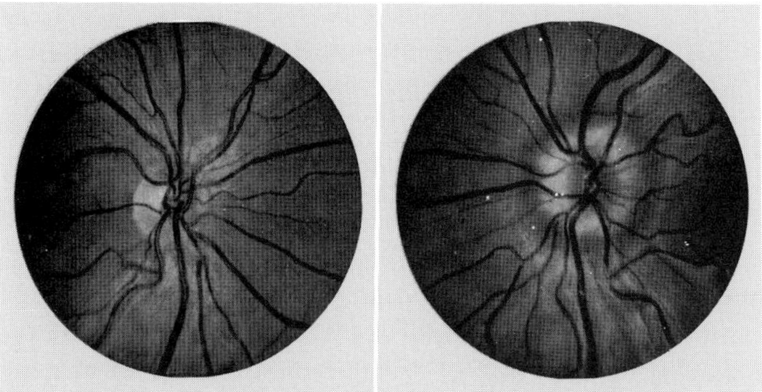

Fig. 2-14. Unilateral apparent choked disc (left side) with intrapapillary drusen of the papilla in an 18-year-old patient. Left disc enlarged, with indistinct margin and slight prominence over the level of the retina. No venous congestion. The drusen hidden in the depth of the papilla just manifest themselves by the slightly protuberant outline of the left disc margin. They could be visualized by means of posterior slit-lamp microscopy (Goldmann's or Hruby's method). The left illustration corresponds to the right eye; the right illustration, to the left eye.

or an edema of the disc due to central retinal vein thrombosis. One should also keep in mind the possibility of *unilateral drusen of the disc* and, in the presence of hyperopia, a *hyperopic pseudoneuritis* (pp. 140 and 143 and Figs. 2-14 and 2-39).

Unfortunately, on the basis of our observations, a fairly large number of so-called unilateral papilledemas cannot be accounted for in spite of the benefit of all kinds of diagnostic procedures. A certain number of cases that cannot be explained properly must be listed under the unsatisfactory category of unilateral *pseudochoked disc or pseudopapillitis* since they are accompanied neither by increased intracranial pressure nor by orbital or ocular conditions. We have not less than eight cases of so-called unilateral "choked disc" on record, cases that were referred because of a suspected brain tumor and that did not yield a satisfactory explanation for the unilateral fundus changes. One cannot help but consider these pseudoforms (for the ophthalmoscopic appearance see Fig. 2-13) of a unilateral choked disc as insignificant morphologic deviations of a normal disc provided they do not change their original picture during an extended period of observation (p. 112).

There are also transient forms of *unilateral papilledema*. We have examined a 25-year-old man with an unquestionable unilateral papilledema of a 1-diopter prominence, hemorrhages, white dots, and a distinct engorgement of the retinal veins. The neurologic and neuroradiologic findings were negative. There was no increased intracranial pressure. The visual acuity and fields were not remarkable. A year later the choked disc had disappeared completely without leaving any residual or functional disturbances! An etiologic explanation for this transient choked disc could not be found (pseudotumor cerebri?) (see Wagener).

These rather gloomy statements regarding the chances of interpreting a unilateral papilledema (including pseudopapilledema) should not allay the zeal of the examiner. On the contrary, because of the very difficulties just described, every attempt should be made at least to rule out the possibility of increased intracranial pressure. In most instances this will hardly be possible without the assistance of the neurosurgeon and neuroradiologist.

Chronic atrophic papilledema

Sooner or later any papilledema will lead to a *secondary optic atrophy* provided there is no intervention to relieve or at least to decrease the increased intracranial pressure. As a rule, this atrophy appears only a few months after the onset of the papilledema. The time required for the involution or collapse of a severe, fully developed papilledema to complete atrophy is 6 to 9 months, occasionally even up to more than 1 year.

The degenerative changes of the nerve fibers caused by the chronic edema produce a glial proliferation—the typical reaction in every kind of secondary optic atrophy. Ophthalmoscopically, it manifests itself as a *grayish white discoloration involving the peripheral parts of the disc initially but, in the more advanced stages, also its center.* As the atrophy progresses, the prominence and the width of the disc decreases in spite of the persistent intracranial pressure. The disc becomes distinctly paler; the original reddish color gives way to the unmistakable grayish white gliosis as the end result. There is narrowing of the retinal arteries. The veins are less congested and may even approach a normal caliber, actually an indication of constriction of the vessels caused by glial proliferation. In certain stages a fine network of dilated superficial capillaries resembling telangiectasia can be observed overlying the white atrophic nerve head. Arteries and veins sometimes may be completely obscured by the glial proliferation in their course across the disc. Occasionally the picture of the chronic atrophic papilledema results in a whitish sheathing of the vessels, blurring and narrowing the red blood column within them (Figs. 2-15, 2-16, and 2-19). Even in the later stages after the prominence of the disc has receded considerably, the original aspect of the choked disc can be surmised from the increased size of the nerve head, from its blurred and slightly prominent border, from remnants of hemorrhages and white dots on or near the papilla (Fig. 2-8), and from an irregular arrangement of the retinal pigment in the surrounding retina. It still may be difficult to rule out a secondary optic atrophy after a papillitis if one depends only on the ophthalmoscopic findings. The presence of grayish arcuate lines (Paton's concentric lines) surrounding the disc is more suggestive of an atrophic papilledema. The final decision will usually depend on the functional examination, especially the determination of the central and peripheral visual fields (pp. 106 and 119).

The visual functions will be preserved for a relatively long time during the acute stages of the papilledema. In contrast, the chronic atrophic papilledema is marked by significant disturbances of the visual function. It is imperative to detect these disturbances as early as possible and not to take chances with a grave, irreversible impairment. *A decompression for a papilledema should be performed before optic atrophy sets in or at least during its very early stages.* Otherwise a restitution of the visual function is no longer possible. In the later stages of a progressing atrophy, which occasionally develops quite rapidly, a decompression for the relief of the increased intracranial pressure may have catastrophic results for the visual function. We have had occasion to observe a reduction of the central vision and the visual fields amounting to complete bilateral amaurosis, especially in patients with tumors of the posterior fossa with chronic atrophic papilledema. Thus the importance of the *early signs of a beginning atrophy in papilledema* cannot be stressed enough. One should make it a rule to search for them carefully, especially in those patients with papilledema in whom, for some reason, neurosurgical intervention has to be postponed or is not indicated.

One of the first signs of beginning atrophy of a papilledema is a *concentric constriction* of the peripheral and middle isopters, which may be associated with peripheral sector-shaped defects or similar defects extending close to the center (Figs. 2-17 and 2-18). Unless the intracranial space-occupying lesion interferes directly with the visual pathways, thus causing specific field changes, these early signs are enormously valuable and reliable. It is essential to search

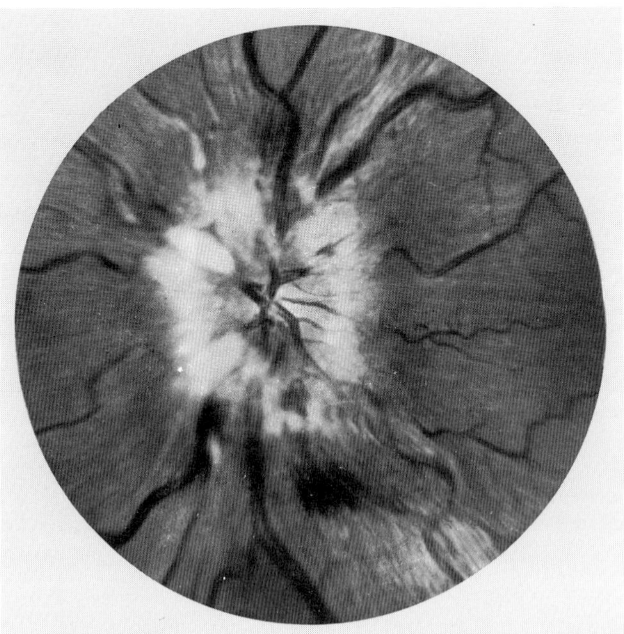

Fig. 2-15. Chronic papilledema of the left eye showing beginning atrophy in a 29-year-old patient with hemangioma of the right cerebellar hemisphere with internal hydrocephalus. Papilla markedly enlarged with blurred outline and an elevation of 4 diopters. Grayish white discoloration of the peripheral as well as the central parts of the disc, indicating beginning atrophy of the nerve fibers and secondary glial proliferation. Numerous white dots ("cytoid bodies") on the disc. Dilation and tortuosity of the veins which, in areas, are buried in the disc. White accompanying stripes of the vessels (dilation of the perivascular lymph spaces). Hemorrhages at the disc margin within the nerve fiber layer. Visual acuity, 1.0. Slight concentric contraction of the visual field.

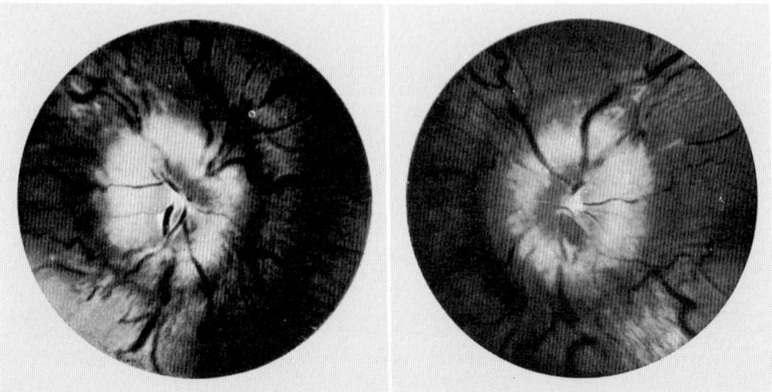

Fig. 2-16. Chronic atrophic papilledema in a 36-year-old patient with syphilis of the central nervous system and syphilitic meningitis. Conspicuous pallor of the edematous discs showing a prominence of 4 to 5 diopters. The veins are barely dilated and the arteries rather narrowed. White perivascular sheaths. Visual acuity, O.D. 1.0; O.S. 0.7. Both visual fields intact. The left illustration corresponds to the right eye; the right illustration, to the left eye.

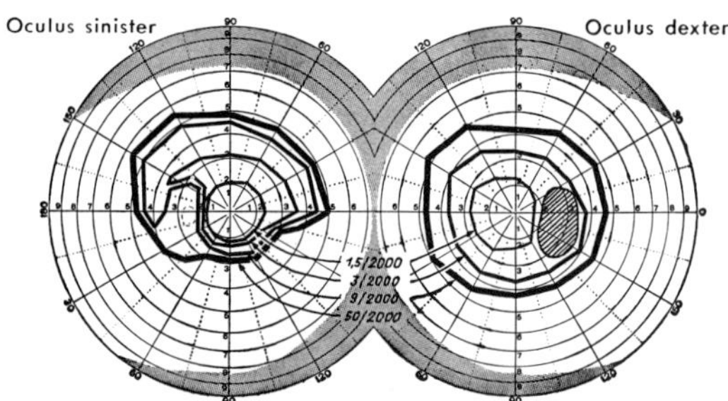

Fig. 2-17. Visual field changes in a 19-year-old patient with chronic atrophic choked discs due to astrocytoma of the fourth ventricle (see Fig. 3-110, p. 274). Concentric constriction of the right field, particularly of the peripheral isopters, and marked enlargement of the blind spot. Contraction of the inferior nasal and inferior temporal quadrants of the left peripheral field, with formation of sector-shaped defects protruding toward the center, as determined on the tangent screen. There is a history of numerous amblyopic attacks. Visual acuity, O.U. 0.6.

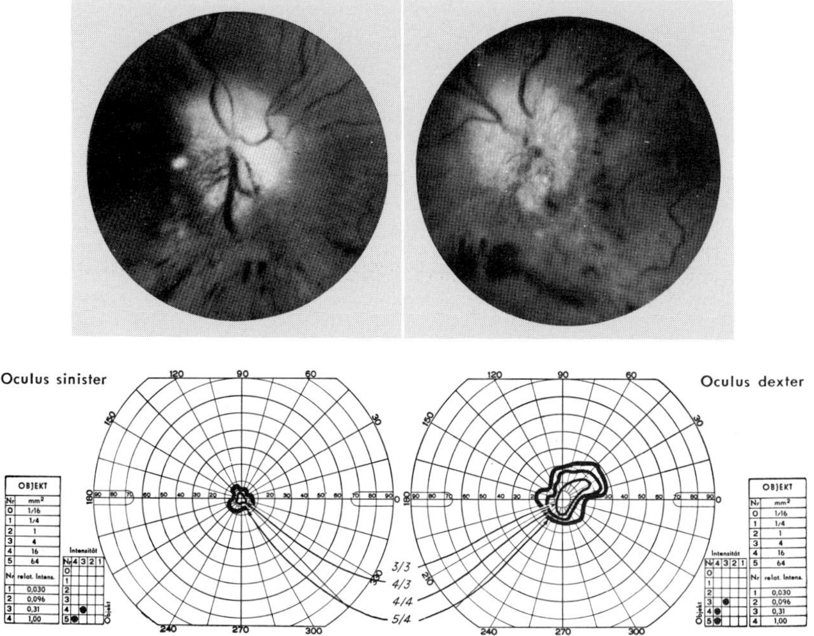

Fig. 2-18. Chronic atrophic papilledemas in an 18-year-old patient with a tumor of the corpus callosum and the septum lucidum in the late stages. Bilateral prominence of the discs of 4 to 5 diopters. Marked pallor and gliosis of the discs. Superficial network of dilated capillaries; arteries partially obscured by glial proliferation. The left illustration corresponds to the right eye; the right illustration, to the left eye. Below, Corresponding visual fields taken on the Goldmann perimeter. Extensive concentric contraction of the peripheral isopters, especially of the left field. The right field shows predominantly a loss of the nasal parts. The central visual acuity is reduced to finger counting at 1 meter O.S. and at 0.8 meter O.D.

for them with the most minute perimetric targets (2/330 and 1/330). Here again, the Goldmann perimeter is particularly well suited. It will indicate a *contraction of the peripheral visual field,* which may occur at a time when the more central isopters still appear relatively normal except for disturbances of the law of summation mentioned previously. Parallel with the increasing concentric contraction of the visual field is a *decrease of the central visual acuity.* As mentioned before, it may have been impaired already by the edema extending into the macular region (resulting in a cecocentral relative scotoma) and further impaired by hemorrhages or by a fan-shaped arrangement of droplets of fluid. As a rule, the concentric contraction of the visual fields progresses more rapidly in the nasal, especially inferonasal, halves of the field, thus simulating some sort of a *binasal hemianopia* during certain stages. In the end there remains a more or less centrally located island of the field, which may or may not include the blind spot and which may enable the patient barely to find his way around. In these advanced stages it may be impossible to determine a field defect caused directly by the pressure of the tumor (for instance, a homonymous hemianopia). Unfortunately, even this small island may be lost, giving way to a complete amaurosis with wide, fixed pupils. (This is in contrast to cortical blindness with a completely intact pupillary reaction!)

In the presence of a chronic atrophic papilledema the size of the peripheral visual field, not the ophthalmoscopic aspect, is of prognostic importance with regard to postoperative loss of vision. When the cause of a persistent increased intracranial pressure cannot be eliminated for some reason or other, the visual field has to be examined repeatedly to detect any progression of the peripheral contraction. This could necessitate an orbital or temporal decompression. In the atrophic stage of papilledema the *electrophysiologic responses* are also altered: decrease of the subjective frequency of fusion, increase of the retinocortical time, and delay or even absence of the specific occipital response. Thus the electrophysiologic examinations may help to confirm an otherwise suspected tendency of a papilledema toward atrophy.

Not without reason did we give such a detailed description of the picture of chronic atrophic papilledema. *We know of numerous instances in which the patient initially consulted an ophthalmologist because of rapidly deteriorating vision and the ophthalmologist diagnosed a chronic atrophic papilledema and was the first to suspect a brain tumor.* Such tumors are mostly in a relatively silent part of the brain and have an early tendency to cause an internal hydrocephalus (cerebellum, third and fourth ventricle). As mentioned previously, surgical removal was not followed by any functional improvement. On the contrary, in some of the patients there was a catastrophic drop of vision, even resulting in total bilateral blindness. Anybody who has experienced such complications once will appreciate the importance of recognizing a choked disc in its early stages—not only in view of the brain tumor but also in consideration of the possible functional impairment of the optic nerve. From this point of view

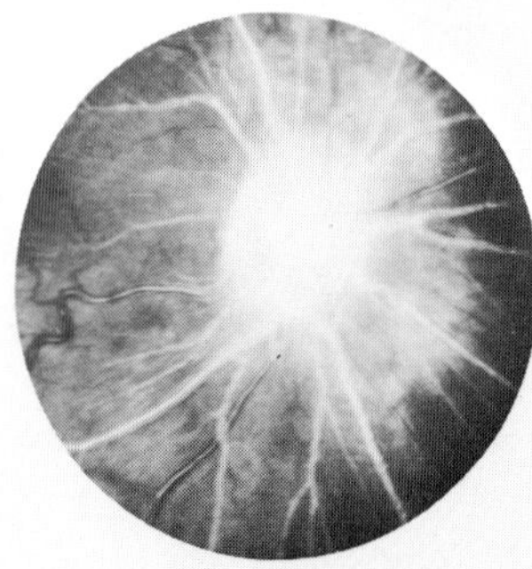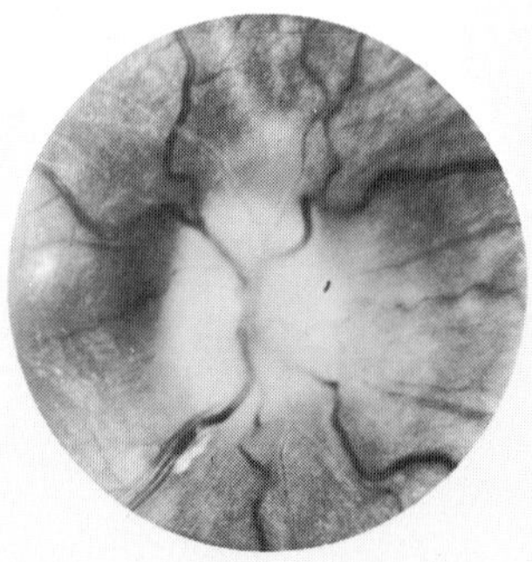

Fig. 2-19. Optic atrophy and gliosis following chronic untreated papilledema in a 28-year-old patient with bilateral acoustic neuroma in neurofibromatosis. Amaurosis of both eyes. The right disc is flat, enlarged, and shows permanent gliosis with folds in the nerve fiber layer. Note the perivascular sheathing extending from the disc far out on the periphery. Optic cup completely obscured by glial proliferation. The left disc is still prominent, but also atrophic, with gliosis and distinct perivascular sheathing. Arteries practically invisible within area of atrophic disc. The optic cup begins to be obscured by the gliosis.

one should almost regret that the papilledema remains asymptomatic for such a long time and that it becomes noticeable only when the dreadful signs of atrophy have already developed, indicating a condition that generally will not improve with active neurosurgical interference but frequently will even become worse.

Under certain circumstances (for instance, meningioma of the olfactory groove), papilledema and optic atrophy may develop at the same time. The optic atrophy (either unilateral or bilateral) depends on the compression of the optic nerves or the chiasm by the tumor. The papilledema is a function of the increased intracranial pressure but may be handicapped in its evolution by the degree of the preceding optic atrophy. In such cases it may become extremely difficult to decide whether one has to deal with a chronic atrophic papilledema or with a combination of simultaneous papilledema and optic atrophy (Fig. 2-19). The functional disturbances and their character can help in making the differential diagnosis: if loss of central vision and defects of the peripheral visual field occur early, one has to think of a primary optic compression; if headaches or other manifestations of increased intracranial pressure precede the visual disturbances, there is a great probability that papilledema has developed into secondary optic atrophy.

Pathologic anatomy of the disc in papilledema

It may be of interest to touch briefly upon the pathologic-anatomic changes of the disc in papilledema and their relationship with the clinical phenomena just discussed. We follow more or less the description given by Duke-Elder in the third volume of his *Text-Book of Ophthalmology.* In principle, there is a *simple edema of the nerve head with an edematous swelling of the nerve fibers and a corresponding exudative infiltration of the rest of the tissues.* This edema is usually limited to a relatively small area and extends from the disc margin to the

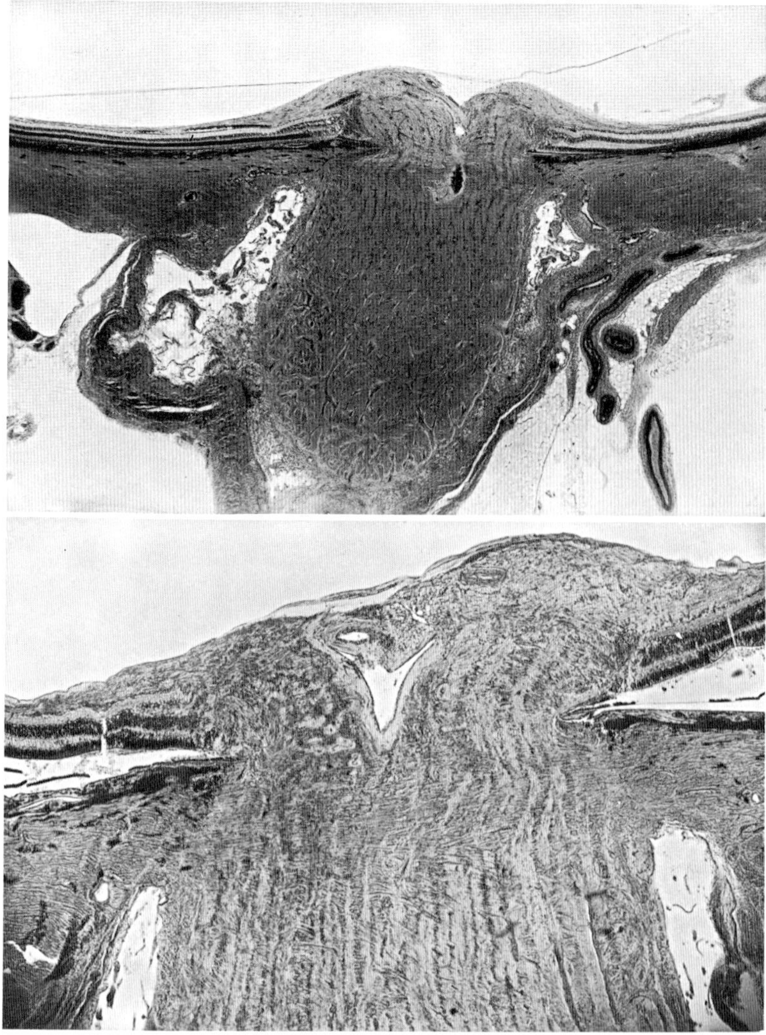

Fig. 2-20. Histologic section of papilledema (above, low power, about ×14; below, higher power, about ×30). Edema of the nerve head with edematous swelling of the nerve fibers and the interstitial tissue. Cup in the lower illustration is filled. Lateral displacement and folding of retina. Dilation of veins and capillaries, especially in the lower illustration. Dilation of the perivascular lymph sheaths. Ballooning of the intervaginal spaces by the transudate.

region in the trunk of the nerve where the retinal vessels leave the nerve. The edematous infiltration and swelling involve the nerve fibers as well as the interstitial tissue. There is an anterior convex displacement of the glial fibers of the lamina cribrosa, and the physiologic excavation appears to be filled. The entire disc mushrooms into the vitreous, displaces the adjacent retina laterally, and causes it to form fine folds. These folds can be recognized ophthalmoscopically as arcuate stripes concentric with the disc (Paton's concentric lines). The retina adjacent to the disc also shows a slight edema with separation of the fibers of the nerve fiber layer (ophthalmoscopically, there is an increase in the radial pattern of the retina near the disc) and accumulation of exudate between the sensory retina and the pigment epithelium (also contributing to the enlargement of the blind spot in the visual field). The veins and capillaries in the area of the disc are markedly dilated. Hemorrhages on the disc or at the margin appear predominantly in the nerve fiber layer. Ophthalmoscopically, they appear as "flame-shaped" hemorrhages. The subarachnoid spaces of the sheaths of the optic nerve are also filled by the edematous fluid, are markedly broadened, and show some ballooning at their scleral end (Fig. 2-20).

The nerve fibers show characteristic changes as a result of the edematous infiltration. The edema first involves the periphery of the optic nerve and gradually progresses toward the axial area. This process has its functional equivalent in the progressive concentric constriction of the visual field. In the early stages there is swelling of the nonmedullated fibers of the disc, especially those near its margin. They show typical varicose dilations. Within these varicosities the fine neurofibrils, separated by the edematous fluid, can be recognized distinctly. As the papilledema progresses, these varicosities multiply and eventually fill the entire disc. After some time, *degenerative changes develop*—the equivalents of the chronic atrophic papilledema! The neurofibrils within the varicosities disappear and are replaced by fine granules. In the end these granules change to a homogeneous mass. The varicosities lose their connection with the nerve fibers. They are now called "cytoid bodies." These cytoid bodies undergo a lipoid degeneration and finally disappear altogether. In the final stages of chronic papilledema there is an ascending and descending degeneration of the nerve fibers and at the same time a neuroglial proliferation. Ophthalmoscopically, this corresponds to the grayish white discoloration of the disc. The degenerated neural elements are phagocytized by microglial cells and replaced by proliferating fibrous astrocytes. Inflammatory alterations play an insignificant and subordinate role during all of these processes.

Pathogenesis of papilledema

From the multiplicity of theories proposing an explanation for the pathogenesis of papilledema one can draw the occlusion that the mechanism is still controversial and by no means settled. It can be definitely stated that the simple increase in the intracranial pressure is not a sufficient pathogenic mechanism to explain completely the origin of papilledema. That the mechanism may not be the same in every case, even if associated with increased intracranial pressure, is indicated by the variety of clinical appearances of the swollen nerve head. Undoubtedly, an *impairment of the return flow of the venous blood* in the optic nerve due to increased pressure in the cavernous sinus and the ophthalmic veins and the resulting congestion play a major role. It also seems possible that there is a certain blockage of the central retinal vein at the point at which it emerges from the optic nerve and enters the intervaginal spaces. It is also imaginable that the fall in the pressure in the retinal arteries, which Bailliart found at the time of the appearance of the papilledema, might be an additional factor in the impairment of the outflow of the venous blood. There is no doubt that *alterations in the intravascular pressure relationship* may be of importance in the pathogenesis of the papilledema. Whenever the ratio of arterial to venous pressure in the central retinal vessels at the nerve head decreases, the disc shows a tendency to swell. Another fact to be considered is an *interruption of a potential flow of the tissue fluid* that normally occurs along and through the optic nerve from the disc centripetally. Under normal conditions the tissues anterior to the lamina cribrosa are subject to the intraocular pressure, which exceeds the intraorbital and intracranial pressure. This sharp drop in tissue pressure is responsible for such a flow of tissue fluid in an anteroposterior direction along the optic nerve. An increased intracranial pressure tends to change this pressure gradient and thus an opposite forward flow of tissue fluid may lead to papilledema. It also has been suggested that *cerebrospinal fluid is forced under increased pressure from the subarachnoidal space into the corresponding intravaginal spaces* of the optic nerve or its perivascular lymph spaces. It was further assumed that an edematous process in the brain spreads forward to the disc. Papilledema, in this view, thus represents merely a local expression of a general cerebral edema.

Probably there are also *colloidal-chemical factors* involved: the water-binding faculty of the tissue colloids caused by the enlarging brain tumor may be transferred to the optic nerve.

It may also appear simultaneously in the optic nerve as an expression of a diathesis in a generalized edema of the white matter. In the interpretation of the pathogenesis of papilledema such physicochemical and colloid-chemical factors must be considered in addition to those of a purely mechanical nature. Furthermore, one has to realize that the nerve head, much more readily than other parts of the optic nerve, and the nerve fiber layer of the retina have a *special tendency to swell and to produce edema* (probably in connection with the intense capillary vascularization and the special structure of the tissues). Finally, one has to mention that a fundamental factor of papilledema is the *disturbance of the capillary permeability*, which on the other hand, depends on the intracranial lesion. Brain and optic nerve are embryologically and functionally identical tissues and therefore undergo similar reactions. Such a reaction is the *vasodilatation that occurs in the course of an increased intracranial pressure*, a vasodilatation that takes place at the same time not only in the brain but also in the optic nerve and especially in the region of the nerve head where there is an abundance of capillaries. For more details concerning theories on the pathogenesis of papilledema refer to textbooks of ophthalmology and special monographs (Duke-Elder; Brégeat; Bonamour, Brégeat, Bonnet, and Juge; Walsh and Hoyt.)

Differential diagnosis of papilledema

The differential diagnosis is important considering the fact that a *papilledema does not indicate a brain tumor in all patients*. Seventy-five percent of all cases of papilledema are caused by brain tumors. For the remaining 25%, other possible factors must be considered in evaluating a papilledema. It is these 25% that occasionally make a definite diagnosis of a papilledema so difficult, especially in its early stages. These difficulties exist not only as a statistical fiction but also in everyday practice. This is clearly demonstrated by the personal communication of a British ophthalmologist who is consultant for a large neurologic and neurosurgical clinic. During the past 12 years not less than 110 patients were referred with the diagnosis of bilateral papilledema in whom neither a brain tumor nor an increased intracranial pressure could be proved with all the up-to-date methods of examination. In addition to the choked disc caused by increased intracranial pressure not related to a brain tumor, papilledemas resulting from a variety of noncerebral conditions with a striking resemblance to a true choked disc have to be considered. Finally, certain congenital anomalies of the disc have certain ophthalmoscopic aspects that may easily lead to confusion with a choked disc. Thus in the differential diagnosis three groups have to be considered: (1) choked disc due to increased intracranial pressure other than that caused by brain tumor, (2) papilledema of noncerebral origin, and (3) congenital anomalies of the disc resembling choked disc.

Choked disc due to increased intracranial pressure other than that caused by brain tumor

In discussing this first group the term "brain tumor" must be defined more precisely. We limit it to true neoplasms; the percentage of 75% cited previously is based on this premise. As "brain tumor" in the broader sense of the word, the following conditions may naturally cause an increased intracranial pressure and thus a choked disc: *brain abscesses,* occasionally with only an ipsilateral choked disc that would be of localizing value; *tuberculous or syphilitic granu-lomas; epidural and subdural hematomas,* perhaps also *subarachnoidal or intra-cerebral* hematomas if they assume large enough proportions (Fig. 5-2); and quite rarely, *aneurysms, cysticercus cysts, and cerebral phakomas.*

In addition to an increase in the volume of the brain tissue, a raised intra-cranial pressure may also be the result of an increase in the volume of the cerebrospinal fluid. Thus occasionally we find a choked disc in patients with

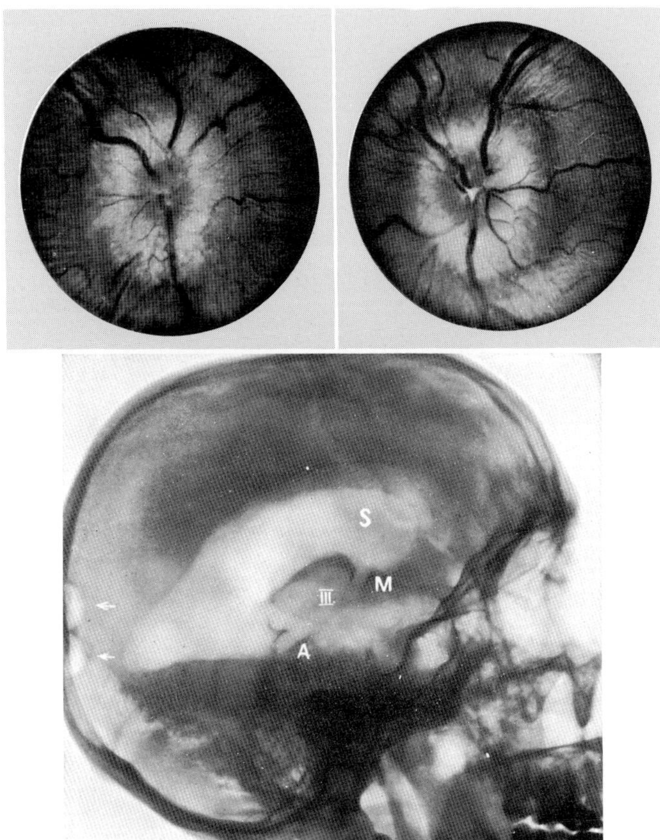

Fig. 2-21. Bilateral papilledema showing beginning atrophy in a 34-year-old patient with in-ternal hydrocephalus secondary to a serous meningitis complicating an attack of flu. Enormous dilation of the disc diameter. Prominence of the right disc, 4 diopters; of the left disc, 3 diopters. Grayish pallor of the discs as a sign of beginning atrophy. Peripapillary retinal edema with radial folding. Veins dilated and tortuous. Hyperemic congestion of the small vessels of the disc, especially on the right side. Diastolic pressure of arteries, O.U. 70 to 80 grams (Bailliart). Visual acuity, O.D. 1.0; O.S. 0.1 (atrophy!). The left illustration corresponds to the right eye; the right illustration, to the left eye. Below, Ventriculogram of a patient with internal hydrocephalus due to occlusion of the Sylvian aqueduct (arrows mark bur holes). **S,** Lateral ventricle; **III,** third ventricle; **M,** interventricular foramina (Monro); **A,** Sylvian aqueduct.

internal hydrocephalus, encephalitis, infectious, serous, or purulent meningitis, syphilitic meningitis, tuberculous meningitis, infectious meningitis secondary to a trauma, further infectious polyneuritis (Guillain-Barré syndrome), poliomyelitis, and infectious mononucleosis (Figs. 2-21 and 2-22).

The rather frequently seen choked discs in patients with septic or aseptic *thromboses of the dural sinuses* are due to an increase in pressure of the cerebrospinal fluid caused by an impairment of the circulation through the main effluent channels of the intracranial circulation (Walsh; Wohlwill; Dill and Crowe; Weber).

Attention should be called here to those somewhat obscure cases in which the patients have bilateral papilledema and increased intracranial pressure, but negative neurologic and general physical findings. Characteristic is the finding of normal-sized ventricles on ventriculography (therefore no signs of obstructing hydrocephalus). A multitude of names has been used for this syndrome, among which are *"pseudotumor cerebri," "*benign intracranial hypertension," "serous meningitis," *"meningeal hydrops,"* and "otitic hydrocephalus." This disease occurs in young to middle-aged adults and is characterized by headaches, sixth nerve palsy, papilledema, occasionally field defects, and rarely blindness. The affection lasts for weeks, sometimes for months, and generally has a good prognosis as long as damage does not result in optic atrophy. The pathogenesis of the syndrome is either a hypersecretion or an obstruction of resorption of cerebrospinal fluid. There are indications, at least in some of the patients, that thrombosis of the sagittal and the lateral sinuses is involved. Otitis media, especially in children, may precede such a thrombosis.

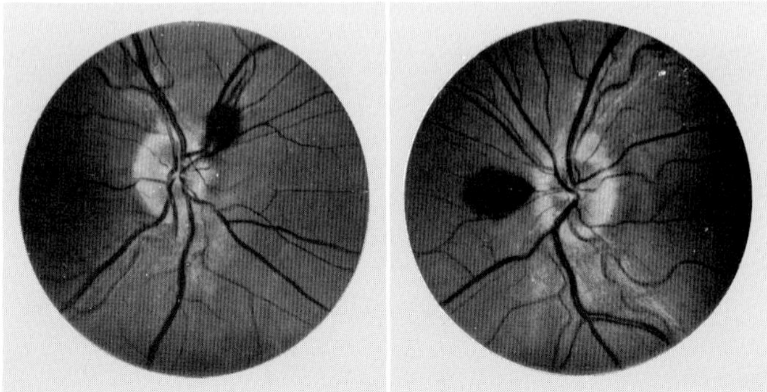

Fig. 2-22. Incipient papilledemas in a 53-year-old patient with internal hydrocephalus due to cerebral arteriosclerosis. The disc margins are blurred throughout. Slight enlargement of disc diameters. Discrete elevation, especially of the nasal disc margin, with minimal deflection of vessels. Minute venous congestion. Large blotches of retinal hemorrhages in the nerve fiber layer on both nasal borders of the discs. Diastolic pressure of ophthalmic arteries, O.U. 80 grams (Bailliart). The left illustration corresponds to the right eye; the right illustration, to the left eye.

Tumors of the spinal cord may be the cause of a bilateral papilledema after blockage of the circulation of the cerebrospinal fluid and formation of a secondary hydrocephalus (Fig. 2-23). Such instances have been described in neoplasms of the cervical portion of the spinal cord near the foramen magnum. However, the tumor may be also situated in the thoracic or lumbar spinal cord (Love, Wagener, and Woltmann). Choked discs have been observed even with *herniated intervertebral discs of the cervical spine;* they disappeared, together with the paresthesias of the hands, when traction was applied to the cervical spine (Girard, Devic, and de Givigney) or surgical intervention performed.

An increase in the intracranial pressure may also result from a dispropor-

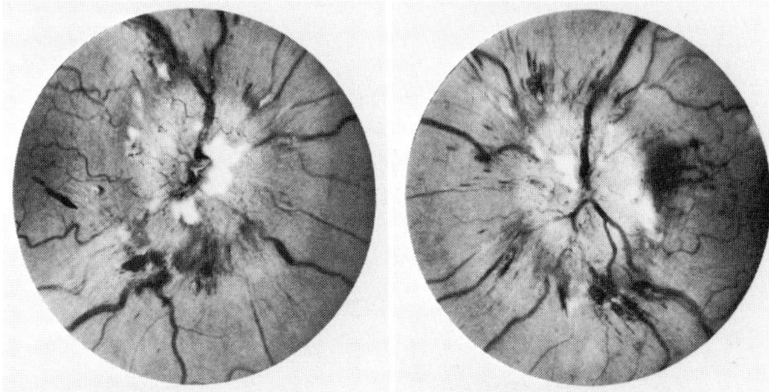

Fig. 2-23. Bilateral papilledema in a 49-year-old patient with reticulum cell sarcoma of the spinal cord (cauda equina) and a concomitant increase of intracranial pressure. Both discs show distinct prominence, enlargement, blurring of the margins, capillary venous stasis, and nerve fiber layer hemorrhages at the disc margins. White "exudates" can be observed among the nerve fibers of the disc and within the retina at the disc margins.

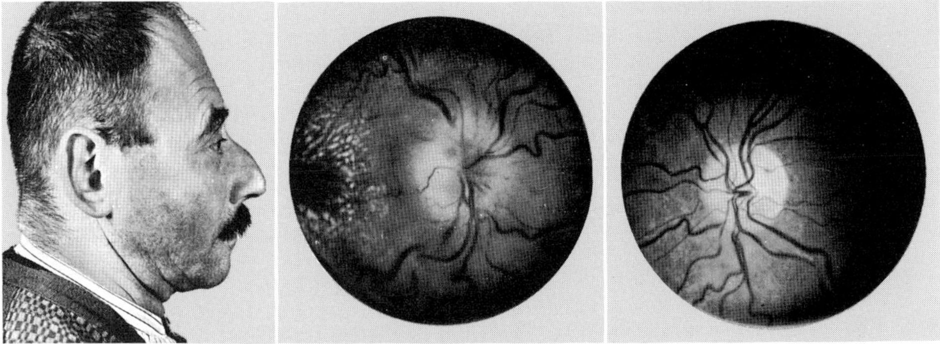

Fig. 2-24. Unilateral papilledema in 53-year-old patient with tower skull. The right disc shows pronounced swelling and prominence with signs of beginning atrophy, venous congestion, macular star, and circumpapillary retinal edema. The left disc shows only slight blurring of the nasal disc margin. It is slightly prominent and somewhat pale. The middle illustration corresponds to the right eye; the right illustration, to the left eye.

tion between the cranial vault and its contents. We have seen a few cases of unquestionable bilateral papilledema in infants with *craniosynostosis* (that is, a premature closure of the cranial sutures). The papilledema and eventually simultaneous atrophic changes of the optic disc (either in unilateral or bilateral form) may be observed with other *malformations of the skull,* especially *tower skull* (oxycephaly), craniofacial dysostosis (Crouzon's disease), basilar impression, and Paget's disease (Fig. 2-24).

The differential diagnosis of cases in the group just discussed is made purely on an etiologic basis. The common mechanism for all of them is increased intracranial pressure. The resulting ophthalmoscopic picture is more or less identical with that of a choked disc caused by a brain tumor and need not be described further. The same principles described for the incipient, the fully developed, and the chronic atrophic papilledema can be applied to these cases.

Papilledemas of noncerebral origin

When increased intracranial pressure has been ruled out as an etiologic factor, we no longer apply the term "papilledema" but use the noncommittal term "edema of the papilla" in accordance with the terminology outlined previously. A first group includes *inflammatory* edemas and a second group includes *vascular* edemas of the papilla.

Inflammatory group. The inflammatory edemas include papillitis, juxtapapillary chorioretinitis, and isolated tuberculosis of the nerve head.

Papillitis (intraocular optic neuritis). In the differential diagnosis of every case of papilledema a *papillitis* (that is, an inflammation of the nerve head or the immediate area surrounding it) must be considered first of all. First it should be emphasized that *it is impossible to differentiate a papillitis from a papilledema only with the aid of the ophthalmoscope.* In both instances an edema of the papilla manifests itself by blurred margins, an increased disc diameter, and a prominent nerve head. Common to both also are hemorrhages and white dots on the papilla or its margin (Fig. 2-25). Dense opaque intraretinal punctate exudates deposited around the disc and in the macula (where they may aggregate into a macular star figure) characterize a particular type of papillitis common in children with signs of acute monocular visual loss (Hoyt and Beeston).

Of great importance in the differential diagnosis is the examination of the disc and the posterior pole of the eye using the *slit lamp,* preferably with the aid of the fundus contact glass. There are prepapillary vitreous changes that undoubtedly speak in favor of papillitis: vitreous haze (increased prepapillary Tyndall phenomenon) and prepapillary and preretinal inflammatory cells, eventually aggregating as precipitates at the posterior surface of a detached vitreous. In certain cases of papillitis the anterior uvea may participate and manifest a positive Tyndall phenomenon or even pathologic cells in the aqueous.

In the fluorescein angiogram, papillitis shows a characteristic vascular pattern of the disc (Fig. 2-26). This can be of great importance in the differential diagnosis since it differs from the angiographic picture of papilledema and pseudopapilledema. In papillitis we see a regular meshwork of radial capillaries connected by anastomoses covering the disc. This network extends beyond the disc margin and goes far into the retina, especially in the area of the large vessels. The capillary stasis, which is so conspicuous in papilledema (Fig. 2-9), is absent or only minimal in papillitis, perhaps confined to one segment. In the venous phase the disc tissue will also fluoresce. This fluorescence, however, is

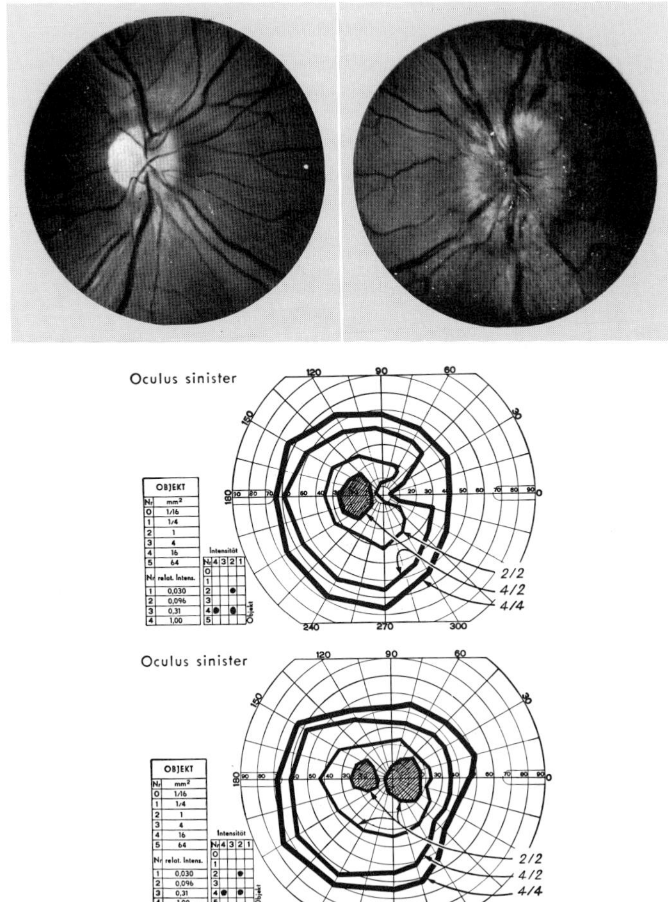

Fig. 2-25. Left acute papillitis (intraocular optic neuritis). Increased diameter of the disc. Blurred disc margin. Grayish yellow edema of the disc. Pronounced prominence of about 2.5 diopters. Deflection of the vessels at the disc margin. Slight venous congestion. Hemorrhages at the 1 o'clock position of the disc margin. This picture is very similar to that of a choked disc! The left central visual acuity is reduced to 0.06 The left illustration corresponds to the right eye; the right illustration, to the left eye. Originally the visual field showed a nasal sector-shaped defect in addition to enlargement of the blind spot (upper field). One week later (lower field) a massive central scotoma had developed.

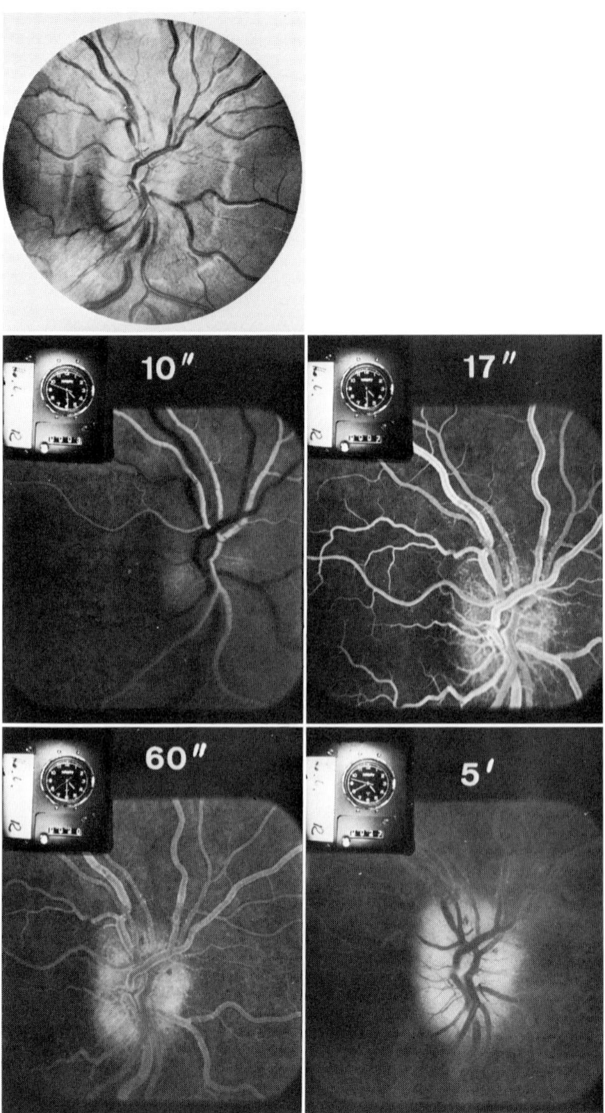

Fig. 2-26. Fluorescein angiogram of the fundus and disc of a patient with acute papillitis. In the arterial phase (10 seconds) no capillary stasis is visible. There is diffuse fluorescence of the papilla in the venous phase (17 seconds). In the late venous phase (60 seconds) there is still fluorescence of the disc, but no capillaries can be detected. This angiographic pattern is in distinct contrast to that of papilledema (Fig. 2-9) in which the capillary stasis, together with aneurysmatic dilatations of the capillaries, represents one of the most pathognomonic signs.

never as intensive as in papilledema. At the same time, it lacks the conspicuously visible capillaries.

True, the prominence of the edema of the papilla, as a rule, is less distinct in papillitis than in papilledema. Also, the dilation of the veins, their tortuosity, and the degree of their deflection at the disc margin are more pronounced in a choked disc than in an inflammation of the nerve head. However, such ophthalmoscopic details are deceptive, especially in the early stages, where such distinctions are missing altogether. *Unilaterality*, in many instances, will weigh in favor of a papillitis. Yet we personally observed several cases of bilateral papillitis (which occurred simultaneously) whose appearance was indistinguishable from a choked disc. Nevertheless, a simultaneous bilateral papillitis is a rare event. In our experience the onset of the inflammation in the second eye is observed much more frequently after an interval of several weeks or even months. The ophthalmoscopic examination may reveal a receding stage in one eye and an incipient stage in the other eye. It is relatively easy to differentiate a papillitis from a papilledema by testing the visual functions. In most instances one can follow the rule that there are *no disturbances of the visual function in papilledema in contrast to papillitis.*

These disturbances involve the central visual acuity as well as the visual fields. It has already been mentioned that, apart from the attacks of obscuration and the enlargement of the blind spot, the visual acuity and the visual field remain intact for weeks and even months in papilledema. The patient is unaware of his papilledema! The situation is entirely different with a papillitis. Perhaps even before fundus changes can be recognized ophthalmoscopically, the patient complains of a sudden foggy, blurred, or veiled vision that, in contrast to the transient obscuration of papilledema, is permanent. Within hours or a few days it may lead to a *progressive loss of vision and possibly to a complete amaurosis.* Actually, this rapid deterioration of the visual function is pathognomonic for a papillitis, which, due to its inflammatory nature, causes an early disturbance in the conductivity of the nerve fibers. Only in rare cases does papillitis develop without impairment of central vision (leaving the macular bundle of the optic nerve intact). In the early stages, *pain in or behind the eye, especially on lateral movement, is characteristic for a papillitis.* The disturbance in the conductivity of the optic nerve may also be demonstrated by means of the so-called *Marcus Gunn pupillary phenomenon.* As a result of the impaired pupillomotor stimuli, the pupil on the involved side will be distinctly wider if the other eye is occluded than the pupil on the opposite side if the involved eye is occluded (p. 26). Such a sign can never be demonstrated in a papilledema—unless there is already evidence of a beginning atrophy.

Visual field defects are of cardinal importance in the differential diagnosis. The edema of the papilla and the immediate area surrounding it frequently causes enlargement of the blind spot and a disturbance of the law of summation for the surrounding isopters (as in papilledema). Important and decisive, however, is the *demonstration of the early occurrence of central and paracentral scotomas.* Their form is not typical and is quite variable. Due to an early impairment of the papillomacular bundle, these scotomas usually are of the *cecocentral* type: they originate from the point of fixation and tend to merge with the blind spot. A number of authors still stress the importance of a *scotoma for red* for the early diagnosis of such disturbances of conductivity. With the more refined methods of perimetry, but especially on the tangent screen, a carefully performed examination should demonstrate similar defects for small white targets even during these stages. The scotomas may also show an arcuate shape and originate from the upper or lower pole of the blind spot, expanding toward the periphery to merge with sector-shaped or quadrantic peripheral defects. This phenomonon is the so-called "breaking through" of a central scotoma. We have also seen central and even cecocentral scotomas in cases of papilledemas. Here they are the result of an edema extending into the macular area. They occur only in the more advanced stages. They are relative scotomas, according to Traquair, especially for blue.

In addition to the central scotomas, the early occurrence of a *peripheral sector-shaped or quadrant-shaped field defect,* indicating damage to the peripheral nerve fibers, is typical for a papillitis and may precede the loss of central visual acuity. These defects may even break through toward the central field and merge with central or paracentral scotomas to form one large defect. In order to give a complete description of these characteristic disturbances of the visual function, it should be added that there may be a *complete or almost complete recovery of the central vision and the field defects* within a few days or weeks. This is in contrast to the defects found in chronic atrophic papilledema. This is not the place to go into detail concerning the numerous etiologic factors that may be responsible for a papillitis except for a reminder that a great percentage of all cases are due to multiple sclerosis. We do want to stress that an acute or chronic *iridocyclitis* is occasionally accompanied by a papillitis, which may resemble a choked disc. We have personally observed such cases that had been referred as a suspected brain tumor—primarily because of an alarming unilateral loss of vision.

Electrophysiologic examinations are an additional aid in the differential diagnosis of choked disc and papillitis. In papillitis there is a distinct decrease of frequency of subjective fusion, an increase of the retinocortical time, an absent or delayed specific cortical response on illumination of the affected eye, and sometimes a supernormal ERG.

In summary, it can be stated that every case of edema of the papilla— regardless of how much the ophthalmoscopic appearance resembles a papilledema—must be interpreted as a papillitis if there are early disturbances of the central vision, central scotomas, or sector-shaped or quadrant-shaped visual field defects or both.

Juxtapapillary chorioretinitis (Jensen). This is a *peculiar type of chorioretinitis,* a simultaneous inflammation of the choroid and retina, that does not involve the entire fundus but shows a predilection for the immediate surroundings of the disc and may come in actual contact with it and the nerve fibers. There is an *edematous swelling of the papilla* and its immediate surroundings that could be mistaken for a papilledema or a papillitis. The similarity with the latter can be even more striking because damage to some nerve fiber bundles may produce either a central scotoma or a sector-shaped field defect. The parapapillary lesion has a moldy grayish green color in its early stages, which changes to a bright yellow and finally becomes white with vicarious heaping of black pigment. This typical picture of a choroidal atrophy, together with its location, should help to clarify the situation. There may be other similar foci, either in the acute or in more advanced stages (the latter pigmented!), scattered over the fundus that will also reveal the true nature of this lesion. More specific diagnostic signs such as posterior vitreous opacities and keratic precipitates ("KP's"), cells, and an aqueous flare in the anterior chamber are of interest only to the ophthalmologist. They may aid greatly in establishing the diagnosis. Toxoplasmosis seems to be a frequent cause of this type of chorioretinitis. We know from personal experience that juxtapapillary chorioretinitis is not only of theoretical interest in the differential diagnosis of the papilledema. We have seen a patient who had been referred because of a suspected unilateral papilledema. The ophthalmoscopic findings of a subsequently developing atrophic parapapillary choroidal scar that showed an extensive marginal accumulation of pigment enabled us to make the proper diagnosis (Fig. 2-27). Sometimes juxtapapillary chorioretinitis, even after healing, may lead to a chronic disc swelling (reactive glial tissue growth) that resembles congenital pseudoneuritis or pseudopapillitis (Hoyt and Beeston).

Juxtapapillary chorioretinitis has a typical fluorescein angiogram. Corresponding to the clinically visible edema, the disc and the focus close to the disc fluoresce even before the dye enters the retinal vessels. The retinal vasculature fills and empties the dye in a normal fashion. The fluorescein in the disc and the peripapillary area increases and the peak of fluorescence occurs in the center of the inflammatory focus. This results in an irregular area of fluorescence lasting for a considerable period of time. This area shows a peculiar feathery, blurred margin and encompasses the entire clinically visible edema, including the disc.

Isolated tuberculosis of the nerve head. This is a rare condition which, however, may be easily mistaken for a papilledema. There is a more or less pronounced edema of the papilla. In contrast to a true papilledema, it is of a striking whitish, spotty appearance. Furthermore, the swelling is quite uneven, with nodular prominences of different sizes. The papillary border

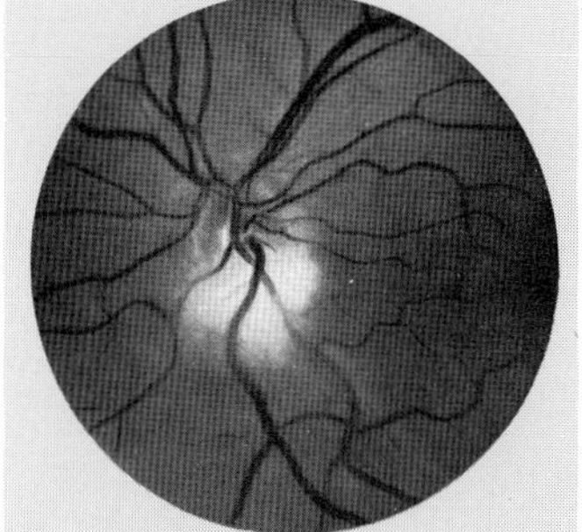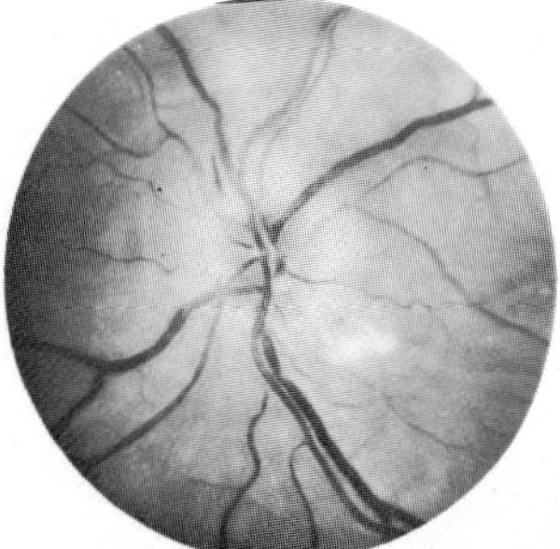

Fig. 2-27. Juxtapapillary chorioretinitis (Jensen) with disc edema. The left photograph shows a small chorioretinitic patch that extends to the lower part of the disc. The right photograph shows a diffuse edema of the disc with venous engorgement, small linear hemorrhages in the nerve fiber layer between 5 and 6 o'clock, and as a probable cause, a chorioretinitic patch at the inferior temporal margin (marked loss of vision, sectorlike defect in the visual field).

is conspicuous by its uneven outline. There may be tonguelike projections of the edema into the retina. The disc is covered by a veillike membrane with several perforations that allow visualization of more or less indistinct parts of the nerve head in addition to sharply outlined areas. The veins are dilated. There are numerous hemorrhages on the disc and in the retina. Characteristic is a rapid breakdown of the visual functions. Duke-Elder considers the diagnosis of a tubercle of the nerve head very difficult, even in the presence of a simultaneous granulomatous iridocyclitis or an active tuberculosis elsewhere in the body. He observed several cases where enucleation became necessary. Only the histologic examination of the globe revealed the true nature of the condition.

We have the record of a patient with Boeck's sarcoid who showed abnormalities of the disc very similar to papilledema. Boeck's nodules buried in the depth of the disc may have been responsible for this misleading ophthalmoscopic picture. Sometimes the disc forms one yellowish white mass with an irregular surface, pronounced neovascularization, and a prominence of 5 diopters or more. The differential diagnosis of a real tumor or a tuberculoma of the nerve head may be extremely difficult.

Vascular group. The vascular edemas include malignant hypertensive retinopathy, occlusion of the central vein, ischemic papillitis, edema of the disc in emphysema, and edema of the disc in anemias and leukemias.

Malignant hypertensive retinopathy. Aside from papillitis, edema of the papilla in patients with malignant systemic hypertension probably is the most important and most common entity to be considered in the differential diagnosis of choked disc (Fig. 2-28). One of the most important reasons is that the *edema of the disc in patients with hypertensive retinopathy practically always occurs bilaterally.* A number of systemic symptoms common to both conditions (for instance, headache, vomiting, vertigo, and cerebral vascular accidents causing hemiplegias, aphasias, and other neurologic complications) are responsible for edema of the papilla in patients with malignant hypertensive retinopathy too frequently being diagnosed as a choked disc. Quite frequently patients suffering from malignant hypertension with bilateral edema of the papillae are referred to the department of neurosurgery as having suspected brain tumors. Reasons for these referrals are based either on the ophthalmoscopic findings alone or on additional cerebral complications.

Disc swelling in malignant hypertensive retinopathy may closely resemble acute or chronic papilledema in brain tumor: under both circumstances, increased diameter of the papilla, blurring of the disc margins, prominence of the nerve head, deflection of the vessels at the disc margin, exudates, and hemorrhages within the papilla or the adjacent retina may occur. In contrast to papilledema in brain tumors, the disc swelling in malignant hypertension manifests a lighter, even anemic color, but no capillary dilatation and no telangiectasias. The retinal veins seldom appear significantly dilated in disc edema due to malignant hypertension. Narrowing of the arteries and arterioles occurs only in late atrophic stages of papilledema, but is a constant and important sign in all stages of hypertensive edema of the papilla. The elevation of the optic disc is not a criterion of diagnostic value: it can go as far as 6 diopters in hypertensive retinopathy! Like papilledema, hypertensive disc swelling after appropriate treatment may disappear completely and may, if left untreated for long

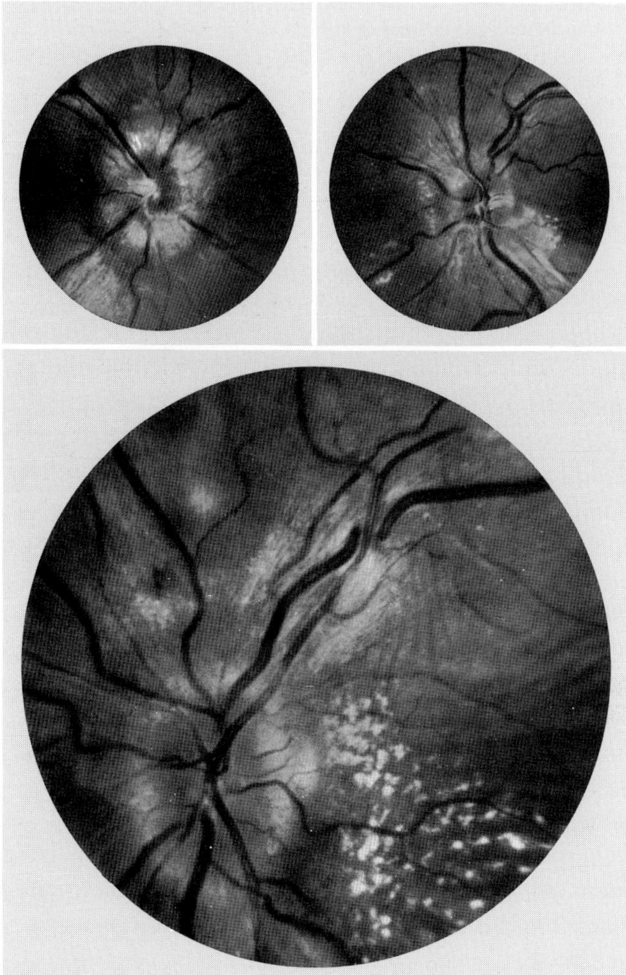

Fig. 2-28. Edema of the papilla in a patient with malignant hypertensive retinopathy (severe systemic hypertension). Disc swelling similar to chronic papilledema. Prominence and edematous imbibition of the discs with enlargement of their diameter. Blurring of disc margins and deflection of the retinal vessels at the disc margins. Slight congestion and tortuosity of the veins. The arteries show extensive attenuation with irregularities in their caliber; in places they are threadlike with silvery reflex stripes (silver wire arteries). Distinct Gunn's signs at the arteriovenous crossings. Peripapillary retinal edema, with patches of ischemic cloudiness of the retina. Soft, fluffy, white cotton-wool exudates in addition to a yellowish white foci of lipoid and fatty nature, especially in the macular region. Isolated streaklike and punctate hemorrhages. Diastolic pressure of ophthalmic arteries, O.U. 120 grams (Bailliart)! The left illustration corresponds to the right eye; the right illustration, to the left eye.

periods, also lead to severe optic atrophy. In contrast to papillitis, hypertensive disc swelling and papilledema can be differentiated with the ophthalmoscope.

The diagnosis should be based on the *unmistakable vascular* changes resulting from hypertonia and the renal damage as well as the secondary retinal lesions representing the *retinopathy* proper. These pathologic signs are missing in patients with papilledema but are absolutely pathognomonic in those with malignant hypertension. The vascular changes apply primarily to the *arteries* and consist of generalized *narrowing* of caliber with considerable localized irregularities. The angiospasm may become so severe that the arteries can scarcely be recognized. The increasing sclerosis of the intima and the proliferation and hypertrophy of the media cause marked thickening of the vessel wall, with narrowing and obscuring of the blood column. The wall reflexes increase in brightness till the stage of *silver wire arteries* or even segmental obliteration is reached. Because of the rigidity and the increased tension of the arteries, *Gunn's sign* will appear at the arteriovenous crossings: the vein appears indented at the point of the crossing, with a tapering of both the distal and proximal adjacent sections; there may even be some arching (Salus' sign). In even more advanced stages the angiospasm may extend to the veins, with caliber variations if they are only partially involved. The retinal changes proper, which are only rarely missing (if so, perhaps only in the initial stages) in patients with malignant hypertension, are just as important in making the diagnosis. The increased arterial pressure leads to extravasations of varying sizes and forms, mostly near the vessels. In contrast to the choked disc, the *hemorrhages* are not only on or near the disc margin, but quite characteristically extend way out into the fundus periphery. Some of the retinal changes are soft, grayish white, fluffy *cotton-wool* exudates in the superficial retina which, in contrast to the white spots of the papilledema, do not appear on the disc but in the surrounding retina. There are also yellowish white punctate exudates deep within the retina that have a fatty and lipid nature, that show a predilection for the macular area, and that appear as sharply outlined, pointed, brilliant dots. Frequently they form a macular star or show a wreathlike arrangement around the macula (circinate retinopathy). If such a macular star is incomplete, it may quite closely resemble the fan-shaped arrangement of the intraretinal droplets seen in papilledema (Fig. 2-8).

Ophthalmodynamometry is quite helpful for differentiating papilledema and edema of the papilla in hypertensive retinopathy. It has been stressed already that the blood pressure of the ophthalmic arteries in fully developed papilledema shows either a normal or even an abnormally low value (p. 110). *In malignant hypertension an abnormally and even discordantly high diastolic pressure in the ophthalmic arteries is a constant sign.*

In summary, it can be stated that *edema of the papilla in hypertensive retinopathy* (although it closely resembles a choked disc) *can be differentiated with some degree of certainty from papilledema on the basis of the characteris-*

tic vascular and retinal changes. Sometimes the retinal changes and the hemorrhages may be minimal or even absent in spite of pronounced disc swelling. In these cases the presence of the characteristic hypertensive arteriolar changes will help to distinguish disc edema in malignant hypertension from papilledema due to brain tumor. One still has to consider the possibility of the coexistence of malignant hypertension and brain tumor! If the ophthalmoscopic findings leave any doubt, the findings of a general physical examination, including blood pressure, urinalysis, nonprotein nitrogen determination, and tests for kidney function, should confirm the diagnosis. We recall a number of hemiplegic patients in whom the question of whether or not there was a papilledema due to cerebral tumor or due to malignant hypertension had to be considered quite seriously. In every patient the exact evaluation of the ophthalmoscopic findings, especially proof of arterial and retinal changes, clarified the situation and eliminated the need for neurosurgical intervention.

Together with hypertensive malignant retinopathy, some quite similar ophthalmoscopic pictures should be mentioned—for instance, in *Kimmelstiel-Wilson disease* in patients with diabetes, *Fahr's malignant sclerosis, retinopathy of pregnancy, and in secondary renal hypertension after chronic glomerulonephritis, pyelonephritis, amyloidosis, and periarteritis nodosa.* Frequently in such patients we find pronounced edema of the disc resembling a choked disc. In the differential diagnosis the same considerations as in malignant hypertensive retinopathy are valid.

Occlusion of the central vein. The process usually is unilateral. A blockage of the venous drainage may at times cause quite a severe edema of the disc, with prominence of the latter and numerous other signs that are common in papilledema. The diagnosis is not difficult if

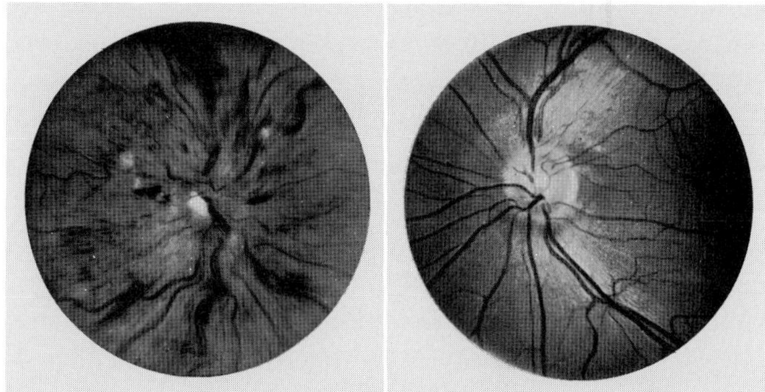

Fig. 2-29. Right central vein occlusion in a 41-year-old patient. The right papilla shows edematous cloudiness, swelling, and blurring of the entire disc margin with an insignificant prominence of 0.5 diopter. Edema of the surrounding retina. Veins markedly dilated and engorged with an enormous tortuosity, in some areas protruding over the level of the retina. Massive hemorrhages on the disc extending into the fundus periphery. Isolated white exudates on the disc and in the retina. The left illustration corresponds to the right eye; the right illustration, to the left eye.

one is cognizant of the *enormous venous congestion* and the distinct corkscrewlike tortuosity of the veins, but most of all the *massive extravasations extending far into the periphery of the fundus.* The dominant factor in the ophthalmoscopic picture of an occlusion of the central retinal vein is the hemorrhages over the swollen disc and the surrounding retina, which mostly show a radial arrangement. Also typical and unmistakable are the immediate disturbances of the visual function caused by macular hemorrhages and interference with the arterial circulation (Fig. 2-29). A chronic disc swelling resembling unilateral papilledema and sometimes persisting for months may result from a primary inflammatory central vein occlusion (papillophlebitis) that occurs in young adults and manifests only minimal loss of vision. Characteristic is the enormous dilation and tortuosity of the veins and the presence of retinal hemorrhages in and around the disc and especially along the veins. The affection leaves discrete partial sheathing of the veins and sometimes newly formed venules on the surface of the disc, which may resolve the edema completely (Hoyt and Beeston).

Edema of the disc seen in hemorrhages into the sheaths of the optic nerve after fractures of the base of the skull (especially if the optic canal is involved) should also be mentioned. Its probable cause is an impediment of the venous circulation. We have observed one case. The appearance of the disc differs little from that of a choked disc (Fig. 2-30). There are, however, early disturbances of the conductivity of the involved optic nerve. In some patients there is an immediate or early amaurosis following the trauma. After a few weeks there is invariably an optic atrophy with partial or complete loss of vision, unless such an event has been prevented by neurosurgical interference. In most cases, however, a hemorrhage into the sheaths of the optic nerve does not produce an edema of the disc but merely functional disturbances of the optic nerve with no primary changes of the disc. Only after weeks is there a primary optic atrophy with a sharp outline of the disc margin.

To the vascular group of disc swellings of noncerebral origin also belongs *ischemic papillitis* (apoplexia papillae, vascular pseudopapillitis, ischemic edema of the disc), which occurs especially in elderly patients with arteriosclerosis and hypertension but sometimes also in younger individuals with collagen vascular disease (polyarteritis), temporal arteritis, or carotid occlusion. The disc swelling resembles that of ordinary inflammatory papillitis, but there are some characteristics that cannot be overlooked and help to confirm the diagnosis: the color of the disc is pale from the beginning and changes within a few days to a whitish, nearly avascular pallor; feathery hemorrhages between the nerve fiber bundles; no capillaries

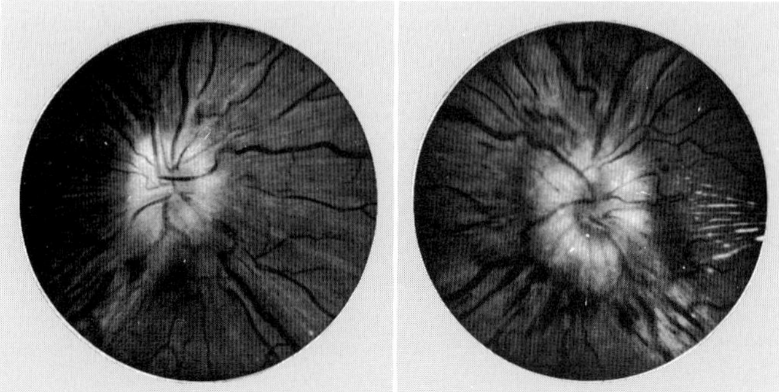

Fig. 2-30. Bilateral papilledema with beginning atrophy in a 19-year-old patient with hemorrhage into the sheaths of the optic nerve due to fall on a concrete floor. Contusion of the brain. Conspicuous pallor of the edematous, swollen discs. Radial folding of the surrounding retina. Beginning left "macular star." Severe impairment of the sensory functions. Reduced visual acuity to 0.1 O.U. Bilateral concentric contraction of the visual fields. The left illustration corresponds to the right eye; the right illustration, to the left eye.

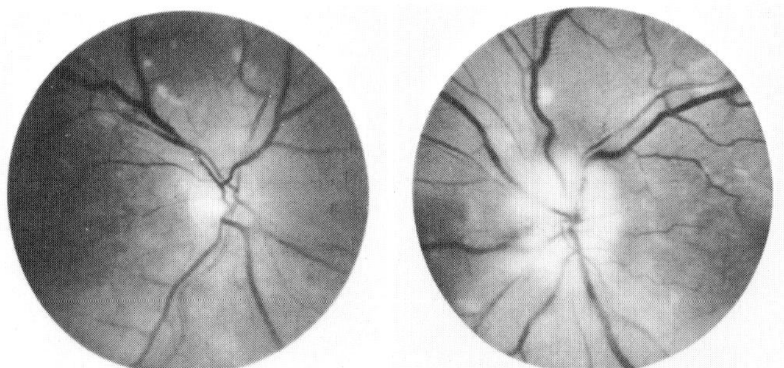

Fig. 2-31. Left ischemic "papillitis" in an 85-year-old patient with general arteriosclerosis. Pale avascular swelling of the disc with no capillary stasis, narrowed sclerotic arteries, slight prominence, and completely blurred margins (infarction of the nerve head). After a few weeks the edema disappeared, leaving a snow-white disc; visual acuity remained at the original level of 0.1.

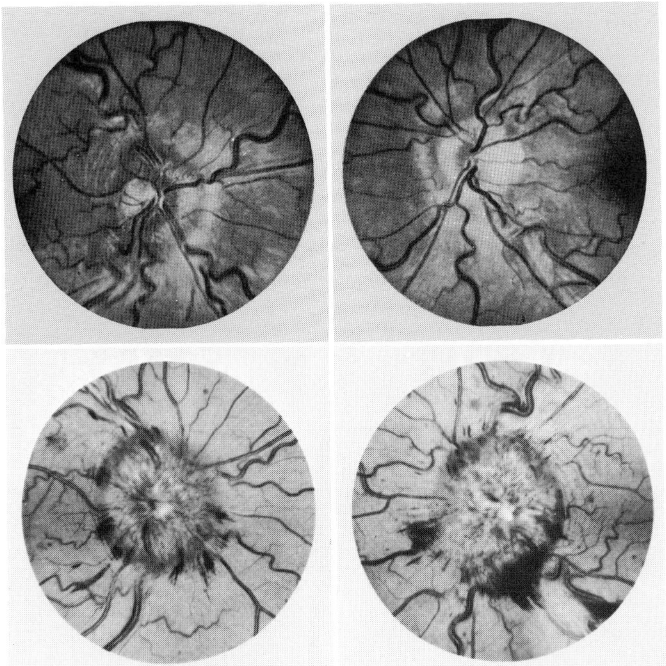

Fig. 2-32. Upper photographs show edema of the discs in a 40-year-old patient with bronchiectasis and cardiac insufficiency (marked cyanosis and polycythemia). Lower photographs show massive disc swelling in a 32-year-old patient with chronic pulmonary heart disease due to bronchial asthma. Note striking resemblance to papilledema due to increased intracranial pressure (capillary stasis, venous engorgement, hemorrhages, white "exudates," etc.)

visible on the disc; distinct narrowing of the arterioles at their exit from the disc; and moderate dilation of the major veins (Fig. 2-31). Ischemic papillitis, the result of total or segmental infarction of the optic nerve head (peripapillary circle of Haller), usually produces rapid impairment of central vision and fasicular visual field defects (especially inferior quadrantanopia or inferior altitudinal hemianopia), which shows a poor or no chance of recovery. Bilateral forms of ischemic papillitis may occur, most often in the sense that the second eye becomes involved after an interval of some weeks or months or even later. The disc swelling on the one side and the optic atrophy on the other side (as a product of the earlier attack, when the edema of the nerve head has subsided) may produce a *pseudo-Foster Kennedy syndrome.* The leading symptoms, described previously (narrowed arterioles in both fundi, visual loss in both eyes, the field defects indicating bilateral optic nerve involvement, and also the history of the affection), speak against a brain tumor genesis. Thus by careful ophthalmologic examination, further neuroradiologic examination or even neurosurgical intervention can be avoided (Bonamour, Brégeat, Bonnet, and Juge; Rintelen).

Edema of the disc in emphysema. There have been reports of cases of emphysema in patients with chronic bronchitis (usually during the stages of cardiac insufficiency) that show a bilateral edema of the disc, together with an increased pressure of the cerebrospinal fluid (Simpson; Cameron). The appearance of such cases is similar to that of choked discs. We observed such a picture in one of our patients with a similar condition (Fig. 2-32). The cause of the elevated pressure of the cerebrospinal fluid supposedly is a decrease in the concentration of oxygen and an increased concentration of carbon dioxide in the blood (hypercapnia), which may cause dilation of the cerebral veins.

Based on a similar mechanism are the edemas of the disc (usually not very pronounced) seen occasionally in patients with congenital or acquired *valvular defects of the heart* with a severe venous congestion. The enormous dilation of the veins as well as the unmistakable cyanosis of the retina should make the diagnosis easy (Fig. 2-32).

Edema of the disc in anemias. We have seen a case of bilateral edema of the papilla resembling choked discs in a patient with acute anemia following a hemorrhage. This hemorrhage resulted from anticoagulant therapy. No signs of increased intracranial pressure could be demonstrated. Similar edemas of the disc have been described in leukemias, chlorosis (Ohashi; Bondi; Hegner; Jaensch; Huber; and others), macroglobulinemias, and polycythemias.

The disc swelling with *leukemias* manifests three different forms which, with regard to the ophthalmoscopic appearance, have no special characteristics. All types and grades of

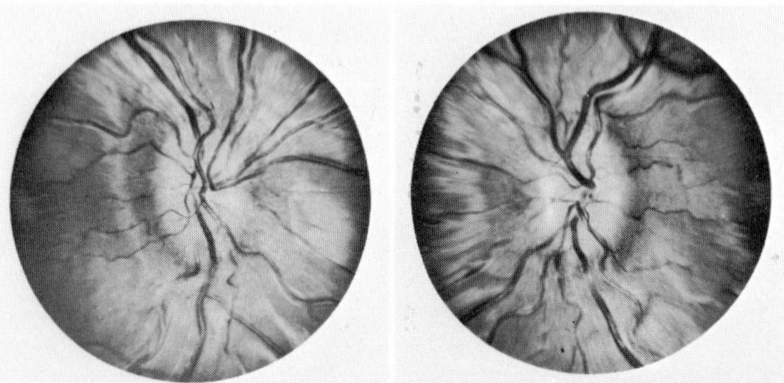

Fig. 2-33. Papilledema (bilateral) in a 12-year-old boy suffering from leukemia (leukemic meningitis with increase of intracranial pressure). Prominence, blurring, venous engorgement of both discs, but no signs of leukemic retinopathy. The latter finding favors a diagnosis of true papilledema due to increased intracranial pressure and not local infiltration of disc tissue by leukemic cells. The left illustration corresponds to the right eye; the right illustration, to the left eye.

swelling may be observed. The first form represents the disc swelling accompanying leukemic retinopathy; it is characterized by scattered retinal and preretinal hemorrhages, an edematous retina, and typical venous engorgement. The second form represents local infiltration of disc tissue by leukemic cells and is accompanied by more or less pronounced visual defects as a sign of affection of the nerve fibers. The third form is a real papilledema due to increased intracranial pressure in connection with a "meningoneuroleukosis" that develops after treatment with antimetabolites, even in the phase of clinical and hematologic remission of the disease (Huber; Hamard) (Fig. 2-33).

Congenital anomalies of the disc resembling papilledema

Neither the ophthalmoscopic appearance nor the pathologic-anatomic findings indicate an edema of the disc or papilledema. There are only certain similarities in the morphologic appearance of the optic disc, which anybody acquainted with these anomalies should recognize as being different from papilledema. Unfortunately, *these elevated disc anomalies are sometimes confused with real papilledema, and thus the suspicion of a brain tumor may lead to unnecessary and unjustified diagnostic procedures or even neurosurgical interventions.* On the other hand, one must not forget that there is always the possibility of a coexistence of such congenital elevated disc anomalies and a brain tumor (Fig. 2-34); however, in such cases the general nonocular symptomatology will most often help make the right diagnosis. In doubtful cases in which a neurosurgical investigation is not yet indicated because of lack of neurologic signs, examining the patient at regular time intervals and if possible taking fundus photographs, possibly combined with fluorescein angiography of the retina, are recommended.

Drusen of the optic disc ("hyaline bodies"). The deposition of granular hyaline-like substances in the disc may lead to a swelling, enlargement, and

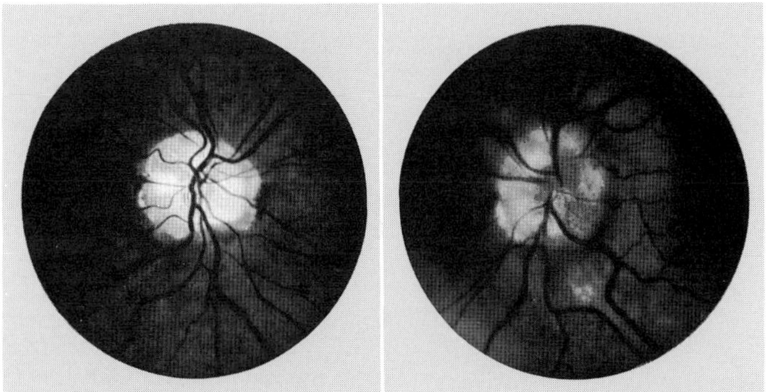

Fig. 2-34. Combination of drusen of the discs with papilledema; giant tumor of the right side of the frontal lobe in a 37-year-old patient. The right disc shows a few exposed drusen in the lobulated surface of the atrophic disc. The left disc is edematous, prominent, and enlarged, with isolated hemorrhages at the disc margin. Absence of an edema on the right side and choked disc on the left side represent a type of Foster Kennedy syndrome! The left illustration corresponds to the right eye; the right illustration, to the left eye.

prominence of the papilla that could be mistaken for a papilledema. This is especially true in those cases in which the hyaline drusen are situated in the depth of the disc and cannot be recognized with the ophthalmoscope (Fig. 2-35). An occasional enlargement of the blind spot, arcuate scotomas, or peripheral, mostly nasal field defects or obscurations may contribute to the possibility of such a mistake. Also, these drusen of the optic disc usually involve both sides (Fig. 2-14). The following observations should help to overcome some of the diagnostic difficulties.

In distinct contrast to the reddish discoloration of papilledema, drusen give the disc a yellowish color. There is a complete absence of venous and capillary dilation, exudates, and hemorrhages*—an important and absolutely reliable sign. As a result of the deposition of the granular hyaline substances, the surface of the disc frequently assumes an irregular, nodular appearance, which is

---

*Only exceptionally will papillary or peripapillary bleeding occur.

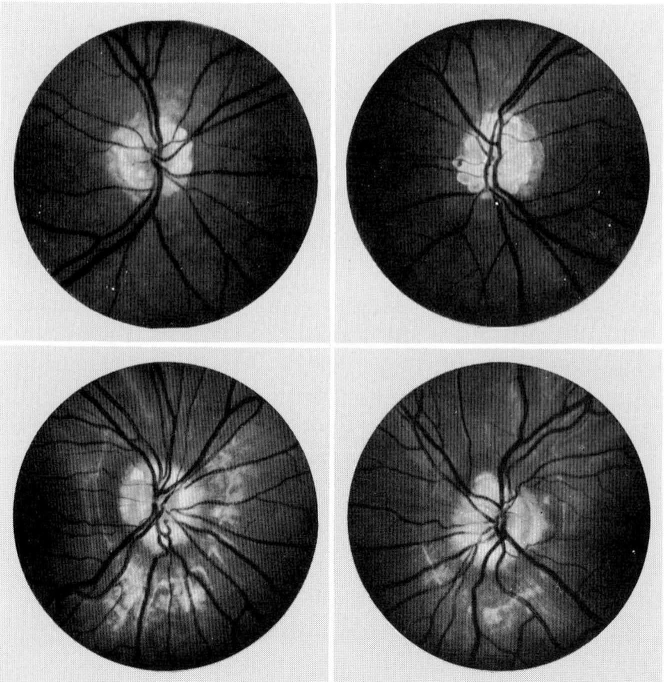

Fig. 2-35. Bilateral drusen of the discs. The upper photographs show exposed drusen that appear as yellowish white, sagolike granules and cause a lobulated surface of the disc. In the lower photographs the drusen are buried in the depth of the discs except for a few superficial ones. Swelling, enlargement, and slight prominence of the discs. Absence of venous congestion! (See Fig. 2-14.) The left illustrations correspond to the right eye; the right illustrations, to the left eye.

especially pronounced if isolated drusen extend to the disc margin and lie exposed on the *surface* of the disc. *In such a case the disc margin shows nodular excrescences, with the drusen being visible as glassy, yellowish white, sago-like grains—findings usually sufficient to make the correct diagnosis.* With several drusen at the disc margin, the nodular appearance of the surface is characteristically described as "mulberry-like." If one suspects drusen, a search should always be made for such superficial translucent hyaline grains, which are usually at the disc margin. Both fundi should always be examined, for the drusen may be completely hidden on one side but may be quite superficial on the other side. By means of posterior slit-lamp microscopy and with the use of a contact lens, it is occasionally possible to visualize hidden drusen of the disc. These drusen of the optic disc are frequently hereditary (irregular dominance).

The hyaline bodies of the optic disc can usually be recognized with fluorescein angiography. The intensive primary fluorescence of these drusen, which can be seen in the preliminary photographs taken before the injection of the dye, is important (Fig. 2-36). Drusen do not take the dye, but show their sharp polygonal outlines during the entire angiogram. The hypoplastic glial tissue of the disc may take the dye to a moderate degree, usually with irregular intensity. This, however, is never as intensive or as long lasting as the fluorescence that occurs in papilledema or in papillitis. There is absolutely no capillary stasis. Deep-lying drusen are often better visualized with fluorescein angiography than with the usual ophthalmoscopy.

Drusen of the optic disc are also seen in tuberous sclerosis (hyaline bodies also in the retina!), optic atrophy, pigment degeneration of the retina, angioid streaks (associated with pseudoxanthoma elasticum), or high hyperopia. Ac-

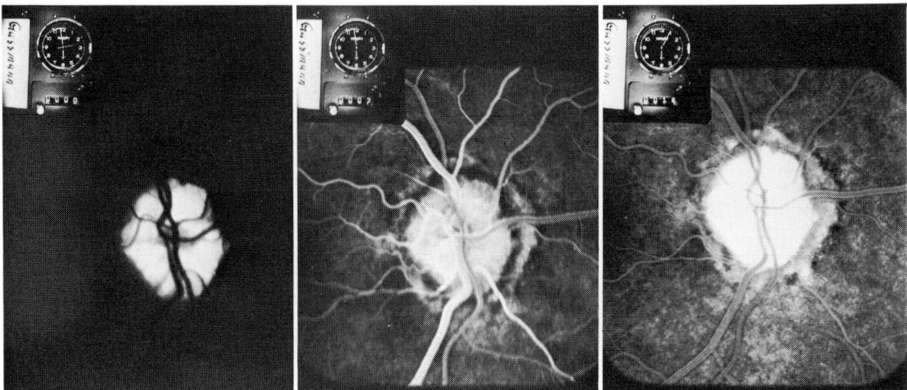

Fig. 2-36. Fluorescein angiogram of a patient with exposed intrapapillary drusen. Before the injection of fluorescein, there is intensive apparent "spontaneous" fluorescence of the drusen (left). In the arterial and venous phase (middle and right), no capillary stasis is visible; rather a lack of small vessels in the disc tissue. Moderate staining of the disc tissue can be observed. Note the characteristic polycyclic margins of the slightly enlarged disc.

cording to the investigations of Chalmers and Walsh, a brain tumor in the presence of drusen is very unlikely. Among several hundred cases of brain tumors, they found only one patient with drusen of the optic disc. Our own observations are in agreement with the findings of these authors (Fig. 2-34). It is interesting that intrapapillary drusen occur especially in children (Fig. 2-36). With increasing age, the hyaline bodies seem to expose themselves first near the border and later on the whole surface of the disc as well, although some drusen remain deeply situated and invisible with the ophthalmoscope throughout life. These remarks explain the frequent occurrence of manifest exposed drusen in the parents and of anomalous disc elevations due to submerged intrapapillary drusen in their children (Hoyt and Beeston).

A 22-year-old girl had a pronounced swelling of one optic papilla, although without venous congestion or hemorrhages. The suspected drusen of the optic disc (hidden drusen) could be confirmed by the finding of manifest bilateral drusen in the mother (see upper photographs in Fig. 2-35).

In summarizing, we can state that *hidden intrapapillary drusen of the optic disc could be mistaken for papilledema.* The absence of dilation of the veins or of hemorrhages as well as the yellowish color of the disc are valuable criteria in the differential diagnosis. Frequently drusen manifest themselves by the nodular appearance of the surface of the disc. Their typical appearance corroborates the diagnosis. Examination of other members of the same family may furnish valuable diagnostic data. In the discussion of *chronic papilledema* (p. 115), yellow, hyalinized exudates simulating congenital drusen were mentioned. These *drusenlike bodies* near the margins of the elevated, pale, atrophic disc are easily differentiated from real drusen if one considers the typical aspect of chronic papilledema with capillary telangiectasias, pallor and gliosis of the disc, narrowing of the arterioles, perivascular sheathing, and finally concentric contraction of the visual fields.

Pseudoneuritis and pseudopapilledema (with or without tortuosity of the retinal vessels). The terms "pseudoneuritis" and "pseudopapilledema" are used for anomalies of the disc that may resemble either a papillitis or a papilledema. Since our previous discussion should have made it clear that there is great similarity between the morphologic picture of papillitis and papilledema, we would like, for simplicity's sake, to dispense with a distinction between these pseudoforms and to discuss pseudoneuritis and pseudopapilledema as one entity. Practical experience has taught us that *such a distinction* is artificial unless one reserves the term "pseudoneuritis" for the description of elevated anomalies of the optic disc combined with mildly defective visual functions and the term "pseudopapilledema" for anomalous elevation of the disc without any visual field defects and signs of visual impairment.

The terms "pseudoneuritis" and "pseudopapilledema" indicate a congenital anomaly consisting of excessive glial tissue proliferation. As a consequence, the nerve fibers appear raised, and the ensuing picture is that of a blurred, enlarged, and occasionally elevated disc. *Two forms* can be distinguished—*one without and one with vascular anomalies.* If the vessels appear normal, the pseudopapilledema (or pseudoneuritis), in contrast to the genuine papilledema, shows a more grayish white opaque discoloration. Edema of the nerve fibers and congestion of the capillary network are characteristically missing. The disc is usually not prominent and appears to be rather firm and compact. In the other form, the one with the vascular changes, in addition to the characteristic findings just described, there are a marked *tortuosity and anomalous early branching of the vessels,* the arteries as well as the veins. The veins, however, are of normal caliber or only moderately dilated (Fig. 2-37). In spite of this tortuosity, their course remains in the same layer of the retina. This is the reason why the hue of their color does not change, in contrast to the choked disc, in which bright sections of the vessels alternate with dark ones. Hemorrhages and white spots are missing entirely. Papilledemas show a certain relationship between the degree of elevation and blurring of the disc on the one hand and the tortuosity of the vessels on the other hand. In the pseudoforms there is always a more pronounced tortuosity with regard to the swelling of the disc. *The size of the blind spot is completely normal, a characteristic sign that differentiates it from a true choked disc!* The visual functions are usually intact, although pseudoneuritis is sometimes accompanied by mild amblyopia. Pseudopapilledema is a congenital anomaly that may frequently involve several members of a family or that may even occur as a hereditary trait (Fischer). We have observed pseudoneuritic changes in both eyes of *identical twins;* both had a bilateral myopia of 4 to 5 diopters. Ophthalmoscopic examination revealed a distinct blurring and elevation of the discs (especially on the nasal side), with tortuosity of the vessels (Fig. 2-38). The myopia has been mentioned purposely:

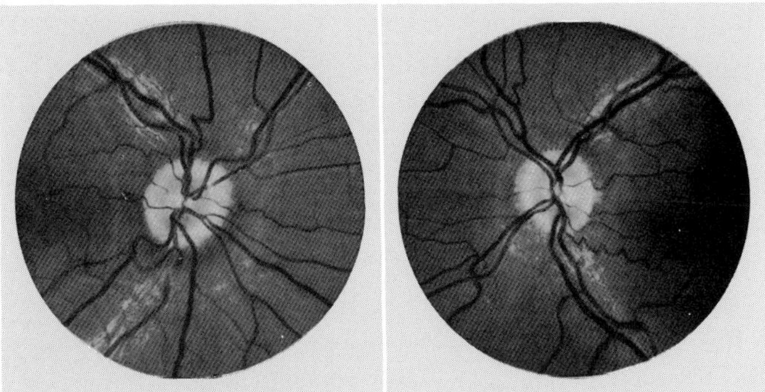

Fig. 2-37. Pseudopapilledema (pseudoneuritis) of the right eye with tortuosity of the vessels. The right disc is barely enlarged, compact, and not edematous and has relatively distinct margins. The veins, like the arteries, are tortuous but not congested. Premature branching of arteries and veins. No hemorrhages. The size of the blind spot is normal! The left illustration corresponds to the right eye; the right illustration, to the left eye.

frequently one hears the opinion expressed that these anomalies occur only in persons with high hyperopia. This is only a conditional truism. We have records of eleven cases of pseudoneuritis in which nine of the patients show *hyperopia* of between 2 and 15 diopters. The hyperopia is not necessarily one of a high degree. It may be quite moderate. In two patients we found a unilateral pseudoneuritis (Fig. 2-39). In one patient with pseudoneuritis we

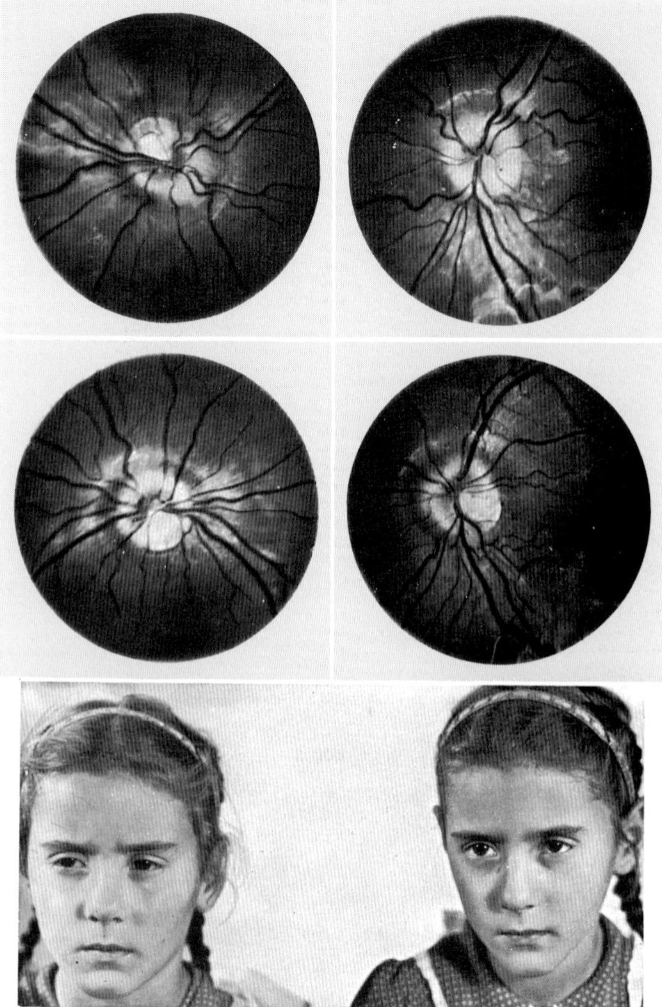

Fig. 2-38. Pseudoneuritis or pseudopapilledema (congenital blurring of disc margins; "super-traction" of papilla) in both eyes of identical twins. The upper photographs belong to the twin on the left, the lower ones to the twin on the right. Blurring and elevation, especially of the nasal halves of the discs. Tortuosity of the vessels. There is an amazing similarity of the twins to the smallest morphologic details of the eyegrounds! The upper left illustrations correspond to the right eye; the upper right illustrations, to the left eye.

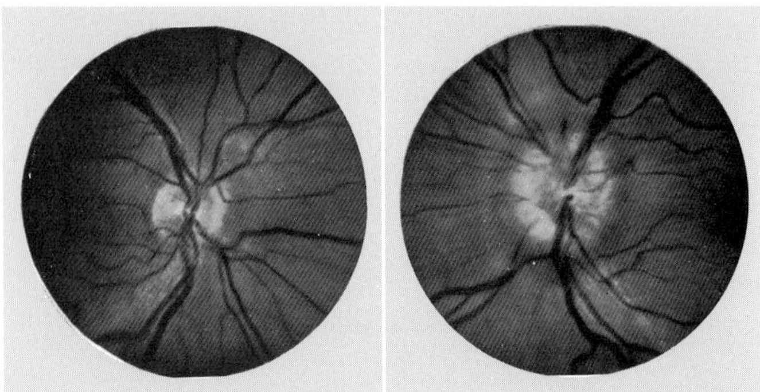

Fig. 2-39. Hyperopic pseudoneuritis ("congenital glial hyperplasia") of the left eye. The right eye is emmetropic and the left shows a hyperopia of 5 diopters. The picture is similar to that of papilledema; however, there is no venous congestion! The left illustration corresponds to the right eye; the right illustration, to the left eye.

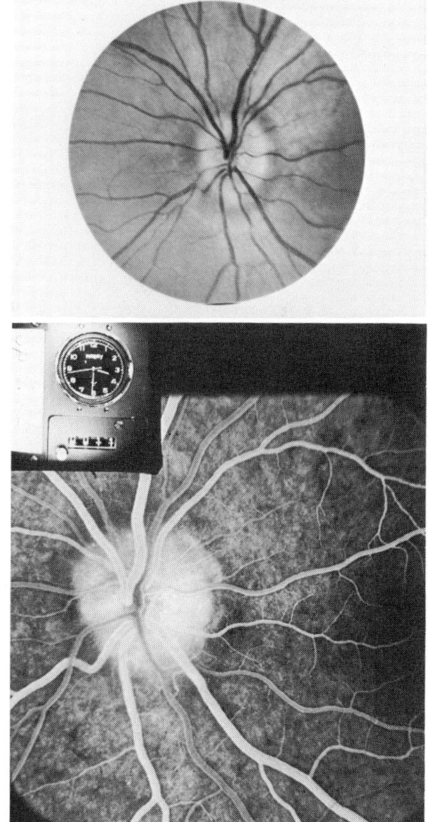

Fig. 2-40. Fluorescein angiogram in a patient with pseudoneuritis ("pseudopapilledema") of the papilla. The upper photograph shows the appearance of the pseudoneuritic disc with slightly blurred margins, discrete prominence, but no capillary stasis and no venous engorgement. The lower photograph is the fluorescein angiogram of the same disc; it behaves like a normal disc and manifests no capillary stasis and no signs of increased permeability of the vessels as does real papilledema.

found a pronounced brachycephaly and an arched palate. Here we would like to call attention to a certain relationship between pseudoneuritis and *malformations* of the skull. Also, Uhthoff's observation that pseudoneuritis occurs twice as often in patients with mental disease as in normal individuals should be mentioned. The occasional unilateral occurrence of a pseudoneuritis plays an important part in the differential diagnosis of unilateral choked disc (p. 143).

Pseudoneuritis and pseudopapilledema in general show no pathologic change on fluorescein angiograms (Fig. 2-40). This differentiates them unequivocally from papilledema or papillitis. The fluorescein picture is that of a normal fundus. There are only a few vessels visible on the disc and these are of normal caliber. The late fluorescence is minimal and lasts only a short time after the vessels have emptied. It is advisable to repeat the fluorescein angiogram at regular intervals when beginning papilledema has to be differentiated from a pseudopapilledema.

The diagnosis of pseudoneuritis or pseudopapilledema is not always easy. It is no coincidence that there are so many cases of this anomaly in our series. The patients were referred to the neurosurgical clinic as suspected of having choked discs and brain tumors. A thorough examination revealed completely negative results. These cases of pseudoneuritis have to be *carefully observed for some time* before the correct diagnosis can be made. Repeated perimetric examinations of the blind spot and photographs of the eyeground at certain intervals aid enormously in making the correct diagnosis. *One should always take heed of the principle that every patient with pseudoneuritis or pseudopapilledema should be suspected of having papilledema until proved otherwise —that is, that there is no increase in the intracranial pressure.*

Medullated nerve fibers. Normally the medullary sheaths of the fibers of the optic nerve begin immediately behind the lamina cribrosa. Occasionally they may extend distally to the papilla and the adjacent retina. They can be recognized as snow-white, flame-shaped stripes

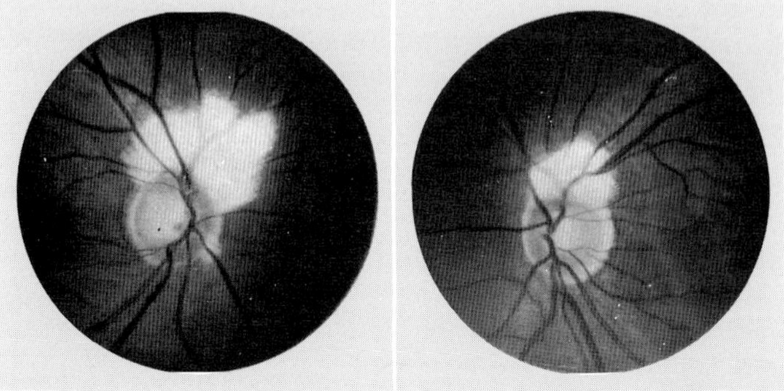

Fig. 2-41. Medullated nerve fibers of both eyes of the same patient. The feathered edges of the myelinated patches are especially obvious on the right side. The left illustration corresponds to the right eye; the right illustration, to the left eye.

that seem to originate from the upper or lower disc border. Thus the disc may appear enlarged and its outline blurred. This picture conceivably could be mistaken for a papilledema. However, anybody who has seen the extraordinarily characteristic appearance of medullated nerve fibers even once is not likely to make such a mistake (Fig. 2-41). It is well known that medullated nerve fibers are often associated with ocular malformations such as epipapillary membranes, persistent prepapillary glial tissue, congenital tortuosity of the vessels, and prepapillary vessel loops. Of course, a real papilledema due to increased intracranial pressure may be superimposed on medullated nerve fibers and the primary condition of the altered disc manifests itself only after corresponding treatment.

Anomalies of the retinal vessels. Although *tortuosity and anomalous early branching of the retinal vessels* have been mentioned previously in connection with pseudoneuritis and pseudopapilledema, it actually should be included as a congenital anomaly in this discussion.

There are congenital vascular anomalies of such extraordinary morphologic variety that some of them might simulate a papilledema, especially if they are localized very close to or directly on the disc. Most of all, we should mention the *arteriovenous aneurysms (racemose aneurysms or cirsoid aneurysms) of the retina,* which show an incredible variety in their ophthalmoscopic aspects, as pointed out so very well by Wyburn-Mason. *The finding of a visible connection between an artery and a vein* (either a direct connection or via a distinct, markedly dilated capillary network) is important and decisive for the diagnosis. Such an arteriovenous shunt causes widening and tortuosity of the veins in one form or another, which contrasts distinctly with the caliber of the arteries. If such an arteriovenous rete mirabile is located near the disc or directly on it, it may cause an edematous blurring, an indistinction of the disc margin, and some prominence of the papilla—signs that may be similar to a papilledema. The protrusion of such a disc usually is minor. The venous congestion and, last but not least, the dilated capillary network in the area of the disc are very closely related to the changes that occur in papilledema. The arteriovenous communication, however, can always be found with careful ophthalmoscopic examination. This should confirm the diagnosis of an arteriovenous aneurysm (Figs. 2-42 and 2-43). Sometimes some larger vessels are sheathed by white tissue neoformation, and in some cases the periphery of the fundus may be covered by grapelike aneurysms surrounded by hard, white exudates (Hoyt and Beeston).

We personally observed an instructive case of a 24-year-old girl who had been referred by a neurologic clinic because of a suspected meningioma of the olfactory groove and a unilateral "choked disc." All neurologic findings were negative. An EEG revealed nothing unusual. After careful ophthalmoscopic examination, the changes of the disc that had been interpreted as "choked disc" turned out to be a typical arteriovenous aneurysm of the retina or rather the papilla (Fig. 2-43) with a distinctly dilated capillary network interpolated between an artery and a vein, in addition to several arteriovenous shunts.

It may be of interest to mention that such *arteriovenous aneurysms of the retina are occasionally associated with similar lesions of the optic nerve, the orbit, or the brain* (for instance, the midbrain). This may explain the occurrence of supranuclear conjugate gaze palsies and bilateral ptosis that has been described in cases of racemose aneurysms of the retina (Wyburn-Mason syndrome).

Tumors of the disc. In rare cases it can happen that tumors of the disc (gliomas, meningiomas, neurinomas, neurofibromas, metastatic tumors) are misinterpreted as a papilledema. The predominantly unilateral occurrence should caution against such a diagnosis. The same is true for malignant melanomas of the choroid near the disc, with secondary invasion of the latter. We have observed a few such malignant melanomas of the disc. The conspicuous pigmentation of such a prominent structure, its size, and its eccentric position with regard to the vessel trunk usually lead to the correct diagnosis (Fig. 2-44). In rare instances of *von Hippel–Lindau disease* an angiomatous hamartoma (hemangioma) may be located directly in the center of the disc and thus simulate papilledema. One should always search for other angiomas far peripherally in the retina and not forget the possibility of the coexistence of similar angiomatous hamartomas in the cerebellum and other organs (kidney, pancreas, etc.).

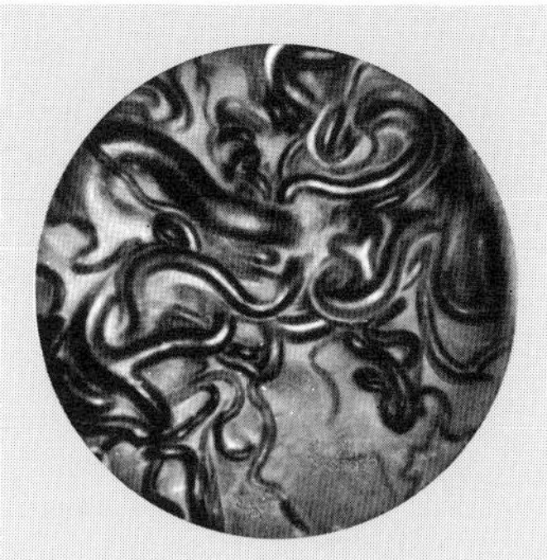

Fig. 2-42. Arteriovenous aneurysm of the retina in the left eye of a 12-year-old girl. Enormous dilation of veins and arteries forming a convolution covering the disc. There was also an arteriovenous plexus within the orbit. (Wyburn-Mason syndrome.)

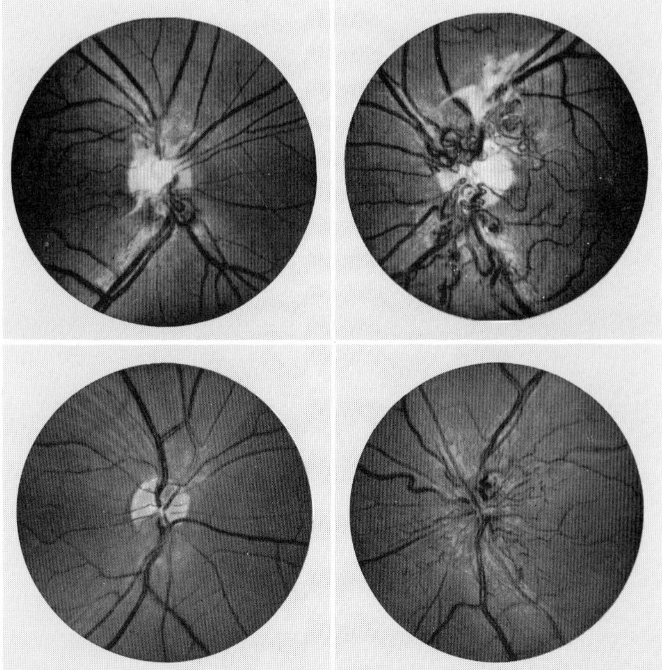

Fig. 2-43. Arteriovenous aneurysms of the retina. Above, Racemose aneurysm at the papillary margin with distinctly visible arteriovenous shunts in a 43-year-old patient. Below, Racemose aneurysm with very fine vessels immediately above the left disc of a 24-year-old patient; it resembles a low degree of papilledema. The left illustrations correspond to the right eye; the right illustrations, to the left eye.

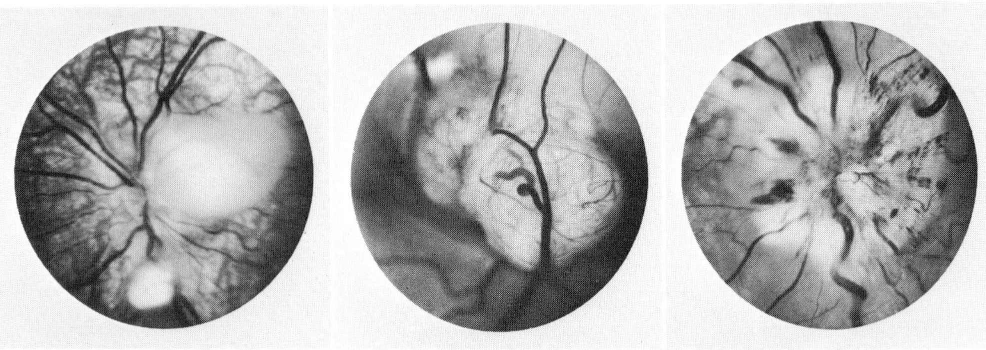

Fig. 2-44. Tumors of the disc simulating papilledema. Left, Congenital tumorlike malformation of the disc (not verified histologically). Middle, Malignant melanoma involving the disc. Right, Bronchogenic carcinoma metastatic to the optic disc and nerve (first diagnosed as acute papillitis because of sudden functional loss, unilateral disc swelling, and edema; only the association with other intraocular and extraocular metastases led to the right diagnosis!).

PARESES OF EXTRAOCULAR MUSCLES

In addition to papilledema, increased intracranial pressure caused by a brain tumor may cause a few other ocular signs which, likewise, are of no localizing significance. One should be cautious not to use them as a basis for a hasty conclusion regarding the seat of the space-consuming cerebral process. Cushing (1910) actually called a *unilateral paresis or the more rarely occurring bilateral pareses of the sixth nerve a "false localizing sign."* It seems that the sixth nerve, due to its long intracranial course, is particularly vulnerable. Displacement of the brain stem by the tumor may easily cause it to be tugged in exposed areas. Cushing has called attention to an important anatomic deviation. Normally the sixth nerve lies between the dura and the branches of the basilar artery after it emerges from the pons. In a certain number of cases, however, it lies for a short stretch between these vessels and the pons. An increased intracranial pressure may cause a compression of the sixth nerve between the pons and branches of the basilar artery, especially the inferior anterior cerebellar artery. The result is a partial disturbance of the conductivity of the nerve with a corresponding *weakening in the action of the lateral rectus muscle of the ipsilateral eye.*

If the paresis is severe, diplopia will occur even in the primary position but will be noticeable only on lateral rotation of the involved eye in the case of a mild paresis. The diplopia is horizontal and homonymous. (For details about diagnosis and symptomatology of sixth nerve paresis see p. 38.) If the intracranial pressure decreases, the nerve may recover its function and the paresis may recede. These variations in the intracranial pressure explain the daily or even hourly "fluctuations" of a sixth nerve paresis as occasionally observed in

1, Superior orbital fissure
2, Trigeminal nerve (first and second branches)
3, Oculomotor nerve
4, Trochlear nerve
5, Posterior cerebral artery
6, Cerebellum
7, Vertebral artery
8, Upper ridge of petrous portion of temporal bone
9, Sixth nerve

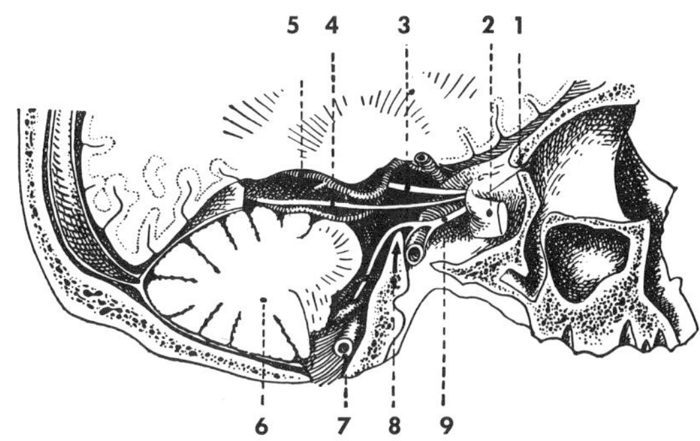

Fig. 2-45. Vertical section through the base of the brain and skull to demonstrate the intracranial course of the sixth nerve. The bend over the sharp upper ridge of the petrous bone is marked by an arrow.

patients with brain tumors. Pathologic-anatomic investigations have demonstrated conclusively that the compression of the sixth nerve between the pons and the branches of the basilar artery causes some damage to the nerve (Fig. 2-45).

In addition to the point at which the sixth nerve leaves the pons, there is another vulnerable area during its extended course along the base of the skull. If a tumor displaces the brain stem inferiorly, superiorly, or posteriorly, it may tug on the nerve at the sharp superior ridge of the petrous bone. The displacement of the brain stem can be caused by supratentorial as well as by infratentorial tumors and has no relation to the side where the space-consuming lesion is located. Therefore a paresis of the sixth nerve (which usually is unilateral) will not permit any conclusions regarding the site of the tumor causing the increased intracranial pressure.

In summary, it can be stated that *a unilateral sixth nerve paresis occurring during the development of a brain tumor is without value whatsoever for purposes of localization; it is merely a general sign of increased intracranial pressure.* However, a bilateral sixth nerve paresis and one that occurs during the early stages of a space-consuming lesion is of decisive value and is highly suggestive of an intrapontine focus (nuclear paresis).

A paresis of the trochlear nerve does not occur merely as a nonspecific general sign of an increased intracranial pressure. Also, nonspecific pareses of the extrinsic muscles supplied by the oculomotor nerve are rare. They may occur only in the terminal stages of a severely increased intracranial pressure (p. 37). There are, however, nonspecific pupillary disturbances caused by increased intracranial pressure. They are most likely to occur if the oculomotor nerve is pressed against the clivus ridge of the tentorial notch. They will be discussed in the following section in connection with the clivus ridge syndrome.

CLIVUS RIDGE SYNDROME

Similar to a sixth nerve paresis resulting from increased intracranial pressure is a *unilateral lesion of the oculomotor nerve that is manifested by a fixed mydriatic pupil.* It is part of the so-called clivus ridge syndrome—another general sign of increased intracranial pressure, especially of a high or an acute degree. Actually, the basic lesions of the sixth and oculomotor nerves are of quite similar nature, with only a gradational difference in the resulting pictures. In the case of the sixth nerve there is only a paresis of the corresponding lateral rectus muscle since there are no other fibers. In the case of the oculomotor nerve, on the other hand, the pupillomotor fibers are impaired first due to their particular vulnerability. Occasionally this is actually the only sign of the impairment. At least in the early stages there is no paresis of any of the extrinsic muscles, but only a unilateral fixed mydriasis. However, in contrast to unilateral sixth nerve paresis, which has no localizing significance, the *unilateral fixed mydriatic pupil is highly suggestive of an ipsilateral or a predominantly unilateral space-consuming supratentorial process.*

Fischer-Brügge stresses in his publications that the ipsilateral mydriatic fixed pupil is only one sign of a broader and more comprehensive clinical picture, the so-called clivus ridge syndrome—in addition to the somatic components (that is, the oculomotor paresis, cardiorespiratory signs, pyramidal tract symptoms), there is a *psychic component* in the form of varying degrees of *disturbances of consciousness.* Thus we have a combination of ocular and systemic signs of increased intracranial pressure in the clivus ridge syndrome. Fischer-Brügge believes that the disturbances of consciousness are brought about by pressure damage and circulatory disturbances in the diencephalon and mesencephalon similar to those found in the oculomotor nerve. We shall discuss the latter in some detail, mostly because this picture is generally not familiar.

Hutchinson (1867) was among the first to describe unilateral fixed mydriasis in connection with cerebral traumas. Since the original observations, a number of investigators have become interested in the origin and mechanism of this clinical sign and its importance in the recognition of the site of an acute intracranial space-consuming lesion (McKenzie). There is almost unanimous agreement among the various authors that the ipsilateral fixed mydriasis is caused by *pressure on the peripheral oculomotor nerve* (particularly the vulnerable pupillomotor fibers) and not by stimulation of the sympathetic fibers (Scharfetter). The site and mechanism of such an oculomotor lesion, however, were quite obscure for a long time.

How can an increased intracranial pressure cause a specific lesion of the oculomotor nerve? On the basis of extraordinarily revealing observations (together with especially instructive pathologic-anatomic specimens), Fischer-Brügge considers two possibilities that may occur in a number of combinations: (1) *the third cranial nerve may be pushed against the ipsilateral clivus ridge*

U, Uncus of hippocampal gyrus (**Hip.**) of temporal lobe (**Tp.**)

R, Gyrus rectus of frontal lobe with orbital sulcus (**F. orb.**)

H, Oblique position of pituitary stalk (displaced from left to right side)

Car., Internal carotid artery

Co. post., Posterior communicating artery

Cb. sup., Superior cerebellar artery

Impr. T., Tentorial notch

B, Pons

Bas., Basilar artery

C.P., Posterior cerebral artery

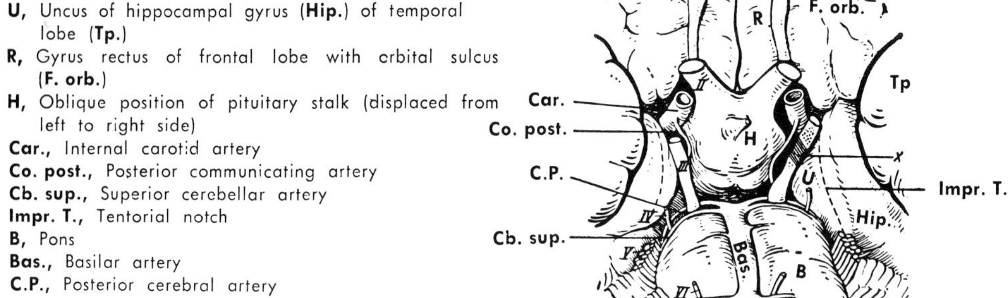

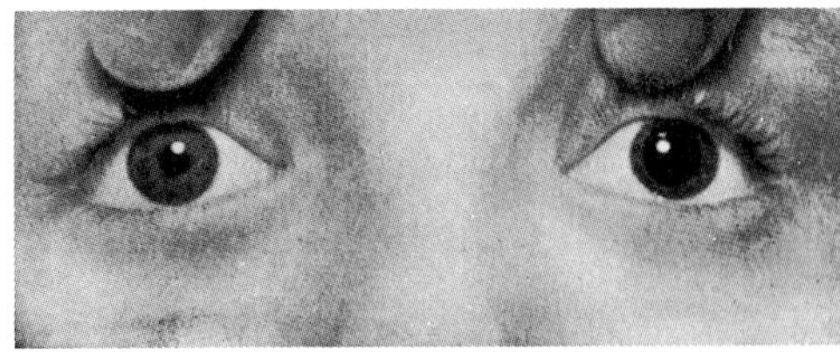

Fig. 2-46. Clivus ridge pressure groove (Fischer-Brügge) of the left oculomotor nerve in a 25-year-old patient with petechial hemorrhages distal to the area of pressure. Abscess of left side of the frontal lobe with breakthrough into the ventricle. Below, Clinical picture of the clivus ridge syndrome. Fixed mydriatic pupil of the left eye, with slight outward deviation of the left eye, a sign of a paresis of the left medial rectus muscle.

of the tentorial notch by increased intracranial pressure to one side and (2) *the oculomotor nerve may be compressed immediately after its emergence from the inner surface of the cerebral peduncle either by the posterior cerebral artery or, more rarely, by the superior cerebellar artery.* The anatomic-pathologic manifestation of such a compression of the nerve consists of a pressure groove (Fig. 2-46), either in the form of the so-called clivus ridge groove (on the caudal surface of the nerve) or, more rarely, of a vessel groove (cranial surface of the nerve).

Just as important as the groove formations are circulatory disturbances. Frequently there are patches of extravasation in the area of the oculomotor nerve distal to the point of compression, manifest signs of interference with the blood supply. *It is just this interference with the intraneural and perineural blood circulation that explains satisfactorily* the frequently only transitory nature of *the clinical signs of the clivus ridge syndrome.* This is especially true for the fleeting and reversible pupillary signs. They are known to recede completely, for instance, immediately following a decompression or even subsequent to an intravenous injection of a hypertonic saline, glucose, mannitol, or urea solution in the treatment of a cerebral edema, together with a simultaneous clearing of the consciousness.

In the clinical picture of the clivus ridge syndrome, Fischer-Brügge distinguishes four stages:

The *first stage* shows a *narrowing of the ipsilateral pupil,* which might be considered an irritation of the oculomotor nerve preceding the lesion (Tönnis). This miosis on the side of the lesion is, however, quite transient and for that reason quite easily overlooked.

The *second stage* is by far the most frequent and the one phase that can be observed longest during the course of increased intracranial pressure. *The pupil is dilated but will still react to light and in convergence.* Occasionally a slight ptosis can be seen. The patient's sensorium is either completely free or already shows a suggestion of somnolence.

The transition to the *third stage* manifests itself by an increasingly *sluggish reaction of the dilated pupil* and a beginning loss of the pupil's regular shape. With the increase in the intracranial pressure, the somnolence and apathy of the patient become more severe and terminate in complete *coma.* The *pupil* becomes *mydriatic, irregular, and completely fixed to light.* During this stage not only the oculomotor fibers supplying the pupil but also those supplying the extrinsic muscles may be impaired. There may be *pareses of the extraocular muscles supplied by the third cranial nerve.* The eye shows a mydriatic fixed pupil and assumes a divergent position (Fig. 2-46).

During the *fourth* (terminal) *stage,* the contralateral pupil becomes mydriatic and fixed to light, obviously also on the basis of a clivus ridge reaction.

In general, the unilateral fixed mydriatic pupil is a nonspecific sign of a generally increased intracranial pressure of higher degree or acute nature. For this reason it has been discussed with the general symptoms. The cause may be a space-consuming process (tumor, abscess, or hemorrhage, especially a subdural or epidural hematoma) in the region of a hemisphere or an edema accompanying a hemispheric tumor. This is definitely a remote effect of a space-consuming process on the oculomotor nerve. The expansion of the hemisphere causes downward displacement of the brain stem and the compression of the third cranial nerve previously described.

In *tumors of the temporal lobe* there is a *local effect* in addition to the remote effect. Frequently the hippocampal gyrus on the tumor side is forced into the tentorial notch. The result is a so-called *herniation of the temporal lobe into the tentorial notch,* which may cause a compression of the oculomotor nerve by direct pressure. We were able to observe unilateral fixed pupils particularly frequently in our patients with temporal lobe tumors.

We can assume a direct pressure effect on the oculomotor nerve in tumors close to the diencephalon, especially craniopharyngiomas.

Although the clivus ridge syndrome with its striking sign of a unilateral fixed mydriatic pupil should be primarily considered a general sign of increased intracranial pressure, it may be of localizing value for certain tumor sites (for instance, tumors of the temporal lobe, p. 172). Thus the discussion of this syndrome leads to discussion of focal symptoms in the next chapter.

In conclusion, we can state that a unilateral mydriatic fixed pupil usually is a general sign of increased intracranial pressure caused by pressure damage to the oculomotor nerve at the clivus ridge. This pupillary sign is seen on the ipsilateral side with unilateral supratentorial space-consuming lesions. With temporal lobe tumors it may be either a general sign of an increased intracranial pressure or a local sign (herniation of the temporal lobe into the tentorial notch). The pupillary disturbances are part of a more extensive picture, the so-called clivus ridge syndrome, which includes, apart from cardiorespiratory signs and pyramidal tract symptoms, also psychic factors, especially disturbances of consciousness.

EXOPHTHALMOS

In Chapter 1 it was stated that a *bilateral exophthalmos* (more rarely, a unilateral exophthalmos) may be a general sign of increased intracranial pressure (p. 17). This is less a simple protrusion of the globes than an "extrusion" of the entire orbital contents—obviously the result of the transmission of the increased intracranial pressure through the orbital fissures into the orbits or the result of pressure effect on the cavernous sinus (Skydsgaard). Among our cases, bilateral exophthalmos as a general sign of increased intracranial pressure has been observed especially in patients with expansive parasagittal meningiomas, meningiomas of the falx, and tumors with a tendency to early hydrocephalus formation (for instance, cerebellar tumors and tumors of the third and fourth ventricles). This exophthalmos usually is quite inconspicuous and can be easily overlooked in certain cases, perhaps because it is of no diagnostic and particularly no localizing value. A bilateral exophthalmos also cannot be evaluated properly with the customary instruments intended for relative exophthalmometry.

A *unilateral exophthalmos* may result from the remote effect of an intracranial growth (Elsberg, Hare, and Dyke) *but should be primarily considered a local symptom.* One should consider here primarily orbital processes, but also brain tumors of the anterior and middle fossas, which may cause an exophthalmos by direct invasion of the orbit or indirectly by interference with the drainage of the ophthalmic vein or the cavernous sinus (Figs. 1-4, 1-5, 3-74, 3-80, and 3-87).

In conclusion, it can be stated that a bilateral exophthalmos occurs occasionally as a general sign of increased intracranial pressure but that it is irrelevant for clinical diagnostic purposes.

 Local symptoms of brain tumors

The classification of the symptomatology of brain tumors into general symptoms of an increased intracranial pressure and into local or focal symptoms was suggested in the introductory remarks to Chapter 2 (p. 86). *The local symptoms are those that originate from localized damage to a definite brain area and thus assume definite significance for purposes of localization.* It has been stressed that, as a rule, the chronologic sequence of the two groups of symptoms is as follows: the focal symptoms, according to the lesion of a certain part of the brain, appear first, to be followed by the general signs of an increased intracranial pressure due to the remote effect of a growth gradually increasing in size and to the reactive edema and swelling of adjacent structures. This chronologic development is true for the majority of brain tumors. Thus the local symptoms assume exceptional diagnostic importance. If the local symptoms per se are not absolute proof of a neoplasm of the brain, *their steady progressiveness (with possible periods of quiescence, but seldom remissions or improvements) is, in fact, one of the principal signs of an intracranial tumor.* In 1904 Collier postulated that *local signs and symptoms occurring in the late stages of a brain tumor are unreliable.* Thus one should make an assiduous search for early symptoms in patients with progressive cerebral symptoms and, if possible, elicit such symptoms by taking a careful case history. In a certain number of patients with brain tumors the local symptoms are so conspicuous and alarming that they may motivate an examination or even surgical interference before general symptoms develop. It has also been mentioned that there are some tumors in silent zones of the brain and others with a tendency to cause an early internal hydrocephalus with general symptoms of an increased intracranial pressure as the first, or perhaps even the only, symptom.

What has been said in a general way in regard to the local symptoms of brain tumors is also true in a stricter sense for the ocular local symptoms in brain tumors. The nature of the visual sensory organ may cause even small circumscribed lesions of the visual pathway to develop distinct disturbances of vision. Early ocular local signs and symptoms in patients with brain tumors are therefore of decided significance for purposes of localization. The importance of some of these signs, however, varies and depends to some degree on the seat of the tumor, that is, its relation to the visual apparatus. Thus pareses of the extrinsic muscles supplied by the oculomotor nerve are usually quite significant, whereas pareses of the lateral rectus muscle (especially if they are unilateral) are of no value for purposes of localization but have to be considered as a general sign of increased intracranial pressure. Visual field disturbances in patients with brain tumors usually represent defects that are valuable for the localization of the lesion. It is rare for remote effects of brain tumors to cause visual field changes that could be mistaken for those of a focal nature. If at all, this usually occurs in the late stages. It should be clear that the ocular local symptoms in patients with brain tumors assume diagnostic significance only in their relationship with other neurologic local signs. *We ophthalmologists should therefore appraise ocular local signs only as part of the entire clinical picture.* The importance of the ocular local symptoms in the overall picture varies a great deal with the location of the tumor. The ocular symptom may be the outstanding feature, it may be of equal significance with other symptoms, or it may be a subordinate accompanying symptom of other neurologic symptoms. It certainly would be a mistake to discuss only ocular local symptoms in this presentation. *We consider it our duty to assign the ocular local symptoms in brain tumors to their proper place within the frame of the entire clinical picture.* For this reason we plan to give a brief summary of the other neurologic symptoms for each symptom complex according to its topographic basis and to point out their proper relationship with the ocular symptoms.

The ocular local symptoms and their diagnostic significance will be discussed from the topographic point of view (Fig. 3-1). The brain tumors will be classified into supratentorial and infratentorial groups. The *supratentorial* group consists of the hemispheres with their frontal, temporal, parietal, and occipital lobes, the pituitary gland and its surrounding area, the anterior and middle fossa, the diencephalon, including the third ventricle, the thalamus, the basal ganglia, and the midbrain. The *infratentorial* group includes tumors of the cerebellopontine angle, the cerebellar hemispheres, the cerebellar vermis, the pons, and the medulla oblongata. Tumors in the mesencephalon may expand in a supratentorial or infratentorial direction. They will be discussed in the concluding discussion of the supratentorial group.

The classification of the tumors into supratentorial and infratentorial groups is used for didactic reasons. It is also based on the symptomatologic difference between tumors of these two groups and, last but not least, on the difference of their surgical accessibility. Also, the so-called tentorium cerebelli forms a definite natural barrier between the supratentorial and infratentorial spaces of the cranium. In its middle section the cerebellar tentorium shows an opening of the dura, the so-called tentorial notch. The stalklike brain stem passes through this opening. Even though the tentorial notch causes a certain discontinuity of the tentorium, the latter presents a mechanically effective barrier that creates a fairly sharp separation between the supratentorial and infratentorial contents of the cranium. A growth in the infratentorial area is not likely to extend into

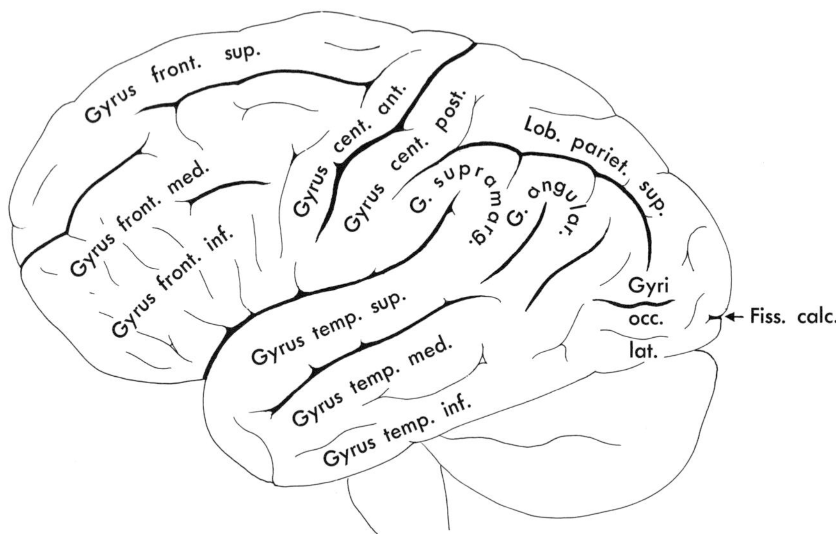

Fig. 3-1. Topography of the cerebral lobes, gyri, and sulci: convexity of the left hemisphere.

the supratentorial space. The membrane of the tentorium protects the occipital lobe against direct invasion by a tumor as well as, to some extent, against the pressure effect from cerebellar or paracerebellar tumors. The tentorium especially forms the lateral dividing line between the infratentorial and supratentorial region. It lies between the cerebellar hemispheres and the poles of the occipital lobes. In the region of the brain stem the separation line coincides with the isthmus, that is, approximately the area where the mesencephalon passes into the rhombencephalon. This area roughly corresponds to the anterior cranial part of the fourth ventricle.

A topographic discussion of the ocular local signs of brain tumors is essential because in numerous patients the ocular signs suggest a definite site for the tumor. In other words, they are of localizing value.

In the diagnosis of brain tumors it is equally important to determine the *nature* of the tumor, especially in view of possible surgical intervention. Certain histologic types of intracranial tumors show a specific and characteristic symptomatology. Some of them (for instance, pituitary adenomas, craniopharyngiomas, and some meningiomas—meningiomas of the tuberculum sellae and the sphenoid ridge) also cause specific ocular local signs and symptoms. In the majority of patients with brain tumors, however, ocular symptoms, if present at all, do not depend on the type of tumor. Following the topographic discussion of the ocular signs, we propose to touch briefly on the relationship between the histologic structure of the tumors and their ocular local symptoms. This chapter should aid in the diagnosis and differential diagnosis of the various types of tumors (p. 282).

SUPRATENTORIAL TUMORS
Tumors of the frontal lobe

In patients with tumors of the frontal lobe the mental changes dominate the picture. Such changes may manifest themselves either in the form of a more or less characteristic psycho-organic syndrome or occasionally as local signs of a frontal lobe syndrome. The latter may be described as lack of impulse (for movement, speech, and thinking), apathy, striking changes in social behavior, euphoria, and occasionally a peculiar silly jocularity (moria). In frontal lobe tumors, with their pronounced tendency to relatively early increase of intracranial pressure, the mental changes may be mixed with disturbances of consciousness in such a way that their differentiation is sometimes almost impossible. Obviously due to loss of inhibitory action on the frontopontine tracts, a motor phenomenon appears that is known as forced grasping or groping—a sign that is almost pathognomonic for frontal lobe lesions, especially those of large size with extension into the corpus callosum or the basal ganglia. It consists of a violent grasping movement of the hand and the fingers if an object is placed into the hand. This sign can also be produced by stroking the palm at the proximal end of the fingers. Frontal lobe ataxia is characterized by a tendency to fall forward or toward one side. A lesion in the left lower frontal lobe gyrus (Broca's center) in a right-handed patient causes motor aphasia. One of the most frequent signs of a frontal lobe tumor is a central paresis of the facial nerve that occurs on the site opposite the tumor. If the damage caused by the tumor extends to the motor cortex of the anterior central region, there may be either signs of irritation in the form of local (jacksonian) or generalized seizures, or there may be signs of paresis or paralysis in the form of a contralateral spastic monoplegia which, dependent on the seat of the process, are more prominent in the face, the arm, or the leg. The focal motor epilepsy (with turning of the head and the eyes to the side opposite the tumor) as well as the paralyses actually belong to the *syndrome of the anterior central gyrus* but have been mentioned intentionally because we include the anterior central gyrus in the discussion of the frontal lobe. Of interest to the ophthalmologist is the fact that in frontal lobe epilepsy there is a specific motor discharge from the frontocortical adversive field that consists of a spasmodic horizontal deviation of both eyes to the side opposite the lesion (possibly with turning of the head in the same direction). The corresponding signs in damage of the frontal lobe are a horizontal gaze paresis or palsy to the side opposite the tumor, frequently resulting in a conjugate deviation of the eyes and turning of the head to the side of the lesion (Fig. 1-11). Frontal lobe tumors, especially glioblastomas, and metastatic processes, with their exquisite tendency to increased intracranial pressure, produce acute, chronic, or intermittent temporal or cerebellar herniation syndromes characterized by unilateral or bilateral pyramidal tract signs, anisocoria, and sixth nerve and hypoglossal nerve pareses. As just mentioned, these are all signs of a general increase of intracranial pressure, but may in certain cases occur so early or be so prevalent as to lead to severe diagnostic errors.

Because of the relatively distant position of the frontal lobe from the visual pathways, ocular localizing symptoms are rather rare in patients with tumors of this region. They actually appear only if the tumor lies at the base of the frontal lobe and extends into the area of the chiasm. Accordingly, symptoms in patients with frontal lobe tumors are usually uncharacteristic and are frequently manifestations of a generally increased intracranial pressure. Patients occasionally complain of double or foggy vision, glittering before the eyes, dark spots suddenly occurring on one side, or formed figures in a certain part of the visual field. The topical symptoms, however, are considerably more important! These are complaints ranging from *unilateral loss of visual acuity to amaurosis.* We have seen this in a number of patients whose conditions assumed various forms of the Foster Kennedy syndrome. However, it should be stressed that in a little more than one half of the patients with frontal lobe tumors no ocular symptoms were present.

To draw a true picture of the ocular signs, we want to emphasize that numerous tumors, even if they reach considerable size, may show no signs whatsoever that would implicate the visual apparatus. We have forty-six verified cases of frontal lobe tumors on record. In two thirds of the patients (thirty-one) there are absolutely no visual field changes—an observation that is in agreement with the statistics of Guillaumat and Robin. In about one third (seventeen) the fundi of both eyes were completely negative (that is, not even a papilledema could be found as a general sign of increased intracranial pressure). Such figures speak for themselves. We should be aware *that the ocular symptomatology of frontal lobe tumors is frequently disappointing.* (See Fig. 3-2.)

In the literature the localizing value of the *Foster Kennedy syndrome* is often stressed. It consists of a primary optic atrophy with a central scotoma on the side of the tumor and a papilledema on the opposite side (Fridenberg; Glees). It must be assumed that the unilateral optic atrophy results from direct pressure on one optic nerve between the optic foramen and the chiasm by a tumor situated at the base of the lobe with a backward extension. It is a well-known fact that an atrophic optic nerve will not develop edema. Hence an increased intracranial pressure will manifest itself only on the side of the intact nerve. *Such a textbook case of a fully developed Foster Kennedy syndrome did not occur in any of our cases.* In the discussion on unilateral papilledema (p. 112) it was stated that there are numerous incomplete forms of this syndrome that occur more frequently than the typical picture. Thus we found a unilateral primary optic atrophy with sharp borders of the disc, concentric constriction of the peripheral visual field, and a central or cecocentral scotoma in four of forty-six patients with tumors (Figs. 3-3 and 3-4). The atrophic papilla was always on the side of tumor. On the opposite side there was either a completely normal or only slightly blurred nerve head but never a pronounced choked disc. The tumor causing these incomplete forms of the Foster Kennedy syndrome was always situated on the side with the optic atrophy— more precisely at the base of the lobe, with a deep basal expansive growth. In one instance a meningioma originating from the roof of the orbit was seen to surround and compress the ipsilateral optic nerve. *The optic atrophy is the pathologic basis for the subjective sensation of a unilateral loss of vision,* which usually does not escape the patient's attention. In the case of a frontal suprasellar cholesteatoma causing such an incomplete Foster Kennedy syndrome, loss of vision was noted first in the ipsilateral eye 5 years prior to surgical intervention! Originally a retrobulbar optic neuritis had been suspected. In the discussion of the basal tumors, especially those of the chiasm and its surroundings, we shall meet frequently with instances in which damage to the optic nerve caused by tumor pressure in the initial stage cannot be differentiated from an ordinary retrobulbar neuritis. *Every case of retrobulbar neuritis should*

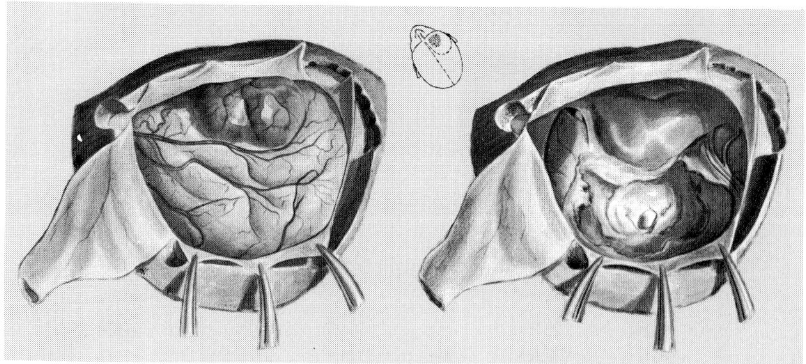

Fig. 3-2. Malignant astroblastoma of the right frontal lobe in a 47-year-old patient before and after extirpation (frontal lobectomy).

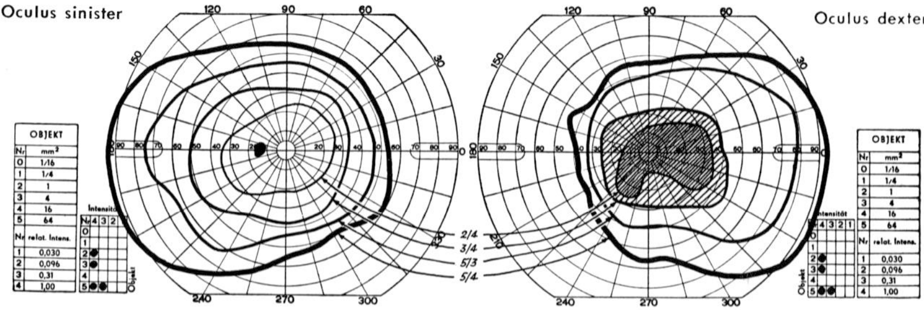

Fig. 3-3. Hemangioblastoma of the right frontal lobe with compression of the right optic nerve (optic atrophy!). Visual field on the side of the tumor shows slight constriction of the peripheral isopters as well as a large relative central scotoma. (Case of Dr. Guillaumat; from Dubois-Poulsen.)

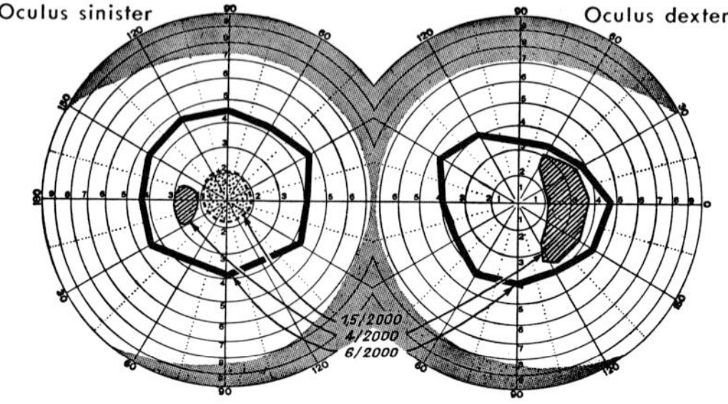

Fig. 3-4. Glioblastoma multiforme of the left frontal lobe. Slight concentric constriction of the peripheral visual fields. Enormous enlargement of right blind spot (choked discs). Left relative central scotoma.

*be suspected as being caused by tumor compression of the optic nerve if re-
covery of the visual function is only temporary or does not take place at all
and if there is further deterioration of the visual acuity as well as of the visual
field.* By the way, it is interesting to note that Foster Kennedy himself talked
about a retrobulbar neuritis and a central scotoma with a subsequent primary
optic atrophy on the side of the lesion and a papilledema on the other side
when he described his syndrome in 1911. We are in agreement with David and
Sourdille; Dubois-Poulsen; Walsh and Hoyt; Tönnis; François and Neetens; and
others that the value and the topographic significance of the Foster Kennedy
syndrome generally is somewhat overrated. We want to state quite emphatically that this syndrome is not pathognomonic for frontal lobe tumors. It is
found, for instance, with meningiomas of the olfactory groove (p. 225), meningiomas of the sphenoid ridge (p. 231), and other basal tumors (for instance,
suprasellar meningiomas)—all of them tumors that may involve the optic nerve.
The value of the Foster Kennedy syndrome is lessened by the fact that even
tumors of the third ventricle or the cerebellum can cause similar symptoms by
remote effect (for instance, via an internal hydrocephalus and dilation of the
third ventricle) and that there are also nontumoral causes (arteriosclerosis,
optochiasmatic arachnoiditis). In the very early stages it is actually not always
easy to differentiate a unilateral retrobulbar neuritis from a beginning Foster
Kennedy syndrome, particularly since signs of papilledema are usually missing
on the opposite side at this stage. Only the bilaterality of the visual field changes
will give actual proof of an intracranial lesion. One has to search for the first
minute signs of visual field changes in the other eye. Such signs usually consist
of minimal losses in the superior temporal quadrant. Dubois-Poulsen is right in
pointing out how important it is to search for such field defects of the better
eye by repeated tests. In addition to the central scotoma in the eye with the
reduced vision, here usually occur during this stage bitemporal superior or
perhaps inferior quadrant defects as signs of damage to the crossing fibers of
the optic nerves in the area of the median parts of both optic nerves or the
anterior inner recess of the chiasm (Fig. 3-5).

Thus an extension of the tumor into the chiasmal region may cause the
development of a *bitemporal hemianopia.* We have observed a tumor of the
base of the frontal lobe with extension into the suprasellar region that caused
a bitemporal superior quadrantanopia (Fig. 3-5).

An actual homonymous hemianopia or homonymous quadrantanopia with
distinctly incongruous field defects occurred in our series only twice. The tumor
causing these defects expanded from the frontal area into the temporal direction. Guillaumat and Robin recorded such defects in 8% of their patients with
frontal lobe tumors. We are convinced that in such cases tumor pressure from
above causes a tract hemianopia (Fig. 3-6).

Tumors located in the frontal region may also cause a homonymous hemianopia by remote rather than by direct effect. This is clearly demonstrated by two

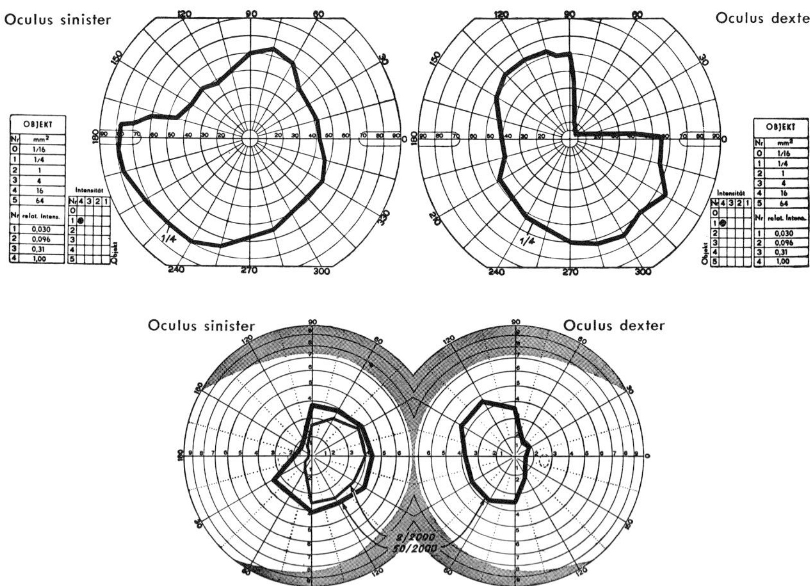

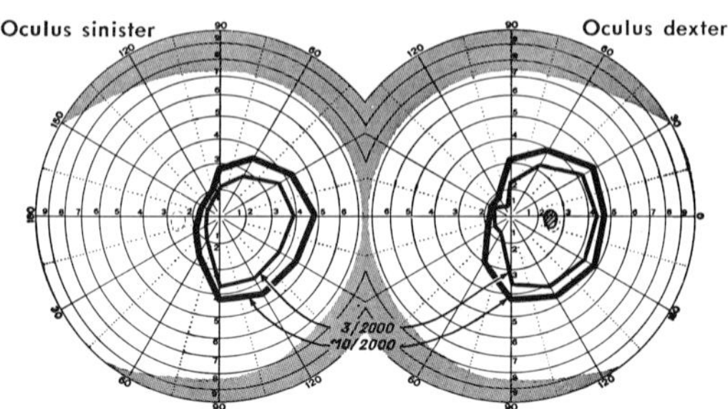

Fig. 3-5. Glioblastoma multiforme at the base of the left frontal lobe with expansion into the suprasellar region. Above, Bitemporal asymmetrical superior quadrantanopia on the Goldmann perimeter. Below, Almost complete asymmetrical bitemporal hemianopia on the tangent screen. These signs indicate interference of the tumor with the anterior chiasmal angle.

Fig. 3-6. Malignant glioma of the right frontal lobe with expansion into the temporal region. Left homonymous hemianopia with sparing of the macula (tract hemianopia?)

patients in our series (one with a falx meningioma of the middle third and the other with a parasagittal meningioma of the anterior third of the sinus), both with a distinct homonymous hemianopia. The field defect must be interpreted as a propagation of the pressure caused by the tumor on the optic tract. The mechanism of this propagation is still quite obscure. Possibly, compression against an artery at the base of the skull may lead to such a lesion of the optic tract (Fig. 3-27).

In spite of normal fields on the perimeter and tangent screen, one may be able to demonstrate a so-called *pseudohemianopia* (Silberpfennig) in frontal lobe tumors. As a rule, it is associated with a gaze palsy toward the side opposite the tumor or a conjugate deviation toward the side of the tumor. This pseudohemianopia consists of a loss of the visual attention toward the side of the gaze palsy. It is assumed that objects in the missing "field of gaze" do not attract attention in spite of an intact "visual field" on this side. The asymmetry of this phenomenon alone indicates that it is not due only to a general lack of attention, as frequently seen in patients with frontal lobe tumors. In such patients the fact that, in addition to the voluntary and the command movements, the optically induced movements (including the optokinetic nystagmus) are impaired in the field of the gaze palsy suggests the following mechanism: the occipital optomotor centers probably function properly only if they are adequately coordinated with the frontal oculomotor centers. An object must stimulate some interest before the eyes will fix it. This interest originates in the frontal centers. It is noteworthy that a gaze palsy as well as a pseudohemianopia can be temporarily improved by an increase in the general attention, by a stimulation of the visual attention, and by a stimulation of the vestibular apparatus. We have observed a patient with an astrocytoma of the frontal lobe in whom the visual fields for very small white and red targets were normal. However, a crude simultaneous test of both visual fields with the hand demonstrated a distinct visual inattention on the side opposite the tumor, although without an accompanying disturbance of the conjugate eye movements.

The *scotomas* are on the side of the optic atrophy and may be central, paracentral, or cecocentral (either relative or absolute), as previously mentioned. Especially in the early stages, they can be the very first signs of a unilateral lesion of the optic nerve (Fig. 3-3).

A markedly *concentric contraction* or *spiral visual fields* in frontal lobe tumors should be interpreted with caution. Usually they are the consequence of the general apathy and lack of impulse on the part of the patient (Ecker) (Figs. 3-4 and 3-7).

A unilateral exophthalmos occurred three times in our series. It should be considered a sign of the direct transmission of tumor pressure into the ipsilateral orbit.

It is known that optomotor centers for the horizontal voluntary schematic gaze movements are located in the paracentral part of the second frontal con-

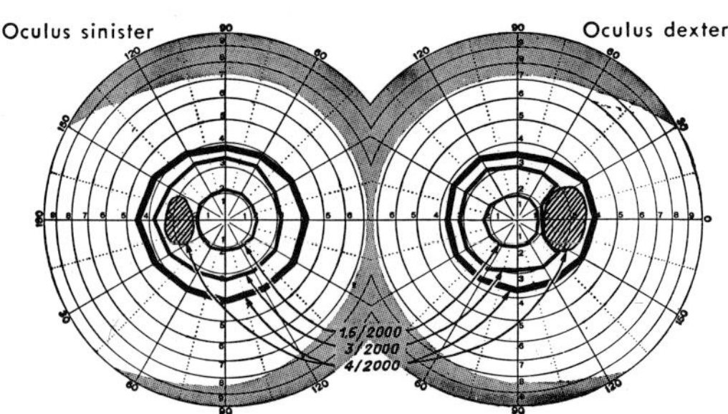

Fig. 3-7. Astrocytoma of the frontal lobe (edge of pallium) with expansion through the falx and corpus callosum toward the right side. Enormous bilateral enlargement of the blind spot (bilateral choked discs of 5 diopters prominence; amblyopic attacks!). Insignificant concentric constriction of the peripheral fields (patient became exhausted rapidly).

volution (area 8 of Brodmann). Nevertheless, we were unable to observe gaze palsies, with or without conjugate deviation, in our patients with frontal lobe tumors. However, during epileptic attacks we found two instances of *turning of the eyes toward the side opposite the tumor*, together with a corresponding torsion of the head into the same direction (frontal adversive attack). It is striking that gaze palsies in patients with frontal lobe tumors are so rare and of such short duration. It is possible that the missing conjugate functions are vicariously taken over by the opposite hemisphere after a brief interval.

There are reports in the literature (Silberpfennig) of cases of gaze palsies accompanied by conjugate deviations toward the side of the tumor. Some of the patients show an isolated impairment of voluntary and command movements, with intact optically induced pursuit and vestibular movements. Others show impairment of the schematic (frontal) and reflex (occipital) movements associated with pseudohemianopia (see above). Since a lesion in the occipital area is not likely with a tumor in the frontal region, one must assume in such a case that the occipital gaze mechanism does not function properly because the connection with the frontal oculomotor centers (whose function, among others, is to stimulate interest for objects in the field of fixation) is missing.

Gaze palsies due to frontal lobe lesions are practically always horizontal; vertical gaze palsy in cortical frontal lesions is extremely rare and therefore is indicative of a lesion in the collicular region. For further information on the theories concerning these facts refer to Kestenbaum and de Recondo.

In diffuse frontal lobe lesions, in addition to a horizontal gaze palsy, there may also be an impairment of optokinetic response. The fast phase of the optokinetic nystagmus is impaired, whereas the slow phase (identical to a sort of pursuit movement) remains intact (p. 58).

A true gaze palsy must be differentiated from a motility disturbance due to lack of attention as seen frequently in patients with frontal lobe tumors. The latter, as a rule, is equal on both sides. In addition to an impairment of the schematic movements, the pursuit movements characteristically may be more affected and involved earlier than optically induced movements (Kestenbaum).

Nystagmus in frontal lobe tumors is rare. We have observed it in only four patients. The amplitude was fine, medium, and even coarse. The direction in one patient was toward the side of the tumor and in another toward the opposite side. It should be kept in mind that minor gaze pareses occasionally manifest themselves in the form of nystagmoid movements.

The general symptoms of increased intracranial pressure in patients with frontal lobe tumors are quite pronounced. We found bilateral papilledema in one half of the patients, mostly with a prominence of from 1 to 3 diopters and frequently with numerous hemorrhages. More developed papilledema on the side of the tumor occurred in about 20% of the cases, especially in meningiomas. A unilateral papilledema without optic atrophy or central scotoma on the opposite side (in other words, without a Foster Kennedy syndrome) occurred twice, both times on the tumor side. There is not an absolute but a considerable correlation between increased intracranial pressure found during surgery and an existing papilledema. Nevertheless, it should be stressed that there are frontal lobe tumors with markedly increased intracranial pressure that will not produce papilledema. Of twenty-seven patients in whom increased intracranial pressure was demonstrated, ten did not show papilledema. Interestingly, almost none of the meningiomas caused papilledema. Unilateral sixth nerve paresis with a corresponding subjective diplopia has been noted only in one tenth of the patients. Bilateral paresis occurred only twice. A unilateral mydriasis as part of the clivus ridge syndrome previously described was always on the side of the tumor. However, it is a relatively rare sign and occurred only eight times among a total of forty-six patients.

In summary, it can be stated that the symptoms as well as the ocular signs are frequently missing in patients with frontal lobe tumors. Further, the Foster Kennedy syndrome, which is frequently considered almost pathognomonic for an involvement of the frontal lobe, is practically never found in its typical form. It occurs occasionally in an incomplete form, with a unilateral optic atrophy and a contralateral normal or blurred disc if the tumor lies at the base of the lobe and expands toward an optic nerve. Bitemporal field defects result if a tumor expands toward the chiasmal region. Homonymous defects are the direct result of pressure by a tumor on the optic tract from above, or they must be interpreted as a remote effect of a frontal lobe tumor on the tract. Unilateral or bilateral scotomas in frontal lobe tumors are the first signs of a pressure lesion of one or both optic nerves. Frequently they precede the Foster Kennedy syndrome. The rather rare gaze palsies, with or without conjugate deviation, are occasionally accompanied by a pseudohemianopia (that is, a lack of visual attention in the missing "field of gaze"). General signs of increased intracranial pressure are frequent and pronounced, especially papilledema. Nevertheless, it happens that very large tumors remain completely "silent" as far as ocular general signs of an increased intracranial pressure are concerned. Concentric contractions of the visual fields should be evaluated with caution. Usually they are an indication of a general lack of attention of the patient.

Tumors of the temporal lobe

The so-called uncinate fits are important as an indication of an irritation of the temporal lobe. These are paroxysmal, usually unpleasant olfactory or gustatory hallucinations of brief duration, which occasionally are accompanied by peculiar sniffing movements or smacking of the lips. These uncinate fits may occur as isolated symptoms or occasionally in association with strange dreamy states. They may even present themselves as the aura of epileptic attacks. These epileptic convulsions dominate the initial clinical picture of temporal lobe tumors and manifest an aura (olfactory, gustatory, optic, acoustic) in about 40% of the cases. Other symptoms of temporal lobe epilepsy are the "déjà vu" or "déjà raconté" phenomena, psychomotor epileptic fits, and paroxysmal optic hallucinations in the form of entire scenes and episodes, which will be described in more detail below. If the tumor is localized in the dominant hemisphere, there may be a sensory aphasia in the form of a disturbed comprehension of speech or, if the anterior part of the temporal lobe is involved, a difficulty in finding the proper words (Fig. 1-2). However, it must be stressed that also in temporal lobe tumors these local symptoms may often be partially or totally masked by the general signs of increased intracranial pressure, producing not only headache, vomiting, and vertigo but also herniation syndromes with consequent compression of the brain stem, characterized by unilateral mydriasis, ptosis, oculomotor or sixth nerve pareses, and pyramidal tract

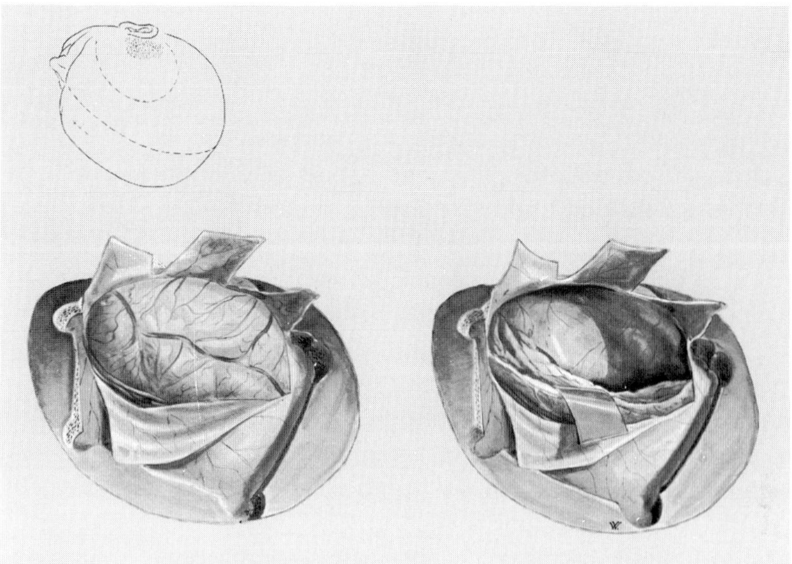

Fig. 3-8. Astrocytoma of the right temporal lobe in a 37-year-old patient before and after extirpation (total right temporal lobectomy).

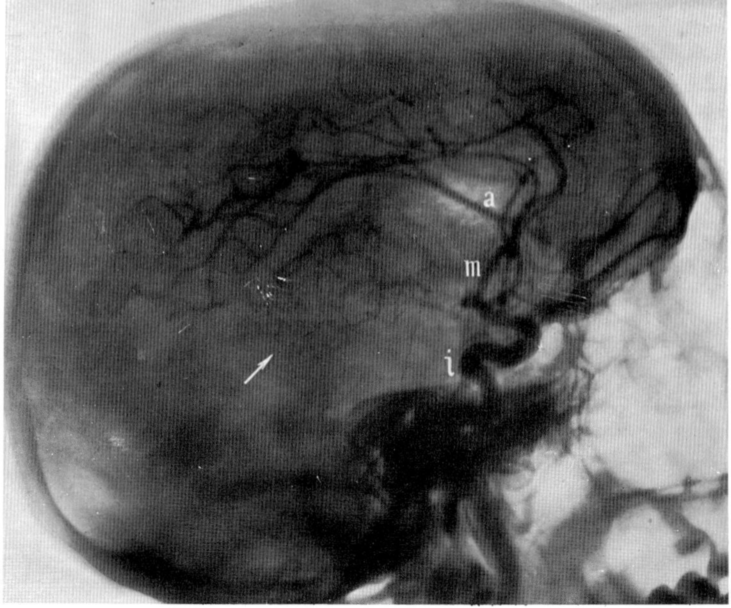

Fig. 3-9. Glioblastoma multiforme of the left temporal lobe. Arteriogram of the left carotid artery. Abnormal vessels in the tumor area (arrow). Upward displacement of middle, **m,** and anterior, **a,** cerebral and internal carotid, **i,** arteries.

signs. Sometimes, especially in lesions of the right temporal lobe, the signs of increased intracranial pressure represent the whole neurologic symptomatology (Figs. 3-8 and 3-9).

The number of localizing ocular symptoms associated with frontal lobe tumors is disappointingly small. This situation, however, is quite different in temporal lobe tumors. Even the *subjective* symptoms, if present, may furnish important localizing cues. The major or minor epileptic fits or dreamy states so typical in patients with temporal lobe tumors may be accompanied by *visual hallucinations* in addition to olfactory and gustatory hallucinations. The patient has sensations (usually on the side of the field defect, provided such a defect is present) of objects, scenery, people, animals, or plants—hallucinations of a definite, formed, and highly organized nature, often reenacting scenes in his past life, with little relation to the immediate surroundings. Their duration is limited to a few seconds. There are also disturbances of the sensorium of varying, but usually slight, degree. Sometimes the visual hallucinations represent an aura of idiopathic or jacksonian epileptic attacks. It is assumed that these hallucinations are the result of an irritation of the optic radiation, which is spread out in the temporal lobe.

In our own series of thirty patients with temporal lobe tumors we found olfactory hallucinations relatively frequently (30%). Visual hallucinations were quite rare. Actually, they occurred only twice. In both instances there were undifferentiated, unformed, and primitive hallucinations in the form of glowing rings, bright stars, multicolored circles, or snowflakes. In the literature it is emphasized time and again that undifferentiated visual hallucinations are characteristic for lesions of the occipital lobe, whereas complex and differentiated visual sensations suggest a temporal seat. Actually, both of our patients show a distinct expansion of the temporal tumor toward the occipital lobe. *It is correct that temporal lobe tumors are more frequently the cause of differentiated and complex visual hallucinations* (Horrax). On the other hand, this type of hallucination is not exclusively found in association with temporal lobe tumors and is not absolutely pathognomonic for such a location. There may be unformed primitive photopsias in patients with tumors limited entirely to the temporal lobe without occipital expansion. According to a number of publications (Paillas, Boudouresques, and Tamalet; Weinberger and Grant), complex visual hallucinations may also originate from a parietal or occipital location. Nevertheless, in the presence of highly differentiated visual hallucinations one should consider primarily the temporal lobe. However, in making a topical diagnosis, and especially in differentiating between the temporal and occipital lobes, one should not depend entirely on these symptoms but should consider the entire neurologic status of the patient. It certainly would be a mistake to base conclusions regarding the seat of the tumor on the nature of the hallucinations. Visual hallucinations generally occur in the blind half of the visual fields, and if there is no hemianopia, they occur in the fields opposite the side of the tumor. There are, however, records of visual hallucinations in the seeing half of the fields.

Ocular symptoms are found in about one half of the patients with temporal lobe tumors. Since visual hallucinations occur quite rarely, these symptoms are limited to manifestations of a generally increased intracranial pressure in the form of flickering or fog before the eyes, temporary blackouts, or unilateral or bilateral decrease of vision. Occasionally there is photophobia or diplopia.

The visual field changes must be counted among the most important *visual signs* in temporal lobe tumors. They could be demonstrated in two thirds of the patients in our series. Of course, the reduced general condition or somnolence of some patients will not permit an exact determination of the visual fields. Actually, this was the case only in one tenth of our patients. If an exact testing of the fields on the perimeter or tangent screen is impossible, the confrontation method or the determination of a relative hemianopia (Thiébaut) may still furnish important clues.

The visual field changes in patients with tumors of the temporal lobe may be labeled with the password *homonymous hemianopia*. We must differentiate between a complete homonymous hemianopia involving both the superior and inferior quadrants and a quadrantanopia. Our own series as well as a number of reports (for instance, Guillaumat and Robin) indicate that a *complete homonymous hemianopia is considerably more frequent than a quadrantanopia.*

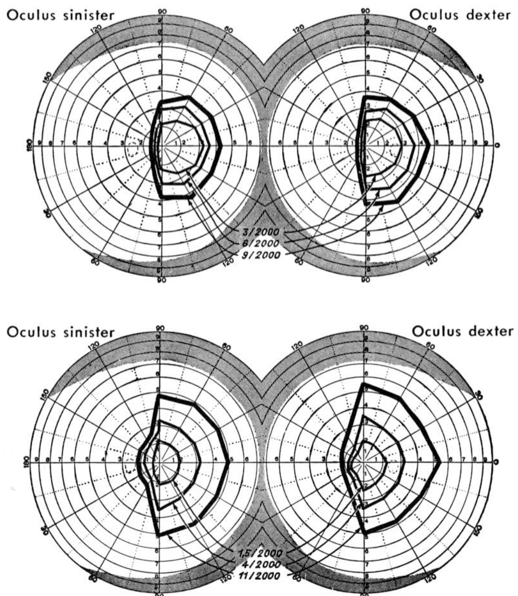

Fig. 3-10. Above, Astrocytoma of the right temporal lobe. Below, Granuloma in the posterior part of the right third temporal gyrus. In both cases there was a complete homonymous left hemianopia with sparing of the macula and a slight incongruence of the visual field defects.

The former occurs in about one third and the latter in about one sixth of all cases. This greater frequency of complete homonymous hemianopias (Fig. 3-10), in our opinion, is due to the expansive growth of the tumors, especially into a posterior direction. We found the macula spared in all types of complete homonymous hemianopias. There were no instances of division of the halves of the visual fields through the point of fixation in patients with tumors of the temporal lobe (described by Penfield as the result of excision of the posterior portion of the temporal lobe), as is seen so frequently in those with occipital lobe tumors. In contrast to hemianopias caused by occipital lobe tumors, those due to tumors of the temporal lobe are characteristic in our series because of a striking *incongruence of the two field defects* or the remaining islands of vision. Time and again this incongruence is stressed by numerous authors (Harrington; Traquair) as typical for the temporal location (especially on the anterior part). *This incongruity of the visual fields can go so far that occasionally there is a complete hemianopia on one side with only a quadrantanopia on the other side* (Fig. 3-11). There is a difference of opinion as to whether the larger defect is on the side of the lesion or on the opposite side. Our own opinion is that this sign does not permit any conclusions as to the seat of the tumor. It should be stressed emphatically that the complete homonymous hemianopia associated with temporal lobe tumors does not differ in principle from that associated with occipital lobe tumors. Even an incongruence of the visual fields is no definite proof for a temporal location, if for no other reason than that tract hemianopias must be considered in the differential diagnosis. However, it is a fact that the congruence of the field defects becomes increasingly more complete the farther posterior the tumor is located in the radiation.

In summarizing his experience, Kestenbaum states that a pronounced incongruity of homonymous hemianopic field defects (a positive "incongruity sign") for all practical purposes rules out a lesion in the parietal or occipital part of the visual radiation. On the other hand, congruence may be found in any loca-

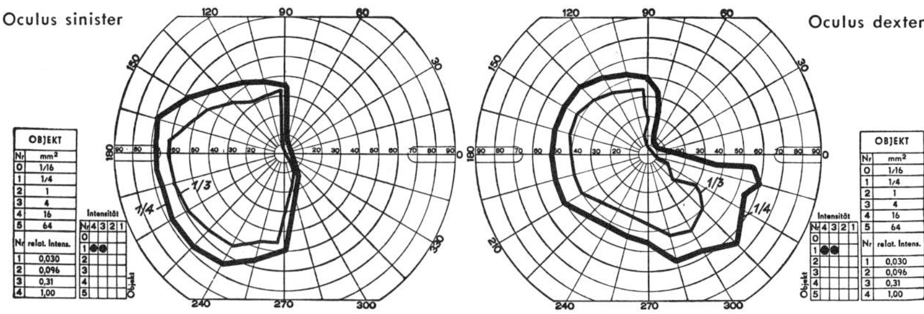

Fig. 3-11. Glioblastoma multiforme of the left temporal lobe. Pronounced incongruence of the field defects: left eye with subtotal hemianopia of the right part of the field; right eye with a quadrant defect of the right superior part of the field. The patient reported formed visual hallucinations!

tion (that is, the optic tract as well as the optic radiation). A homonymous hemianopia merely indicates an interruption of the visual pathways above the chiasm. Incongruence of the visual fields is useless for the differentiation of lesions in the infrageniculate or suprageniculate tracts. Traquair even suggests that the incongruity of the field defects, as seen in association with temporal lobe tumors, is indicative of damage to the optic tract below the temporal lobe. The fact that temporal lobe hemianopias usually leave the macular function intact seems to refute this opinion because optic tract hemianopias result relatively early in the formation of central scotomas (isolated or splitting of the macula by a complete homonymous hemianopia). This opinion has been generally abandoned since Meyer published his fundamental investigations on the course of the optic radiation in the temporal lobe. His anatomic findings demonstrated that the incongruity of the visual fields, especially in processes of the anterior part of the temporal lobe, can be explained by the evidence of a dissociation of the fibers of corresponding retinal points in this region (Fig. 3-14).

A superior homonymous quadrantanopia or sector-shaped defect is typical and almost pathognomonic for tumors of the temporal lobe (Figs. 3-12 and 3-13). We have observed them in one sixth of our patients. In the majority the

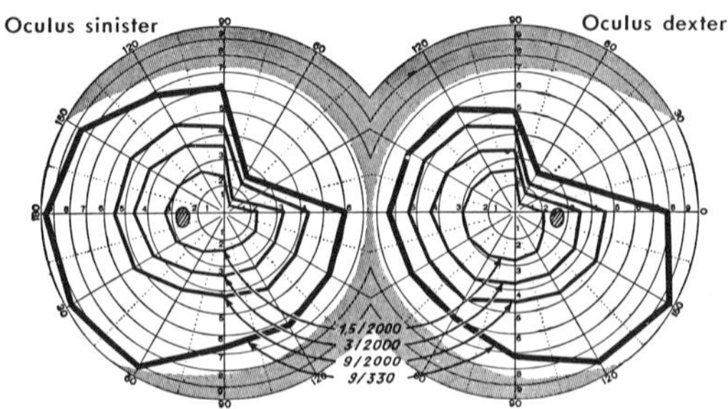

Fig. 3-12. Protoplasmic astrocytoma of the left temporal lobe the size of a walnut. Tumor situated in the depth of the third left temporal, fusiform, and hippocampal gyri. Typical right superior homonymous quadrantanopia, a sign of tumor interference with Meyer's loop. Congruence of field defects.

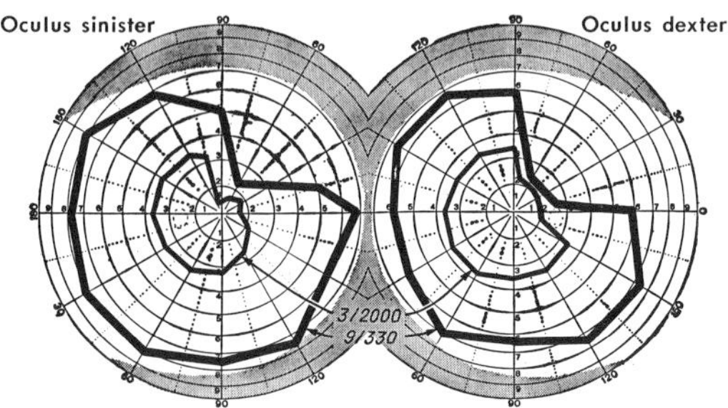

Fig. 3-13. Astrocytoma of the left temporal lobe. Right superior homonymous quadrantanopia. Incongruence of quadrant defects.

defect was rather congruous and in a smaller number incongruous. This defect is the *result of damage to Meyer's loop,* the ventral bundle of the secondary optic radiation from the external geniculate body, which swings forward in the temporal lobe to form a loop around the apex of the lower horn and to run from there ventrally into the anterior part of the calcarine cortex (Fig. 3-14). Cushing was the first to investigate the superior homonymous quadrant defects in temporal lobe tumors and to point our their localizing value. The incongruity of these quadrant defects can be easily explained as damage to the optic radiation, particularly if it involves the anterior part of the optic radiation where the crossed fibers still have an anterolateral and the uncrossed fibers a posteromedian position in their course. If one remembers that only tumors of the posterior part of the temporal lobe area interfere with the optic radiation, one will understand the relatively rare event of this type of field defect. *But if present, a superior homonymous quadrantanopia definitely has localizing value.* The importance of the most careful perimetry, the use of very small targets, and the plotting of a number of isopters is urgently stressed. Otherwise such field defects may go unnoticed (Krayenbühl). Frequently they start merely with a slight indentation of the outer isopters in the superior quadrants and are likely to escape detection (Fig. 3-15).

In agreement with the experience of other authors we have also observed an *inferior quadrantanopia* (involvement of the superior part of the temporal lobe) in addition to a superior quadrantanopia (involvement of the inferior part of the temporal lobe). This defect occurs much less frequently (one-fifteenth). If a superior quadrantanopia or sector-shaped defect increases in size, the incongruity becomes less distinct, and the defect changes to a complete homonymous hemianopia. This is always a sign that the tumor process expands into a superior and particularly a posterior direction (Fig. 3-16).

In addition to these homonymous field changes, we have also observed concentric constrictions of an unspecific nature that had to be interpreted as signs of a beginning optic atrophy secondary to a papilledema. Of course, they have no localizing value.

Other specific ocular signs of an involvement of the temporal lobe may be particularly helpful in a complicated diagnostic situation when complete homonymous hemianopias are difficult to differentiate from those caused by occipital lobe tumors. Such a sign is the *homolateral mydriatic fixed pupil described in the section on general signs of the increased intracranial pressure as a component of the clivus ridge syndrome.* It has been stated (p. 23) that a unilateral mydriatic fixed pupil in patients with temporal lobe tumors may be caused by direct pressure of the neoplasm on the oculomotor nerve as well as by such pressure of the temporal lobe herniating through the tentorial notch. Among thirty patients with temporal lobe tumors, we have observed this phenomenon of a unilateral mydriasis six times on the side of the tumor and three times on the opposite side. Once there was miosis in place of mydriasis which,

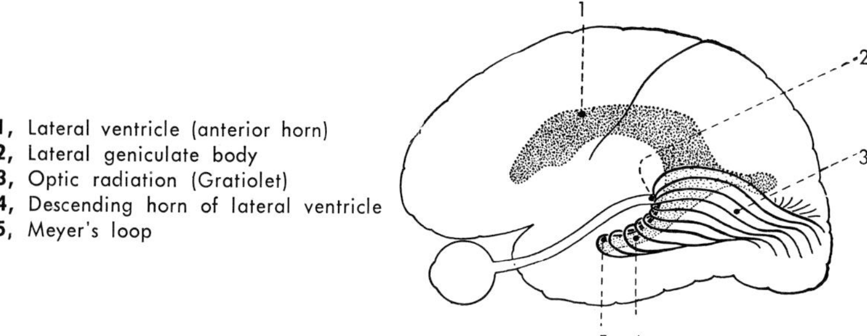

Fig. 3-14. Diagram of Meyer's loop. The ventral fibers of the optic radiation pass from the lateral geniculate body anteriorly into the temporal lobe to form a loop around the tip of the descending horn.

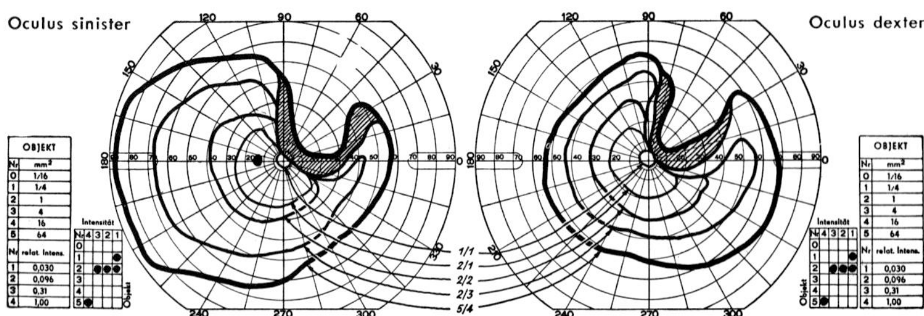

Fig. 3-15. Tumor of the left temporal lobe. Typical right superior homonymous quadrantanopia. This is the initial stage, with involvement of only part of the right superior quadrants. The shaded areas indicate the progressive deterioration of the visual field during the time of observation and before surgical intervention. (Case of Dr. Guillaumat; from Dubois-Poulsen.)

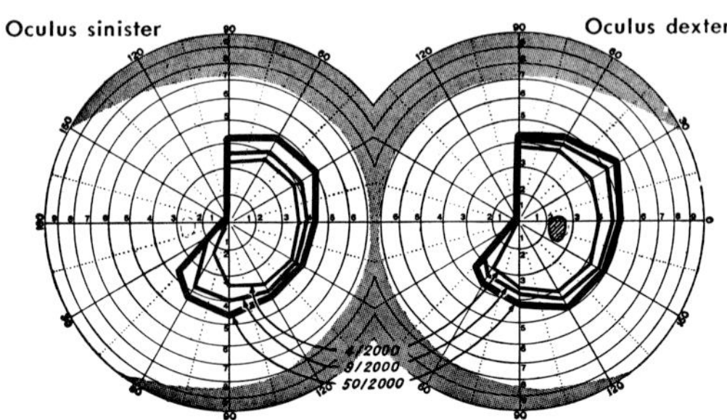

Fig. 3-16. Glioblastoma multiforme of the right temporal lobe (second and third basal temporal gyri). Left partial homonymous hemianopia. The left superior quadrants are missing completely. The central parts of the left inferior quadrants are still preserved. This is an intermediate form between a superior homonymous quadrantanopia and a complete homonymous hemianopia. The neoplasm expanded from the front and below in a posterosuperior direction.

according to Fischer-Brügge and Tönnis, we took as an early sign of irritation on the oculomotor nerve (p. 24). However, as the clivus ridge syndrome is a general sign of increased intracranial pressure, unilateral mydriasis should be interpreted with caution as a localizing sign in temporal lobe tumors.

Pareses of the oculomotor nerve, especially in the late stages of tumor growth, may form part of the clivus ridge syndrome. We have seen such oculomotor pareses in one sixth of our patients. Usually such pareses can be recognized by the divergent position of the eyes (due to a paresis of the medial rectus muscles) and by unilateral ptosis. We regard the pareses of the oculomotor nerves as components of the clivus ridge syndrome. In the case of temporal lobe tumors there is also the possibility of a direct pressure effect on the nerve, as described previously.

A *contralateral central facial palsy* in patients with temporal lobe tumors has been seen more frequently than pareses of the extrinsic ocular muscles—in approximately two tenths of our patients. In a central facial palsy the frontal branch of the nerve is not involved; thus no lagophthalmos occurs. In contrast to the ordinary supranuclear paralysis of the facial nerve, the emotional motility of the mimic muscles, according to Kennedy, is impaired in addition to the spontaneous motility (pressure on the thalamus).

Occasionally an *ipsilateral disturbance in the area of distribution of the trigeminal nerve with hyporeflexia or areflexia of the corresponding cornea* (Krayenbühl) is found in patients with temporal lobe neoplasms.

At times a unilateral *exophthalmos* is observed in patients with temporal lobe tumors. We have seen three such cases among thirty patients. It occurred twice in bilateral form as a general sign of increased intracranial pressure. A unilateral exophthalmos in the temporal lobe syndrome is mostly the manifestation of an extracerebral growth in this region (for instance, a meningioma).

Disturbances of motility have already been discussed in connection with the clivus ridge syndrome. They are oculomotor pareses as the result of a direct pressure effect of the temporal lobe herniating through the tentorial notch. We have never observed supranuclear gaze disturbances, although there are reports of disturbances of convergence and conjugate vertical movements (remote effect on the brain stem, especially the quadrigeminate plate) in the literature (Walsh and Hoyt; Bing). Relatively often (in about one third of our patients) we found a horizontal *fine or medium,* easily exhaustible *nystagmus* of the gaze nystagmus type. This also must be a remote effect of the tumor on the brain stem.

The *general signs of increased intracranial pressure* are usually quite distinct. Bilateral papilledema, frequently with hemorrhages and with a prominence of from 1 to 4 diopters, was found in about one half of all patients (Tönnis found them in 82% of his cases). Among thirty patients, unilateral papilledema occurred three times—twice on the side of the tumor and once on the opposite side. There was considerable correlation between an increased intracranial pressure determined during surgery and the presence of papilledemas. Nevertheless, in one sixth of the patients with increased intracranial pressure, no papilledema was seen. In our series of thirty patients with temporal lobe tumors, a sixth nerve paresis as a general sign of increased intracranial pressure was observed only once.

The significance of the ocular symptoms in patients with temporal lobe tumors can be summarized as follows: highly differentiated, formed hallucinations suggest a temporal location, whereas primitive, unformed photopsias are more typical for an occipital site. This differentiation is only conditionally, not absolutely, correct—differentiated as well as undifferentiated hallucinations may be triggered both by occipital and temporal lobe tumors. Superior homonymous quadrantanopia must be considered as pathognomonic for a temporal seat of the tumor. However, it is rare compared with the complete incongruous homonymous hemianopia. Practically, however, the latter cannot be differentiated from a similar lesion caused by an occipital tumor. Of localizing importance are an ipsilateral mydriatic fixed pupil and oculomotor pareses as components of a clivus ridge syndrome as well as the frequent contralateral central facial palsy. There are no specific general signs of an increased intracranial pressure typical for the temporal lobe.

Tumors of the parietal lobe

The posterior central gyrus, by definition, is part of the parietal lobe. Lesions of this anterior part of the parietal lobe produce a characteristic syndrome known as the *syndrome of the posterior central gyrus*. An irritation of the postcentral convolution manifests itself in the form of sensory jacksonian fits. The region of the initial sensory sensation points out the localization of the focus. The destructive loss of this specific cortical area results in a disturbance of the epicritic sensibility. According to the seat of the lesion, the arm or leg on the contralateral side is involved. Kinesthesia, position sense, vibration sense, and two-point discrimination are decreased or abolished. Recognition of objects by touch is limited or impossible (astereognosis). Lesions of the medulla of the supramarginal gyrus cause apraxia in the form of bilateral motor apraxia (ideokinetic apraxia or Liepmann's apraxia)—inability to carry out purposeful movements in the absence of motor paralyses or sensory disturbances. Affections of the various areas of the angular gyrus and its adjacent regions in the temporal or occipital lobe on the dominant hemisphere result in loss of comprehension of the written word and consequent inability to read (alexia), often associated with loss of ability to write (agraphia); in inability of calculation (acalculia); in disturbances of the body scheme; and in right-left disorientation. *Gerstmann's* angularis syndrome consists of finger agnosia, right-left disturbance, agraphia, acalculia, and right homonymous hemianopia. Lesions in the region of the angular gyrus of the nondominant hemisphere cause disturbances in spatial orientation (topographic agnosia) and a left homonymous field defect. Biparietal lesions may produce global visual object agnosia (inability to identify objects and to recognize persons), eventually associated with lack of voluntary control of eye movements (Balint's syndrome). (See Figs. 1-2 and 3-17.)

Parietal lobe tumors may produce ocular symptoms of localizing significance that should not escape the ophthalmologist during his examination. Some of the most important sensory cortical centers are situated in the parietal lobe. As mentioned previously, defects caused by a parietal tumor involve important functions such as writing, reading, calculation, stereognosis, and others, in ad-

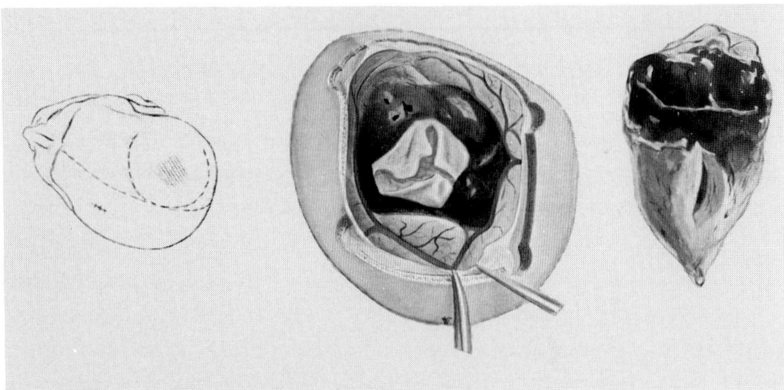

Fig. 3-17. Malignant glioma of the right parietal lobe in a 41-year-old patient before and after surgical extirpation (right parietal osteoplastic craniotomy).

dition to sensory disturbances. A tumor in the region of the angular gyrus of the dominant hemisphere may cause *alexia,* or *word blindness—the patient is unable to read (in other words, he does not comprehend the significance of written words, musical notes, or figures).* In our series of twenty-seven proved cases of parietal lobe tumors we found complete alexia three times and an incomplete form (dyslexia) once. Alexia is usually combined with *agraphia*—the inability to write spontaneously or on dictation. Alexia without agraphia occurs in patients who have a unilateral occipital lobe lesion of the dominant side and a lesion of the splenium, but an intact angular gyrus. Alexia can be observed with or without hemianopia (usually right-sided homonymous hemianopia); it can occur independent of any lesion of the optic radiation. In our own patients the alexia and agraphia usually were combined with other sensory-aphasic and with apractic disturbances. We found no cases of visual object agnosia (that is, *mind blindness*—an inability to recognize and identify objects although they are seen). Visual object agnosia is also considered a result of damage to the angular gyrus or to its association pathways. Even though we have not seen a typical case of visual object agnosia in our series of patients with tumors of the parietal lobe, we have observed one patient with findings suggestive of such a condition. This patient stated that he recognized his physician visually but that he was unable to comprehend the entire significance of the situation he visualized. We were of the opinion that this was an incomplete form of visual object agnosia. So much for the subjective ocular symptoms concerning the higher visual centers in the parietal lobe (see p. 8 and Fig. 1-1).

In two patients in our series we were able to observe an aura before jacksonian fits, which always occurred in the form of *undifferentiated primitive visual hallucinations* such as stars, flickering, or dots before the eyes. Such visual sensations occur infrequently with parietal lobe tumors. They suggest an expansion of the tumor toward the optic radiation. As can be expected from the prox-

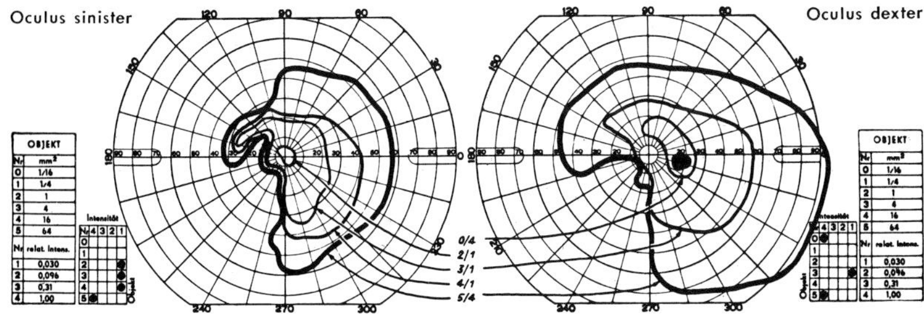

Fig. 3-18. Tumor of the right parietal lobe. Left inferior homonymous quadrantanopia with distinctly incongruous field defects. Enlargement of the right blind spot due to papilledema. (After Dubois-Poulsen.)

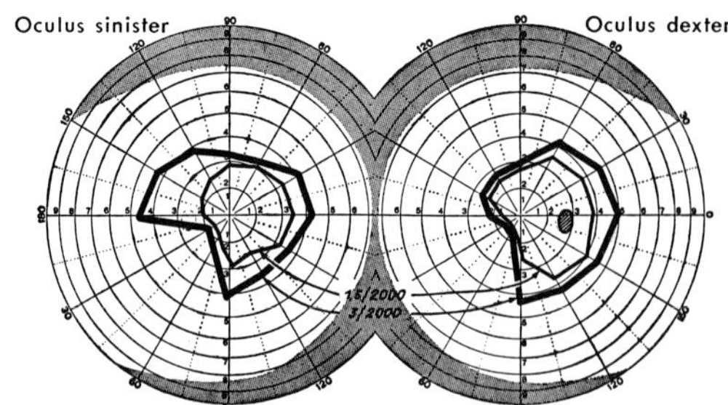

Fig. 3-19. Cortical and subcortical sarcoma of the right parietal lobe. Rather congruous left inferior quadrantanopia.

imity of the occipital lobe, these hallucinations usually are unformed and undifferentiated in character.

Parietal lobe tumors will interfere with the optic radiation only in case of a relatively extensive downgrowth. This is the reason for the *rather infrequent occurrence of visual field changes.* We observed them in only one third of our patients; Guillaumat and Robin reported them in only one fourth of their series. Some reports (Traquair; Dubois-Poulsen) indicate that *inferior homonymous quadrant defects* are characteristic for parietal lobe tumors (Figs. 3-18 and 3-19). This type of defect occurred in our series only once. In another patient there was a bilateral superior homonymous quadrant defect. We observed a *complete homonymous hemianopia* much more frequently (especially in gliomas, most seldom in meningiomas). It occurred in seven of twenty-seven patients with parietal lobe tumors. A *relative hemianopia* in the form of "visual inattention" that could not be demonstrated on the perimeter or tangent screen was found twice. Our own experience leads us to conclude that inferior homonymous quadrantanopia, which supposedly is characteristic for parietal lobe lesions, actually occurs quite rarely. Other visual field changes (especially complete hemianopias, mostly with sparing of the macula and with rather congruous defects) are difficult to differentiate from those caused by occipital lobe tumors. This differentiation actually is quite irrelevant because numerous parietal lobe tumors extend into the occipital lobe, and vice versa. *It is impossible to distinguish a "parietal" from an "occipital" homonymous hemianopia on the basis of the visual field.* Lack of awareness of the homonymous field defects (anosognosia) and a strange sensation of persistence of an afterimage in the blind half of the field (paliopsia) accompany many parietal lobe lesions.

We have never observed the disturbances of the conjugate lateral movements reported by David and Hecaen in patients with parietal lobe tumors. *A rotation of the eyes toward one side* (usually opposite the tumor) *during an epileptic seizure* must be considered as a parietal adversive phenomenon—the "associated adversive looking" of Penfield. We have seen two such cases in our series. In view of the cortifugal efferent tracts of the occipital fixation mechanism in the middle and posterior thirds of the optic radiation, one should always search for asymmetrical disturbances of optically induced movements and pursuit movements (cogwheel movement!) and optokinetic nystagmus (p. 58) in patients with suspected parietal lobe tumors. Kestenbaum, by the way, emphasizes the strikingly constant manifestation of a positive optokinetic nystagmus sign (optokinetic nystagmus to the side of the hemianopia or to the side opposite the lesion strikingly less than to the other side) in parietal lobe lesions. Lesions of the parietal lobe are by far the most common lesions seen in association with abnormal optokinetic responses. The most important feature is the *loss of the fast phase of the response to the opposite side* with retention of the pursuit response. Voluntary gaze and the fast phase of nystagmus induced by caloric stimulation are normal with the patient's eyes closed. The abnormal optokinetic response does not depend on the hemianopia: it may be positive with dorsal parietal lesions even when there is no affection of the visual radiation.

Pursuit movements, also, are frequently disturbed. They may show a cogwheel phenomenon (to the side contralateral to the hemianopia). However, patients with a tumor of the parietal lobe in whom this sign is missing completely are seen time and again. Stadlin made some interesting investigations bearing on this subject in a case of a parieto-occipital astrocytoma which are referred to in the section on occipital lobe tumors (p. 185). A gaze palsy of the up movement was seen once. Since there was simultaneously a diminished pupillary reaction to light, we interpreted this vertical conjugate impairment as a remote effect of the tumor on the midbrain. A fine or medium, easily exhaustible horizontal gaze nystagmus was noted in about one fourth of the patients.

In addition to a relatively *frequent contralateral central facial palsy,* we found *hypesthesia of the contralateral cornea* in one fourth of our patients with parietal lobe tumors—usually associated with a corresponding disturbance of the other trigeminal branches. This phenomenon is an analog to other sensibility disturbances in patients with parietal lobe lesions.

General signs of increased intracranial pressure occur in about one half of the patients in the form of bilateral papilledema. It is noteworthy that increased intracranial pressure is frequently found during the operation (in about one fourth of the patients) in spite of the absence of papilledema. Generally, the prominence of the choked disc is rather minor (that is, between 1 and 2 diopters). Signs of optic atrophy secondary to papilledema occur rarely. Other ocular signs of increased intracranial pressure were surprisingly rare. In three of our patients the nonspecific sixth nerve paresis and a unilateral mydriasis were both on the side opposite the tumor!

In summarizing, the localizing symptoms in patients with parietal lobe tumors are the subjective phenomena of disturbances of the higher visual centers that occur in the form of the visual agnosias, especially alexia or dyslexia. An inferior homonymous quadrantanopia, which is somewhat pathognomonic for parietal lobe lesions, is rare. Much more frequent are complete homonymous hemianopias. The latter type, however, can hardly be distinguished from similar defects caused by occipital lobe tumors. Noteworthy are contralateral disturbances of the corneal sensitivity, usually associated with disturbances in the area of distribution of other branches of the trigeminal nerve. The general signs of increased intracranial pressure in the form of choked discs are rather rare or frequently completely absent in spite of a proved increased intracranial pressure.

Tumors of the occipital lobe

Except for the ocular symptoms, the occipital lobe is actually a neurologically silent zone. Manifestations of tumor-induced irritation in this area are epileptic seizures occasionally preceded by unformed visual hallucinations or ocular involuntary deviations. Defects involve primarily the visual field, in the form of complete or incomplete homonymous hemianopia. (However, we know today that small lesions of the striate cortex may occur without a defect in the visual field.) If an occipital tumor extends forward and involves the peristriate and neighboring parietotemporal areas, there may result alexia with or without agraphia (dominant hemisphere), topographic agnosia (nondominant hemisphere), visual object agnosia, achromatopia (inability to recognize colors), and visual irreminiscence (inability to call up visual images from the past). Bilateral lesions of the occipital lobes may lead to total amaurosis, occasionally accompanied by a denial of the blindness (Anton's syndrome). In reality, about 70% of the occipital lobe tumors start with symptoms of increased intracranial pressure (Tönnis) and because of disturbing neighborhood symptoms (parietal, temporal) and early herniation syndromes (tentorium notch, foramen magnum) occasionally lead to a false diagnosis. Pressure atrophy and enlargement of the sella, especially pronounced in occipital lobe tumors, may, in connection with visual disturbances, even simulate a lesion of the sellar region.

Thus the *results of a proper ophthalmologic diagnosis will assume utmost importance* and supply the neurosurgeon with valuable data for a tumor localization (Fig. 3-20).

Occipital lobe tumors occur relatively infrequently. Nevertheless, we are able to report on more than twenty cases limited more or less exclusively to the occipital lobe.

To state the most important fact, *the subjective ocular symptoms are related primarily to the homonymous hemianopia.* However, it must be emphasized that quite a few patients are not aware of their hemianopia and will not

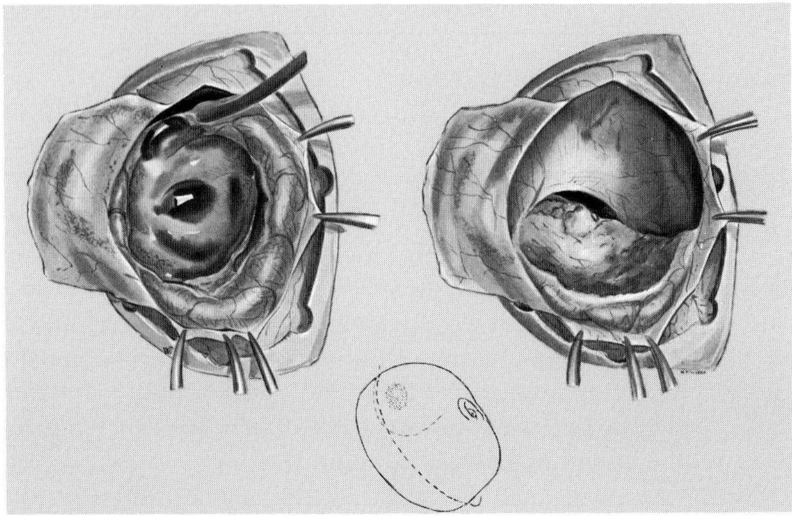

Fig. 3-20. Astrocytoma of the left occipital lobe in a 37-year-old patient before and after extirpation (left total occipital lobectomy).

mention it spontaneously (p. 4). Less intelligent patients, for instance, complain of a vaguely defined loss of vision, especially toward the side of the hemianopia. Much more frequently patients will make *statements that imply a hemianopia:* the patient strikes a door post; the motorist misses a pedestrian or cyclist on one side; the cyclist cannot see the curb correctly; when reading, the patient sees only half a word; or when writing, he always writes past the margin of the paper (especially disturbing with hemianopias on the right side; with hemianopia on the left side there is difficulty in finding the next line of print). Occasionally a patient will only state he cannot see very well with one eye. Another will say that "there is no depth" toward the side of the hemianopia. Much rarer are peculiar deformations of objects: they appear reduced in size (micropsia) or change their size. One patient with an extensive glioblastoma multiforme of the occipital lobe experienced attacks of micropsia associated with visual hallucinations. Similarly, another patient complained of undefined "deformed vision" (metamorphopsia).

Visual hallucinations of an unformed primitive pattern, mostly in the blind half of the visual fields, are surprisingly rare in patients with occipital lobe tumors. Such hallucinations may assume the form of fiery globes, stripes, discs, zigzag lines, or flashing lights of purple, golden, or other bright colors and occur as transient attacks—even with the eyes closed. We observed visual hallucinations only three times in twenty proved cases of occipital lobe tumors. All of them were of the primitive type. Although there are occasional reports that more highly organized forms of visual hallucinations can originate not only in the temporal but also in the occipital lobe (Weinberger and Grant), it can be stated on the basis of reports by numerous authors (Parkinson, Rucker, and

McCraig; Penfield and Rasmussen) as well as our own experience that *occipital lobe tumors usually produce simple primitive visual hallucinations*. If, during the course of the disease, the latter give place to more complex visual sensations, this may be an indication of an expansion of the tumor into the temporal region. Although visual hallucinations are important localizing symptoms in the diagnosis of occipital lobe tumors, their importance unfortunately is diminished by their rare occurrence.

The subjective complaints of reading difficulties may make it difficult at times to decide whether they are caused by only the hemianopic field defect or by an additional alexia. In addition to field defects, twice we observed an unquestionable *alexia* and twice *agraphia* manifested by disturbances of writing. *Alexia without agraphia* occurs in patients with unilateral left occipital lobe lesions and associated lesions of the splenium. Such patients are able to write but are unable to interpret what they see. If the tumor of the occipital lobe responsible for such symptomatology involves also the angular gyrus of the adjacent parietal lobe, it will produce the better known and more frequent syndrome of alexia with agraphia (p. 9).

In our series of patients with occipital lobe tumors there were no instances of associated disturbances of the higher visual centers such as visual object agnosia or mind blindness—the inability to recognize and identify objects although they are seen (Fig. 1-1).

Nor have we observed *cortical blindness* as the functional equivalent of bilateral tumor damage to the occipital cortex, especially the calcarine fissure. There was, however, one case of necrosis of both occipital lobes caused by x-ray treatment after an operation to remove a temporal lobe tumor. Another case occurred after a bilateral ventriculography performed through the occipital lobes. Obviously this caused a localized edema, which proved to be reversible. The characteristic signs of cortical blindness were discussed in detail in Chapter 1 (p. 8). Briefly, cortical blindness is characterized by complete loss of all visual sensations, by loss of reflex lid closure, by intact pupillary reactions on light and convergence, and by intact motility of the eyes. Cortical blindness may occur as an indirect consequence of any supratentorial cerebral tumor causing herniation of the brain through the tentorial notch and thus producing compression of the posterior cerebral arteries and their branches supplying the occipital cortices. Sometimes patients with cortical blindness deny their defect, using the most extravagant confabulations (Anton's syndrome). Cortical blindness must not be confused with *cerebral blindness or double homonymous hemianopia*. In such a case there is complete loss of vision at the onset, but generally only for a relatively short transient interval, and vision reappears in a small central field around the point of fixation (corresponding to both areas of the spared maculas!). Cerebral blindness is usually of vascular origin (arteriosclerosis, hypertension) and not related to cerebral tumors.

Some of the most typical and important symptoms of occipital lobe tumors are the visual field changes, which may be summarized with the phrase *complete or incomplete homonymous hemianopia on the opposite side. Visual field changes were not missing once among our twenty patients with occipital lobe tumors.* Ten (that is, one half) of the patients showed complete homonymous hemianopia with distinct sparing of the macula. In six (approximately

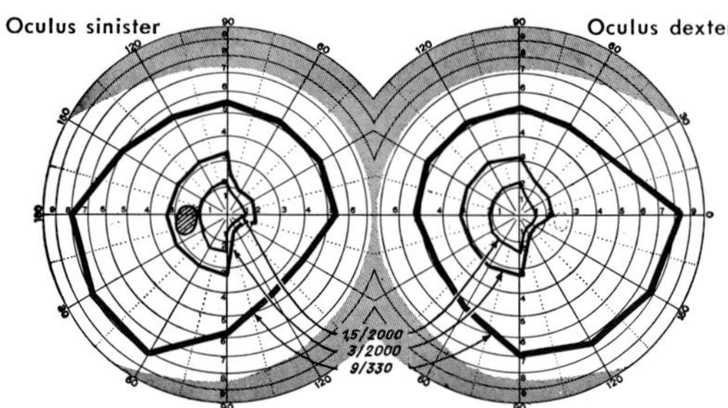

Fig. 3-21. Cherry-sized parasagittal metastatic hypernephroma of the left occipital lobe. Right homonymous hemianopia of the central isopters with distinct sparing of the macula. The 9/330 isopters determined on the perimeter do not reveal the homonymous hemianopia at all!

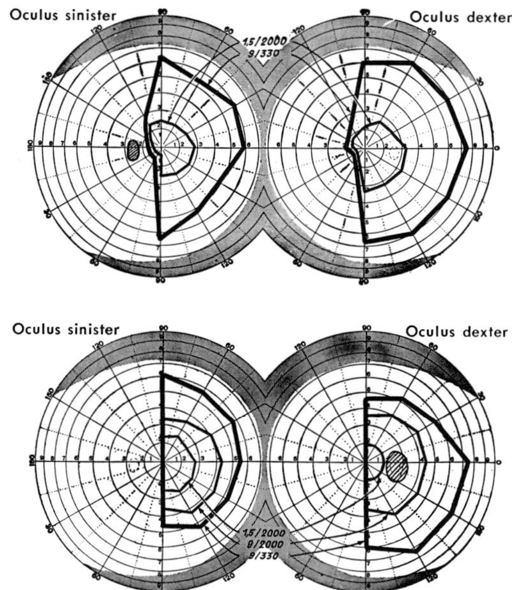

Fig. 3-22. Above, Glioblastoma multiforme of the right occipital lobe. History revealed unformed visual hallucinations! Below, Oligodendroglioma of the right occipital lobe. In both cases there is a complete homonymous hemianopia which is characteristic for occipital lesions. The upper fields reveal sparing of the macula. This is missing in the lower fields— the splitting of the macula is a sign of total interruption of the optic radiation. The field defects and the remaining parts of the fields are congruous. The hemianopia could be demonstrated on the perimeter (peripheral isopters) as well as on the tangent screen (central isopters).

one third) the complete homonymous hemianopia occurred with splitting of the macula (that is, with the dividing line between the seeing and the blind part of the field running directly through the macula). *There is a conspicuous congruence of the field defects and the remaining parts of the fields of these patients with complete hemianopia—both those with and those without sparing of the macula* (Figs. 3-21 and 3-22).

Incomplete hemianopia was seen only once in the form of an inferior homonymous incongruous quadrantanopia (Fig. 3-23). In two instances we observed a concentric constriction on one side in the presence of an incongruous hemianopic field defect on the other side. In agreement with numerous authors (Dubois-Poulsen; Guillaumat and Robin; Walsh and Hoyt; Horrax and Putnam; and others) we feel that complete homonymous hemianopia is characteristic for tumor lesions of the occipital lobe. There is always a lively discussion concerning sparing of the macula. We have examined the visual fields after occipital lobectomy in fifteen patients operated on for occipital lobe tumors (Huber). We demonstrated that the *removal of the occipital lobe (that is, the complete interruption of the optic radiation) results in complete homonymous heminopia without sparing of the macula* in the real sense of the word (that is, less than 3 degrees). However, there exists in most cases a definite sparing of the central fovea, which is rather difficult to explain and raises the problem of a possible bilateral representation of the fovea. On the basis of these findings we come to the conclusion that the absence of sparing of the macula is suggestive of a complete or nearly complete interruption of the fibers of the optic radiation or of the visual cortex. Sparing of the macula (that is, more than 3 degrees) occurs in homonymous hemianopia if the fibers of the optic radiation, particularly those of the macula, are not interrupted in their entirety. Since the optic radiation spreads out in its posterior part toward the calcarine fissure (Fig. 1-30), one can draw the conclusion that very extensive tumors or those situated in a rather anterior part of the occipital lobe are more likely to produce homonymous hemianopia without sparing of the macula. And

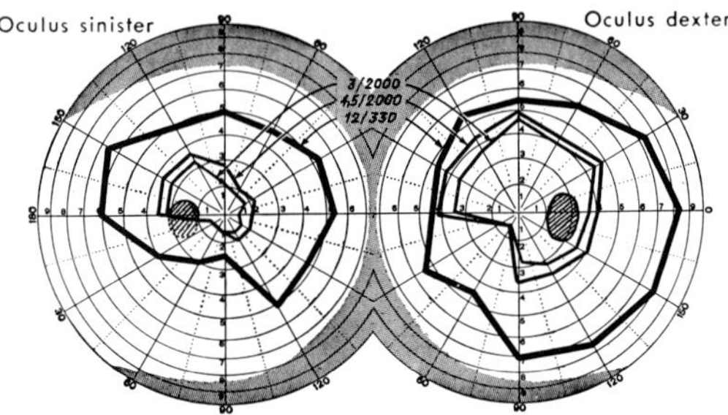

Fig. 3-23. Astrocytoma of the right occipital lobe. Left asymmetrical inferior quadrantanopia. This picture is similar to the one seen in parietal lobe tumors. Enormous bilateral enlargement of the blind spot due to bilateral choked discs of 4 diopters elevation.

thus we are in agreement with Kestenbaum, who also considers *macular sparing as a sign that points to a lesion in the most posterior part of the optic pathway, that is, the hindmost part of the optic radiation or the calcarine cortex itself.*

Occasionally the temporal part of the peripheral visual field, the so-called *"temporal crescent or half-moon,"* may be spared in cases of occipital lobe tumors producing homonymous hemianopia. In other instances a unilateral temporal crescent defect may be the only visual field sign in a cerebral tumor affecting the optic radiations (Bender and Strauss). We agree with Walsh and Hoyt that sparing of the temporal crescent indicates an occipital localization, whereas an isolated temporal crescent defect is of no value in topical diagnosis.

We agree with Dubois-Poulsen that it is not possible to differentiate lesions of the optic radiation from those of the cortex on the basis of the visual fields. Furthermore, such a differentiation is irrelevant because many tumors involve the cortex as well as the gray matter. *Central or paracentral homonymous scotomas (with apices at the points of fixation) are generally regarded as signs of affection of the occipital cortex, especially in the area of the occipital pole.* One of our own observations concerns a case of hemorrhagic pachymeningitis over the right occipital pole with paracentral homonymous scotomas in the left side of the fields (Fig. 3-24). It is noteworthy that the patient had epileptic seizures with a decided visual aura during which he experienced flashing concentric circles and sparks of light synchronous with the pulse in addition to certain noises. In case a tumor develops in the space between the two occipital poles, it may press on the upper or lower calcarine lips, thus creating a *vertical hemianopia.* Such cases have been described in the literature. We personally have never observed one.

Next to visual field changes, other ocular signs play a lesser role. In a little more than one fifth of our patients we noticed a fine or medium horizontal *gaze nystagmus.* It is probably due to a remote effect on the cerebellum or

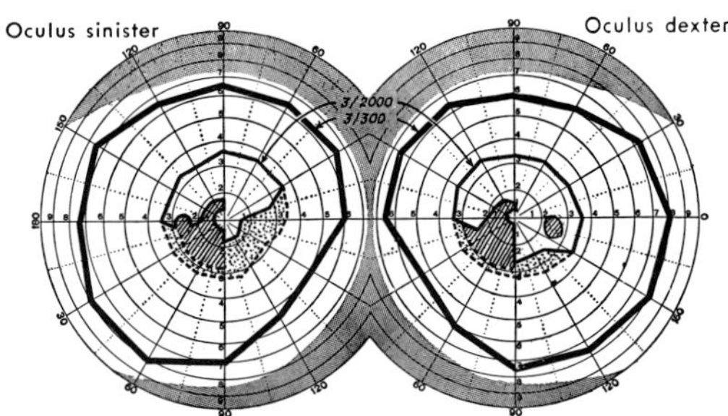

Fig. 3-24. Hemorrhagic pachymeningitis in the region of the right occipital pole after pneumococcal meningitis. Jacksonian epilepsy with visual and acoustic aura. Left paracentral homonymous hemianopic scotomas with sparing of the macula in the left inferior quadrants. Vision, O.U. 1.0. The scotomas are absolute in the shaded area. The dotted areas indicate relative scotomas (also involvement of left occipital pole?).

the brain stem. *Gaze palsies are extremely rare.* We observed only one patient with a gaze paresis up and toward the side opposite the tumor.

Let us add some remarks regarding the usefulness of *pursuit movement* and its change to cogwheel movement as well as *optokinetic nystagmus* for the differential diagnosis of a tract hemianopia and an optic radiation hemianopia (p. 59). In a patient with a parieto-occipital astrocytoma with homonymous hemianopia, Stadlin was able to demonstrate that following the extirpation of the tumor the optokinetic nystagmus toward the side of the tumor was missing. Before surgery it had been symmetrical to both sides. In the case of a homonymous hemianopia with an absent or weakened optokinetic nystagmus toward the side of the hemianopia, or an impaired pursuit movement toward the opposite side, one must assume a lesion in the middle or posterior part of the optic radiation (in other words, the parietal or parieto-occipital region). It is important, however, that the optokinetic nystagmus as well as the pursuit movement remains normal in patients with lesions of the cortex of the calcarine fissure.

Unilateral or bilateral sixth nerve pareses resulting from an increased intracranial pressure manifest themselves a little more often in occipital lobe tumors than in those of the parietal lobe (one fourth of the patients). A unilateral fixed mydriasis also occurred in one fourth of the patients. Usually it is homolateral—occasionally associated with a ptosis. Sometimes there is also homolateral miosis (sign of irritation of the oculomotor nerve).

There was a *conspicuous frequency in the occurrence of choked discs* (that is, in seventeen of our twenty patients with occipital lobe tumors). We always found a *pronounced prominence of the disc of up to 5 diopters as well as numerous retinal hemorrhages and exudates*—findings that are in agreement with those of Walsh and Hoyt; Bailey; and Parkinson and McCraig. The frequency and the pronounced character of these choked discs may be related to the location of the neoplasms near the posterior fossa. Tumors here lead to a relatively early impairment of the circulation of the cerebrospinal fluid and formation of an internal hydrocephalus. Consequently, there is an almost complete parallelism between the development of a choked disc and the finding of increased intracranial pressure during surgery (Fig. 3-25).

Finally, it should be mentioned that a brain tumor, regardless of its location (that is, in the frontal or temporal lobe), may occasionally simulate an "occipital picture." Evidently a compression of the posterior cerebral artery against the tentorium may cause an *ischemic or hemorrhagic infarction of one or both occipital lobes.* (See discussion of cortical blindness, p. 7.)

Following is a summary of the important ocular symptoms of occipital tumors. Frequently either clear or disguised statements in the patient's history suggest hemianopia. Signs of irritation in the form of primitive undifferentiated visual hallucinations are relatively rare. An expansion of the tumor toward the parietal region leads to disturbances of the higher visual centers in the form of

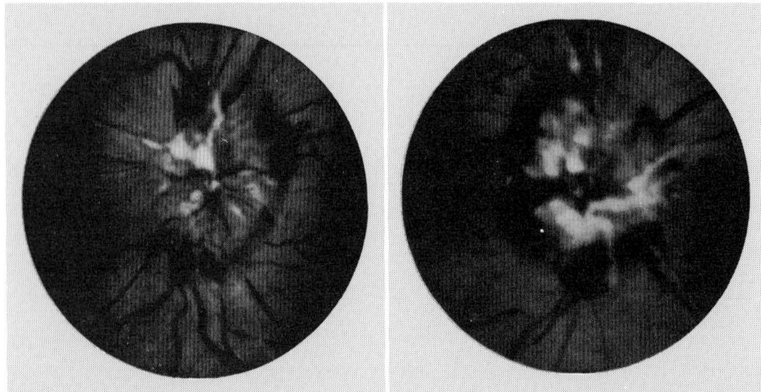

Fig. 3-25. Bilateral choked discs. Advanced chronic stage in glioblastoma multiforme of the right occipital lobe. Pronounced elevation of 3.5 to 4 diopters. Numerous hemorrhages and white exudates (cytoid bodies) on the disc and in adjacent retina. The left illustration corresponds to the right eye; the right illustration, to the left eye.

visual object agnosia, alexia, agraphia, and perhaps also micropsia or meta-morphopsia. The most frequent visual field change is a complete homonymous hemianopia, with or without sparing of the macula, the former perhaps being the more frequent one. Another typical sign is a striking congruence of the remaining halves of the visual fields. Quadrantanopia as well as a concentric constriction is quite rare. A differentiation between cortical and subcortical lesions cannot be based on the visual fields. Signs of an increased intracranial pressure in the form of choked discs are remarkably pronounced and practically always present. Characteristically, the choked discs are relatively prominent and present numerous hemorrhages and exudates. In the presence of alexia and other central disorders of visual integration it may be quite difficult to differentiate occipital lobe tumors from those of the parietal lobe. It may be irrelevant in cases of expansive growth of an originally occipital tumor. Generally, temporal lobe tumors can be differentiated from those of the occipital lobe on the basis of other neurologic symptoms. Superior homonymous quadrant defects and highly organized visual hallucinations suggest a temporal site. Complete homonymous defects and primitive hallucinations signify an occipital focus.

Tumors of the corpus callosum

A *unilateral apraxia* of the left hand generally is considered as the only, though rare, characteristic neurologic symptom of a tumor of the corpus callosum. In addition, such tumors usually cause grave psycho-organic disturbances accompanied by marked hypokinesia or akinesia, lack of initiative, and urinary and fecal incontinence. The clinical diagnosis of such tumors often is quite difficult—mostly because general signs of increased intracranial pressure may appear at an early stage and dominate the whole symptomatology (including signs of herniation into the tentorium and the foramen magnum).

The general ophthalmologic signs of increased intracranial pressure, especially choked disc, may be seen in patients with tumors of the corpus callosum as the first symptoms. We found a papilledema, although lacking characteristic details, in every one of our patients. The tendency to secondary optic atrophy with loss of vision and concentric contraction of the visual fields is pronounced. Sixth nerve pareses and central facial palsies are rather common signs (Tönnis). Pupillary disturbances such as anisocoria and fixed pupils have been described (Bailey). We have not observed them in any of the patients in our series. Occasional field defects must be considered as a pressure effect on the neighboring structures of the frontal and temporal lobes.

In summary, it can be stated that the ocular signs associated with tumors of the corpus callosum have no localizing value. They are merely manifestations of a general increase in the intracranial pressure, especially in the form of a papilledema. With the tumor situated in the most posterior part of the corpus callosum near the posterior commissure, an expansion may cause pressure on the roof of the midbrain (that is, the quadrigeminate plate). In such a case a Parinaud syndrome may appear (p. 260).

Parasagittal meningiomas and meningiomas of the falx

Meningiomas of the falx originate in the falx itself. The parasagittal meningiomas develop in the recess between the superior longitudinal sinus and the falx. One differentiates between meningiomas of the anterior, middle, and posterior thirds of the sinus according to their location and symptomatology. The neurologic signs depend on the effect on the adjacent part of the cerebral hemispheres. In meningiomas of the anterior third of the sinus, compression signs of the frontal lobe will predominate; those of the middle third of the sinus will show signs typical for the central region. Tumors of the posterior third of the sinus are associated with compression signs of the parietal and occipital lobes. Meningiomas of the falx and bilateral parasagittal meningiomas will result in signs involving both hemispheres (Fig. 3-26).

The ocular signs of parasagittal meningiomas and meningiomas of the falx depend on the part of the brain compressed by the tumor. For this reason we merely refer to the detailed discussion of the appropriate regions of the cerebral hemispheres. In our series of patients, parasagittal meningiomas and meningiomas of the falx *almost invariably caused papilledemas of from 2 to 3 diopters prominence, although without any peculiarities characteristic for this type of tumor.* The ocular symptoms usually are related to the general increase in intracranial pressure. They manifest themselves as flickering before the eyes, blackouts, photophobia, and double vision. Visual field changes are relatively rare. If present, they must be interpreted as remote effects of the tumor, with the exception of homonymous hemianopias caused by meningiomas of the posterior part of the falx or the sinus with direct involvement of the occipital lobes. A tumor in the falx between the occipital lobes may cause *bilateral homonymous hemianopia,* for instance, in the form of a bilateral inferior homonymous quadrantanopia, which looks like an altitudinal hemianopia with a line separating the seeing and the blind half of the field in the horizontal meridian (Rucker and Kearns). Klingler and Condrau have called attention to the *misleading localizing signs of the visual field changes caused especially by meningiomas of this type.* For instance, a deep-seated parietal meningioma

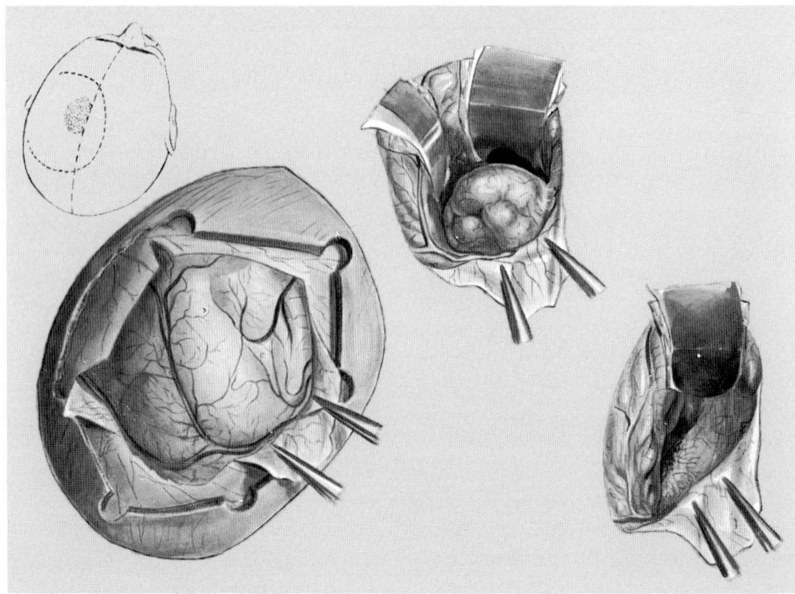

Fig. 3-26. Deep-seated meningioma of the falx in the left parietal region of a 31-year-old patient before and after extirpation.

of the falx on the left side caused a central scotoma of the left eye, with concentric contraction of the peripheral field. In such a case one must assume a remote effect of the tumor on the optic nerve; the exact mechanism cannot be explained (Custodis). We have observed a homonymous hemianopia of an incongruous type in a patient with meningioma of the middle third of the sinus as well as in a patient with a parasagittal meningioma of the frontal third of the sinus (Fig. 3-27). These homonymous hemianopias also must be interpreted as a remote effect on the optic radiations or possibly on the optic tract. A satisfactory explanation is difficult. In one patient with meningioma of the middle third of the falx we even observed a defect resembling bitemporal hemianopia. We cite this case as an illustration of how complex the task of interpretating these visual fields may become in such meningiomas and how misleading the field changes may be.

In summary, it can be stated that parasagittal meningiomas and meningiomas of the falx are generally associated with distinct general signs of increased intracranial pressure in the form of choked discs. It is of importance that they may cause visual field changes, mostly in the form of a homonymous hemianopia. Since these tumors are quite distant from the optic radiation and the optic tract, they must have a remote effect (except for those in direct contact with the occipital lobes). This type of tumor seems to be disposed to produce visual fields with misleading features for an exact localization because of their slow growth and their relatively large size.

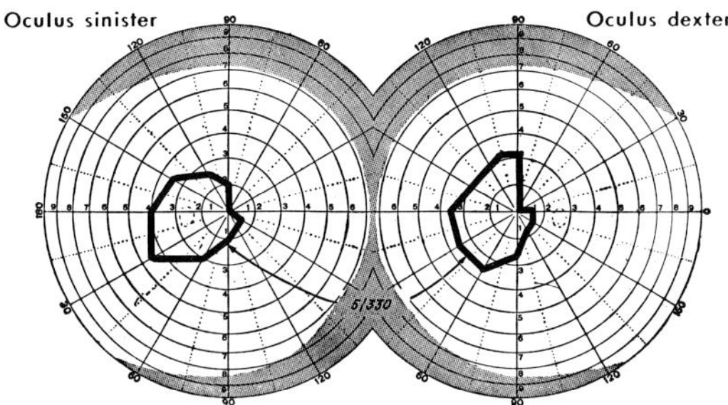

Fig. 3-27. Meningioma of the falx in the middle third of the sinus (left frontoparietal region). Right incongruous homonymous hemianopia, possibly the remote effect of the tumor on the optic tract. This visual field is of misleading localizing value.

Intrasellar, suprasellar, and parasellar tumors

After the discussion of tumors of the cerebral hemispheres, we must consider the space-consuming lesions of the anterior fossa, especially those of the *region of the chiasm.* We proceed, so to speak, from the hub of this area, the sella and its contents, in radial directions to the suprasellar structures situated anterosuperiorly as well as to the parasellar area on both sides. Thus we plan to set forth a logical method to differentiate the various types of space-consuming sellar lesions (Fig. 3-28).

Pituitary adenoma is considered the prototype of an intrasellar neoplasm because it originates within the sella in spite of its frequent extrasellar expansion during later stages. *Craniopharyngioma* shows an intrasellar and extrasellar growth, with the extrasellar expansion usually proceeding in a suprasellar direction. *Meningiomas of the tuberculum sellae* and *olfactory groove meningiomas* represent decidedly suprasellar types of growths. Finally, *meningiomas of the sphenoid ridge* will be discussed as typical examples of parasellar tumors. In the broadest sense of the word, tumors of the middle fossa and the cavernous sinus should be considered as growths of the parasellar space. The latter will be discussed in a special section because of the characteristic cavernous sinus syndrome they produce.

Statistics tell us that *one in four of all intracranial tumors arise in the region of the chiasm.* In the majority, visual symptoms represent their initial manifestations. As most of these sellar and parasellar tumors are of a benign nature, their early recognition is of utmost importance. Successful treatment therefore depends largely on the correct and well-timed diagnosis by the ophthalmologist!

Pituitary adenomas (intrasellar type of tumor)

The neuro-ophthalmologic examination reaches a maximum of importance with tumors of the pituitary gland, not so much because it permits exact localization but because the symptoms produced by these tumors, in the initial

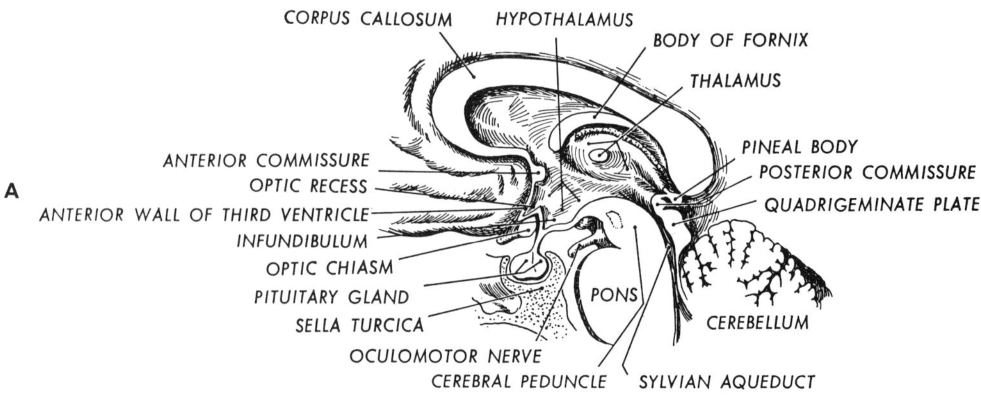

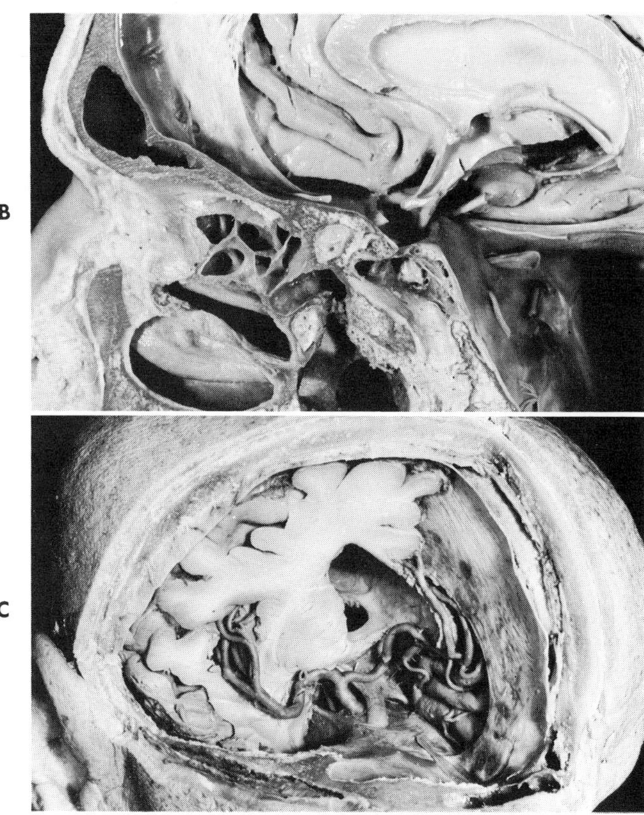

Fig. 3-28. A, Semidiagrammatic illustration (longitudinal section) showing the pituitary gland, sella, optic chiasm, third ventricle, thalamus, hypothalamus, and brain stem. **B,** Median section through the skull showing the sella, pituitary gland, optic chiasm, third ventricle, thalamus, hypothalamus, and bony structures of the forehead, nose, and epipharynx. **C,** Transverse section of the right anterior part of the skull and the brain showing the optic chiasm with its anterior rim and its relation to the internal carotid, arterial circle of Willis, and adjacent brain areas (especially the third ventricle). (**B** and **C** specimens of Prof. Kubic, Anatomic Institute of the University of Zurich.)

as well as the fully developed stage, are ocular symptoms (Fig. 3-29). Sooner or later these symptoms will bring the patient to the ophthalmologist. It is entirely the responsibility of the general practitioner or the ophthalmologist to detect pituitary tumors and arrange for early treatment. The neurologist and especially the neurosurgeon usually see these patients at a later date.

There are three principal types of pituitary tumors, all of them originating from elements in the anterior lobe. (1) The most frequent ones (75 to 80%), the *chromophobe adenomas* (Fig. 3-30), are generally marked by signs of endocrine deficiency such as sexual disturbances in the form of amenorrhea, impotence, and loss of libido, a decreased basal metabolism with weight gain and adiposity, skin changes in the form of straw-colored discoloration and wrinkling of the skin, scanty growth of the beard in males, and less frequently and manifestly, hypoglycemia, reduced excretion of 17-ketosteroids, and collapse under stress situations. Only a marked intracranial expansion of the adenoma may cause additional hypothalmic signs such as disturbances of the water balance (diabetes insipidus with polyuria and polydipsia) and disorders of the thermoregulation (hyperthermia, poikilothermia). (2) Next in frequency (20%) are the *acidophil adenomas,* especially frequent in females, which according to a pattern of endocrine hyperfunction, produce in young individuals during the growth period pituitary *gigantism* and in adults *acromegaly* (enlargement of hands and feet, prominence of jaw, thickness of tongue, deepening of voice, insulin-resistant diabetes mellitus, amenorrhea, impotence, loss of libido). (3) The third form is represented by the *basophil adenoma* (2% of all pituitary adenomas), which induces hyperadrenalism known as *Cushing's syndrome* and is marked by adiposity, moon facies, hypertension, osteoporosis, steroid diabetes mellitus, and hypogonadism. From the neuro-ophthalmologic point of view, only the chromophobe and acidophil adenomas are of interest because they are the ones that produce a chiasmal syndrome due to compression of the chiasm and because basophil adenomas cause chiasmal disturbances only rarely and usually after they have manifested themselves by their endocrine symptomatology.

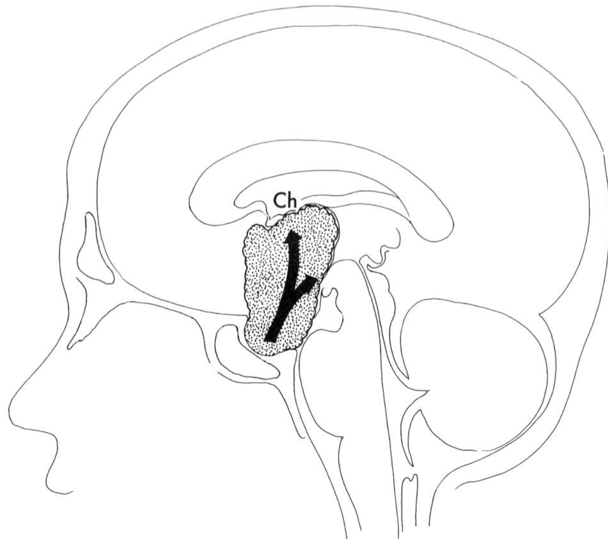

Fig. 3-29. Median section through the skull showing a pituitary adenoma growing out of the sellar region. Note the compression and upward displacement of the optic chiasm, **Ch,** as well as the narrowing of the third ventricle in comparison with those structures shown in Fig. 3-28.

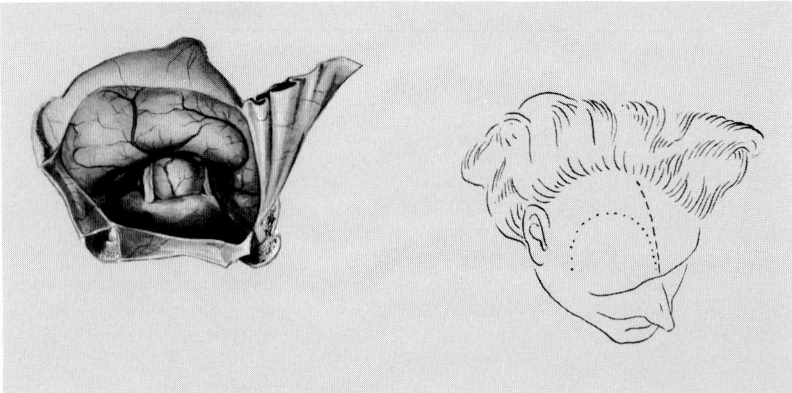

Fig. 3-30. Solid chromophobe pituitary adenoma (mostly intrasellar) with beginning suprasellar extension in a 29-year-old patient. The tumor was approached via a right frontal osteoplastic craniotomy. Note the separation and compression of the optic nerves caused by the neoplasm growing upward from the sella.

In this connection we want to stress also that the acidophil adenomas involve the chiasm far less commonly than the chromophobe type. In our series of twenty-three patients with acromegaly a chiasmal syndrome was observed only five times (a frequency that is still lower than that found by others: Mundinger, Riechert, and Reisert, 30%; Tönnis, 36%; and Bakay, 41%). Furthermore, the compression of the chiasm caused by an acidophil adenoma is always minor. Surgical intervention is usually not necessary; radiotherapy will suffice in most cases. Apart from visual field defects and optic atrophy (described in the discussion of chromophobe adenomas), proptosis and infrequently palsies of the ocular muscles (resulting from a hemorrhage into the adenoma) may occur with acromegaly.

In addition to the above-mentioned three types of pituitary tumors, there are also *mixed tumors* consisting of both chromophobe and acidophil cells and combining the tendency of chromophobe adenomas to expansive growth with the endocrine activity of the acidophil adenomas in respect to acromegaly. Their frequency is about 7 to 8% (Mundinger, Riechert, and Reisert). Sometimes a secretory type of adenoma may change into a nonsecretory type, thus causing a so-called transient acromegaly. It is therefore not surprising that today the classic division of pituitary adenomas into three functional types of tumors must be accepted with reservations, and there is a strong trend toward a newer classification based on functional principles (for instance, adenomas with and adenomas without endocrine activity).

The cardinal sign of pituitary tumors can be expressed in two words: *chiasmal syndrome.* The signs of this syndrome are primarily *disturbances of the visual field, loss of vision,* and *eyeground changes* (in other words, the consequences of an impaired conductivity of the chiasm, the optic nerves, or the optic tracts caused by the tumor). The functional endocrine changes of the pituitary gland previously mentioned produce secondary or accompanying signs that may assume some importance for differentiation of other tumors in the sellar area.

The subjective complaints of the patient are strikingly similar and may in themselves lead to the right diagnosis. Time and again the complaints are a *decrease of vision* (in 70% according to Tönnis)—*often at first on one side but occasionally on both sides.* This decrease of vision is described as a veil, as fog before the eyes, as vision through a dirty or dull spectacle lens, as a cloud, or as vision through a tinted lens. Frequently patients ask for new glasses. With the appearance of the visual disturbances, headache that may have been present for years and that was localized in the frontal region, in the temporal region, or in the eyes themselves may cease. Unfortunately, we have seen numerous patients who sought medical aid only after one eye had already become amaurotic and the other eye had begun to lose sight (24% of cases according to Tönnis). It is characteristic and valuable that some patients assert that they notice a decrease of vision only at dusk or after physical exertion—in some cases also during pregnancy, with a transient improvement of vision post partum! Among the thirty-three patients with chromophobe adenomas of the pituitary gland in our series, the more or less sudden onset of a unilateral loss of vision in three caused the consulting ophthalmologist to make *the mistaken diagnosis of a retrobulbar neuritis*. The subconjunctival or retrobulbar injections were credited with a temporary improvement, which actually seemed to confirm the wrong diagnosis in these cases. This improvement, however, was only short-lived and was followed by final deterioration of vision. In the end the diagnosis of a pituitary tumor was made, although after considerable delay.

Although the loss of vision in pituitary tumors is usually gradual, extremely rapid loss of vision or onset of amaurosis in one or both eyes may occur (see pituitary apoplexy and differential diagnosis of aneurysms of the sellar region, p. 242). Patients complain only occasionally of attacks of transient amaurosis lasting from a few seconds to an hour or more.

Hallucinations among patients with adenomas of the pituitary are unusual, but according to Weinberger, Adler, and Grant, do occur either in the form of flashes of light or in the form of complex figures and complicated geometric patterns. They are explained on the basis of a secondary involvement either of the cerebral peduncles (peduncular hallucinosis) or of the temporal lobes. One extraordinary interesting observation concerns a patient who noticed rapid bilateral deterioration of his visual acuity following a stellate ganglion block for trophic disturbances in one arm. A permanent impairment on the side of the injection caused annoying clouding and fogging of vision. Subsequently, an acidophil pituitary adenoma was discovered.

Although patients with pituitary tumors practically always complain of loss of vision, *they will less frequently mention disturbances of the visual field*. To some extent, this is due to the fact that they are not conscious of the superior temporal field defects in the early stages of the development of a pituitary tumor—defects which, because of their peripheral situation, are not disturbing. In the later stages, however, complaints relating to visual field changes are quite characteristic. The patient states that he collides with people approaching from the side; that he drives his car over the sidewalk because he does not see the curb; that when reading he sees only part or perhaps only one half of the line; that he cannot see out of the corner of his eyes and must turn his head too much; that he feels as if he has *blinders* on both sides of his eyes (a motorist noticed these blinders for the first time after a strenuous drive); that at times he sees only the middle of people's faces (that is, only the nose); and that either one or both of the outer sides of his field appear in the shade. Such more or less spontaneous descriptions of field defects so characteristic for pituitary adenomas were recorded in more than one fourth of the patients in our series (28% according to Mundinger, Riechert, and Reisert). Occasionally

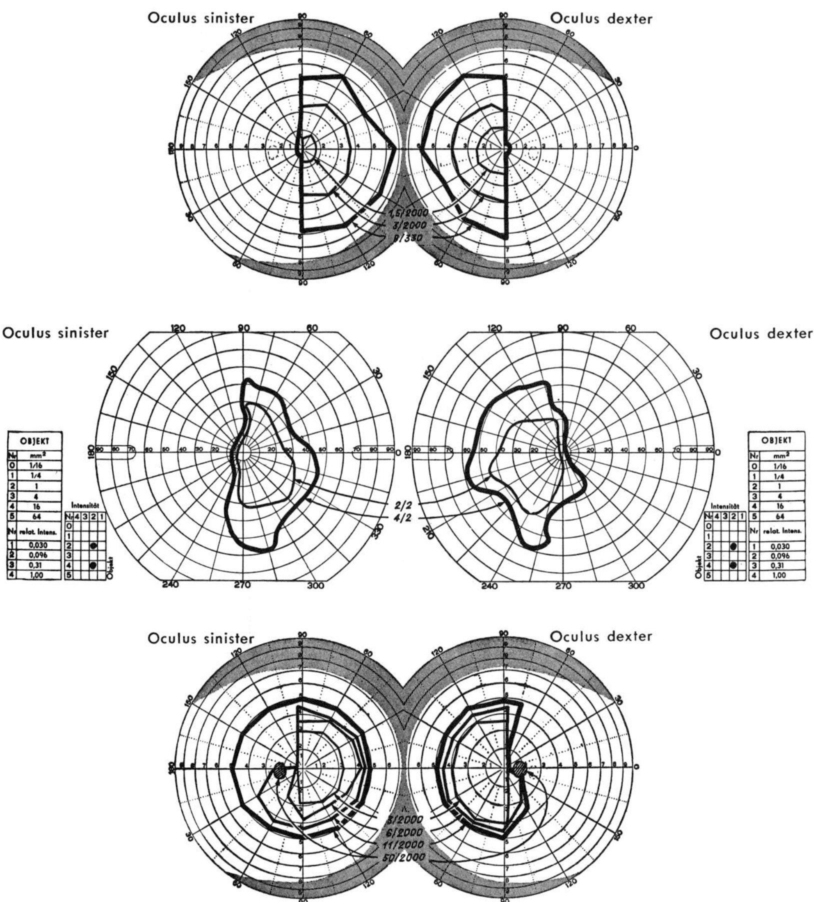

Fig. 3-31. Three cases of intrasellar and suprasellar chromophobe pituitary adenomas with typical chiasmal syndrome: bitemporal hemianopia, loss of vision, and optic atrophy. Above, Symmetrical bitemporal hemianopia with vertical line of separation and slight sparing of the macula. The peripheral and central isopters are involved. Middle, Bitemporal hemianopia with slight asymmetry of the missing and remaining parts of the visual fields. Below, Symmetrical, slightly incomplete bitemporal hemianopia with preservation of islands of vision in both inferior temporal quadrants. The left peripheral isopter (50/2000) does not reveal the hemianopia!

patients complain of transient or permanent diplopia, which usually can be traced to a corresponding extraocular muscle paresis.

The *ocular signs of the chiasmal syndrome* consist of changes of the *visual field* and the *eyegrounds*.

The characteristic field change is *bitemporal hemianopia.* We have seen such a characteristic field in a large number of cases (about one half of our patients, 60% according to Tönnis, 63% according to Bakay) during the first examination. There has always been a more or less vertical line of separation either through the point of fixation or more frequently just sparing it. In such patients the subjective visual disturbances, especially symptoms of a hemianopia, have been pronounced. *Symmetry of the bitemporal field defects is characteristic for pituitary tumors. It is important for their differentiation from extrasellar conditions.* This symmetrical bitemporal hemianopia usually involves the peripheral as well as the central field (Fig. 3-31). In some of our patients, peripheral isopters indicated perfectly normal limits and only the central isopters demonstrated the bitemporal defect (Fig. 3-32); in others only the peripheral field was defective.

The presenting visual field changes in patients with pituitary tumors are not always so typical. The classic fully developed bitemporal hemianopia actually is only an advanced stage in the gradual development of the field changes. *The very first defects in a pituitary tumor start in the superior temporal quadrants* (Cushing and Walker; Traquair). Frequently these defects are bilateral and symmetrical, indicating an impairment of the inferior surface of the body of the chiasm by the tumor protruding from the sella, in other words, an involvement of the crossing fibers originating from the inferior nasal quadrants of both retinas. Frequently, however, the field defect in a temporal upper quadrant is limited to one eye in the early stages (Figs. 3-33 and 3-34). Only at a later date will the other eye show a similar change. It is often necessary to search for such peripheral superior temporal field defects with the most

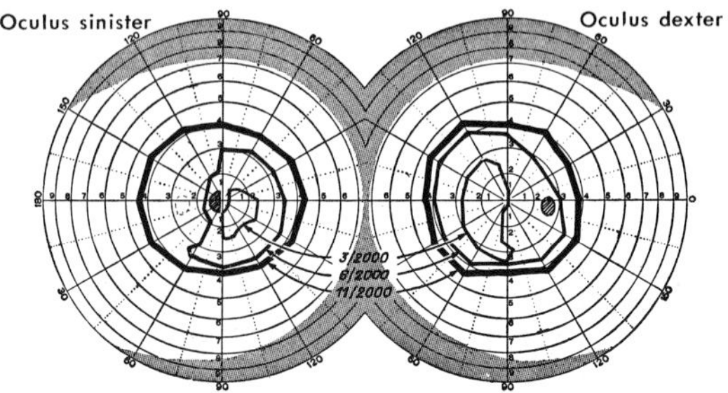

Fig. 3-32. Macrocystic chromophobe pituitary adenoma with marked compression of the chiasm and optic nerves. Beginning bilateral optic atrophy. Corrected visual acuity, O.D. 0.6; O.S. 0.2. Slightly asymmetrical bitemporal hemianopia of the central isopters with normal peripheral (11/2000) isopters. Left paracentral hemianopic absolute scotoma temporally to vertical meridian ("junction" scotoma of Traquair).

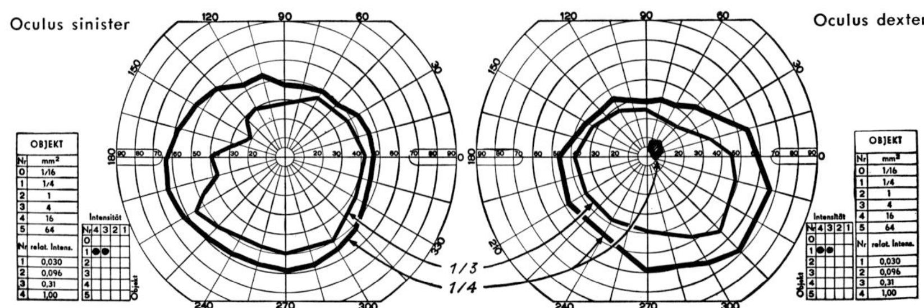

Fig. 3-33. Chromophobe pituitary adenoma with beginning extrasellar extension up to the right optic nerve. Temporal pallor of the right disc with visual acuity of 0.3. Normal left disc with visual acuity of 1.0. Very early beginning bitemporal constriction of superior fields, left somewhat more than right. Absolute paracentral scotoma of the right eye temporally and superiorly to point of fixation.

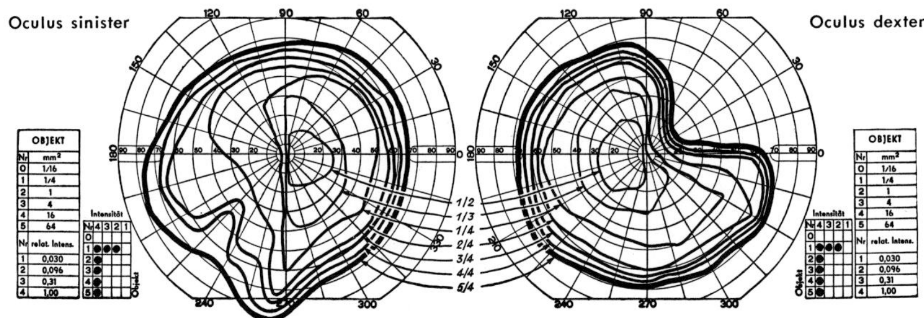

Fig. 3-34. Intrasellar chromophobe pituitary adenoma. Slight atrophy of the left disc. Temporal pallor of the right disc. Visual acuity, O.S. 0.9; O.D. 1.0. Subjective complaint of veil before the eyes. Beginning loss of field in right superior temporal quadrant (peripheral and central isopters). At least the central isopters of the left field already show a complete hemianopic defect. (Case of Prof. Brückner.)

painstaking methods of perimetry and campimetry (Chamlin and Davidoff). Again we want to call attention to the signal value of kinetic perimetry performed with targets of varying light intensity and size on the Goldmann instrument, preferably combined with static perimetry performed on the Harms perimeter (p. 79).

It is imperative to determine a fairly large number of peripheral and central isopters because experience shows that the peripheral isopters show either no change or merely a superior temporal indentation, whereas the central isopters already indicate a distinct bitemporal hemianopia. In one sixth of the patients in our series there was evidence of such a (mostly bilateral symmetrical) quadrant loss or indentation in the superior temporal regions starting from the periphery. Here again, the usual course is a more or less pronounced symmetry. *During subsequent stages the field defects progress from the superior temporal to the inferior temporal quadrants* (crossing fibers from the superior nasal retinal quadrants!), leading to the full bitemporal hemianopia just dis-

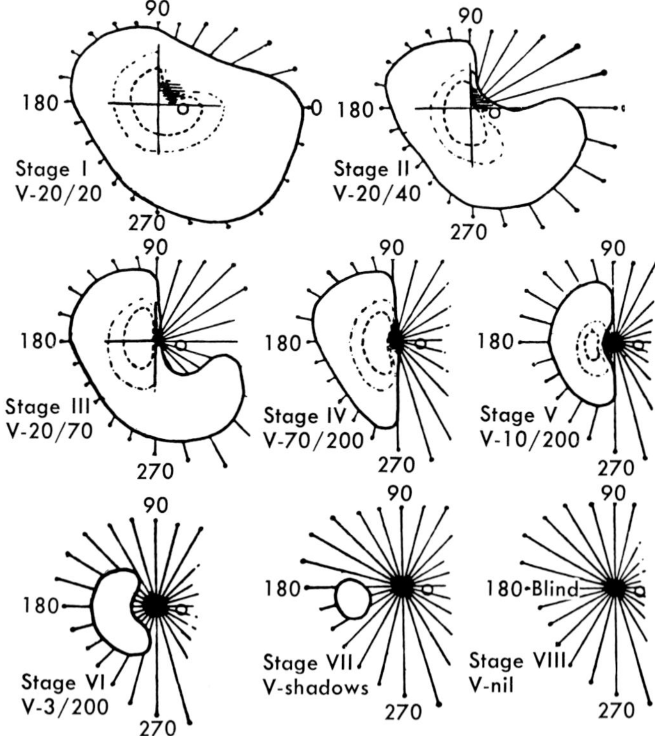

Fig. 3-35. Eight stages in the progression of a primarily temporal field defect in the right eye with progressive growth of a pituitary tumor. (After Cushing and Walker.)

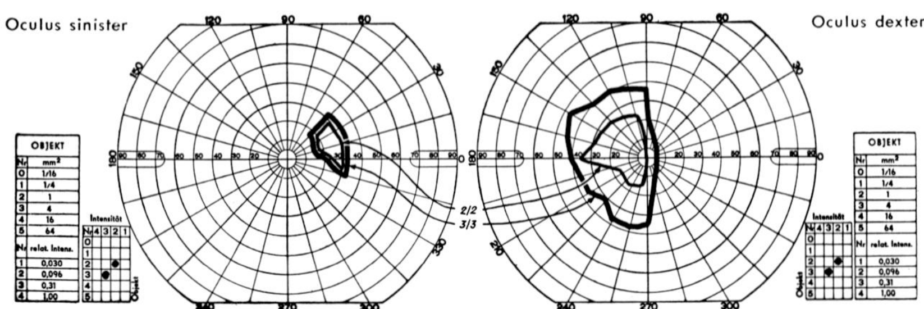

Fig. 3-36. Chromophobe pituitary adenoma with marked suprasellar extension. Bilateral temporal pallor of the disc. Visual acuity, O.D. 0.3; O.S. 0.1. Temporal hemianopia (stage IV according to Cushing) of the right eye. Island of vision in the left superior nasal quadrant (stage VII according to Cushing). The tumor has reached the left outer rim of the chiasm!

cussed. The latter remains stationary for some time before a deterioration of other areas sets in. *Eventually, there follows a breakdown of the nasal halves of the fields, beginning with the inferior quadrants; the superior quadrants* (Figs. 3-35 and 3-36) *frequently remain the most enduring ones* (Cushing and Walker). Such course is obvious if one considers that the tumor must reach a fairly large size before it can strike at the noncrossing fibers which represent the nasal field and which, due to their position at the lateral aspects of the chiasm, are better protected than the fibers at the inferior surface. *It should be emphasized again that the rather symmetrical progression in the deterioration of the visual fields of the two eyes is characteristic of pituitary tumor.*

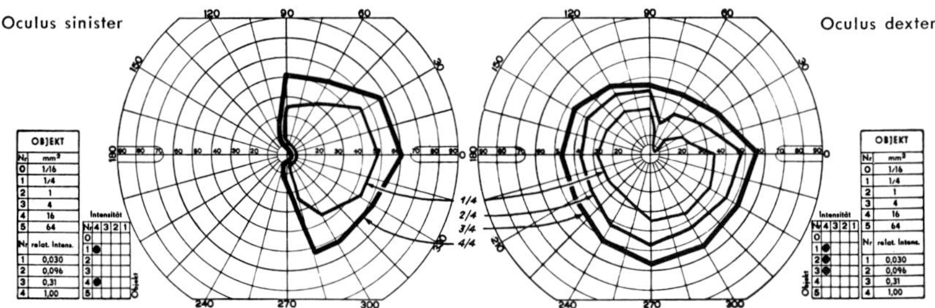

Fig. 3-37. Chromophobe pituitary adenoma with suprasellar extension, especially toward the left optic nerve. Advanced left optic atrophy. Normal right disc. Visual acuity, O.S. finger counting at 50 cm.; O.D. 1.25. Asymmetrical visual fields: left temporal hemianopia with loss of macula; right beginning superior temporal quadrantanopia.

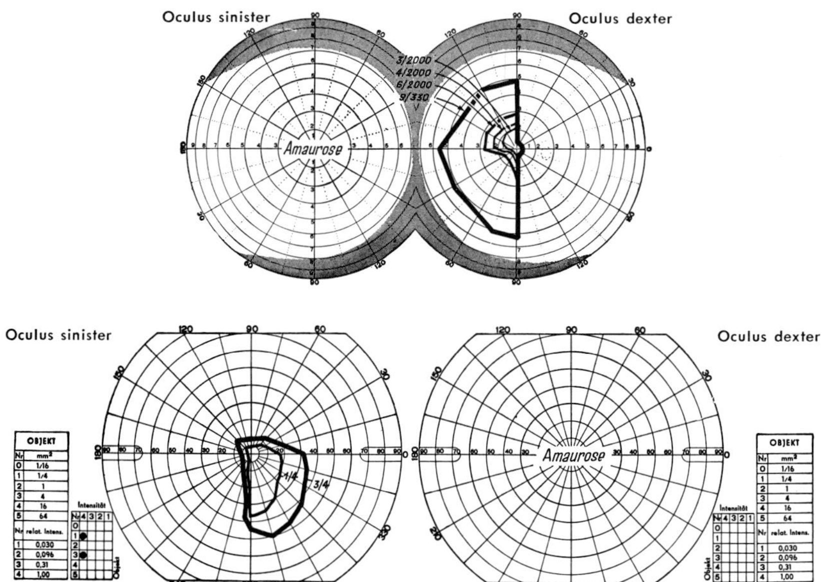

Fig. 3-38. Above, Chromophobe pituitary adenoma with suprasellar extension and invasion of the third ventricle. Bilateral optic atrophy. Visual acuity, O.D. 0.8. Amaurosis of the left eye with temporal hemianopia of the right eye. Below, Enormous chromophobe pituitary adenoma with suprasellar extension. Temporal pallor of both discs, right more than left. Visual acuity, O.S. 0.5. Amaurosis of the right eye with island of vision in the inferior nasal quadrant of the left eye.

In addition to these characteristic features and stages of the field changes, there are other, less pathognomonic, forms (Fig. 3-37). The field loss, for instance, may progress more rapidly in one eye than in the other. It may lead to complete amaurosis in the first eye, whereas the other eye still shows a temporal hemianopia. Such an *amaurosis of one eye with a temporal hemianopia of its fellow* occurred in one tenth of the patients in our series (16% according to Tönnis, 15.5% according to Bakay) (Fig. 3-38). Blindness of one eye and a temporal field defect in the contralateral eye can be explained either by a frontal and lateral extension of the pituitary adenoma from the chiasm to one optic nerve or by a temporal extension from the chiasm to one optic tract (the blind eye is always on the side of the extension of the tumor). However, it should be stressed here that the same picture can be produced by extrasellar tumors such as meningiomas of the tuberculum sellae (p. 218) or of the sphenoid ridge (p. 231) that extend into the chiasm either as primary *prechiasmal* tumors (first blindness of one eye and later temporal field defect in the contralateral eye) or as primary *postchiasmal* tumors (first homonymous hemianopia and later loss of temporal field in the eye on the side of the lesion). Thus an *amaurosis of one eye and temporal hemianopia of the other eye should be evaluated with great caution in diagnosing a pituitary tumor.* The most extensive destruction of the visual fields occurred in one of our patients with amaurosis on one side and preservation of only the inferior nasal quadrant on the other side (Fig. 3-38). In general, however, a patient with amaurosis on one side will notice disturbances on the other side relatively early.

In addition to disturbances of the peripheral field, we also find *scotomas* of varying sizes and types (scotomatous type of chiasmal syndrome) associated with pituitary tumors. Traquair mentions them especially in rapidly growing tumors. Supposedly they are not found if the tumor grows slowly or is stationary. In a review of our own cases we found relatively few instances of central or paracentral *scotomas,* an observation in agreement with the statistics of Guillaumat and Robin. (Tönnis found them in 3.5% of patients with adenomas.)

The scotomas are important as localizing signs if they are the only manifestations of a pituitary tumor in its early stages—a type of field defect that occurs only infrequently. The quadrant or hemianopic shape of these scotomas as well as their location temporally to the vertical meridian is characteristic. Before a quadrant scotoma of the superior temporal quadrant extends into the inferior temporal quadrant, it usually merges with the blind spot. The development of such scotomas in pituitary adenomas must be interpreted as an interference of the tumor with the posterior angle of the chiasm since the crossing macular fibers are situated near the posterior rim of the chiasm. Actually, neoplasms that attack the posterior rim of the chiasm (suprasellar tumors, craniopharyngiomas, tumors of the third ventricle) most readily cause the scotomatous type of chiasmal syndrome. In one patient with a pituitary tumor we were able to

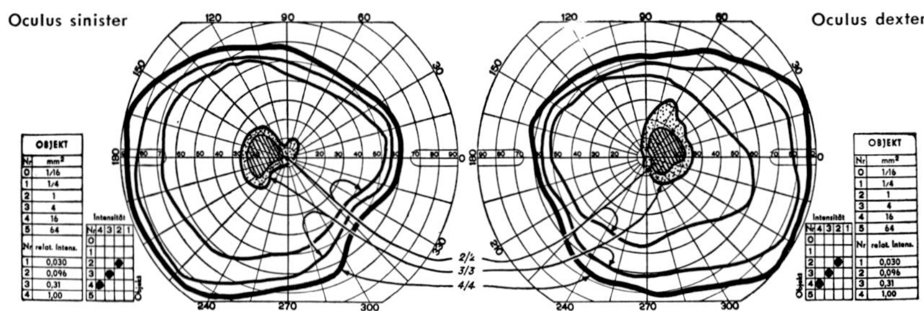

Fig. 3-39. Partly cystic, partly necrotic intrasellar chromophobe pituitary adenoma with compression of the optic chiasm. No optic atrophy. Visual acuity, O.D. 1.25; O.S. 0.4. Bitemporal hemianopic paracentral scotomas: interference of tumor with posterior angle of the chiasm. Peripheral isopters are completely intact. Scotomas are the only manifestation of field defect!

demonstrate bitemporal hemianopic scotomas without impairment of the peripheral isopters (Fig. 3-39). Thus a central or paracentral scotoma in one or both eyes may not only be the first evidence of a pituitary tumor but also remains the only ocular manifestation. In general, however, the development and progress of central and paracentral scotomas parallel those of the peripheral field. Thus their diagnostic importance should not be overestimated, except as an indication of tumor interference with the posterior angle of the chiasm.

Arcuate scotomas, especially below the horizontal meridian, mostly unilateral, and exactly resembling those in chronic simple glaucoma, have been described in pituitary tumors by different authors (Kearns, Salassa, Kernohan, and MacCarty; Rucker; Dubois-Poulsen; Hoyt). They are interpreted as a pressure effect of the anterior cerebral and anterior communicating arteries on the superior part of the optic nerves and chiasm. Hoyt concludes from these observations that some arcuate fiber organization must persist even in decussating fiber projections within the chiasm. Central and paracentral scotomas require very careful perimetry or campimetry for early detection and exact follow-up; in this connection we call special attention to the cardinal value of *static perimetry* (Harms; Aulhorn).

One should be mindful of the fact that some authors talk about "scotomas" of pituitary tumors if the peripheral isopters are still completely intact while the intermediate or central isopters already show a distinct quadrant or hemianopic defect. Some patients under our observation showed an initial indentation of the central isopters in the superior temporal quadrant. During the course of growth of the tumor a gradual corresponding indentation of the peripheral isopters resulted in a regular quadrant defect (Fig. 3-40). Also, parts (especially temporal) of the field may become separated to form isolated islands ("îlot temporal"). During the later development of the tumor they also become extinct.

In our patients we observed only a single case of an asymmetrical *homonymous hemianopia* (according to Tönnis, in 2% of patients with chromophobe adenomas) with the entire half of the field of one eye and three quadrants of

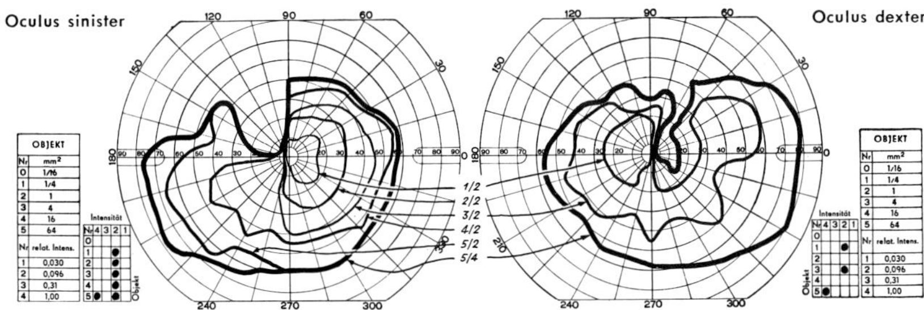

Fig. 3-40. Intrasellar and suprasellar chromophobe pituitary adenoma with large cyst. Bilateral temporal pallor of discs, more on right side. Visual acuity, O.D. 0.1; O.S. 1.0. Asymmetrical bitemporal superior quadrantanopia of the peripheral as well as the central isopters as initial stage of a chiasmal syndrome. The right upper temporal field remnant may break off to form an "îlot temporal."

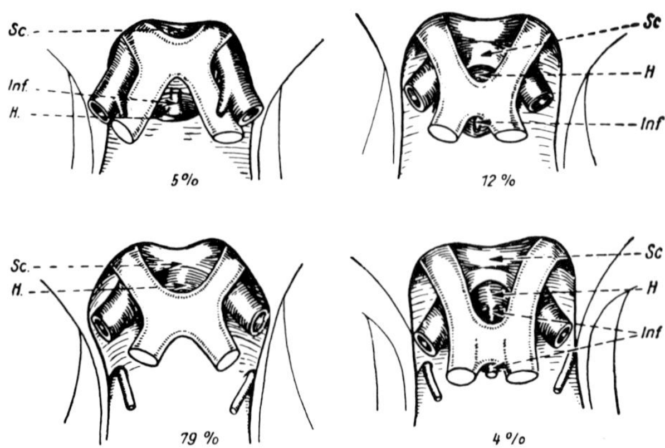

Fig. 3-41. Normal variants in the topographic relationship between the chiasm on the one hand and the pituitary gland and the sella turcica on the other hand. **Sc,** Chiasmal sulcus; **Inf,** infundibulum; **H,** pituitary gland. **5%,** Chiasm above the tuberculum sellae; **12%,** Chiasm of the sellar diaphragm; **79%,** posterior rim of the chiasm above the dorsum sellae; **4%,** chiasm above and behind the dorsum sellae. (After de Schweinitz; from Kyrieleis.)

the field on the other side missing. It occurred in a patient with a tremendous pituitary tumor with pronounced extension into the retrochiasmal space. It must be assumed that mainly one optic tract was damaged in this case.

It may be of interest to mention that there are chromophobe adenomas that cause *no defect in the visual fields*. Mundinger, Riechert, and Reisert gave a figure of about 4%, whereas Cushing and Walker and also Tönnis cited a figure of 20% in their total cases of pituitary adenomas. This is in agreement with the statistics of Guillaumat and Robin, who found in seventy-eight cases of pituitary adenomas fourteen cases of normal visual fields. These high figures for normal fields might well be reduced in the future due to more exact and more refined perimetry (for instance, static perimetry)!

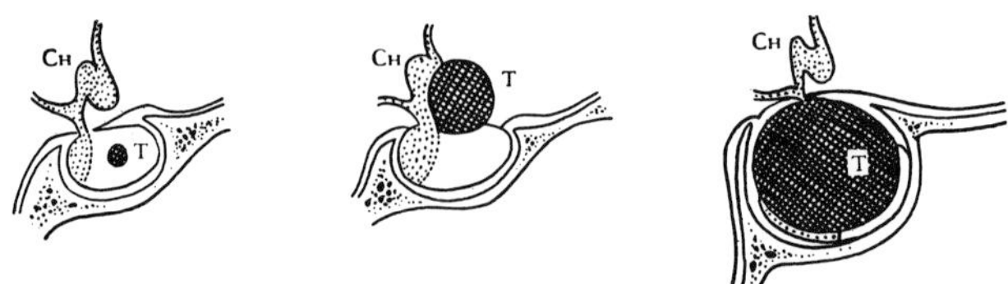

Fig. 3-42. Three different types of growth of a pituitary adenoma. Left, Small asymptomatic adenoma, **T**, within sella of normal size. Center, Larger adenoma with extrasellar extension through the sellar diaphragm causing compression of the chiasm, **Ch**, and chiasmal syndrome. Right, Large intrasellar adenoma with enlargement of the sella. No impairment of the chiasm. The original part of the pituitary gland (white area) is compressed into a small space. (After Cushing.)

There is a certain relationship between the size of the field defect and the extent of the tumor. However, it should be emphasized that *in the case of pituitary adenomas the relationship between the size of the tumor and that of the field defect depends on the position of the chiasm relative to the diaphragma sellae* (Fig. 3-41). Under normal conditions a pituitary adenoma must expand approximately 2 cm. in the superior direction before its pressure against the chiasm produces symptoms of a field defect. If the chiasm is fixed anteriorly or posteriorly, the tumor must reach an even larger size in order to produce visual field changes (Fig. 3-42). In some instances chromophobe pituitary adenomas may assume an enormous *extrasellar extension* (Jefferson). They may expand into one of the following various directions: toward the sphenoid sinus, toward the hypothalamus, toward the temporal region, toward the cavernous sinus, posteriorly into the infratentorial region, and anteriorly into the frontal region. The various neurologic symptoms characteristic for each particular area thus are added to the chiasmal syndrome. Whereas a suprasellar extension of the pituitary adenoma is quite frequent (in 70% of the cases according to Tönnis) (it actually is responsible for the chiasmal syndrome), a *parasellar extension through the thin wall of the cavernous sinus* is rare (Weinberger, Adler, and Grant; Cairns). In such cases the result may be *extraocular muscle pareses,* especially those supplied by the oculomotor and sixth nerves, with the patients occasionally reporting a corresponding diplopia. We have observed this in about one fourth of the patients with pituitary tumors (10% according to Tönnis, 7% according to Bakay). Twice the extrasellar lateral expansion resulted in an actual *cavernous sinus syndrome* (p. 39) with involvement of the third, fourth, fifth (facial neuralgia or anesthesia), and sixth cranial nerves. If the tumor produces a stasis within the cavernous sinus, it causes a more or less pronounced *exophthalmos* on the same side. In one of our patients a pituitary tumor extended into the third ventricle, causing an internal hydrocephalus with occlusion of both foramina of Monro and, at the same time, a severe psycho-organic syndrome.

Sometimes combined with bitemporal field defects in chromophobe adenomas (and also in other chiasmal lesions) one may observe a so-called *see-saw*

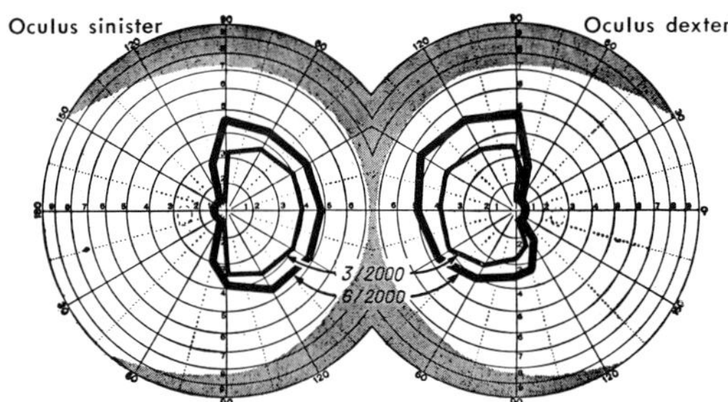

Fig. 3-43. Acidophil pituitary adenoma with suprasellar extension. Acromegaly. Visual acuity, O.D. 0.1; O.S. 0.5. No definite optic atrophy. Symmetrical bitemporal hemianopia with sparing of the macula.

nystagmus, a peculiar dissociated vertical nystagmus consisting of rising of one eye and falling of the other combined with torsional movements of both eyes (Maddox; Mark, Smith, and Kjellberg).

It has already been mentioned that *acidophil adenomas show much less of a tendency for extrasellar extension and an associated compression of the chiasm* (Davidoff). We have observed visual field changes five times among twenty-three patients with acromegaly (36% according to Tönnis, 41% according to Bakay). There were two instances of bitemporal superior quadrant defects and one of a complete bitemporal hemianopia with sparing of the macula (Fig. 3-43).

This may be the opportune place to clarify in a few words the *relationship between the visual field defects and the tumor effect on the chiasm.* It has already been mentioned that the relationship between the size of the tumor and the nature of the field defect depends on the position of the chiasm relative to the diaphragma sellae. If the chiasm is in a normal position (lying above the sella turcica), bitemporal hemianopia results, indicating a lesion of the entire midline of the body of the chiasm. If the chiasm is fixed posteriorly, one or both optic nerves may be involved, and if the chiasm is fixed anteriorly, one or both tracts may be affected.

The fundamental anatomic basis for the clinical sign of bitemporal hemianopia is the separation of crossing fibers (in the midline of the chiasm) and the uncrossed fibers (at the lateral edge of the chiasm). However, regional anatomy, intrinsic nerve fiber organization, and physiology of the chiasm are still incompletely understood even today and therefore our ability to interpret the visual defect on this basis is still limited. In this connection we would like, in agreement with Hoyt, to call attention to the numerous existing diagrams of the chiasm that depict the fiber organization of the chiasm in schematic form but at the same time contain much conflicting and misleading information. Hoyt, an expert in chiasmal pathology, comes to the conclusion that no existing diagram (including his own) is completely satisfactory and has to be applied with caution in correlating clinical data with neuropathologic conditions (Fig. 3-44). Moreover, the mechanism by which pituitary tumors and other tumors originating from structures surrounding the chiasm produce chiasmal damage is still incompletely understood. It certainly must consist of various factors. Deformation, flattening, and stretching of the chiasm are the purely mechanical consequences of extrinsic tumor pressure. Whether such a mechanical distortion of the chiasm is really responsible for the visual field defects is not certain. Such a distortion usually is manifest before defects in the visual fields become evident. It is not even established whether direct contact of the tumor with the chiasm is necessary for the production of visual field defects. It is most probable that factors

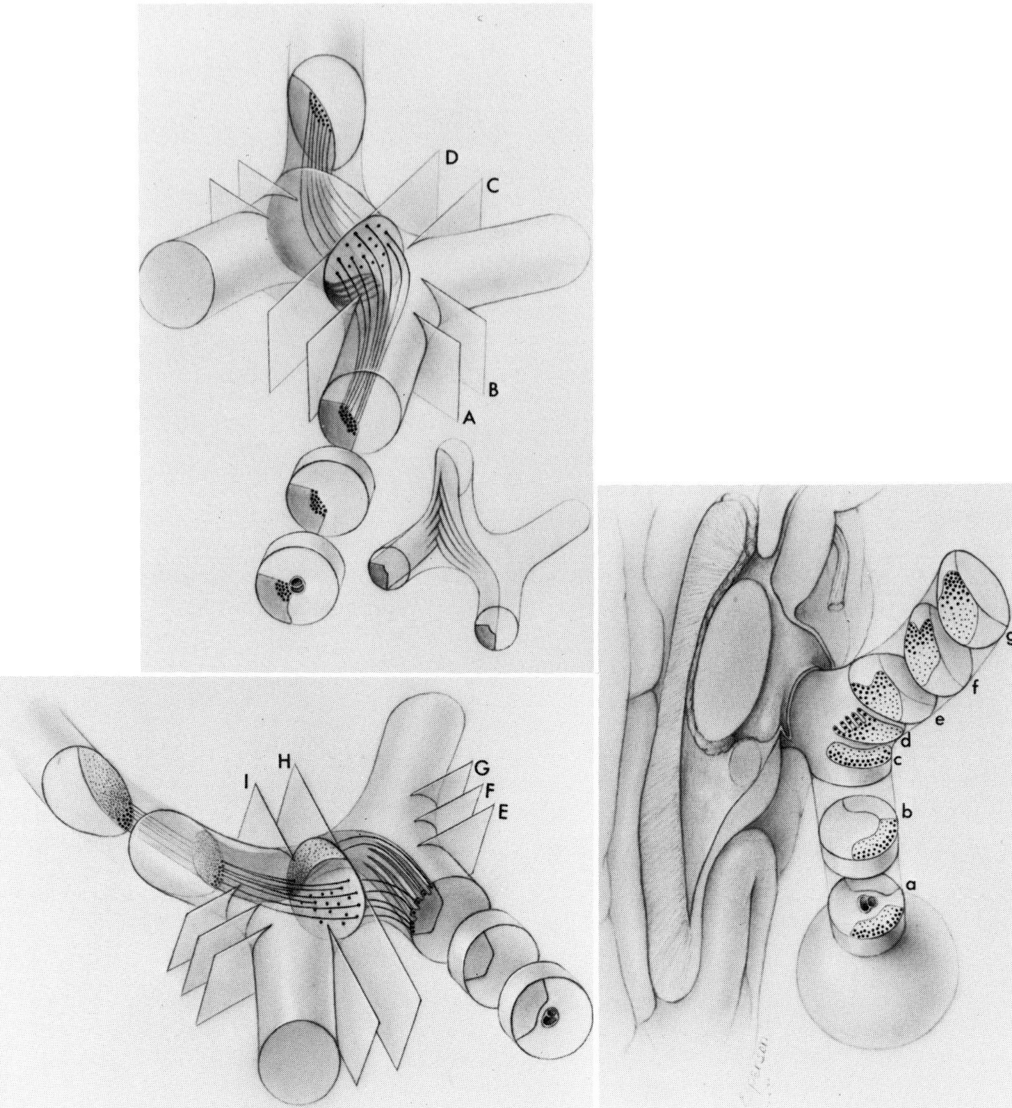

Fig. 3-44. Visual fiber anatomy of the chiasm, as determined from Nauta's axon degeneration studies (Hoyt) in primates. Top, Inferior crossed extramacular retinal projections passing posteriorly into the lateral optic tract of the opposite side. Left, Superior crossed extramacular retinal projections passing posteriorly through the chiasm (forming the posterior inferior notch) and entering the inferior medial optic tract of the opposite side. Right, Uncrossed extramacular retinal projections undergoing an inversion. The superior uncrossed projection takes a medial position in the optic tract. Large-caliber axons drop down into the inferior optic tract (large black dots); smaller-caliber axons rise into the medial dorsal optic tract (small dots). The corresponding position of the superior uncrossed quadrant projection (stippled) and the inferior uncrossed projection (unstippled) are demonstrated at different levels in the optic nerve, **a** and **b**; in the chiasm, **c** and **d**; and in the optic tract, **e, f,** and **g**. (From Hoyt, W. F., and Luis, O.: Arch. Ophthal. **70:**69, 1963.)

other than nerve fiber compression might be involved in the bitemporal hemianopia produced by pituitary tumors. It is a fact that despite numerous variations in the normal anatomy of the chiasm and great variation in the degree and pattern of chiasmal distortion, the resulting visual field defect, bitemporal hemianopia, remains monotonously the same. Also, the often observed rapid recovery of vision and visual fields after decompression of the chiasm suggests that nerve compression cannot be the only mechanism responsible. *Of great and often under-estimated importance seems to be the arterial blood supply and its impairment by the growing tumor.* After extensive autopsy examinations, Bergland and Ray come to the conclusion that the optic chiasm receives an arterial blood supply from a *superior group of vessels* (derived from the anterior cerebral arteries) that spares the central chiasm and from an *inferior group of vessels* (derived from the internal carotid artery, the posterior cerebral artery, and the posterior communicating artery). The decussating fibers in the central chiasm receive their arterial supply only from the inferior group. During pituitary tumor surgery these authors found the inferior group of vessels commonly distorted and compressed (the superior group is not affected) and therefore conclude that the bitemporal field defects caused by these tumors result from ischemia rather than from neural compression (Fig. 3-45).

Classic *bitemporal hemianopia* in its full development indicates a lesion of the entire midline of the chiasm, in other words, a total functional interruption of crossing projections. *Bitemporal superior field defects* indicate involvement of fibers in the anterior chiasmal notch by the expanding tumor and impairment of conduction from the inferior nasal retinas. *Bitemporal hemianopic scotomas* alone or associated with inferior field defects indicate involvement of the posterior chiasmal notch where the macular projections occupy a large area of the transverse bar of the chiasm. The clinical observation that chiasmal field defects can be recognized in the central field at the same time or even before they are present in the peripheral fields is not easily understood and must, apart from diffuse pressure effects, be due to disturbances of the arterial blood supply.

Gradual development of *blindness in one eye* (or central scotoma or temporal hemianopic "junction" scotoma) and *later development of a temporal upper defect in the contralateral eye* point to a lesion of the intracranial segment of one optic nerve and an interruption of the anterior knee fibers at the junction of the optic nerve and the chiasm. Such pictures are also produced by prechiasmal tumors.

A lesion at the junction of the medial optic tract and the chiasm produces an *inferior quadrant defect in the contralateral temporal field* (superior crossing fibers) and a *small ipsilateral nasal field defect,* in other words, a sort of incongruous homonymous hemianopia. Hoyt has found this type of field defect in patients with glioma of the anterior third ventricle. *Homonymous hemianopia with additional loss of the temporal field of the eye on the side of the lesion* indicates the extension of a tumor from the tract into the chiasm and is characteristic for postchiasmal tumors.

Total temporal hemianopia in one eye with no temporal defect in the other eye may also be observed in pituitary tumors (5.5% according to Tönnis) and must indicate a chiasmal lesion. However, such a defect, indicating involvement of all the crossing fibers only from one side, is difficult to explain and must be related to some peculiar pattern of chiasmal distortion by the growing tumor.

Binasal hemianopias are occasionally observed in patients with pituitary tumors. They are no longer regarded as a direct affection of the uncrossed fibers of both lateral chiasmal rims but rather as the result of a compression of these lateral rims by the carotid arteries, anterior cerebral arteries, or their communicating branches. However, they occur mostly in patients with infratentorial tumors (p. 264), tumors of the third ventricle (p. 253), or chiasmal arachnoiditis (p. 243).

Ipsilateral nasal hemianopia combined with an upper temporal defect in the contralateral eye (eventually associated with ipsilateral scotoma and reduction of central vision as a sign of optic nerve involvement) points to an ipsilateral lesion of the lateral margin of the chiasm. In such a case the superior temporal defect can be explained by the lateral position of the inferior crossed fibers from the optic nerve within the chiasm and the proximal parts of the contralateral optic tract.

In extremely rare instances, pituitary tumors may cause *horizontal hemianopias*. So far

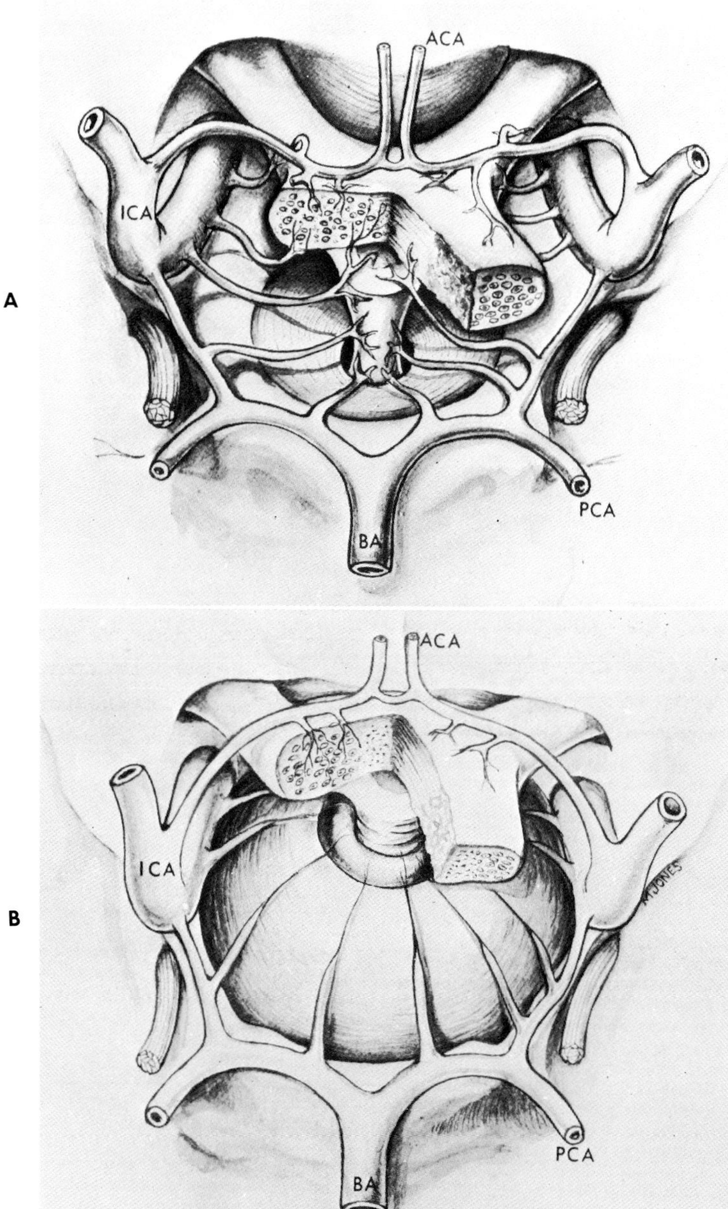

Fig. 3-45. A, Chiasmatic visual pathway and its arterial supply. The arterial supply of the chiasm can be divided into a superior and inferior group. The superior group of vessels is derived from the anterior cerebral arteries, **ACA,** and spares the central chiasm. The inferior group of vessels is derived from the internal carotid artery, **ICA;** the posterior cerebral artery, **PCA;** and the posterior communicating artery. The central chiasm containing the decussating fibers derives an arterial supply only from the inferior group of vessels. **BA,** Basilar artery. **B,** Chiasmatic visual pathways distorted by a pituitary tumor. The inferior group of vessels is distorted and compressed: thus the central chiasm, containing the decussating fibers, becomes ischemic. The superior group of vessels is not affected by the tumor and the lateral portions of the chiasm retain their arterial supply. (From Bergland, R., and Ray, B. S.: J. Neurosurg. 31:327, 1969.)

there has been no satisfactory explanation for the pathogenesis of this phenomenon. They occur much more frequently in patients with chiasmal arachnoiditis or aneurysm of the circle of Willis (p. 299) or following traumas to the skull.

These reflections on some of the relations between the development and type of the visual field defect and the topographic position of the tumor with regard to the chiasm are helpful in suggesting that part of the chiasm in which the tumor exerts a maximum of interference. This area, however, is not necessarily identical with the location of the main body of the tumor (Figs. 3-42 and 3-44).

The eyeground changes associated with pituitary tumors can be best paraphrased as *primary simple optic atrophy*. There are gradual transitions from a partial pallor (sometimes a yellowish tint) to a snow-white, completely atrophic disc (Fig. 3-46). The disc characteristically is sharply outlined in all stages and surrounded by normal vessels. Papilledemas are extremely rare (occurring only in cases of secondary internal occlusive hydrocephalus). This is understandable if one bears in mind that patients with pituitary tumors, because of early visual disturbances, seek treatment long before the tumors reach a size where they could cause signs of an increased intracranial pressure. The atrophy is of the descending type (retrograde axonal degeneration). It is always bilateral. The involvement of the optic nerve may show the following variations: a uniform pallor of the entire disc (the most frequent form), a more pronounced pallor on the nasal side (atrophy of the nasal crossing fibers), and a more pronounced temporal pallor (in the case of a previously existing physiologic temporal pallor or excavation of the disc).

Frequently the degree of atrophy differs in the two eyes. If, in addition to the chiasmal lesion, the optic nerve is involved (for instance, amaurosis of one eye and temporal hemianopia on the other side), the atrophy is particularly

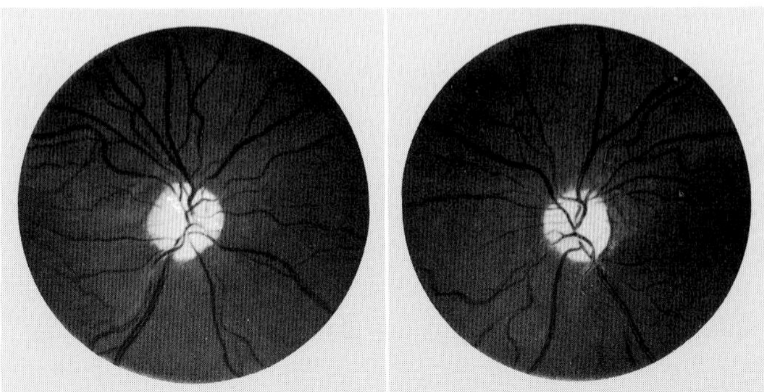

Fig. 3-46. Primary complete optic atrophy with distinct outlines in chromophobe pituitary adenoma with suprasellar extension (same patient as shown in Fig. 3-30). Temporal and nasal halves of both discs appear chalk white, are distinctly outlined, and are at the level of the retina. There is not yet an atrophic excavation. Slight narrowing of the arteries. The veins are normal. Capillaries on the disc are missing completely. The papillomacular fiber pattern cannot be demonstrated in red-free light. The left illustration corresponds to the right eye; the right illustration, to the left eye.

severe on the ipsilateral side. In extreme stages the atrophy causes not only pallor but also an atrophic, pearly white crater, with the cribriform plate visible on its ground (with steep walls and a deep depression but not cupping of the disc margins as in glaucoma!). In the one case this pallor is due to the ischemia of the disc structure; in the other case it is due to the secondary glial tissue proliferation that replaces the destroyed nerve fibers. It should be stressed that *there is no absolute relation between the degree of the optic atrophy and the central vision. It is particularly characteristic for pituitary tumors that the optic atrophy lags behind the loss of vision.* The vision may have already failed considerably, yet the discs may show normal color. Evidently the retrograde atrophy and glial proliferation have not yet extended to the nerve head.

In patients with unquestionable loss of vision and suspected bitemporal visual field losses, examination of the papillomacular bundle in red-free light (Vogt) may already demonstrate signs of atrophy (that is, absence of the pattern of the papillomacular fibers that normally is distinctly visible in red-free light). We have observed *optic atrophy* in the majority of our patients—however, always *with a distinct difference between the two sides,* which usually corresponded to the difference in the visual acuity rather than the field defects. As mentioned before, the field defects in patients with pituitary tumors are rather symmetrical. Although the lag of optic atrophy behind the loss of vision is the rule, the reverse (that is, pallor of the discs without marked loss of vision) is observed occasionally. *A pallor of the disc need not necessarily be identical with a functional loss of the nerve fibers. The morphology of a pale disc may permit certain prognostic conclusions.* If, for instance, the pallor of the disc is combined with atrophic excavation, the outlook for recovery of the visual function after extirpation of the pituitary tumor is generally poor. If there is merely pallor of the disc without marked wasting of the nerve tissue, the chances for restoration of the visual function after surgical intervention are much better.

In order to give the degree of atrophy of the optic nerve a numerical value, Kestenbaum suggests counting the vessels passing over the margin of the disc. In normal eyes the number of vessels passing over the margin is fairly constant: four to five arteries, four to five veins, and about ten small vessels. In primary optic atrophy the number of arteries and veins remains unchanged, but the number of the small vessels is diminished to seven, six, or even as few as three and thus represents a numerical measure of the degree of the pallor, allowing one to judge the development of optic nerve damage.

The *fluorescence angiogram* of primary simple optic atrophy is quite characteristic: the disc remains unstained and dark throughout all phases (which is in contrast to the fluorescence of the surrounding choroid). Even reactive glial tissue may show no staining if the atrophy of vessels of the nerve head has progressed far enough.

Pupillary changes are rare among our patients with a pituitary chiasmal syndrome. *In those with marked unilateral loss of vision the pupil on the involved side shows a diminished direct reaction to light. The Marcus Gunn pupillary phenomenon is decidedly positive* (p. 26). In patients with unilateral amaurosis the pupil in question is often found to be wider than the pupil of the contralateral side and fixed to light. It shows merely a consensual light reaction. It is common experience that the pupillary reaction remains intact longer than the visual function. Thus proof of a still-existing hemianopic pupillary reaction in an amaurotic eye with temporal hemianopia of the other eye may be an important diagnostic cue for a pituitary affection.

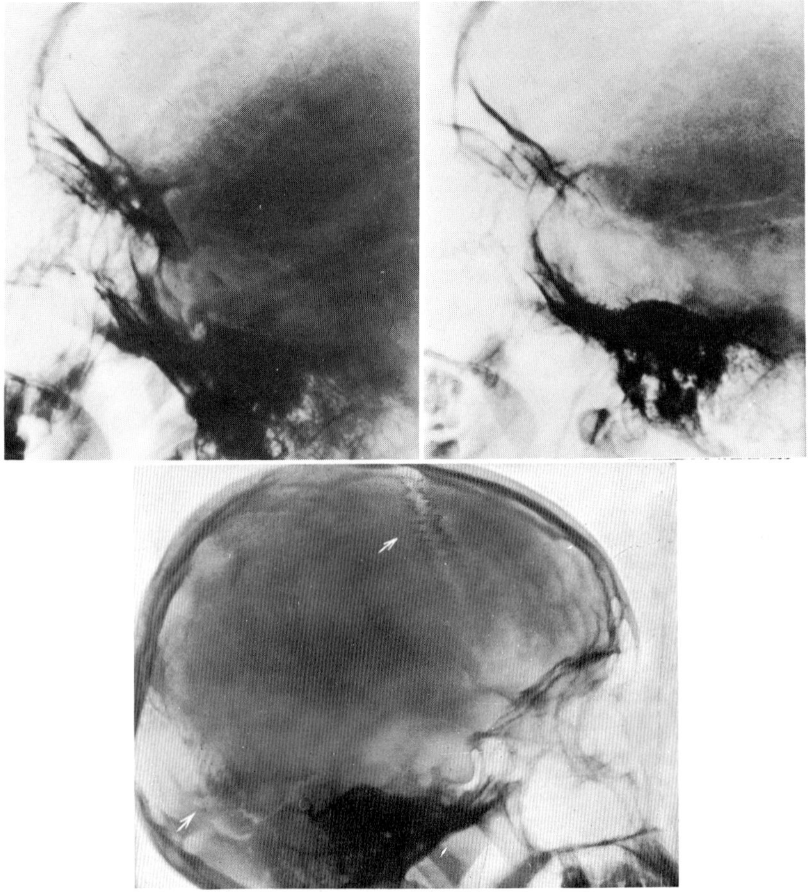

Fig. 3-47. Above, Roentgenogram of the sella (lateral view) of a 50-year-old patient with chromophobe adenoma of the pituitary. Right, Enormous enlargement and ballooning of the sella, thinning of the sellar floor, dorsum sellae evidently destroyed, pointed appearance of the tuberculum. Left, Sella of the same patient 16 years before! Beginning enlargement of the sella already visible. Below, Lateral view of the skull of a 15-year-old patient with astrocytoma of the cerebellar vermis to demonstrate the sellar changes in chronic increased intracranial pressure: slight enlargement of the sella with normal thickness of its floor, posterior clinoid processes missing, marked osteoporosis of the sella contour. Rupture of the coronal and lambdoidal sutures (arrows).

The typical roentgenologic changes of the bony structure of the sella should be briefly summarized—most of all because the general practitioner or the ophthalmologist frequently finds himself in a situation where he is the first to order and interpret a film. The characteristic roentgenologic sellar changes are as follows: *ballooning of the sella* with uniform enlargement in all diameters (observed in about one third of cases, especially in acidophil adenomas); depression and thinning of the floor of the sella, which protrudes into the sphenoidal sinus; and above all, *destructive changes at the dorsum and the anterior and posterior clinoid processes,* ranging from slight decalcification to destruction or erosion (Fig. 3-47). In patients with severe sellar changes and considerable extrasellar growth of the adenoma the optic foramina may manifest decalcification, blurring of their contours, or even destruction (Lombardi). Highly important information about shape, size, and connections of the pituitary adenoma can be gained by *pneumonencephalography* (preferably combined with tomographic techniques), revealing the encroachment upon the suprasellar cisterns and the third ventricle and eventually also an indentation of the anterior horns of the lateral ventricles. It is important that roentgenologic sellar changes occur in patients with intrasellar pituitary tumors before the manifestation of visual field changes. This is in contrast to tumors of a primarily suprasellar growth, which frequently show visual field changes before roentgenologic alterations occur. These x-ray findings are characteristic for pituitary adenoma and should not be confused with the effects of a generally increased intracranial pressure on the bony structures of the sella. The latter, as a rule, produces atrophy of the clinoid processes and the dorsum sellae as well as occasionally a certain widening of the opening of the sella (Fig. 3-47). Other neuroophthalmologic and purely neurologic signs should aid in the differential diagnosis of doubtful cases.

Pituitary apoplexy represents a syndrome characterized by intense headaches, unilateral or bilateral ophthalmoplegia, sudden occurrence of blindness in both eyes, and coma, which without surgical intervention leads to death within hours or days. Such a syndrome develops as a result of hemorrhage into a pituitary adenoma or an infarction within such a tumor. We have observed one such patient: after operation the ophthalmoplegia disappeared within days, but one eye remained amaurotic and the other retained a complete temporal hemianopia. Before the acute onset of the syndrome, the patient had no symptoms whatever, especially no ocular signs of his growing pituitary adenoma!

In summary, it can be stated that pituitary adenomas are characterized by the development of the classic chiasmal syndrome. This includes the distinctive bitemporal hemianopia, which in numerous patients takes on a symmetrical form. The visual field changes start in the superior temporal quadrants, usually also in a symmetrical fashion. The field changes progress from the superior temporal to the inferior temporal quadrants. Only in the late stages is there an involvement of the inferior nasal and, finally, of the superior nasal quadrants. The central or paracentral scotomas in the scotomatous type of chiasmal syn-

drome show a similar development. They may be the first and only ocular sign of a pituitary tumor! Amaurosis of one eye and temporal hemianopia of the other eye is a rather infrequent event. Symmetry of development and the final shape of the visual field defects are characteristic for the pituitary adenoma. The eyeground changes consist of a more or less pronounced optic atrophy which, as a rule, lags behind the loss of vision: there is no direct relation between the degree of the atrophy and the loss of vision. The loss of vision may vary considerably and may progress to complete amaurosis. It is the first and most frequent ocular symptom. Complaints referring to temporal hemianopia are rare. In the case of extensive extrasellar expansion the pituitary adenoma may invade the cavernous sinus, causing extraocular muscle pareses (especially of the oculomotor and sixth nerves) or even an actual cavernous sinus syndrome. Pupillary disturbances are found mostly with severe loss of vision and especially with a unilateral amaurosis. In such cases the ipsilateral pupil is dilated and does not react to light.

After this discussion of pituitary tumor as the prototype of the sellar chiasmal syndrome, we have to consider the *differential diagnosis,* a rather complex problem because of the numerous types of tumors possible in this and the surrounding area. However, from the neuro-ophthalmologic point of view, in principle, we are always dealing with *more or less distinct modifications of a chiasmal syndrome.* Its atypical character can be summed up as *irregularity and asymmetry in its progressive development as well as in its final stage.*

Craniopharyngiomas (intrasellar and suprasellar tumor type)

Craniopharyngiomas originate from epithelial remnants which, as vestiges of Rathke's pouch (craniopharyngeal duct), are scattered throughout the stalk of the pituitary gland (Erdheim). Such epithelial remnants are found above the diaphragma sellae on the anterior part of the infundibulum and below the diaphragm at the immediate area of transition from the stalk of the pituitary gland to the anterior lobe. Accordingly, there are craniopharyngiomas with a *primary suprasellar* and (less frequent) with a *primary intrasellar origin.* The latter may show a suprasellar expansion in the course of their later development and thus form the third group, the intrasellar and suprasellar pharyngiomas. Craniopharyngiomas are predominantly tumors of childhood. However, they do occur in older persons up to the sixth decade. In reviewing our own series of twenty-two cases of craniopharyngiomas, we were surprised to find that only one third involved adolescents and children, whereas all other patients were past 20 years of age (according to Thiebaut, 66%!). The general as well as the neuro-ophthalmologic symptomatology in children differs from that in adults (Figs. 3-48 and 3-49).

In children and adolescents the signs of an increased intracranial pressure, especially headache and vomiting, dominate the clinical picture. Tumor invasion of the third ventricle interferes with the circulation of the cerebrospinal fluid and causes an internal hydrocephalus. The head of such children is out of proportion to the size of the body. On percussion of the head one frequently finds the typical "resonance of a cracked pot," a sign of suture ruptures. During puberty, endocrine disturbances (due to an impaired development of the anterior lobe of the pituitary gland resulting from the tumor growth) produce the clinical picture of *dystrophia adiposogenitalis* (Fröhlich's type), *severe cachexia* (Simmonds' type), or *pituitary dwarfism* (Lorain's type). As a rule, there are also *hypothalamic signs* in evidence (such as, for instance, polyuria, polydipsia, and diabetes insipidus) as well as disturbances of sleep (lethargy) and temperature (hyperthermia).

In *adults* the presenting complaints that bring the patient to a physician are not the symptoms of increased intracranial pressure but usually *visual disturbances.* In addition,

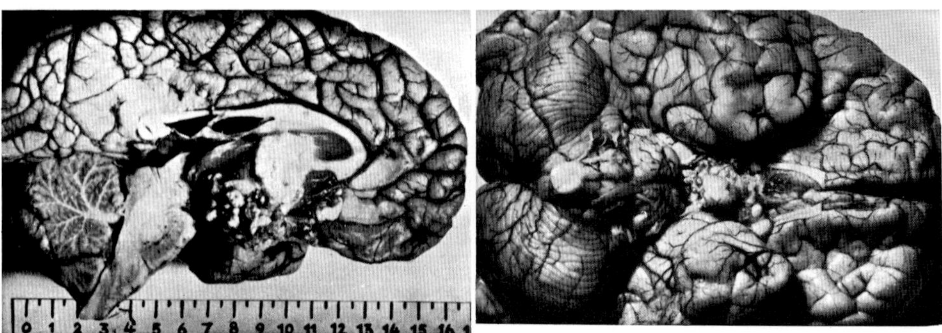

Fig. 3-48. Left, Median section through the brain of a 7-year-old girl with craniopharyngioma invading the third ventricle. Right, Appearance of the tumor as seen from below. Displacement and invasion of the chiasm can be seen.

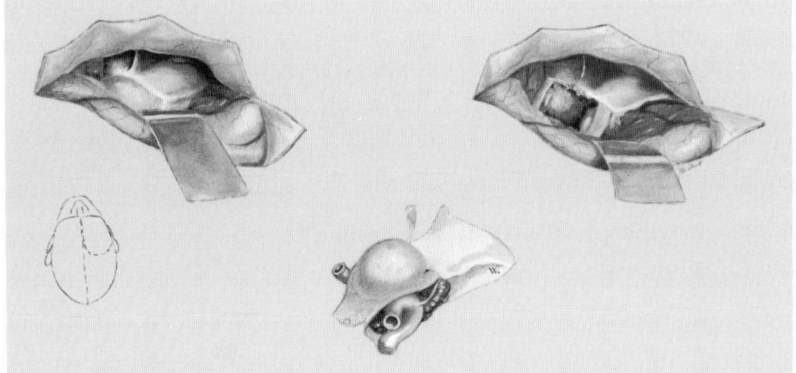

Fig. 3-49. Walnut-sized craniopharyngioma in the region of the tuberculum sellae of a 41-year-old patient before and after surgical extirpation (right transfrontal osteoplastic craniotomy and extradural exposure of sellar region). The illustration in the right upper corner shows the right optic nerve as a flattened cord. The illustration in the center below shows how the tumor compresses the optic nerve and chiasm from above and from behind.

adults frequently show endocrine signs typical for hypopituitarism (sexual disturbances, adiposity, soft skin, loss of axillary and pubic hair, scanty growth of beard in males, etc.). There might also be hypothalamic signs in the form of polyuria, polydipsia, lethargy, hyperthermia, etc.

Some criteria regarding the *ocular symptoms* can already be deduced from these introductory remarks. In *children,* especially babies, signs of increased intracranial pressure in the form of headaches, nausea, and vomiting predominate. Loss of vision is a phenomenon that occurs mostly in the later stage of the disease but, once present, frequently progresses rapidly and even leads to blindness. There is a combined effect of compression of the chiasm and atrophy of the choked discs, with variabilities in the two factors. The parents of a child with severe impairment of vision stated that the child was unable to recognize grain while gleaning, that he ran into nettles, and that he did not

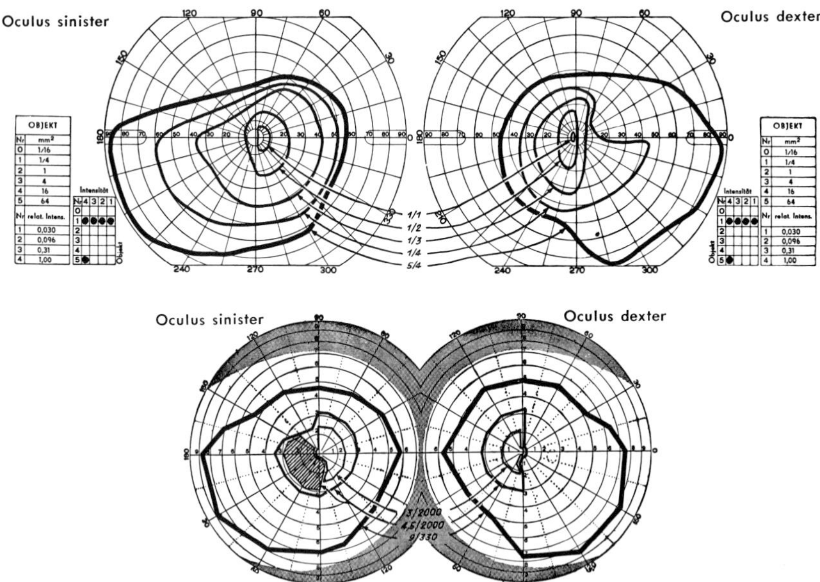

Fig. 3-50. Walnut-sized intrasellar craniopharyngioma near the tuberculum sellae, compressing the chiasm and optic nerve from an anteroventral direction. One year prior to surgical intervention, loss of right visual acuity. Seven months prior to surgical intervention, right temporal hemianopia without symptoms in left eye! First signs of left field defects 2 months prior to surgery. Visual fields (above, Goldmann perimeter; below, tangent screen) taken shortly before surgery show an asymmetrical bitemporal hemianopia of the central isopters, further advanced on the right side. The peripheral isopters demonstrate a similar asymmetrical onset in both superior temporal quadrants. Temporal scotoma extending into the blind spot on the left eye. Visual acuity, O.U. 0.7; no optic atrophy.

look at people while talking to them. The ocular symptoms in *adults* are similar to those caused by a pituitary adenoma. In about 50% of the cases there are *statements about loss of vision, especially on one side* (18% according to Tönnis). *In some cases patients noted a distinct fluctuation of the visual acuity.* The frequently cystic nature of craniopharyngiomas explains such changes and may even simulate successful treatment of the visual loss caused by such a tumor. Such was the case in a patient of Franceschetti and Blum; his vision in one eye improved from 0.2 to 1.0 two weeks after retrobulbar injections of cobra venom, only to show a relapse later on. The *chances of confusing a disturbance of conductivity in the optic nerve caused by a craniopharyngioma with the signs of a retrobulbar neuritis* are great and need special consideration. In addition to making statements indicating loss of vision, the patients will always mention a corresponding constriction of the visual fields, as discussed in connection with pituitary adenomas (p. 195): a narrowing of the field to the side, the sensation of blinders, colliding with people in the street, and reading difficulties with half a word or even half a line missing. Frequently double vision is mentioned. The underlying pareses of the extrinsic muscles are easily explained by the frequently massive extrasellar extension of these tumors.

The *ocular signs,* especially in adults, are marked by the *chiasmal syndrome,* usually with an *asymmetrical development and asymmetry in the end stage.* However, patients with the intrasellar type of craniopharyngioma may show visual field changes, which can easily be confused with those of a pituitary adenoma. In one case we observed a typical onset in the two superior temporal quadrants combined with a temporal scotoma that extended into the

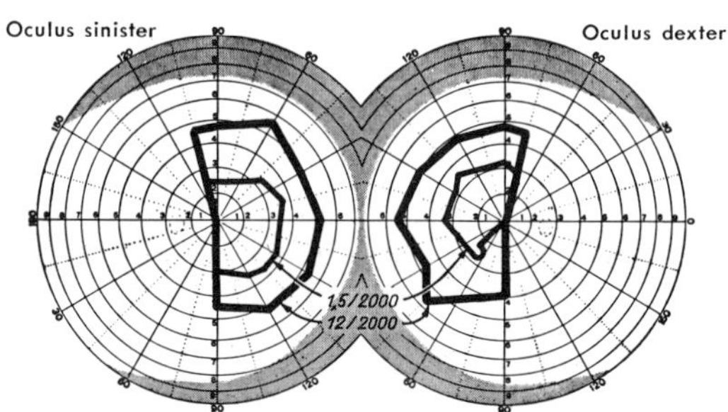

Fig. 3-51. Intrasellar and suprasellar craniopharyngioma of a partly solid and partly cystic structure. Visual acuity, O.D. 0.8; O.S. only hand movement at 3 meters. Left disc slightly paler. Slightly asymmetrical bitemporal hemianopia without sparing of the macula on either side.

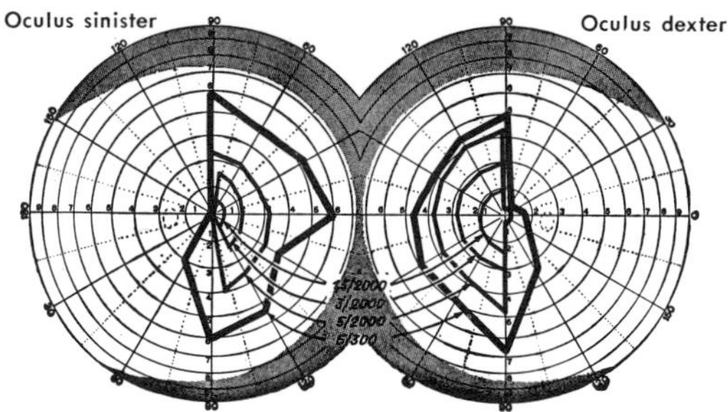

Fig. 3-52. Mostly intrasellar craniopharyngioma. Visual acuity, O.D. 0.25; O.S. 0.25. Bilateral optic atrophy. Bitemporal hemianopia with asymmetrical development. No straight vertical dividing line. Right, Macular sparing. Left, Macular loss.

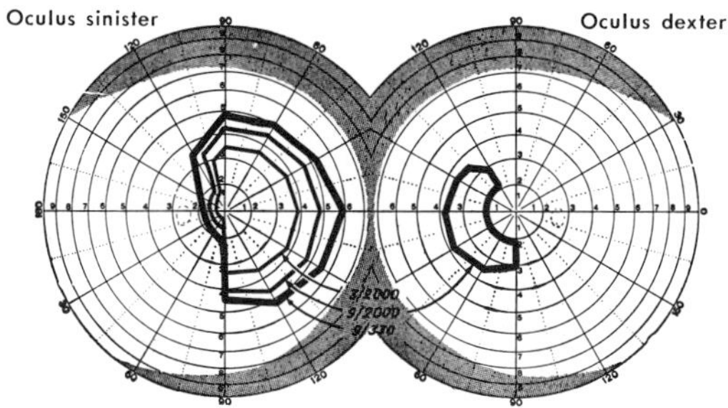

Fig. 3-53. Cystic craniopharyngioma in the third ventricle. Visual acuity, O.D. 0.1; O.S. 0.3. Temporal optic atrophy more pronounced on right side. Bitemporal hemianopia with extreme asymmetry of visual field defects. Left, Macular sparing. Right, Kidney-shaped nasal island of vision hugging macula.

blind spot (Fig. 3-50). This type of field change has been described previously in the discussion of pituitary adenomas. In not quite one fourth of our twenty-two patients with craniopharyngiomas we have observed a typical *bitemporal hemianopia* (40 to 50% according to Tönnis). However, it is important to stress again the *more or less pronounced asymmetry of the visual field defects* (Figs. 3-51 and 3-52). No vertical line separates the seeing and blind halves of the field, but the dividing line usually projects from the seeing into the blind part of the field, sparing the macula. In other cases this dividing line deviates into the seeing part of the field, causing a macular loss. Occasionally we have found kidney-shaped remnants of the field on the nasal side, with a peculiar hugging of the macular area (Fig. 3-53). This asymmetry may also manifest itself with one eye still showing a completely normal field and the other eye already revealing an indentation of the superior or inferior temporal quadrant (Fig. 3-54). Since craniopharyngiomas with primary extrasellar origin tend to expand into a retrochiasmal and suprachiasmal direction and to exert pressure on the chiasma from above and behind, they can be *expected to cause initial field defects in the inferior quadrants.* We made no personal observation of

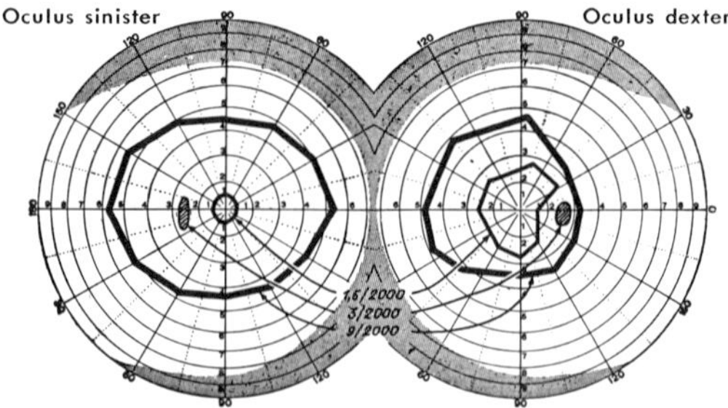

Fig. 3-54. Intrasellar, suprasellar, and retrosellar cystic craniopharyngioma. Visual acuity, O.D. 1.0; O.S. 0.8. Bilateral temporal optic atrophy, further advanced on left side. Beginning right temporal hemianopia with left peripheral isopters normal.

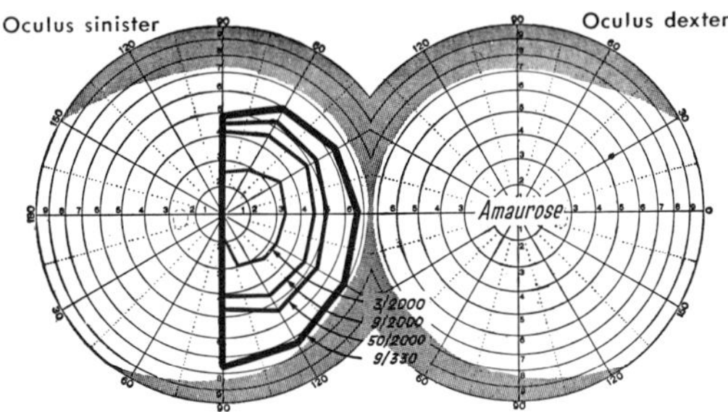

Fig. 3-55. Partly solid and partly cystic intrasellar and suprasellar craniopharyngioma. Left visual acuity, 0.1. Complete bilateral optic atrophy with chalk-white discs. Amaurosis of the right eye with temporal (strikingly symmetrical) hemianopia of the opposite eye.

such a case. We do have records of two patients with amaurosis on one side and temporal hemianopia on the other side (33% in intrasellar craniopharyngiomas according to Tönnis) (Fig. 3-55). One patient showed a distinctly incongruous homonymous hemianopia (8% according to Tönnis) more suggestive of a tract lesion. Some reports in the literature (Wagener and Love; Dubois-Poulsen; Tönnis) stress the frequent and early occurrence of unilateral or bilateral central and paracentral scotomas in patients with craniopharyngioma. These observations are hardly in agreement with our experience: we have seen such scotomas (they are quite similar to the ones seen in pituitary adenomas, p. 189) rather infrequently. It is evident that *it is quite difficult to prove the visual field changes in children,* especially with an increased intracranial pressure. Frequently we have to depend here on a crude confrontation test by hand (p. 71).

In children the *predominant signs of increased intracranial pressure include more or less advanced choked discs.* These choked discs are frequently associated with a secondary optic atrophy (p. 115). Occasionally a distinct choking is missing, but the atrophic discs show blurred or even slightly prominent margins, a quasi "forme fruste" of chronic atrophic papilledema. There are all conceivable intermediate forms and combinations between a pure atrophy and a pure papilledema. Venous and capillary stasis, tortuosity of veins, extravasations near the disc, blurring of the disc margin, and a slight prominence are infallible signs of increased intracranial pressure.

In *adults,* in whom there is a compression of the chiasm rather than the formation of a hydrocephalus, the ophthalmoscopic picture is that of a *primary optic atrophy with sharp borders.* In the section on papilledema (p. 120) we stressed that an atrophic disc, as a rule, is incapable of developing the typical signs of choking. This fact as well as the different tendency to form a hydrocephalus must be taken into consideration in our interpretation of the difference between the fundus pictures of children and adults. Usually there is a distinct dissimilarity in the degree of atrophy between the two eyes. Other phenomena associated with atrophy (p. 207) have been dealt with in the discussion of pituitary adenomas. They are also present here, especially a random or even inverse relation between the extent of the atrophy and the degree of loss of the visual function. In spite of normal optic discs, there may be extensive visual field defects and considerable loss of visual acuity. We found a slight blurring and a trace of prominence of an atrophic disc (manifestation of an increased intracranial pressure; Seidenari) in only a few adults.

Disturbances of motility involved mostly the oculomotor nerve, much more rarely the abducens. The involvement of the *oculomotor* nerve results from the parasellar expansion of the tumor—obviously into the region of the cavernous sinus. The diplopia occasionally mentioned by the patient has its objective equivalent in these muscular pareses. In one patient we even found a binocular oculomotor paresis. Unilateral or bilateral *sixth nerve* palsies are due to the increased intracranial pressure and accordingly may show fluctuations. In this connection we would like to mention that the *onset of a paralytic strabismus in a child sometimes represents one of the first symptoms of a growing cranio-pharyngioma!*

Pupillary disturbances are rather rare and occur in case of unilateral severe impairment of vision or complete amaurosis as a diminished or even abolished direct reaction to light. There are cases of amaurosis in which the reaction to light is paradoxically retained (Walsh and Hoyt).

Once again, it should be stressed that the symptoms, especially the ocular symptoms, associated with craniopharyngiomas may remain stationary or may even improve without any treatment because of the cystic nature of these tumors. In some cases this has caused the understandable but *mistaken diagnosis of a unilateral or bilateral retrobulbar neuritis,* a diagnosis that seemingly was confirmed by the success of whatever therapy was employed. A tireless repeated *search for visual field defects in both eyes* is imperative in such cases: such defects may be missing in the initial stages of development of a craniopharyngioma. We observed a patient with a hemianopic defect in one eye and no change whatsoever in the other eye. Only 6 months later was it possible to demonstrate a similar temporal defect in the other eye and thus substantiate the suspicion of a tumor in the chiasmal region.

The *proof of suprasellar calcifications on the x-ray film* is of considerable value in the diagnosis of craniopharyngiomas. They are present in 90% of the patients under 15 years of age and in only 40% of those over 15 years of age (Lombardi). Some of these calcifications are massive and others are quite delicate, resembling splashes. Usually they are suprasellar, only rarely intra-sellar (Fig. 3-56). As a result of the decalcification and erosion of the clinoid processes and the dorsum, the sella may appear compressed. Only in case of an intrasellar growth will there be ballooning and enlargement of the sella. Its

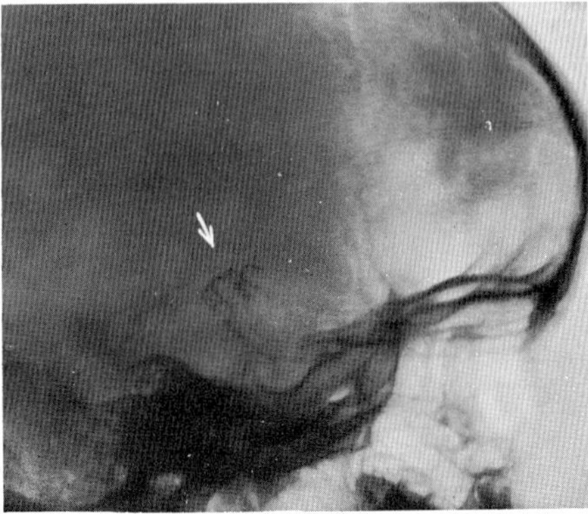

Fig. 3-56. Lateral x-ray film in patient with craniopharyngioma. Suprasellar calcifications (arrow). Compressed sella.

appearance resembles that seen in pituitary adenomas (p. 210). It is important to remember that in tumors with a predominantly suprasellar growth such as craniopharyngiomas *the visual field defects usually precede the roentgenologic signs.* Thus the former are of particularly great diagnostic significance.

The ocular symptoms of craniopharyngioma can be summarized in the following manner. In children the ocular signs of increased intracranial pressure (that is, papilledemas) dominate the picture. The bitemporal loss of the visual fields usually is difficult to demonstrate in children. As in the case of pituitary adenomas, the loss of vision is the most conspicuous sign in adults. Usually it involves one side first, with the other side becoming impaired later. The pathologic-anatomic basis for the loss of vision is a primary optic atrophy, usually more pronounced on one side. Occasionally minor signs of a papilledema are superimposed on the optic atrophy. The chiasmal syndrome is practically always present in adults. On the other hand, an asymmetry during the developing stages and in the end stage is typical. The characteristic visual field defect is a bitemporal hemianopia, usually with a certain asymmetry of the two halves of the field; the dividing line lies rarely in the midline but is displaced either toward the blind or seeing part of the field. Oculomotor nerve pareses are not so rare—at any rate, more frequent than those of the sixth nerve.

Meningiomas of the tuberculum sellae (suprasellar tumor type)

These tumors originate from the dura of the tuberculum sellae and its immediate surroundings (sphenoidal plane, chiasmal sulcus, anterior clinoid process, medial sphenoid ridge).

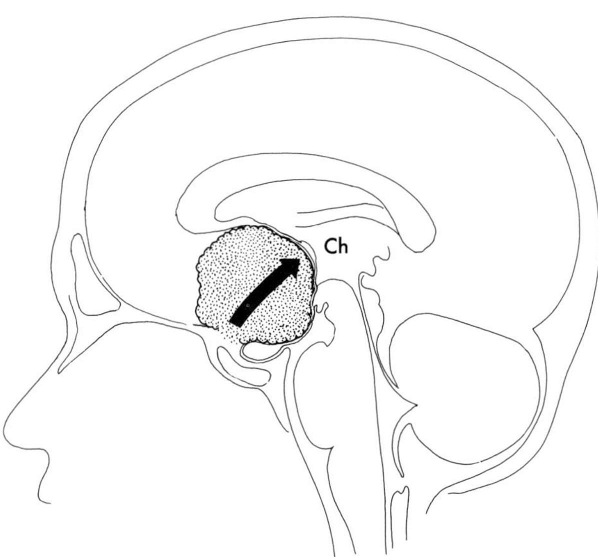

Fig. 3-57. Median section through the skull schematically showing a meningioma of the tuberculum sellae displacing the chiasm, **Ch**, in a superior and posterior direction.

A pronounced suprasellar development results in a superior and posterior displacement of the chiasm (Figs. 3-57 and 3-58). The tumors usually assume an asymmetrical growth, involving at first only part of the chiasm or only one optic nerve. This causes a distinct *asymmetry of the chiasmal syndrome,* especially with amaurosis of one eye and a temporal visual field defect on the other side. The suprasellar meningiomas distinguish themselves by their *exceedingly slow growth,* which may extend over many years. Endocrine pituitary or hypothalmic signs, especially in the initial stages, are extremely rare and occur only in the late stages after the tumor has reached a certain size. Compression of the third ventricle and blocking of the

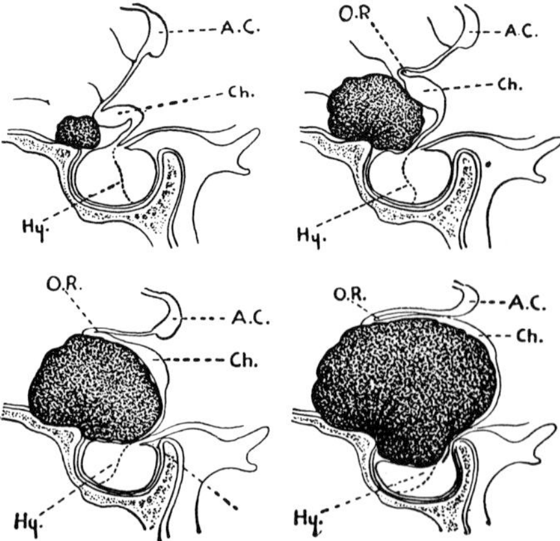

Fig. 3-58. Four stages in the development of a meningioma of the tuberculum sellae demonstrating the progressive deformation and compression of the chiasm, which is pushed upward and backward. **A.C.,** Anterior commissure; **Ch.,** chiasm; **O.R.,** optic recessus; **Hy.,** pituitary gland. (After Cushing and Eisenhardt.)

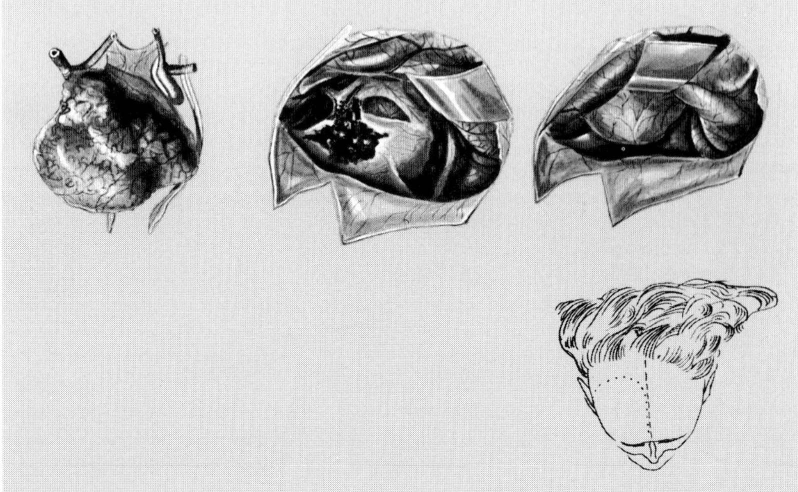

Fig. 3-59. Walnut-sized meningioma of the tuberculum sellae in a 48-year-old patient before and after extirpation (right transfrontal osteoplastic craniotomy). A small stump of the tumor remains at the right internal carotid artery.

intraventricular foramina of Monro may cause an internal hydrocephalus during such late stages (Cushing). In contrast to pituitary adenomas and craniopharyngiomas, there are no important roentgenologic changes of the sella turcica. Demonstration of the frequently occurring hyperostosis of the tuberculum sellae is of diagnostic importance. However, because of the paucity of nonvisual signs, suprasellar meningiomas are often misdiagnosed.

Cushing first described the picture of the chiasmal syndrome as it occurs in meningiomas of the tuberculum sellae: primary optic atrophy in adults with bitemporal visual field defects and a normal sella turcica. In general, this definition is correct. However, it does not take into consideration the *conspicuous and characteristic incongruence of the chiasmal syndrome during its stages of development and in its final form.* Yet this phenomenon is quite pronounced in meningiomas of the tuberculum and still much more distinct than in craniopharyngiomas, as the following considerations should clarify (Cushing and Eisenhardt; Hartmann; Guillaumat) (Figs. 3-57 and 3-59).

The *subjective ocular symptoms already suggest such an asymmetry:* the very first sign is always a significant *unilateral loss of visual acuity,* mostly in the form of a fog, veil, or glimmer before the eye involved first. Another characteristic symptom is a statement that the *unilateral loss of vision in one eye progresses slowly, often to complete amaurosis, yet without manifestation of any signs in the other eye.* Such signs may follow only later. Occasionally it happens that a patient is unaware of the amblyopia or amaurosis in one eye because the vision on the other side is still well preserved. The latent period between the affection of the first eye and the second eye may be years or even decades—20 years in a case we have followed up personally. *The unilateral impairment of the visual function that frequently is accompanied by a central or paracentral scotoma shows symptoms which, again, are quite similar to chronic retrobulbar neuritis*—in the broadest sense of the word.

Among nineteen patients with meningioma of the tuberculum sellae, we observed three in whom the ophthalmologist who was originally consulted diagnosed a retrobulbar neuritis. In one patient a series of retrobulbar injections of tolazoline (Priscol) seemed to cause a temporary improvement of vision. This apparently confirmed the suspected diagnosis of a retrobulbar neuritis. In another patient an optic atrophy associated with the "retrobulbar neuritis" was interpreted as being due to multiple sclerosis. In one patient a unilateral impairment of vision improved for a short time after x-ray treatment. In this case the diagnosis of a tumor was made and an operation finally performed 15 years later! Because of unilateral loss of vision, one patient received irradiation of the pituitary region (without a convincing result) for 5 years. The first signs of loss of vision had occurred 8 years prior to the irradiation. A definite diagnosis of meningioma of the tuberculum sellae and the surgical intervention took place about 20 years after the occurrence of the initial symptoms in the first involved eye! Quite instructive is the history of another patient who complained of loss of vision in one eye for 9 years before a definite diagnosis was made. Throughout this period of time he had been treated for an occupational toxic amblyopia and, accordingly, had received benefits from workmen's compensation. The tragic fact was that during this long latent period the meningioma of the tuberculum sellae resulted in amaurosis of the first involved eye and a temporal hemianopia of the other eye!

The fact that impairment of vision and, with it, the field defects usually occur first only in one eye makes it understandable that subjective statements regarding these visual field defects are rather rare. The symptoms involve the temporal part of the field. The patients have the sensation of shadows, veils, a curtain, a wall, etc. on the side of the involved eye. Frequently the patient remarks about defects only after the second eye has become involved and the first one is already amaurotic. Diplopia is rarely mentioned in the history.

Compared with craniopharyngiomas, the asymmetry of the visual fields in patients with meningiomas of the tuberculum sellae is even more pronounced. Amaurosis of one eye with temporal hemianopia of the other eye, in our experience, *is the most frequent and most characteristic sign.* We have observed this phenomenon in twelve of nineteen patients. Guillaumat and Robin reported a similar ratio, as did Mundinger and Riechert. The explanation for this pattern is that the tumor is limited for a considerable time to one optic nerve and later extends to the median surface of the anterior angle of the chiasm, the area that contains the crossing nasal fibers of the contralateral side. Consequently, a unilateral amaurosis with temporal hemianopia of the field of the other eye is called *the syndrome of compression of the anterior angle of the chiasm.* It is *particularly characteristic for meningiomas of the tuberculum sellae* (Fig. 3-60). The temporal hemianopia of the second eye may either have a symmetrical character with separation of the two halves of the field in the midline or, as

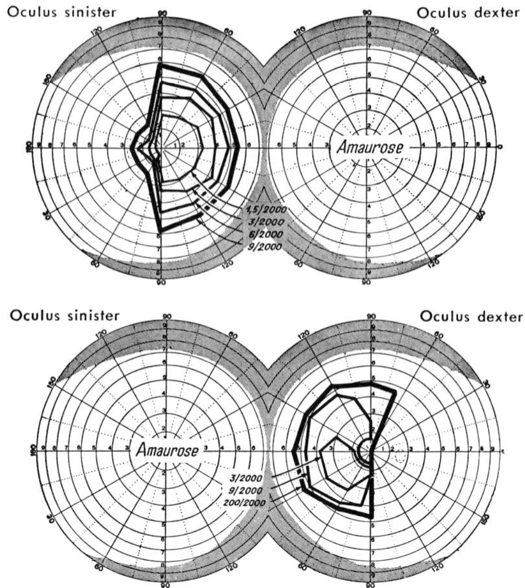

Fig. 3-60. Two cases of meningioma of the tuberculum sellae with typical amaurosis on one side and temporal hemianopia on the opposite side. Above, Meningioma weighing 20 grams. Complete right optic atrophy. Temporal pallor of the left disc. Visual acuity, O.S. 0.7. Left symmetrical temporal hemianopia with sparing of the macula. Below, Meningioma weighing 10 grams. O.S., Complete optic atrophy; O.D., incomplete optic atrophy! Right visual acuity, 0.15. Slightly asymmetrical right temporal hemianopia without sparing of the macula.

happens frequently, there may be asymmetrical temporal defects, with overlapping of the dividing line either into the temporal or nasal part of the field. In advanced stages the second eye may show only quadrant islands of the superior or inferior nasal field. In one patient we observed the onset of the temporal hemianopia in the superior temporal quadrant of the second eye. In another we noted an altitudinal hemianopia, with a horizontal dividing line and a loss of both inferior quadrants (Fig. 3-61).

In four of the nineteen patients with meningiomas of the tuberculum sellae, visual field defects in both eyes could be demonstrated in the form of an extraordinarily *asymmetrical, incongruous bitemporal hemianopia* (Figs. 3-62 and 3-63), occasionally with fairly large central and paracentral scotomas, often of the hemianopic type (Schlezinger, Alpers, and Weiss). For example, the asymmetry manifested itself in such a way that one eye showed a complete temporal hemianopia, with the dividing line through the point of fixation,

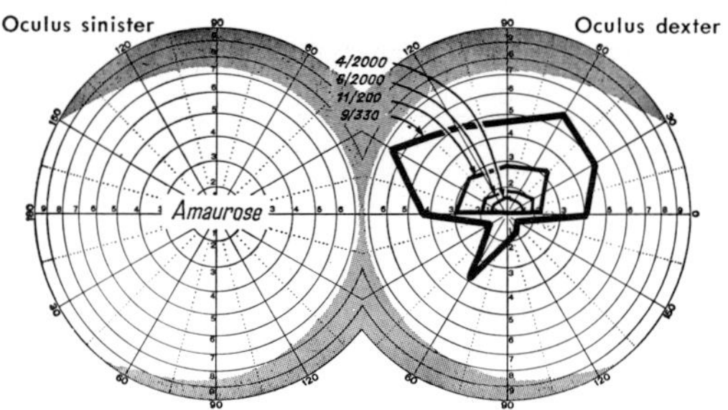

Fig. 3-61. Meningioma of the tuberculum sellae that weighs 5 grams. Incomplete bilateral optic atrophy. Right visual acuity, 0.05. Left amaurosis with right altitudinal hemianopia (partial loss in both inferior quadrants).

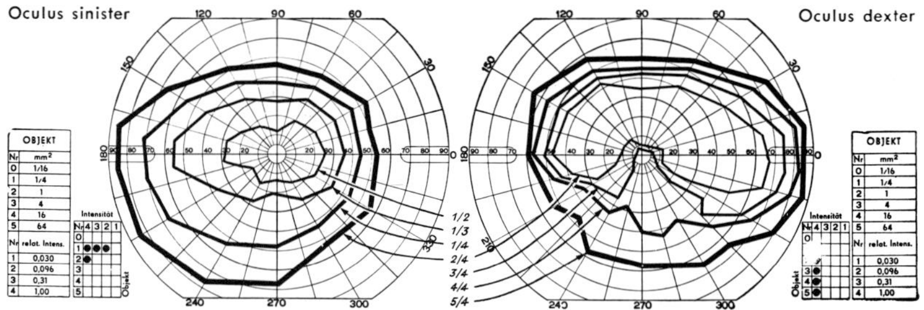

Fig. 3-62. Meningioma of the tuberculum sellae that weighs 6 grams. Right disc shows temporal pallor. Left disc is normal. Visual acuity, O.D. finger counting at 1 meter; O.S. 1.5. Because of loss of right vision, this patient had been treated (without success) for retrobulbar neuritis! There is a suggestion of an asymmetrical bitemporal inferior quadrantanopia. Distinct right temporal inferior quadrant defect. The left most central isopter shows an indentation in the inferior temporal quadrant. The involvement of the right inferior nasal quadrant can be explained by the extensive compression of the right optic nerve confirmed during the operation.

whereas in the other eye only the inferior temporal quadrant was missing (Fig. 3-64). In the early stages of the bilateral field changes the defects of the inferior temporal quadrants seem to dominate the picture. Unilateral central or paracentral scotomas of a midline temporal hemianopic type, which are usually seen in association with peripheral temporal defects of the contralateral eye, correspond to the so-called junction scotomas of Traquair and are typical for a lesion in the area of the optic nerve blending into the chiasm in the anterior angle of the chiasm. As these scotomas are easily overlooked, they have to be searched for by careful (especially static) perimetry.

Just as in patients with pituitary adenomas and craniopharyngiomas, the *eyeground findings* in patients with meningiomas of the tuberculum sellae are characterized by primary optic atrophy with sharply outlined discs. In line with the involvement of mainly one eye in the initial stages, a *unilateral optic atrophy or one more pronounced on one side is the rule.* The atrophy may vary in degree from partial pallor to a completely chalk-white disc with excavation

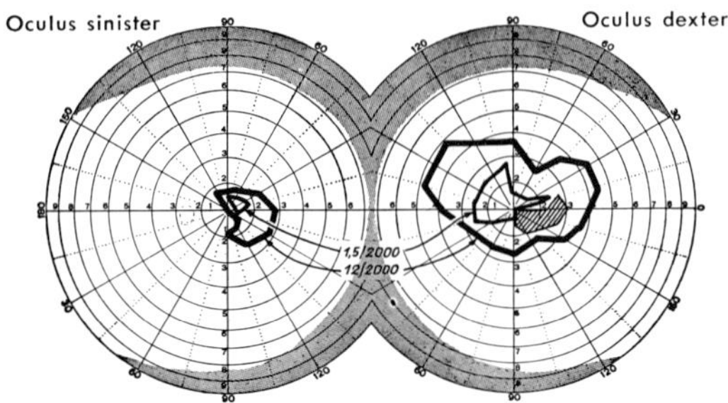

Fig. 3-63. Meningioma of the tuberculum sellae. Primary optic atrophy with distinct margins and excavation of both discs, more on the left side. Narrowing of the arteries, especially on the left side. Visual acuity, O.U. 0.15. Completely asymmetrical, irregular bitemporal hemianopia with bizarre incongruence of remnants of the visual fields. Right central-paracentral temporal scotoma of midline hemianopic character ("junction" scotoma).

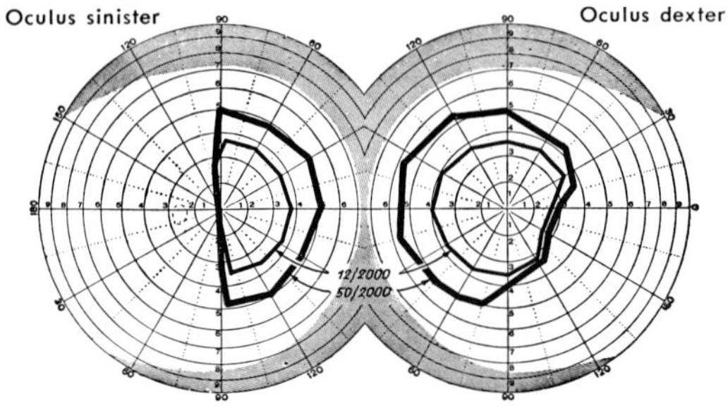

Fig. 3-64. Meningioma of the tuberculum sellae that weighs 70 grams. Visual acuity, O.D. 1.0; O.S. 0.25. Right disc of normal color. Left disc shows temporal pallor. Asymmetrical incongruous bitemporal hemianopia. Left side, Complete temporal hemianopia. Right side, Defect mostly in inferior temporal quadrant.

(Fig. 3-46). As a rule, the atrophy lags behind the impairment of vision and the visual field defects. The latter may be present before the disc changes become manifest. In an extremely severe optic atrophy, especially with an amaurosis, there may be a narrowing of the retinal arteries and later of the veins. In such cases the caliber of the vessels is in contrast to those of the other side. Minor signs of a choked disc in the form of blurring of the margins and a slight prominence of the disc with a simultaneous atrophy of the papilla were noted once among our nineteen patients—a patient with a particularly huge tumor. As a rule, there is no increased intracranial pressure, perhaps because the ocular symptoms are so alarming before the tumor grows too large. Ocular pareses were exceedingly rare in our series—one instance of a sixth nerve paresis among nineteen patients.

Pupillary disturbances occur primarily in the first affected eye—usually as a diminished or abolished reaction to light. In a unilateral amaurosis the pupil characteristically is dilated and does not respond directly to light.

The *x-ray film* in meningiomas of the tuberculum sellae offers important diagnostic clues. The tumors usually do not attack the sella. Thus it appears perfectly normal, at least in the initial stages. Later a decalcification and erosion of the anterior and posterior clinoid processes may be observed. Much more important from a diagnostic point of view is the *proof of hyperostosis of the tuberculum sellae or thickening of the sphenoidal plane. Occasionally there are calcifications of the tumor mass itself* (Fig. 3-65).

Cerebral arteriography is of great diagnostic help: the carotid siphon and the anterior cerebral arteries are displaced upward, forming a sort of arch, and in the venous phase there may be a capillary "blush" staining the meningioma itself. Sometimes the ophthalmic artery is hypertrophic, indicating that it also contributes to the blood supply of the tumor.

In summary, we note that a unilateral onset of ocular signs is quite charac-

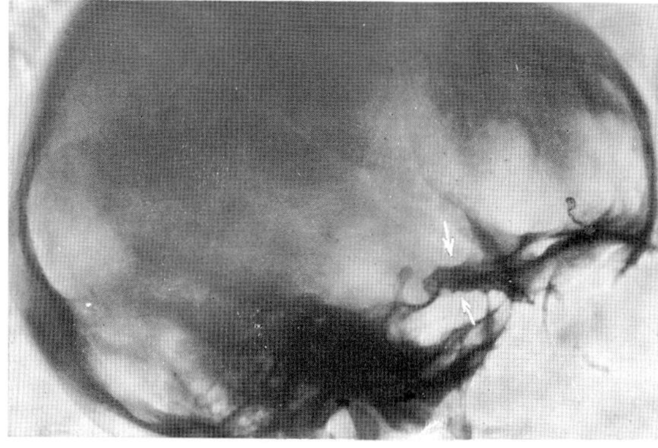

Fig. 3-65. Lateral x-ray view of the skull in a patient with meningioma of the tuberculum sellae. Hyperostosis of the tuberculum sellae (arrow).

teristic for patients with a meningioma of the tuberculum sellae. The history reveals a unilateral loss of vision that may progress to complete amaurosis. This process may extend over several years. The latent period between the involvement of the first eye and of the second eye also may be quite long. Amaurosis of one eye with temporal hemianopia of the other eye is quite characteristic. It is a sign of unilateral involvement of the optic nerve and the anterior angle of the chiasm. In addition, there are asymmetrical, incongruous bitemporal hemianopias, with a preponderant involvement of the inferior temporal quadrants. As a consequence of the unilateral onset of the disease, the ophthalmoscopic appearance of the optic atrophy, likewise, is unilateral or more pronounced on one side. The unilateral loss of vision, visual field changes (central scotomas!), and fundus alterations frequently cause confusion with retrobulbar neuritis. In the differential diagnosis the x-ray findings are of importance, that is, the absence of gross changes of the sella, the proof of hyperostosis of the tuberculum sellae, or actual calcification of the tumor.

Meningiomas of the olfactory groove (presellar tumor type)

These tumors originate from the arachnoid cells near the olfactory groove and the crista galli (Figs. 3-66 and 3-67). Although a meningioma of the olfactory groove arises in the midline, it frequently tends to grow more toward one side. According to their location, we can differentiate three groups: an anterior type with a frontal location, an intermediate type near the crista galli, and a posterior type originating near the tuberculum sellae (Cushing and Eisenhardt; David and Askenasy). The anterior and intermediate types rarely cause signs involving the visual apparatus. The posterior meningioma of the olfactory groove is most likely to press against the chiasm and the optic nerve, thus displacing and compressing these structures backward and downward. *Clinically, the most important symptom is unilateral or, later, bilateral anosmia.* During the course of several years the tumor may reach considerable size, which may lead to a more or less pronounced *psycho-organic syndrome* characterized by disturbances of memory, associations, and affect. Thus the symptomatology of these tumors frequently may convey the impression of frontal lobe tumors. The psycho-organic syndrome must be considered the result of compression of the base of the frontal lobe (Fig. 3-66).

In patients with meningiomas of the olfactory groove the symptoms of anosmia and those of a generalized increase in intracranial pressure in the form of headache, nausea, vomiting, or vertigo usually precede the ocular symptoms. The farther forward the meningioma originates, the later it causes ocular symptoms. As a rule, there are complaints about a unilateral *loss of vision*, with cloudiness, fogginess, and darkness of vision. Usually loss of vision occurs first on one side and later perhaps on the other side but may occasionally show a bilateral onset. We observed one patient who became practically blind in both eyes before undergoing surgery. In one patient, apart from a state of confusion, generalized apathy, and loss of impulse, visual hallucinations could be recorded. There are numerous references in the literature to the fact that pregnancy (Hagedoorn) or trauma may activate the growth of meningiomas of the olfactory groove (like other meningiomas). In one patient in our series the ocular symptoms first manifested themselves shortly after trauma to the skull

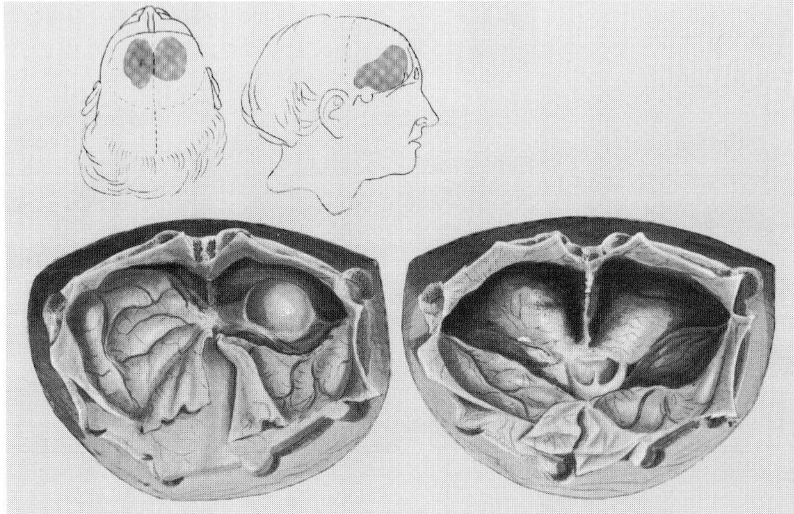

Fig. 3-66. Tangerine-sized bilateral meningioma of the olfactory groove in 53-year-old patient before and after extirpation (bilateral frontal osteoplastic craniotomy).

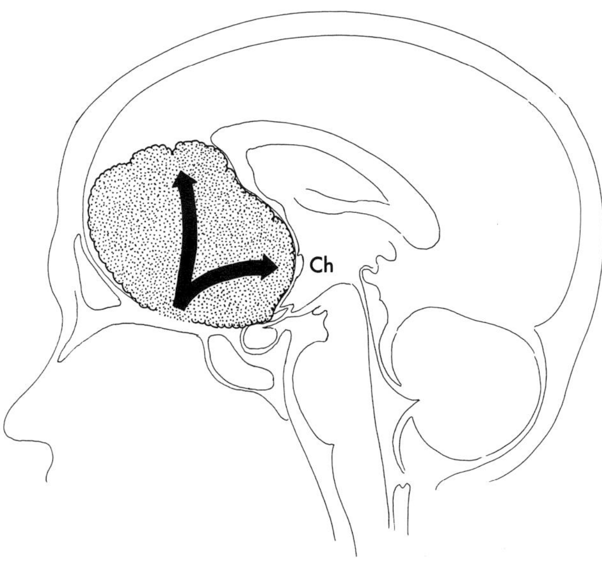

Fig. 3-67. Median section through the skull schematically showing a meningioma of the olfactory groove. Compression and deformation of the chiasm, **Ch,** and the third ventricle.

and progressed in both eyes to the point where there was an inability to read. In one patient the vision improved following delivery of her infant only to show further loss later.

The visual fields and eyeground findings depend to a large extent on the position and size of the tumor. A meningioma of the olfactory groove with an anterior location that does not interfere with the optic nerve or the chiasm may assume a certain size and produce the ophthalmoscopic picture of only a generalized increase of intracranial pressure. Thus we have noted a unilateral or bilateral papilledema in about two thirds of our sixteen patients with meningioma of the olfactory groove. The prominence varied from slight blurring of the disc margins to 3 diopters or more. Hemorrhages of the disc are rare. There was no correlation between the side of the tumor and the side of the more pronounced papilledema.

During the extended course of the increased intracranial pressure the choked discs frequently undergo a chronic atrophy. In addition to the commonly seen enlargement of the blind spot, there will be a progressive constriction of the peripheral visual field and a loss of central vision. One should always consider the possibility that mental and psychic alterations are responsible for concentrically constricted visual fields (p. 81), because a general lack of impulse and apathy may, to a large extent, cause such a defect. In our series of sixteen patients we found papilledema on one side and optic atrophy on the other side, in other words, a *Foster Kennedy syndrome,* only three times. Like the cases reported by Hartmann, David, and Desvignes, our patients did not always show a typical picture. One had bilateral papilledema, with one disc revealing a conspicuous atrophy. Another showed bilateral optic atrophy, but with distinct blurring and a slight prominence of the disc on one side. The third patient was a rather typical example of this syndrome, with

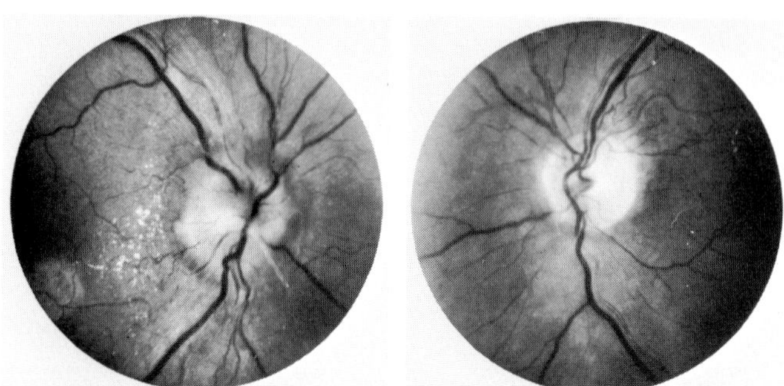

Fig. 3-68. Foster Kennedy syndrome in a 61-year-old patient with bilateral meningioma of the olfactory groove. Distinct papilledema with a tendency to atrophic development on the right side (left) and pallid left disc with sharp margins and no signs of swelling (right). Visual acuity, O.D. 0.2; O.S. below 0.1. One of the first symptoms of the condition was progressive loss of vision on the left side!

optic atrophy on one side and a blurred and prominent nerve head on the other side. *The situation is similar to that in tumors of the frontal lobe, where we also emphasized that the Foster Kennedy syndrome manifests itself rarely in a truly typical form* (Fig. 3-68).

If a meningioma of the olfactory groove extends backward and interferes with the optic nerve and the chiasm, it causes *unilateral* or *bilateral* primary optic atrophy with dissimilar involvement of the two sides. The atrophy may also show all possible gradual variations, ranging from temporal pallor to complete chalk-white atrophy with a corresponding excavation of the disc. A bilateral atrophy of equal degree is not so rare because the tumor originates in the midline and frequently progresses in a symmetrical manner toward the optic nerves and the chiasm.

Except for changes caused by papilledema, the *visual fields* in patients with meningiomas of the olfactory groove are usually quite *bizarre* and *unreliable*. It is entirely possible to find normal fields with anterior meningiomas. Amaurosis of one eye and temporal defects of the other field are not so rare. They are similar to those of suprasellar meningiomas (syndrome of compression of the anterior angle of the chiasm, Fig. 3-69). Actual bitemporal hemianopias are extremely rare in our series. They occur in an incomplete form if at all. *From a diagnostic point of view the central and paracentral scotomas, with or without temporal hemianopia, are more important and characteristic.* Usually they occur first on the side of the tumor but may involve the other eye later. These scotomas must be regarded as the result of a direct tumor pressure on one or both optic nerves. They may be confused with the consequences of chronic retrobulbar neuritis. As a rule, they are quite extensive, involving both the nasal and temporal areas, sooner or later merging with the blind spot (Fig. 3-70).

X-ray films furnish important diagnostic hints. In about one half of the patients there are characteristic *hyperostoses along the crista galli,* sometimes also *calcifications* of the tumor itself, and occasionally a small ossification at the point of origin of the tumor. Angiography demonstrates that the anterior cerebral arteries and their branches are pushed and stretched above the floor of the anterior fossa and sometimes there is a direct tumor stain. Not infrequently the ophthalmic artery is seen to contribute to the vascularization of the meningioma (especially well recognized in the subtraction pictures according to the method of Ziedses des Plantes).

In summary, it must be stated that the ocular symptoms in patients with meningiomas of the olfactory groove have no localizing value. Some of the anterior tumors, especially the larger ones, may lead to papilledema (which usually is bilateral) or rarely to papilledema on one side (contralateral to the tumor) and to optic atrophy on the side of the tumor, that is, a Foster Kennedy syndrome (Fig. 3-71). The latter rarely appears in a typical form. If the tumor grows backward toward the optic nerve and the chiasm, it causes a unilateral or bilateral optic atrophy and loss of vision with the occurrence of large central

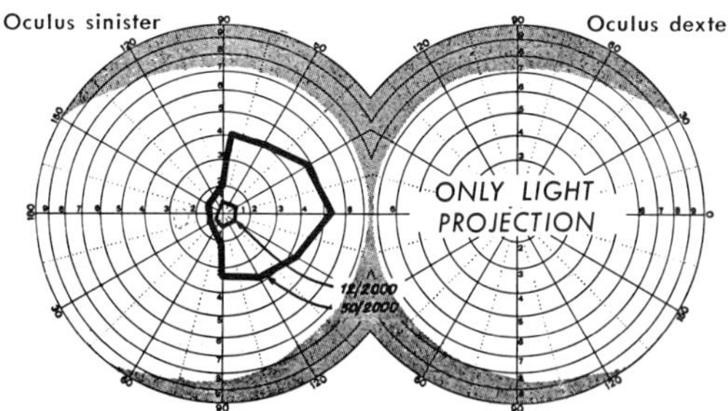

Fig. 3-69. Bilateral meningioma of the olfactory groove with chiasmal compression. Complete optic atrophy on the right side; temporal pallor on the left side. Visual acuity, O.D. light localization; O.S. finger counting at 4 meters. Almost complete right amaurosis with temporal hemianopia on the opposite side (syndrome of compression of the anterior chiasmal angle).

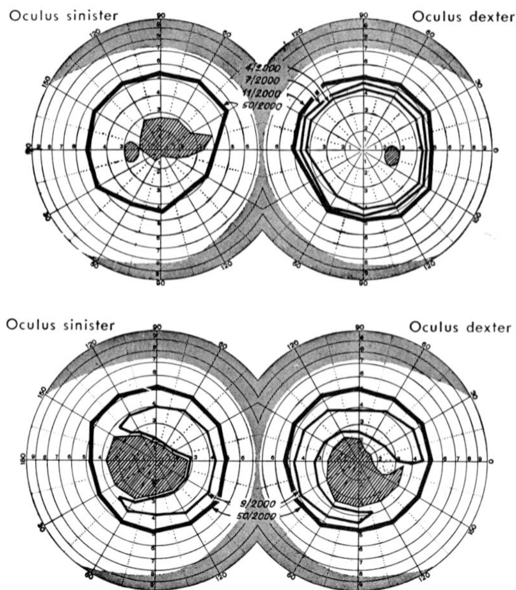

Fig. 3-70. Above, Bilateral meningioma of the olfactory groove that weighs 65 grams and extends mainly toward the left side. Bilateral papilledema of 2 to 3 diopters elevation. In contrast to the right disc, the left one shows distinct pallor; atypical Foster Kennedy syndrome. A large left central and paracentral scotoma extends nasally and temporally, indicating a lesion of the ipsilateral optic nerve. Below, Large bilateral meningioma of the olfactory groove that weighs 120 grams. Bilateral optic atrophy, more advanced on the right side. Visual acuity, O.U. 0.01. Bilateral anosmia. Bilateral large central-paracentral scotomas (targets 50/2000, 9/2000). On the left side the scotoma merges with the blind spot (cecocentral scotoma) but does not include it on the right side. Signs of a lesion of both optic nerves!

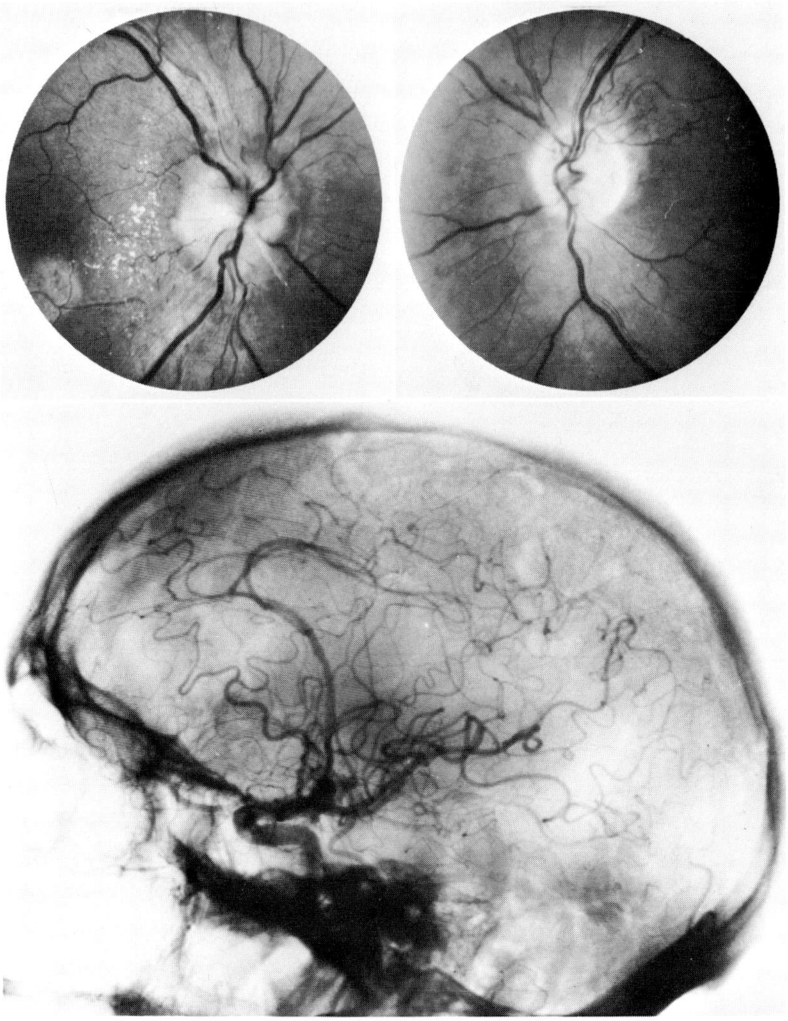

Fig. 3-71. A, Foster Kennedy syndrome with olfactory groove meningioma in a 58-year-old patient. Papilledema with moderate elevation (1.5 diopters), distinct venous engorgement, and obscuration of all disc margins and the optic cup on the right side. Pallid disc with sharp margins and vessels of normal caliber in the left fundus. Note signs of systemic hypertension (arteriovenous crossing phenomena, yellow deep retinal exudates, narrowing of arterioles, etc.) in both eyes. Diastolic pressure of ophthalmic arteries, O.U. 70 grams (Bailliart). Visual acuity, O.D. 0.5; O.S. 0.1. The left illustration corresponds to the right eye; the right illustration, to the left eye. **B,** Corresponding cerebral arteriogram after left carotid angiography. Dorsal displacement of the anterior cerebral artery, typical curvilinear capsular vessels, compression of carotid siphon.

or paracentral scotomas as well as asymmetrical, atypical, and incomplete hemianopias in case of interference with the chiasm. Usually the field defects are quite variable, bizarre, and not very reliable for diagnostic purposes. The diagnosis of such tumors is no longer based on the neuro-ophthalmologic signs but on the two most important symptoms: that is, unilateral or bilateral anosmia and a more or less characteristic psycho-organic syndrome. In differentiating these meningiomas from those of the tuberculum sellae, one should remember that the latter, even at an early stage, lead to ocular symptoms because of an upward displacement of the chiasm. Meningiomas of the olfactory groove, on the other hand, must reach considerable size before pressure from above on the optic nerves or the chiasm will impair the visual function.

Meningiomas of the sphenoid ridge (parasellar tumor type)

Meningiomas of the sphenoid ridge, due to their lateral position, show the greatest variance from the chiasmal syndrome in their symptomatology. Cushing and Eisenhardt distinguish three different types of manifestation of these tumors.

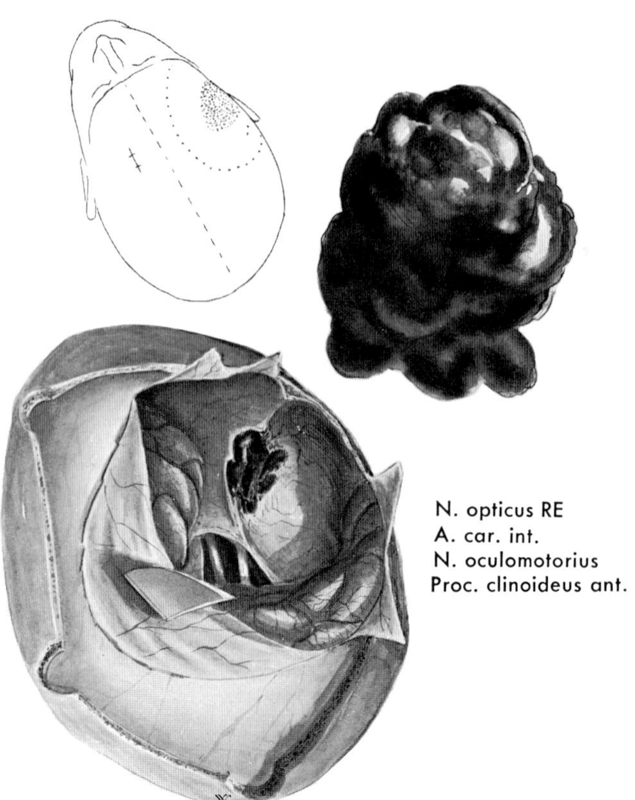

Fig. 3-72. Tangerine-sized meningioma of the right sphenoid ridge, middle third, in a 47-year-old patient after extirpation (right frontotemporal osteoplastic craniotomy).

The first type originates in the *outer third of the sphenoid ridge* (part of the larger wing). It shows two forms with a distinctly separate symptomatology (that is, *pterional meningioma* and *meningioma "en plaque"*). The pterional meningioma grows from the pterional tip of the sphenoid ridge toward the Sylvian fissure, which separates the frontal and temporal lobes. In distinct contrast to the meningioma "en plaque," it reaches considerable size before it causes any symptoms. Except for papilledema and occasional visual hallucinations (temporal lobe!), the symptoms are predominantly of a nonophthalmologic nature (such as headaches, epileptic attacks, uncinate fits, central facial paresis, as well as roentgenologic evidence of changes in the pterional region in the form of abdominal vascularization). In view of its important ocular symptomatology, meningioma "en plaque" will be discussed in greater detail on p. 239.

The second type includes meningiomas originating from the *middle third of the sphenoid ridge* (Fig. 3-72). The third type is of greatest interest to the ophthalmologist. It develops at the *inner third of the sphenoid ridge,* quite close to the clinoid processes and the superior orbital fissure ("clinoid type"). The middle and inner thirds of the sphenoid ridge are part of the lesser sphenoid wing.

Because of their peculiar position, usually a part of the meningiomas of the sphenoid wing extends either into the anterior or middle fossa. For this reason they may interfere with the frontal lobe and the olfactory nerve (in the form of anosmia) or, more often, with the temporal lobe (frequently there are *uncinate fits,* but also other forms of *epileptic attacks* and fainting spells). In spite of its variance with the chiasmal syndrome, the neuro-ophthalmologic symptomatology (David and Hartmann; Kearns and Wagener) is quite decisive for the diagnosis of meningiomas of the sphenoid ridge, especially the inner ridge type: even if small in size, *these tumors interfere less with the chiasm than with the optic nerve and the nerves for the extrinsic eye muscles entering the orbit through the superior orbital fissure.*

Ocular symptoms are almost always mentioned by patients with meningiomas of the lesser sphenoid wing.* In addition to the general signs of increased intracranial pressure such as headache, nausea, and vomiting, there are quite frequently statements about a *unilateral, slowly progressing loss of vision.* The loss of vision frequently is described directly. At times patients complain about cloudy, indistinct, or blurred vision or about the appearance of a fog or shadow. These unilateral types of impaired vision should not be confused with the transient amblyopic attacks or fits of glimmering, which must be regarded as the result of an increased intracranial pressure and of a papilledema (p. 102). *In addition to loss of vision, diplopia is one of the most important subjective symptoms.* Its pathologic basis is a paresis of the extrinsic ocular muscles, which will be discussed below. Because of the frequent involvement of the oculomotor nerve, the *double images often show a horizontal as well as a vertical separation.* Rarely are there *subjective statements concerning an exophthalmos.* In our series of twenty-five cases of meningiomas of the lesser sphenoid wing, patients occasionally complained about a peculiar pressure sensation, about a certain feeling of tension or swelling in the involved eye, or of an impression of pulsation in the eye or the ipsilateral temple. One patient stated that the exophthalmos occurred only during her menstrual period.

---

*Henceforth this abridged term will be used for meningiomas, especially of the inner third of the sphenoid ridge ("clinoidal type"). Middle ridge tumors in their later stages manifest the symptoms of inner ridge tumors. Outer ridge tumors (meningiomas "en plaque") are described separately (p. 239).

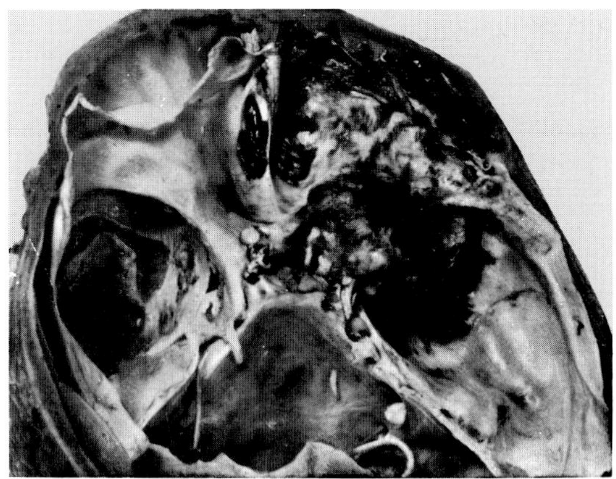

Fig. 3-73. Anterior and middle cranial fossae with meningioma of the inner and middle ridge of the right lesser sphenoid wing. Observe how the tuberous tumor invades the right optic nerve as well as the oculomotor, trochlear, sixth, and trigeminal (ophthalmic division) nerves before their entrance into the superior orbital fissure.

Visual field changes are mentioned rarely. With meningiomas "en plaque" the patient notices a conspicuous swelling of the ipsilateral temporal region.

The *objective symptomatology* of meningiomas of the lesser sphenoid wing (especially the inner ridge type) permits the distinction of *two main groups of phenomena:* (1) the group that is caused by an extension of the tumor into the superior orbital fissure, producing a superior orbital fissure syndrome, and (2) the group that is caused by interference of the tumor with the optic nerve (optic atrophy, visual field changes) (Fig. 3-73).

The *superior orbital fissure syndrome* that may occur in the course of the development of a meningioma of the lesser sphenoid wing is primarily characterized by *disturbances of motility of the extrinsic ocular muscles* (that is, the *sixth, oculomotor,* and *trochlear nerves*). We were able to record such disturbances of motility in about one fifth of the patients in our series. Usually a combination of various nerves is affected, with no apparent predominant involvement of one particular nerve. In addition to the orbital fissure syndrome, there is generally a *unilateral exophthalmos.* Often it is partly responsible for the impaired motility of the involved eye. Frequently it is difficult to decide whether the impairment is due to the affection of the extraocular nerves or the limitation of the muscle action by the exophthalmos. This exophthalmos is an extraordinarily important and frequent (30 to 40% of cases) ocular sign with meningiomas of the lesser sphenoid wing (Elsberg, Hare, and Dyke; Thorkildsen). We have observed it in a quite well-developed form in a little more than one half of our patients (Fig. 3-74). Depending on how far the tumor has progressed, the exophthalmos may range from slight, almost imperceptible degrees to quite severe forms measuring 10 to 15 mm. It may even assume the grotesque form of a protruding "chameleon eye." In extreme cases

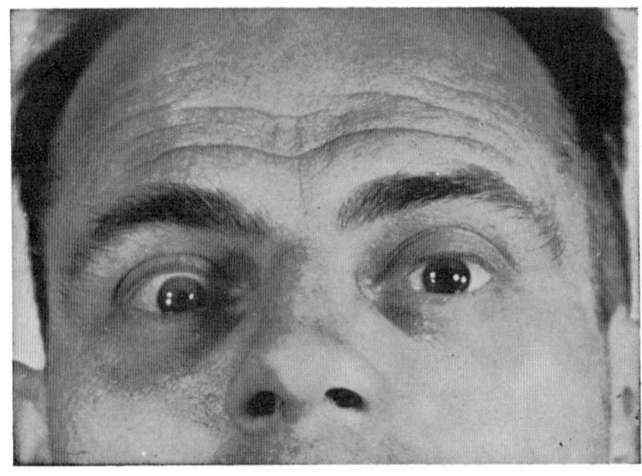

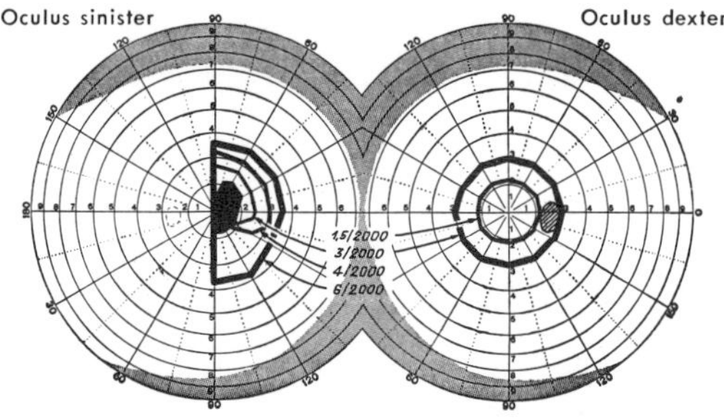

Fig. 3-74. Exophthalmos of the left eye in a patient with meningioma of the left lesser sphenoid wing (middle ridge). Diplopia due to slight paresis of the left sixth and oculomotor nerves. Visual acuity, O.D. 1.0; O.S. 0.1. Left disc shows temporal pallor. Temporal hemianopia of the left eye with central scotoma. Slight concentric constriction of the right isopters.

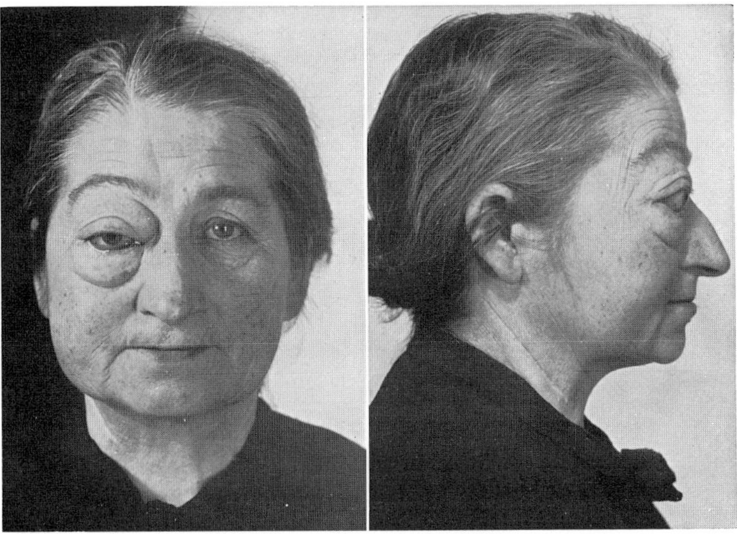

Fig. 3-75. Right unilateral exophthalmos with lid swelling and slight downward displacement of the globe in a patient with meningioma of the inner third of the right sphenoid ridge. Because of optic atrophy, right visual acuity reduced to light perception. Motility of the right eye limited in all, especially temporal, directions.

there may be, in addition to the ordinary proptosis of the globe, a displacement, for instance, downward and outward (Figs. 1-5 and 3-75). The exophthalmos is due to compression of the venous drainage system of the orbit and the cavernous sinus as well as (in many cases) to direct invasion of the orbit by the tumor through the orbital fissure. The venous stasis is frequently evidenced by a corresponding dilatation and tortuosity of the episcleral and conjunctival vessels as well as the vessels of the skin of the upper and lower lids.

Part of the superior orbital fissure syndrome is a lesion of the first division of the trigeminal nerve, with hypesthesia in the area of its distribution, especially a *diminished ipsilateral corneal reflex.* We were able to record a distinctly diminished corneal reflex (mostly in the superior and inferior quadrants) in one third of the patients in our series. *Except for the exophthalmos, the superior orbital fissure syndrome is quite similar to the cavernous sinus syndrome* (p. 249). A meningioma of the lesser sphenoid wing, especially the inner ridge type, may easily infiltrate this area. Thus it is often difficult to decide which of these two syndromes is present. Actually, the difference is only a question of definition and of little importance for a topical diagnosis.

Half of the patients in our series showed pronounced *papilledemas* which, in an overwhelming majority, were bilateral. *Often the progress of these tumors is quite slow: thus the frequent appearance of a chronic atrophic papilledema is easy to understand.* The prominence of the papilledema is often quite pronounced and reaches 4 to 5 diopters, with numerous exudates and hemorrhages on the disc. The latter may have a causal relationship to the impaired drainage of the orbital veins or the cavernous sinus. This mechanism may give rise to the conspicuous congestion of the retinal veins that we observed frequently. We have seen a *Foster Kennedy syndrome* only twice among twenty-five patients. Even these were *not typical in their appearance* but showed a slight blurring or even prominence of the atrophic disc on the tumor side. In the other half of the patients either the discs were normal on both sides or there was a pronounced involvement of one optic nerve with unilateral *signs of atrophy* ranging in degree from a slight temporal pallor to complete optic atrophy with excavation of the disc. Including the two patients with the Foster Kennedy syndrome, we have seen a unilateral optic atrophy in one fifth of the patients in our series.

There is a great variety of visual field changes in patients with meningiomas of the sphenoid wing. This depends on whether and how much the tumor involves the optic nerve. *In slightly less than one third of our patients we found either completely normal visual fields* or merely a slight concentric constriction and enlarged blind spots as a result of the papilledemas (Fig. 3-76). Next in frequency are unilateral visual field changes on the side of the tumor resulting from a lesion of the optic nerve. These field changes may begin either on the nasal side and progress to a *monocular nasal hemianopia* or they may involve the temporal side, leading to a *temporal hemianopia* of the involved eye. We

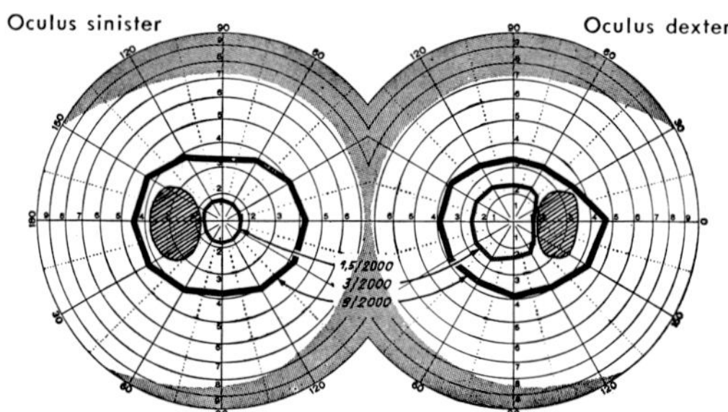

Fig. 3-76. Left meningioma of the lesser sphenoid wing (middle and inner thirds of the sphenoid ridge). Left exophthalmos. Bilateral papilledema of 3 to 4 diopters elevation (amblyopic attacks!). No diplopia. Visual acuity, O.U. 0.7. Enormous bilateral enlargement of the blind spot. Concentric constriction of the visual fields. (See Fig. 1-5, p. 19, recurrence of this same tumor with invasion of the orbit and development of extensive exophthalmos.)

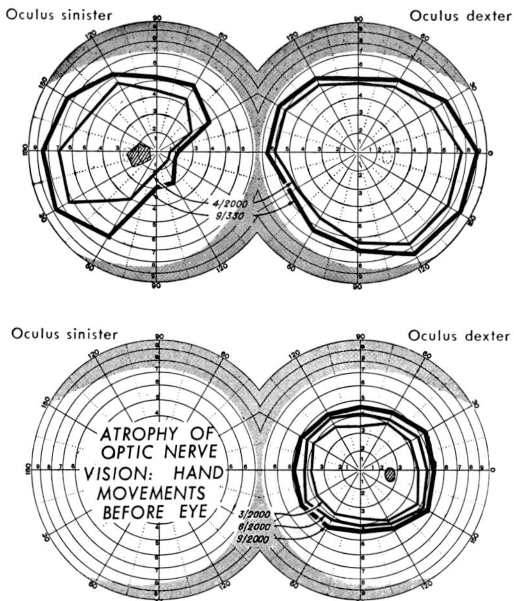

Fig. 3-77. Above, Meningioma of the lesser left sphenoid wing (inner third) extending over the orbital roof to the olfactory nerve. Bilateral papilledema of 3 diopters elevation. Visual acuity, O.D. 1.0; O.S. 0.2. Monocular left inferior nasal quadrantanopia. Right visual field normal. A paracentral temporal scotoma on the left side is a sign of a lesion of the left optic nerve! Below, Meningioma of the inner third of the left sphenoid ridge. Exophthalmos of 4 mm. prominence on the left side. No motility disturbances. Complete left optic atrophy. Right optic disc normal. Visual acuity, O.S. hand movements before eye; O.D. 1.5. Almost complete amaurosis of the left eye without demonstrable visual field. Right visual field normal.

have records of both forms in our series (Figs. 3-74 and 3-77). Occasionally unilateral *central scotomas* (at first relative, later absolute) are seen that must be regarded as the first signs of damage to the optic nerve. The deterioration of the visual field defects may progress to *complete amaurosis on the side of the tumor,* with the field on the other side still intact (Fig. 3-77). In the later stages a temporal field defect may occur in the contralateral eye, indicating a posterior extension of the tumor and pressure upon the anterior angle of the chiasm (Tönnis; Walsh and Hoyt). In one case of recurrent meningioma of the sphenoid ridge we were even able to demonstrate bilateral, inferior, nasal, sector-shaped defects. The binasal defects seem to be based on the bearing of the tumor on the chiasm; however, one also has to consider the possibility of pressure on the lateral aspects of the optic nerves. It is an interesting and noteworthy fact that in four of a total of twenty-five patients with meningioma of the sphenoid ridge there was a distinct *homonymous hemianopia* of a rather congruous form. In one patient the dividing line went through the point of fixation. In another there was definite sparing of the macula (Fig. 3-78). We are in agreement with Dubois-Poulsen, Walsh and Hoyt, and others that these *homonymous hemianopias* (in particular those involving the superior quadrants) *result from an interference of the tumor with the temporal lobe.* However, frequently one cannot ascertain merely on the basis of the visual field findings whether or not a lesion of the optic tract is also a causal factor for such hemianopias.

Pupillary disturbances on the side of the tumor may be due to either a lesion of the afferent fibers of the optic nerve or a lesion of the efferent pupillomotor fibers of the oculomotor nerve. Accordingly, there is only a disturbance of the direct or of the direct and consensual pupillary reaction to light on the tumor side, which may even assume the form of a mydriatic fixed pupil.

The radiologic changes in meningiomas of the sphenoid wing are frequent (40 to 60% of cases) *and quite characteristic* (Fig. 3-79). There may be an

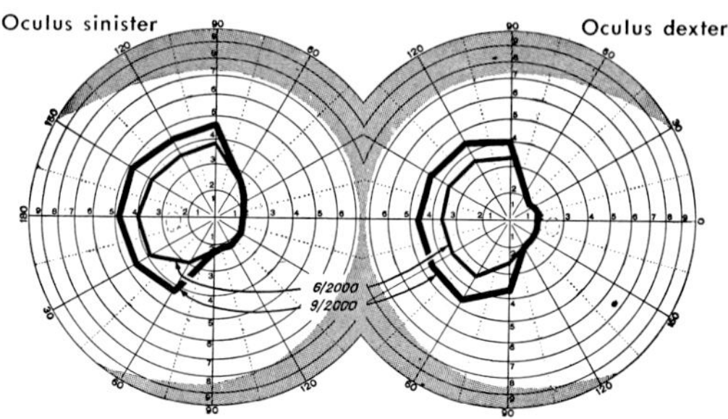

Fig. 3-78. Meningioma of the left sphenoid outer ridge in the pterional region with temporal extension. No exophthalmos. No motility disturbances. Both discs sharply outlined and of normal color. Visual acuity, O.U. 0.4. Right homonymous hemianopia with sparing of the macula, a sign of interference with the left temporal lobe.

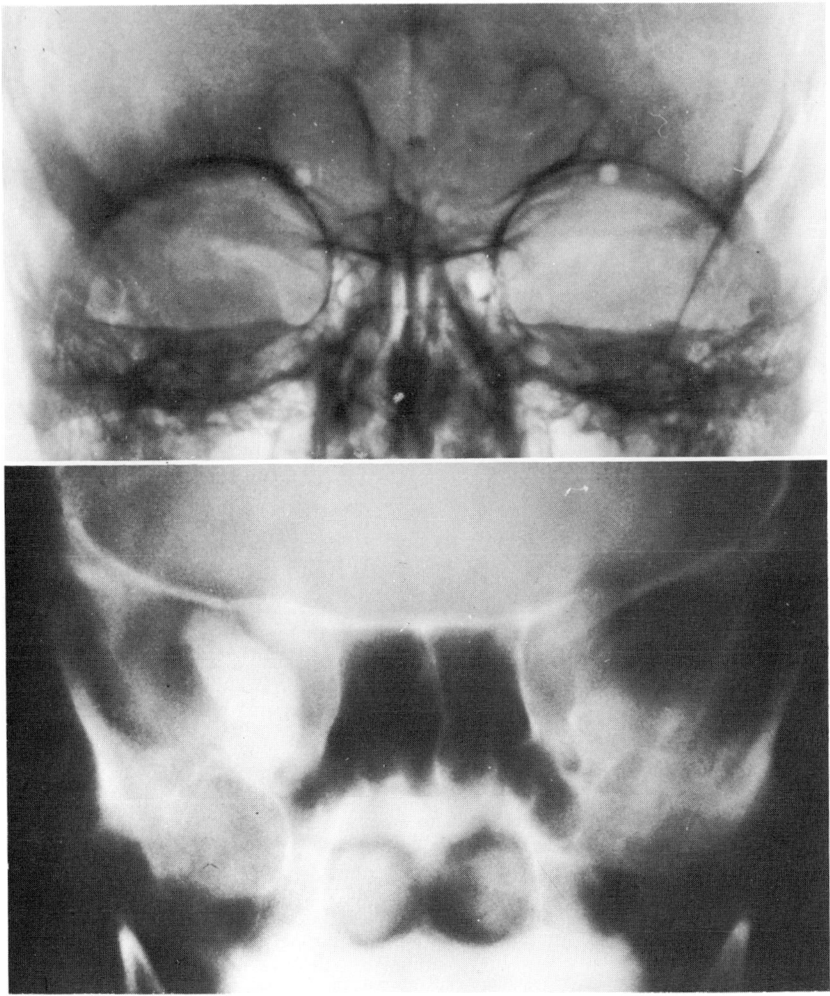

Fig. 3-79. Above, Anteroposterior x-ray view of the orbits. Increased bone density and sclerosis of the lesser and greater sphenoid wings on the right side due to meningioma of the right sphenoid ridge (middle and inner third). Normal size and contour of the superior orbital fissure on the right side. Below, Corresponding orbital tomogram demonstrating even more distinctly the hyperostosis of the bone surrounding the right superior orbital fissure.

osteomatous reaction of the lesser sphenoid wing (especially at the site of the tumor attachment, eventually combined with narrowing of the orbital fissure and of the optic foramen); or one may find decalcification of the sphenoid ridge, pathologic enlargement of the superior orbital fissure, and decalcification of the optic foramen. Angiography helps to confirm the type and site of the tumor; in more than a third of the cases there occurs a more or less intensive tumor staining. Very often the ophthalmic artery will be found to be greatly enlarged, sending off supplying vessels to the meningioma.

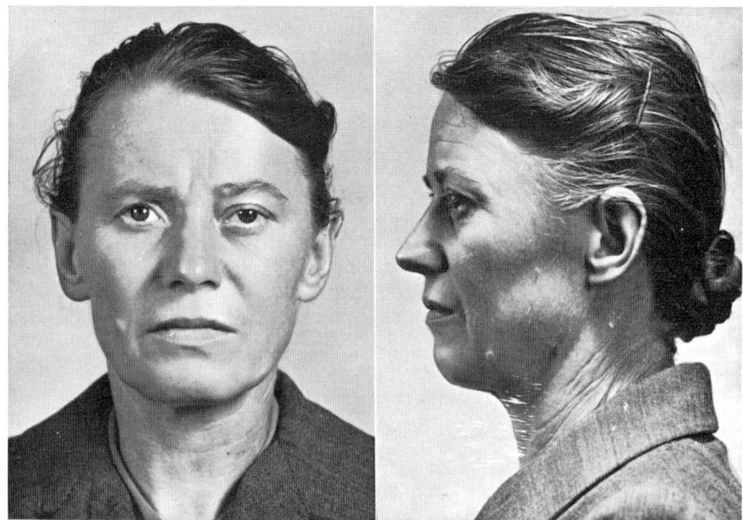

Fig. 3-80. Meningioma "en plaque" in the pterional region of the left sphenoid wing. Left exophthalmos of 5 mm. prominence. No motility disturbances. Minimal left papilledema. Slight blurring of the nasal margin of the right disc. Visual acuity, O.U. 1.0. Palpable swelling in the left temporal region.

In this connection the special value of *serial tomograms of the optic canal* in the diagnosis and differential diagnosis of these tumors must be mentioned. Whereas gliomas of the optic nerve produce only enlargement of the optic foramen and of the canal without decalcification of its walls, meningiomas of the optic nerve may cause osteomatous thickening of the optic canal, going so far as to narrow the foramen concentrically.

The secondary hyperostosis is particularly impressive in cases of *meningioma "en plaque."* Not only does it form a thin layer on the inside of the dura but also infiltrates it and invades the adjacent bone. For this reason these meningiomas, in spite of their lateral origin in the pterional region, may produce symptoms similar to those of tuberous meningiomas of the middle and, especially, the inner third of the sphenoid ridge (that is, a combination of a superior orbital fissure syndrome with a unilateral impairment of the function of the optic nerve and an exophthalmos). A unilateral, slowly progressing, painless, and nonpulsating *exophthalmos* accompanied by a characteristic *swelling in the ipsilateral temporal region* (due to hyperostosis and thickening of the larger sphenoid wing) is one of the most important early signs (Fig. 3-80). In differentiating these from meningiomas of the inner third of the sphenoid ridge, Cushing stresses the more or less pronounced lid edema as well as a combined forward and downward displacement of the globe. Loss of vision and diplopia occur later and are not constant. The rare occurrence of a papilledema, even with exophthalmos of considerable degree, is characteristic for meningioma "en plaque"—a tumor which, according to Cushing, occurs almost exclusively in women between the ages of 40 and 60 years. The radiologic changes may determine the diagnosis. They are both erosion and bone thickening in the area of the lateral third of the sphenoid ridge and its surroundings (that is, the roof of the orbit, the lateral orbital wall, and the superior orbital fissure).

In summary, we state that meningiomas of the sphenoid ridge cause ocular symptoms if they expand in a mesial direction and involve the optic nerve and the superior orbital fissure. This makes it easy to deduce the characteristic signs: unilateral exophthalmos, variable pareses of the abducent, oculomotor, and trochlear nerves, and occasional hypesthesia of the ipsilateral cornea. Damage to the ipsilateral optic nerve is relatively frequent and manifests itself by the subjective sensation of unilateral loss of vision. In such patients the visual fields show unilateral changes beginning on the temporal or nasal side and progressing to unilateral temporal or nasal hemianopia or even to unilateral amaurosis. Occasionally one finds homonymous hemianopias of the congruous type— obviously due to a lesion of the adjacent temporal lobe. A patient's subjective complaints of double vision and diminution of vision in combination with proptosis of one eye always suggest the possibility of a meningioma of the sphenoid ridge, in addition to a basal aneurysm or orbital processes (osteomas, sarcomas, angiomas, dermoids, tumors of the optic nerve, mucoceles, etc.). A painless, slowly developing exophthalmos with a palpable swelling in the ipsilateral temporal region, and a later impairment of vision and ocular motility, indicate a meningioma "en plaque" in the pterional region.

In the differential diagnosis of such a syndrome other types of tumors must, of course, be considered—tumors that occur much more rarely than meningiomas of the sphenoid ridge. We mention dermoids (we observed one case), meningiomas of the middle fossa that expand toward the sphenoid region, and in particular, metastatic tumors. Occasionally one must consider the possibility of a meningioma of the sheaths of the optic nerve or a glioma of the optic nerve (p. 239).

Differential diagnosis of tumors of the sella and its surroundings

The discussion of the various forms of sellar, suprasellar, presellar, and parasellar tumors included the differential diagnosis of tumors of this particular area of the base of the skull. In our opinion, however, it would be incorrect and misleading not to consider other possible causes of a chiasmal syndrome. Although, strictly speaking, they are outside the scope of our discussion, it is essential to mention them at least briefly in order to present a complete differential diagnosis of pituitary and nonpituitary chiasmal syndromes.

In the presence of a chiasmal syndrome one always must consider, in addition to sellar, suprasellar, presellar, and parasellar tumors, those of the chiasm itself, the so-called *gliomas of the chiasm* as well as tumors originating in the sheaths of the optic nerve, the *meningiomas of the optic nerve*. Malignant tumors at the *base of the skull* (nasopharyngeal carcinomas, sarcomas, metastatic

Table 4. Differential diagnosis of chiasmal syndrome*

	Pituitary adenoma	Meningioma of tuberculum sellae	Craniopharyngioma	Glioma of chiasm	Aneurysm	Chiasmal arachnoiditis
Age	30-50	30-50	Most children, also adults	Children	As a rule, adults	All ages
Eyeground	Primary optic atrophy	Primary optic atrophy, often more severe on one side	In children papilledema, in adults optic atrophy	Primary optic atrophy	Normal discs or optic atrophy	Unilateral or bilateral optic atrophy or papilledema
Visual field	Symmetrical bitemporal hemianopia	Asymmetrical chiasmal syndrome, often amaurosis of one eye with temporal hemianopia of the other	Asymmetrical bitemporal hemianopia	Unilateral or bilateral temporal visual field defects	More or less distinct bitemporal hemianopia (fluctuations!)	Central scotomas, concentric constriction, unilateral or bilateral hemianopia
Endocrine signs	Hypopituitarism, hyperpituitarism, or mixed	None	Hypopituitarism, perhaps also hypothalamic signs (polyuria, polydipsia, etc.)	None or hypopituitarism	None	None
Localization	Sellar	Suprasellar	Intrasellar and suprasellar	Suprasellar	Suprasellar (supraclinoid aneurysm), intrasellar (infraclinoid aneurysm)	Optic and chiasmal arachnoid
X-ray findings	Ballooning of sella, thinning of floor of sella, erosion of clinoid processes	Sella normal, hyperostosis of tuberculum sellae	Suprasellar calcifications	Unilateral or bilateral widening of optic foramen	Often negative, occasionally annular calcium shadows	Negative

*Modified from Walsh and Hoyt.

tumors) in an advanced stage may cause compression of the chiasm. As a rule, this is marked by asymmetry combined with an incomplete or complete cavernous sinus syndrome (regarding these tumors, refer to p. 249). In addition to the types of tumors just mentioned, one must also consider the rather rare *suprasellar cholesteatomas* as well as *specific granulomas* (tubercles, gummas), certain *angiomas, chordomas, osteochondromas, dermoids, ectopic pinealomas, carcinomas of the sphenoid sinus,* and *affections caused by parasites.* These clinical conditions are so rare that their detailed discussion would be out of proportion in the overall picture. *Of greater importance in the differential diagnosis of the sellar and extrasellar tumors* are *aneurysms,* especially supraclinoid aneurysms originating from the anterior cerebral artery or the anterior communicating artery (p. 299).

Furthermore, one must always consider the possibility of an *indirect compression of the chiasm,* primarily a remote effect of a *hydrocephalus of the third ventricle* (p. 253). Finally, there is a group of inflammatory chiasmal changes (the so-called *chiasmal arachnoiditis*) that may also produce a more or less typical chiasmal syndrome. Such a differential diagnosis of the chiasmal syndrome may at first appear quite complex and difficult. Actually, it is not. We have intentionally given a very detailed discussion of the most important sellar and extrasellar tumors and their characteristics so that their differential diagnosis should create no difficulty. In addition to the neoplasms mentioned, one must primarily consider, for practical clinical purposes, aneurysms, indirect compression of the chiasm by a hydrocephalus of the third ventricle, gliomas of the chiasm, and inflammatory changes. (See Table 4.)

Aneurysms. Infraclinoid and supraclinoid aneurysms of the internal carotid artery must be differentiated. The former develop in the cavernous sinus and thus usually present the picture of an ophthalmoplegia combined with an affection of the trigeminal nerve. Only by breaking through the wall of the cavernous sinus will they invade the sella and then produce the picture of a chiasmal syndrome and mimic a pituitary tumor or a craniopharyngioma even with endocrine dysfunction. This happens much more frequently with *suprasellar aneurysms* of the supraclinoid section of the carotid artery that originate from the anterior cerebral or anterior communicating arteries. An important symptom that is significant for an aneurysm is the relatively early occurrence of *headaches,* mostly localized in the frontal region. Occasionally they occur in violent paroxysms. Also characteristic is the *more or less sudden appearance of visual disturbances and visual field changes,* which may be subject to *fluctuations.* One of the usual manifestations of the chiasmal syndrome in aneurysms is loss of vision, more pronounced in one eye. Sudden improvement or deterioration of vision is significant. The ophthalmoscopic picture reveals a primary optic atrophy that is usually bilateral but of distinctly different degrees in the two eyes.

A bitemporal hemianopia of a rather asymmetrical type is the most frequent field change in aneurysms (Fig. 5-10). According to Jefferson, an inferior bitemporal quadrantanopia is more suggestive of a suprasellar aneurysm, whereas onset of the field changes in both superior temporal quadrants is more likely due to a pituitary adenoma. There may also be amaurosis of one eye and a temporal hemianopia of the other. In the course of the disease the visual field defects may even vary, which is ordinarily not the case with tumors. In many cases the x-ray findings will furnish important data. Aneurysms may show a destruction of the contours of the sella and the clinoid processes, usually at first only on one side, but no enlargement of the sella. Occasionally there are calcium deposits in the wall of the aneurysm that radiologically appear as annular streaks or small compact circles. However, such findings may be completely missing in aneurysms with a chiasmal syndrome. A definite diagnosis of an aneurysm is made by means of cerebral angiography.

Indirect compression of the chiasm. The mechanism of remote effect of distant tumors on the chiasm usually consists in the formation of a hydrocephalus of the third ventricle, with its dilated anterior wall causing damage to the chiasm by pressure from above or from behind. Such a remote effect is possible with tumors of the third ventricle (p. 253), tumors of the aqueduct of Sylvius, and tumors of the posterior fossa (cerebellar tumors, p. 265). Klingler and Condrau, in their article on misleading signs in visual fields, reported such observations (p. 187). Characteristically, such an indirect compression of the chiasm causes a *more or less distinct bitemporal hemianopia* (Fig. 3-105) *with a more or less advanced unilateral or bilateral loss of visual acuity.* A simultaneous increase in the intracranial pressure frequently causes signs of choking in the eyegrounds. Fluctuations in the cerebrospinal fluid pressure may be responsible for considerable improvement or deterioration of field defects and visual acuity.

In addition to the bitemporal field defects, one observes *unilateral or bilateral central scotomas, with greater frequency in the temporal halves.* The cause for the origin of such central scotomas supposedly is a downward kink of the optic nerve shortly before it enters the optic foramen as a result of the internal hydrocephalus (Mooney and McConnel) or, perhaps, of exceptional pressure by the third ventricle on the posterior rim of the chiasm (Hughes).

Chiasmal arachnoiditis. In spite of numerous publications on this inflammatory type of a chiasmal disease, it must be considered a rare event (this is in agreement with Walsh and Hoyt). In our series we have only two undisputed cases of chiasmal arachnoiditis proved after surgery. We must emphasize that this diagnosis has been made all too often, especially during the last decades. In numerous cases it turned out there was an occult sellar or extrasellar tumor.

Just as in sellar or extrasellar neoplasms, the usual onset is *loss of vision*—first in one eye and later in the other. *Characteristically, the visual field defects are quite irregular.* Likewise, their development is rather pleomorphic. These irregularities are based on the varied extent of the changes in the arachnoid, the circulatory disturbances caused by them, and a spread of the inflammatory process to the optic nerve itself. According to Bollack,

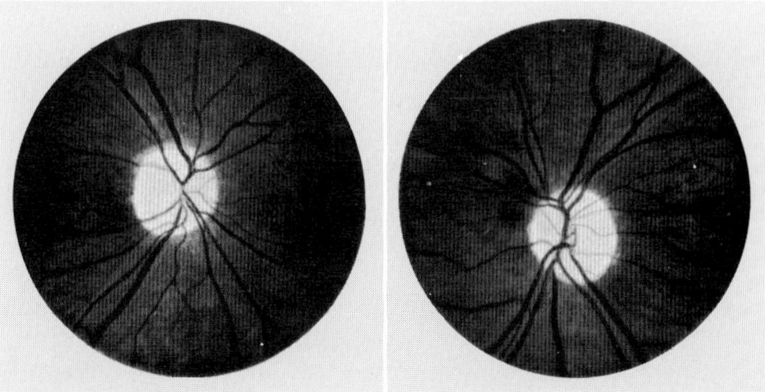

Fig. 3-81. Primary complete optic atrophy with distinct disc margins in chiasmal arachnoiditis, confirmed on surgical exploration of the chiasmal region. Both discs are chalk white and sharply outlined. Slight excavation of the central funnel. Slight narrowing of the arteries. Normal veins. On both sides the pattern of the papillomacular bundle cannot be visualized in red-free light. Visual acuity before surgery, O.D. 1/20; O.S. 1/60. Large bilateral central scotomas. The left illustration corresponds to the right eye; the right illustration, to the left eye.

David, and Puech, the most frequent and earliest signs are—almost always—bilateral *central scotomas.* Next in frequency is a *concentric constriction and a unilateral or bilateral hemianopia.* Much less frequent are binasal, altitudinal, or homonymous hemianopias.

The eyeground changes are also quite irregular and may vary considerably in one and the same patient during different stages of the disease. Most frequently (about 50%) one finds a *primary optic atrophy* with sharply outlined disc margins (Fig. 3-81). Next in frequency is an atrophy with blurred disc margins (mixed atrophy according to Vail). In one tenth of the patients a *papilledema* is seen that suggests a spreading of the process toward the base and an involvement of the cisterna magna. One tenth of the patients show a perfectly normal fundus. A general neurologic examination usually is noncontributory. Other ocular symptoms are quite rare. Nonocular symptoms such as headache, sleeplessness, polyuria, etc. are not very characteristic. X-ray films of the skull generally show a completely normal picture (especially a normal sella)—an extremely valuable fact in differentiating this process from others, especially from tumors in this region. Pneumoencephalography shows an irregular and deformed sellar cisterna or even an obliteration of the chiasmal cistern.

Gliomas of the chiasm; tumors of the optic nerve. Glioma of the chiasm is an instance of a primary tumor of the optic nerve in this particular localization; according to Bürki, tumors of the optic nerve can be classified into two main groups: (1) tumorlike fibrogliomatosis and (2) endotheliomas or meningiomas of the optic nerve. Because of their identical or at least very similar symptomatology, both types can be discussed together.

Glioma of the chiasm, like glioma of the optic nerve (Figs. 3-82 and 3-83), is in many cases (15% according to Walsh) merely a sign of a more general affection of the peripheral and central nervous system—namely *neurofibromatosis (von Recklinghausen's disease)* (van der Hoeve). Bürki is of the opinion that an isolated primary tumor of the optic nerve may be the first or only sign of von Recklinghausen's disease. Glioma of the chiasm occurs mostly in infancy. Most of all, it must be differentiated from craniopharyngioma. As mentioned previously, there may not always be systemic signs of neurofibromatosis (Fig. 3-84). They may be completely missing. At times only "café au lait" spots (a so-called "forme fruste" of von Recklinghausen's disease) can be demonstrated on the skin. Occasionally it may be necessary to search carefully for such skin changes among other members of the family.

Whereas fibrogliomatosis of the optic nerve and the chiasm is primarily a disease of infancy, *meningioma of the sheaths of the optic nerve is more likely to occur in the older age group.* Although we concern ourselves here only with the glioma of the chiasm, it should be mentioned that a primary tumor can occur in any part of the optic nerve (that

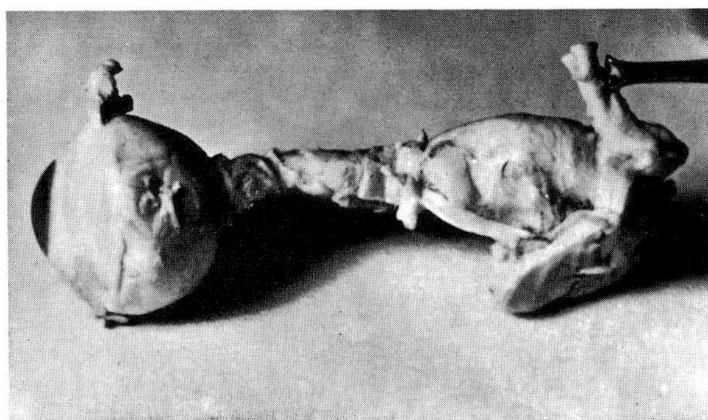

Fig. 3-82. Glioma of the chiasm and left optic nerve. Systemic neurofibromatosis (von Recklinghausen's disease) of the peripheral and sympathetic nervous system as well as glioma in the medulla oblongata in a 19-year-old patient.

is, intraorbital, intracanalicular, or intracranial). Intraorbital optic nerve tumors tend to occur as a single focus (Figs. 3-86 and 3-87), whereas the intracranial form shows multiple foci in the chiasm and the optic nerves (Fig. 3-82).

The chiasmal syndrome, as seen with gliomas of the chiasm (Martin and Cushing), is marked by a *progressive loss of vision in both eyes,* often beginning in one. *The visual field changes take an irregular course and differ from the ordinary type seen in the chiasmal syndrome.* There are unilateral temporal defects. In bitemporal hemianopias a

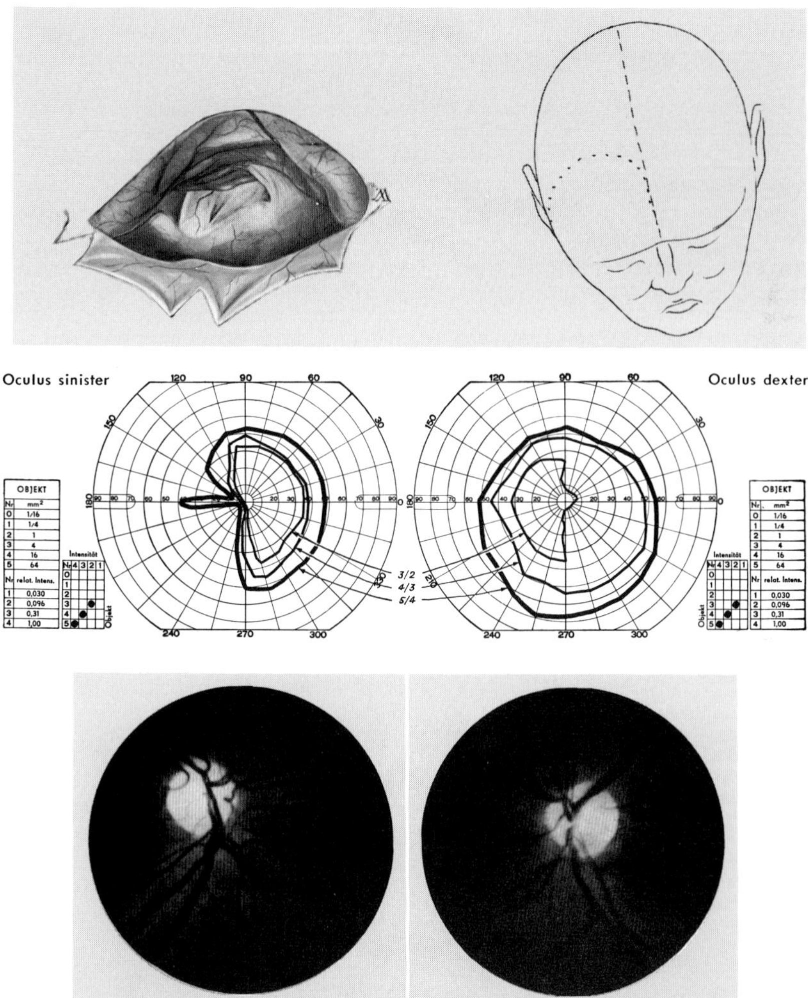

Fig. 3-83. Above, Glioma of the chiasm in situ after exposure of the sellar region by right transfrontal osteoplastic craniotomy in a 25-year-old patient. Center, Asymmetrical bitemporal heminaopia involving only most central isopters of the right field and peripheral as well as central isopters of the left field. There is a peculiar spurlike remnant of the left temporal field. Below, Corresponding fundus photographs, demonstrating bilateral primary optic atrophy with sharp margins. Slight venous congestion. Visual acuity strikingly well preserved: O.U. corrected to 0.9! The left illustration corresponds to the right eye; the right illustration, to the left eye.

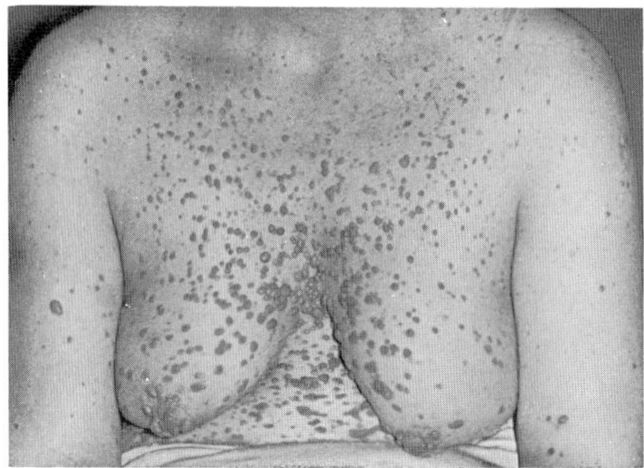

Fig. 3-84. Multiple neurofibromas of the skin in the region of the breasts with neurofibromatosis (von Recklinghausen's disease) and glioma of the chiasm (same patient as in Fig. 3-83).

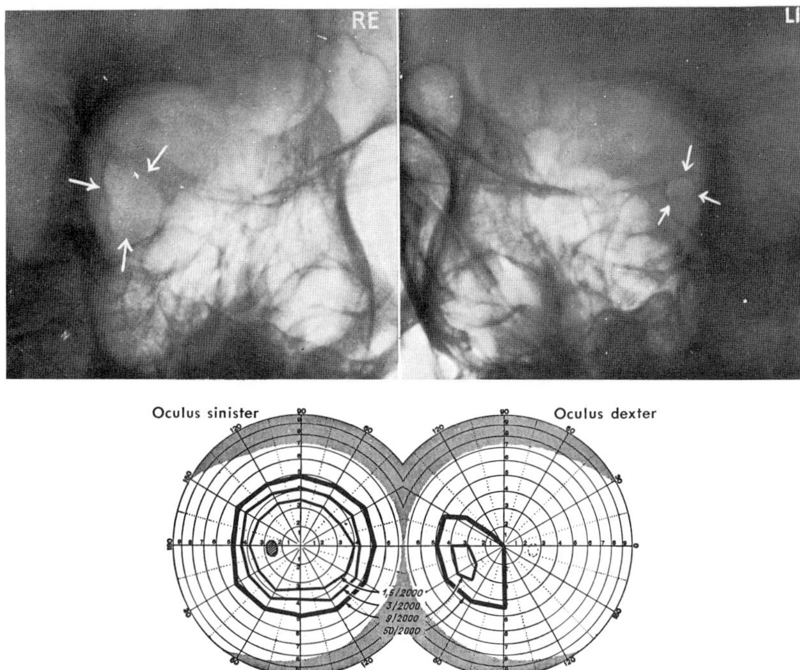

Fig. 3-85. Glioma of the right optic nerve and chiasm in a 13-year-old boy. Slight exophthalmos and primary optic atrophy on the right side. Marked loss of right vision: 0.01. Above, Radiologic visualization of the optic foramina (arrows), Rhese-Goalwin position. Marked unilateral enlargement of the right optic foramen. Below, Corresponding visual fields. Right temporal hemianopia with beginning destruction of the upper nasal quadrant. Three years later, a left temporal hemianopia and thus a true chiasmal syndrome developed.

certain asymmetry is the rule (Fig. 3-85). Occasionally even homonymous hemianopias occur (Brégeat). The ophthalmoscopic appearance usually is that of a simple bilateral optic atrophy with sharp outlines of the discs (Fig. 3-83). If there are superimposed signs of increased intracranial pressure, the margins may be blurred. Marked enlargement of the tumor may embarrass the circulation of the cerebrospinal fluid. It may also produce hypothalamic signs such as drowsiness, adiposity, or polyuria.

Of utmost diagnostic importance are the radiologic findings. If the tumor already has invaded the optic canal, an *enlargement of one or both optic foramina* without decalcification can be easily demonstrated (Fig. 3-85). This sign, of course, is of greater diagnostic significance if it remains unilateral (Bürki), although the bilateral optic foramen is just as frequently observed in cases of chiasmal glioma. Another diagnostically important change is *deformation of the sella turcica* in the form of an *omega or a pumpkin* (also called □-shaped deformity) resulting from destruction of the anterior wall of the sella turcica and the lower part of the anterior clinoid processes by the tumor, which spreads along the optic nerve toward the orbit. These radiologic signs are of extraordinary importance, especially in the differential diagnosis of a craniopharyngioma. The pumpkin shape of the sella and the enlargement of the optic foramina are missing in craniopharyngiomas. The characteristic findings for the latter are the frequent suprasellar calcifications.

Expansion of a glioma of the optic nerve or chiasm toward the orbit leads to the very early development of a unilateral *exophthalmos* (Fig. 3-87). The motility of the globe in fibrogliomatosis is said to remain intact for a long time, whereas a meningioma supposedly causes early disturbances of the ocular movements. We have in our series three cases of primary tumors of the optic nerve, two of them in patients with von Recklinghausen's disease. The third case, a meningioma of the optic foramen with progressive unilateral loss of vision, had been considered as an essential optic atrophy and the patient had been treated by injections. The ophthalmoscopic findings in this patient were a white atrophic

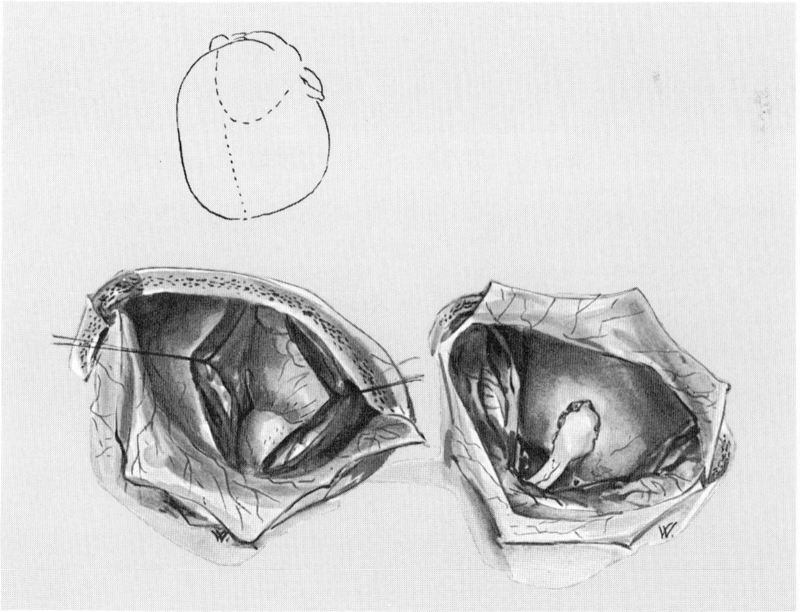

Fig. 3-86. Glioma of the right optic nerve in a 3-year-old girl before (right) and after (left) deroofing of the orbit (approach via right transfrontal osteoplastic craniotomy). (See Fig. 3-87.)

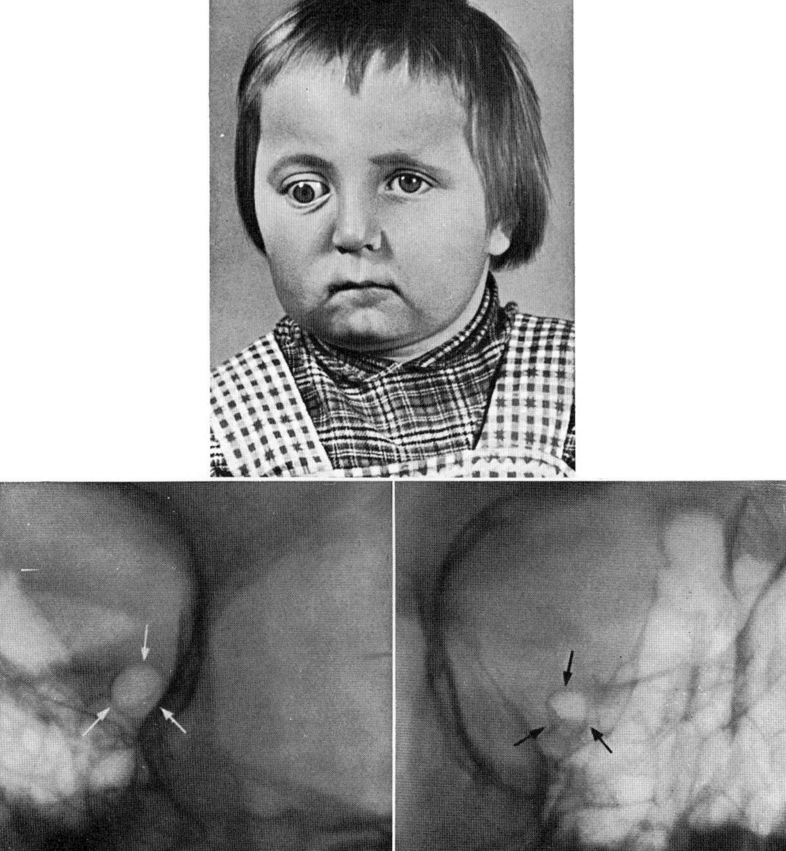

Fig. 3-87. Above, Glioma of the right optic nerve in a 3-year-old girl (see Fig. 3-86). Right exophthalmos with downward displacement of the globe. Motility of the right eye restricted up and out. Beginning congestion of the blood vessels of the upper lid and nasal aspect of the bulbar conjunctiva. Right disc shows temporal pallor, blurred nasal margin, and slight elevation. Left disc normal. Below, Radiologic visualization of the optic foramina, Rhese position. Right optic foramen definitely enlarged but round and of distinct outline (white arrows).

disc, with a normal papilla in the other eye. A distinct dilatation of the optic foramen with thickening of the contours on one side suggested the correct diagnosis. The meningioma was extirpated, although the optic nerve had to be severed. In this connection the special value of *serial tomograms of the optic canal* in the diagnosis and differential diagnosis of these tumors must be mentioned. Whereas gliomas of the optic nerve produce only enlargement of the optic foramen and of the canal without decalcification of its walls, meningiomas of the optic nerve may cause osteomatous thickening of the optic canal, going so far as to narrow the foramen concentrically.

In patients with a chiasmal syndrome a *granuloma* (tuberculous or syphilitic) occasionally must be considered in addition to primary tumors of the chiasm. This is illustrated by the history of a patient with an initial loss of vision of the left eye which rapidly progressed to complete amaurosis, followed by a similar course in the right eye. The left

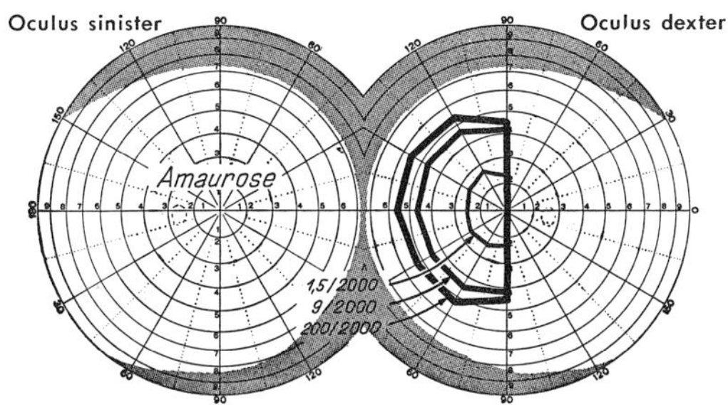

Fig. 3-88. Retro-orbital solitary tubercle of the left optic nerve (intracranial section) in a 44-year-old patient. Blindness of the left eye. Temporal hemianopia without sparing of the macula of the right eye. No chiasmal syndrome but direct pressure of the tumor on the medial part of the right optic nerve!

amaurosis and a right temporal hemianopia could not be explained satisfactorily. The subsequent picture was that of an acute meningitis with sudden death. Autopsy revealed a *solitary tubercle* in the intracranial section of the left optic nerve in addition to a systemic proliferating tuberculosis of various organs. This case is remarkable because of the development of an amaurosis in one eye and a temporal hemianopia in the other eye, yet without direct interference of the tubercle with the chiasm. The tubercle in the intracranial section of the left optic nerve, causing amaurosis on this side, pressed against the medial part of the right optic nerve, thus producing a temporal hemianopia (Fig. 3-88)!

Tumors of the cavernous sinus

In the discussion of meningiomas of the sphenoid ridge the superior orbital fissure syndrome was mentioned. Actually, it is related to the cavernous sinus syndrome. This should cause no surprise because the anatomic structures in the cavernous sinus are identical with those entering the orbit through the superior orbital fissure. *The characteristic sign of the cavernous sinus syndrome* (Foix) *is the more or less simultaneous involvement of the third, fourth, fifth, and sixth cranial nerves* (in other words, the three nerves supplying the extraocular muscles and the trigeminal nerve). In contrast to the superior orbital fissure syndrome (with involvement of only the first division of the trigeminal nerve), all branches of the trigeminal nerve might be involved here. This symptomatology results from the close topographic relationship of these structures within the cavernous sinus and their proximity to the internal carotid artery (Figs. 3-89 and 3-90). According to Jefferson, one can distinguish between a *posterior cavernous sinus syndrome* (involvement of all branches of the trigeminal nerve; frequently only sixth nerve paresis) and an *anterior cavernous sinus syndrome* (only the first division of the trigeminal nerve involved, oculomotor nerve palsy alone or in association with a paresis of the fourth and the sixth nerve). This differentiation is particularly valid for aneurysms of the internal carotid artery within the cavernous sinus (Fig. 3-91) (Krayenbühl).

Our series includes twenty-four cases of cavernous sinus affections. Of these, not less than eighteen were caused by various tumors originating from the base of the skull in this area. Two of these tumors were *neurinomas of the trigeminal*

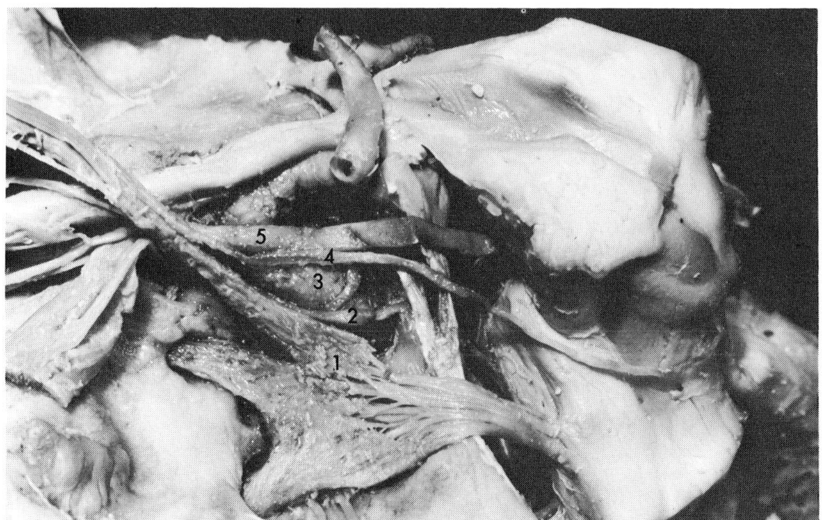

Fig. 3-89. Lateral view of the cavernous sinus. Note the intimate connection between the trigeminal nerve, oculomotor nerve, trochlear nerve, and abducent nerve. 1, Trigeminal nerve (ophthalmic branch); **2**, sixth nerve; **3**, internal carotid artery; **4**, trochlear nerve; **5**, oculomotor nerve. (Specimen of Prof. Kubic, Anatomic Institute of the University of Zurich.)

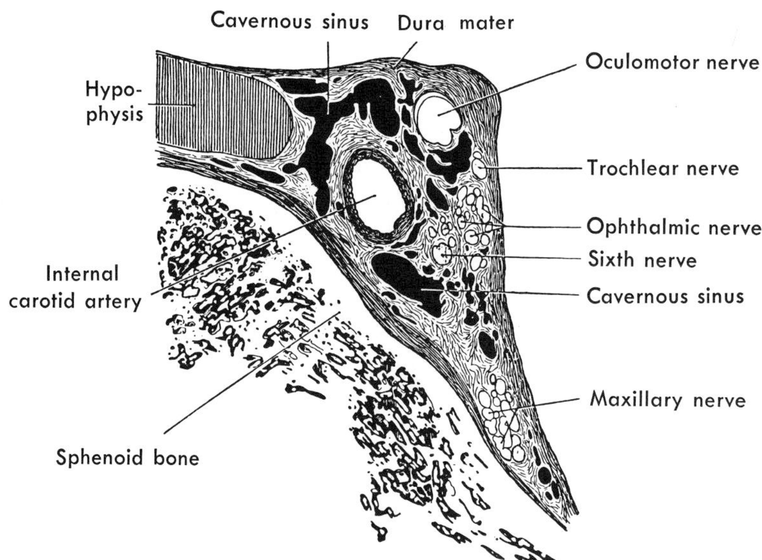

Fig. 3-90. Median section through the cavernous sinus, hypophysis, sella turcica, internal carotid artery, and nerves of the extraocular muscles. (After Corning.)

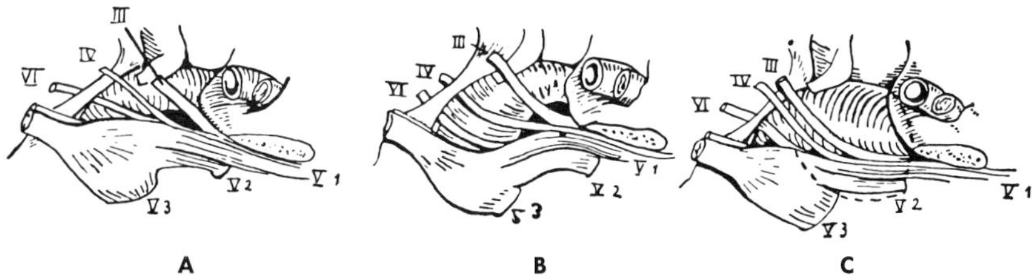

Fig. 3-91. Differentiation of the cavernous sinus syndrome, according to Jefferson, with aneurysms of the internal carotid artery within the cavernous sinus. **A,** Normal. **B,** Posterior cavernous sinus syndrome. **C,** Middle cavernous sinus syndrome.

nerve, with the tumor in contact with the median part of the cavernous sinus. There was a characteristic history of neuralgic pains of the face, sensory disturbances, and later, numbness in the same area as well as diplopia in the later stages. Even loss of vision was mentioned. Both patients showed an incomplete cavernous sinus syndrome—only the abducent and trigeminal nerves were involved, whereas the oculomotor and trochlear nerves remained unaffected. *A total impairment of all branches of the trigeminal nerve with analgesia and anesthesia is pathognomonic for these tumors.* X-ray films in both patients showed decalcification and destruction of the ipsilateral tip of the petrous bone.

Two other cases in our series consisted of *meningiomas of the middle fossa.* These tumors are of importance because they are frequently responsible for a *lesion of the ipsilateral optic nerve.* Their symptomatology can resemble closely that of meningiomas of the sphenoid ridge. In addition to pain, ptosis, and diplopia, *characteristically* there are *complaints of unilateral loss of vision or even of a sudden blindness,* which may be mistaken for a retrobulbar neuritis. The meningiomas in our two patients extended to or even infiltrated the wall of the cavernous sinus. One originated from the sella and the other from the sphenoid ridge. The first case even showed a bilateral cavernous sinus syndrome. The second case was unusual because of an optic atrophy on the ipsilateral side and a papilledema on the contralateral side (that is, a Foster Kennedy syndrome). In both patients there was involvement of all the nerves of the extrinsic eye muscles as well as the first and second branches of the trigeminal nerve.

Two cases of cavernous sinus syndrome caused by *chondromyxomas* or *myxomas* with parasellar or suprasellar localization were less characteristic and not so well defined. The two most severe tumors with perhaps the most malignant course originated in the nasal and retronasal space and extended to the base of the skull. One was a *round cell sarcoma of the retronasal area;* the other, a lymphoepithelial carcinoma of the nasopharyngeal area. The patient showed a complete bilateral cavernous sinus syndrome with bilateral amaurosis! The second showed unilateral amaurosis with protrusion of the ipsilateral globe.

An important group of tumors causing a cavernous sinus syndrome is the large number of metastatic neoplasms (seven cases), *especially metastatic carcinomas in the region of the cavernous sinus.* Naturally these occur mostly in patients in the older age group. Almost all of the patients initially complain of frontal or temporal headaches and also of pain or sensory disturbances in the various areas of distribution of the branches of the trigeminal nerve. In later stages there is almost always diplopia, associated quite frequently with unilateral loss of vision. *The oculomotor nerve is invariably involved* (external and internal ophthalmoplegia and ptosis), *as are all branches of the trigeminal nerve.* In six of our seven patients the ipsilateral corneal reflex was distinctly diminished or absent. The abducent and trochlear nerves were not regularly involved. Noteworthy and, in view of the rapid growth of malignant tumors, not surprising is the relatively frequent impairment of the ipsilateral optic nerve in four of our seven patients. The radiologic findings showed destruction of the sella, the anterior clinoid processes, and especially the tip of the petrous bone. In five patients the metastatic carcinoma in the region of the cavernous sinus could be verified during surgery or autopsy—the primary tumors were in the breast, the bronchi, or the thyroid gland. It is not always easy to make a diagnosis without surgical exploration. However, even without evidence of a primary tumor, one should always consider a metastatic carcinoma in older people if there is a fully developed cavernous sinus syndrome with involvement of the optic nerve, a relatively rapid progress of symptoms, general cachexia, and most of all, marked roentgenologic changes in the region of the cavernous sinus and its neighborhood.

In addition to metastatic tumors, other *tumors at the base of the skull* (we cite as an example in our series a chondro-osteoma) may reach the size that could lead to a cavernous sinus syndrome. In the section on *pituitary adenomas* (p. 189) we cited two examples to discuss the possibility of the development of a cavernous sinus syndrome on the basis of an—admittedly rare—lateral extension of a pituitary tumor (Weinberger, Adler, and Grant).

In the broadest sense of the word, aneurysms are a form of tumor. In our series, of the twenty-four cases of cavernous sinus syndrome, four were due to *aneurysms of the internal carotid artery in the region of the cavernous sinus* (Fig. 5-7). All of the cases in question are examples of the so-called *paralytic type of basal aneurysm of the brain.* Without prodromal signs, they usually set in with *pain or a sensation of numbness of the face, followed within a short period of time by diplopia with or without ptosis.* The pain is localized in the area of distribution of the trigeminal nerve and usually is sharply demarcated. In contrast to the ordinary trigeminal neuralgia, it is not paroxysmal but constant. In addition to the subjective symptoms of irritation, it is always possible to demonstrate more or less distinct defects such as hypesthesia or hypalgesia in the area of distribution of the first, second, or third branch of the trigeminal nerve. *Of special significance, again, is the corneal reflex which, sometimes as the only trigeminal sign, may be diminished or abolished.* Occasionally an aneurysm of unusual size may damage the ipsilateral optic nerve, causing atrophy with a corresponding loss of vision. We have records of two such cases. After their sudden onset, extraocular pareses due to aneurysms usually remain stationary or may even regress during the further course of the disease. Not infrequently the trigeminal pain, which was quite violent at the onset, disappears. Now the hypesthetic or anesthetic zones that are so significant for the diagnosis become evident. This symptomatology actually is so typical that it almost allows a diagnosis. Radiologic proof of bony changes at the base of the skull is valuable. These bony changes consist of destruction or erosion of the mesial part of the ipsilateral lesser wing of the sphenoid and the anterior or posterior clinoid processes or of widening of the superior orbital fissure or the optic foramen. Actual proof of the aneurysm is obtained by cerebral angiography. It is interesting to note that an arteriovenous aneurysm of the internal carotid artery within the cavernous sinus (carotid-cavernous fistula) does not cause a cavernous sinus syndrome and that it rarely leads to pareses of the extrinsic ocular muscles (p. 305).

In addition to tumors and aneurysms, one must consider *sinus thrombosis* as a third etiologic group in the differential diagnosis of the cavernous sinus syndrome. This picture sets in with unilateral or bilateral *proptosis, lid swelling,* and *chemosis.* Usually there is an impairment of the three nerves supplying the extrinsic muscles as well as an involvement of the trigeminal nerve. A *preceding infection in the area of the head* (such as furunculosis, cellulitis of the lid, tonsillitis, maxillary or sphenoidal sinusitis) and signs of a *septicemia* make the diagnosis obvious (Weber). We have two recorded cases of cavernous sinus thrombosis: one was secondary to a septic tonsillitis; the other, to an erysipelas of the face. (See Fig. 1-3.)

Quite rarely a cavernous sinus syndrome may develop secondary to *herpes zoster.*

In summary, it can be stated that in the majority of cases a cavernous sinus syndrome is caused by tumors. Usually these are metastatic carcinomas in the cavernous sinus or malignant tumors at the base of the skull. Sellar, suprasellar, or especially parasellar tumors likewise may produce a cavernous sinus syndrome if their expansion reaches sufficient size. Next to tumors, aneurysms must be primarily considered in the differential diagnosis. A more or less simultaneous involvement of the third, fourth, fifth, and sixth cranial nerves is characteristic for the cavernous sinus syndrome. An impairment of the ipsilateral optic nerve is frequent with malignant tumors. Besides diplopia and unilateral loss of vision, a pronounced incidence of pain and sensory disturbances in the various areas of distribution of the trigeminal nerve are characteristic. X-ray films frequently show destruction of the sella and the tip of the petrous bone.

Tumors of the third ventricle

Tumors of the third ventricle produce early *embarrassment in the circulation of the cerebrospinal fluid* by obliteration of either the foramina of Monro (tumors in the anterior section) or the anterior opening of the aqueduct (tumors of the posterior section, Fig. 3-92). This results in the formation of an *internal*

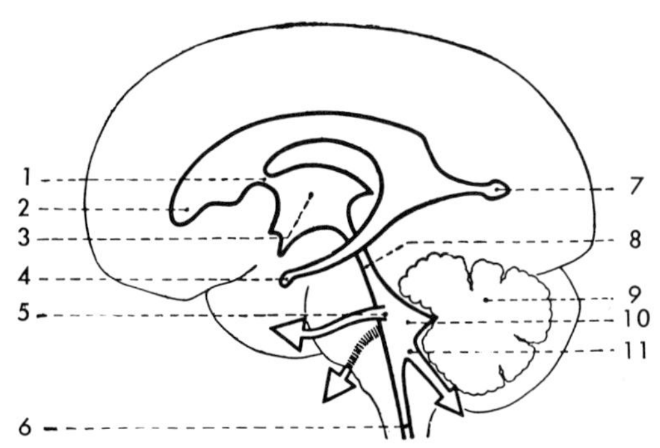

Fig. 3-92. Diagrammatic cross section through the ventricular system indicating the connection openings (arrows) to the subarachnoid space.

hydrocephalus (Fig. 3-94) and, consequently, the appearance of symptoms of *increased intracranial pressure.* It follows that general symptoms of increased intracranial pressure dominate the picture for some time, especially in view of the fact that tumors within the third ventricle cause focal signs relatively late (Oldberg and Eisenhardt; Dandy). Tumors of the infundibulum produce hypothalamic and pituitary signs quite early: actually, they are *craniopharyngiomas* and should be discussed here with other tumors of the third ventricle. However, they have already been discussed together with the sellar processes in view of their importance in the differential diagnosis. The symptomatology of tumors of the third ventricle is, to a large extent, similar to that of craniopharyngiomas. For more details refer to the section on craniopharyngiomas (p. 211).

It is quite characteristic that *tumors of the third ventricle predominantly cause ophthalmoscopic signs of increased intracranial pressure* (Fig. 3-93). Very frequently there are *pronounced* papilledemas with a prominence of up to 6 diopters. *Relatively early they change into the chronic atrophic form, with disastrous consequences for the visual acuity and the visual fields* (Fig. 3-93). *Intermittent attacks of blurred vision or even transient amaurosis,* mostly combined with paroxysmal headache, nausea, or vomiting, occur characteristically in colloid cysts of the third ventricle and its area, often related in a peculiar way to changes in the position of the head (Dandy). With the head bent to the side the vision may suddenly disappear, and with the head in an erect or backward position it may come back gradually. The intermittent headaches can be explained by paroxysmal blocking of the foramina of Monro, but not the loss of vision. The latter has to be interpreted on the basis of intermittent pressure of the dilated third ventricle upon the chiasm and the optic nerves.

In addition to the general symptoms of increased intracranial pressure, there may be *local symptoms referable to the chiasm:* a tumor in the anterior part of the third ventricle and the infundibulum may exert direct pressure on the chiasm, or a tumor in the posterior part may cause a hydrocephalic dilation of the ventricle, whose anterior wall compresses the chiasm (Hughes) (Fig. 3-95). In both

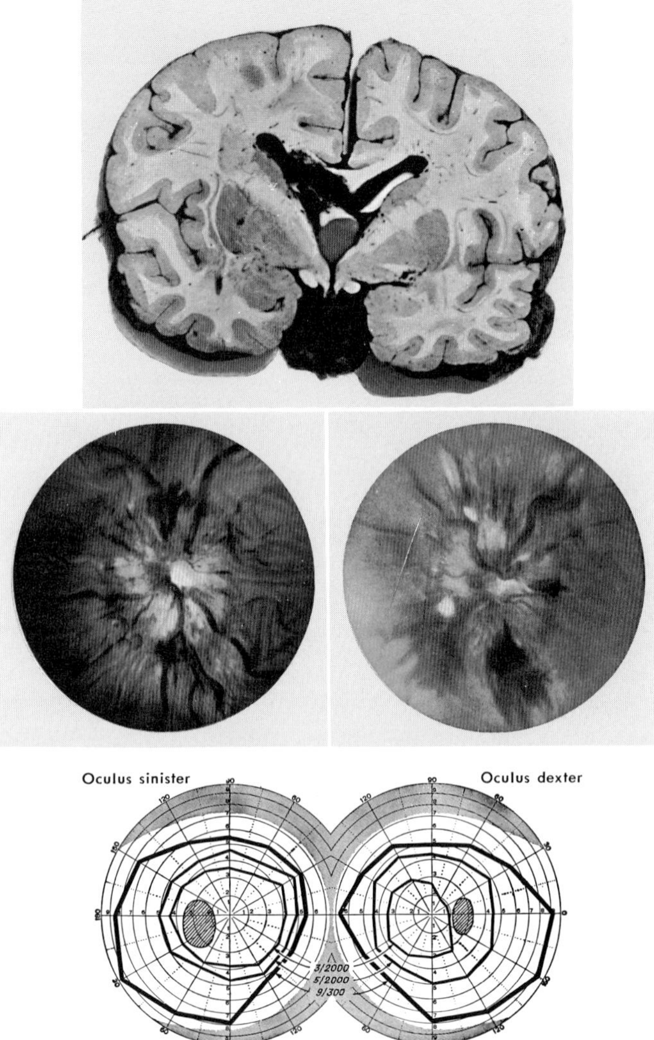

Fig. 3-93. Plexus cyst in the third ventricle with internal hydrocephalus in a 43-year-old patient. Above, Frontal section through the brain. Center, Fully developed papilledemas with marked prominence of 5 to 6 diopters. Numerous hemorrhages. White spots. Extension of edema into the adjacent retina (radial folding). The left illustration corresponds to the right eye; the right illustration, to the left eye. Below, Corresponding visual fields. Enormous enlargement of both blind spots. Suggestion of right temporal hemianopia in the area of the central isopter (beginning chiasmal syndrome?).

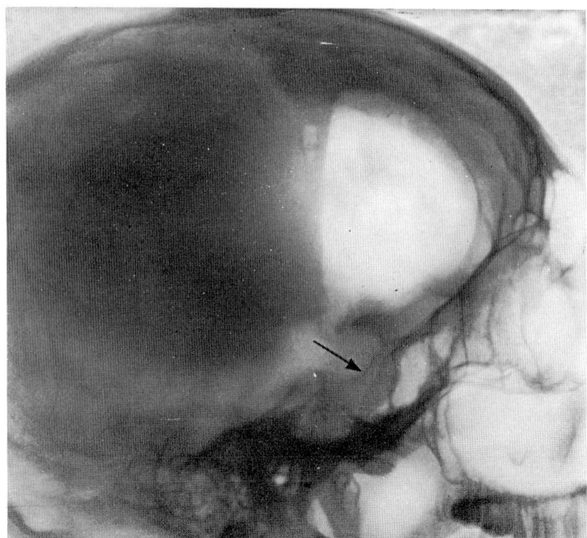

Fig. 3-94. Ventriculogram (lateral view) of internal hydrocephalus due to stenosis of the aqueduct in a 16-year-old boy. The hydrocephalic lateral ventricle (in this illustration, only the anterior part is visible) causes downward displacement of the enlarged third ventricle and extension into the sella turcica (arrow). Chronic atrophic bilateral papilledema. Visual acuity of both eyes reduced to 0.5. Questionable bitemporal hemianopia.

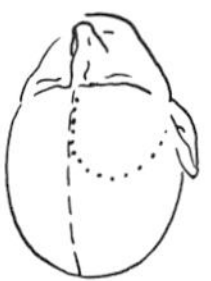

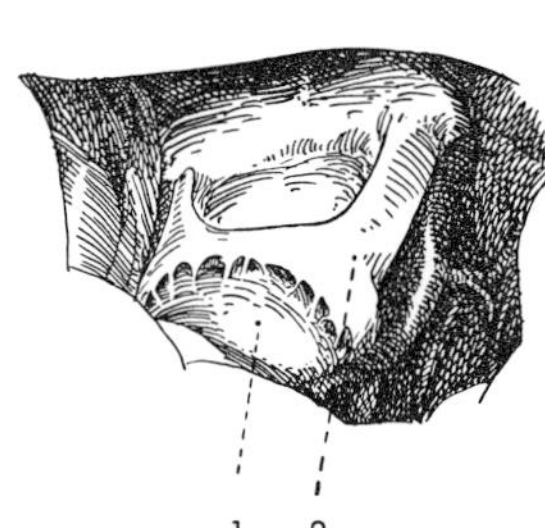

Fig. 3-95. Dilation of the third ventricle and compression of the chiasm and right optic nerve in internal hydrocephalus due to stenosis of the aqueduct in a 48-year-old patient. Distinct chiasmal syndrome. Bilateral optic atrophy. Asymmetrical bitemporal hemianopia. Visual acuity reduced to O.D. finger counting at 1 meter; O.S. 0.6.

instances the result will be a *more or less asymmetrical chiasmal syndrome with bitemporal hemianopia and primary optic atrophy,* or secondary atrophy after papilledema. The visual fields in patients with tumors of the third ventricle show a rather asymmetrical appearance. Usually there is first an involvement of the upper fibers of the chiasm, causing *mostly initial inferior temporal hemianopias.* Frequently the compression embarrasses the posterior rim or the chiasm and thus often causes *central scotomas.* One may also observe vertical or binasal hemi-

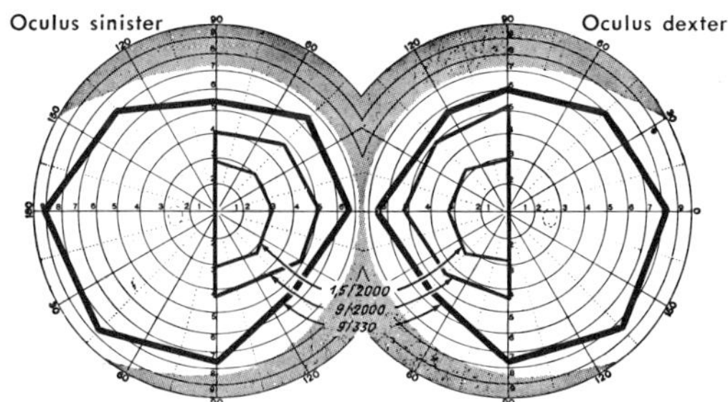

Fig. 3-96. Cystic hemangioma in the rostral part of the third ventricle in a 29-year-old patient. Visual acuity, O.D. 0.15; O.S. 0.3. No optic atrophy. No radiologic changes of the sella turcica. Slight asymmetrical bitemporal hemianopia of the central isopters without sparing of the macula.

anopias with tumors of the third ventricle (François). A tumor at the floor of the third ventricle in one of our patients caused temporal hemianopia on one side and loss of the inferior nasal quadrant on the other side. In another, a hemangioma in the third ventricle caused an almost symmetrical bitemporal hemianopia (Fig. 3-96). A lateral extension of the tumor toward the thalamus or the visual pathways may produce a homonymous hemianopia. If a tumor in the posterior part of the third ventricle grows upward toward the quadrigeminate plate, the result may be a Parinaud syndrome (p. 260), characterized by dilated fixed pupils and vertical gaze palsy.

In summary, tumors of the third ventricle cause general symptoms of increased intracranial pressure and thus papilledemas at a relatively early stage. The latter soon change to the chronic atrophic form, with its disastrous sequelae for the visual functions. Intermittent and paroxysmal headaches produced or relieved by alterations in the position of the head are sometimes accompanied by corresponding intermittent loss and recovery of vision. *A direct pressure effect by the tumor in the anterior part of the third ventricle or an indirect effect due to an internal hydrocephalus leads to asymmetrical chiasmal syndromes with loss of vision and optic atrophy (possibly combined with choking) and to quite irregular visual field defects (mostly in the form of an incomplete bitemporal hemianopia). Such a chiasmal syndrome frequently can only be differentiated from the other forms previously discussed by means of encephalography or ventriculography, which permits proof of an internal hydrocephalus.*

Dilatation of the third ventricle with consequent pressure upon the chiasm and production of a chiasmal syndrome (incomplete bitemporal hemianopia, unilateral central scotoma with temporal hemianopia in the opposite eye, unilateral amaurosis with temporal hemianopia in the other eye, always choked discs) occurs also in the late stages of posterior fossa lesions (tumors of the cerebellum and the cerebellopontine angle), in pinealomas, in obliterations of the Sylvian aqueduct, and in internal hydrocephalus of any origin (Wagener and Cusick; Hughes). (See also pp. 253, 257, 263, and 265.)

Because of their topographic position, tumors of the third ventricle may produce hypothalamic symptoms such as polyuria, polydipsia, or drowsiness. Another hypothalamic disturbance has been described by Penfield as *diencephalic autonomous epilepsy.* It is found especially in association with tumors of the third ventricle. This form of epilepsy is distinguished by attacks that set in with great restlessness and that later lead to reddening of the face and the arms, slowdown of respiration, epiphora, intense perspiration, drooling, hiccough, dilation of pupils, and tachycardia. Certain psychic disturbances associated with tumors in the region of the third ventricle such as lack of impulse, apathy, and general loss of interest should also be mentioned.

Tumors of the thalamus

True tumors of the thalamus actually are rare. More often, an invasion of tumors from the brain stem, the third ventricle, the midbrain, or the pineal body into the thalamus causes neurologic disturbances. For this reason the primary signs of the original tumor predominate in the symptomatology. Thus what has been called the classic *thalamus syndrome* is quite rare in tumors of the thalamus: crossed hemihypesthesia (involving more deep than superficial sensitivity), "main thalamique," hemiastereognosia, incessant or paroxysmal intolerable pain on the involved side (thalamic pain), occasional choreoathetotic movements, hemiataxia, or even a slight transient hemiparesis (Bailey).

The ocular symptoms associated with such tumors are not characteristic and, at any rate, not of localizing value. A papilledema usually occurs only in the later stages (80% according to Tönnis). Important ocular symptoms occur if the tumor expands toward the chiasm and creates a chiasmal syndrome (p. 192). There are reports of pupillary disturbances in tumors of the thalamus (Névin). One of our patients with a tumor in this region showed a difference in the size of the pupils, with a distinct decrease of the pupillary reaction of both eyes to light and in convergence. This patient also showed a distinct decrease of the corneal reflex on the contralateral side. Nystagmus (horizontal or vertical), paralysis of upward or downward gaze, and oculomotor or sixth nerve pareses are described in primary tumors of the thalamus (glioblastomas, astrocytomas, etc.) by McKissock and Paine.

A few of our patients with glioblastoma in the thalamic region show symptoms of motor and sensory hemiplegia as well as a homonymous hemianopia. It must be assumed that the tumor has invaded the internal capsule and the lateral geniculate body or the optic radiation.

Tumors of the basal ganglia

Tumors of the basal ganglia (caudate nucleus, putamen, and pallidum) frequently show an onset with headache, followed by vomiting. The increased intracranial pressure may produce signs of cerebellar herniation, with violent pain in the neck or even extensor spasms. Usually there are pyramidal tract signs (partial or complete hemiplegia)—as a rule on the contralateral side. Hemiparesthesias and other sensory disturbances have been described. Tremor athetosis, rigidity, choreatic convulsions, disturbances of coordination, speech disturbances, ataxia, and memory changes are late manifestations.

It should be emphasized, by the way, that tumors of the basal ganglia frequently cannot be differentiated clinically from those of the thalamus. They are discussed separately to present an overall topographic picture.

The ophthalmologic signs of tumors of the basal ganglia, again, are not characteristic and localizing. Papilledemas occur quite often (in about 80% of all patients). There are practically no visual field changes. Occasionally there is an ipsilateral disturbance of the pupillary reflex. Eye muscle palsies (fourth and sixth nerves) occur in about one half of the cases (Tönnis). In isolated cases a vertical or ipsilateral horizontal gaze paresis or gaze palsy is seen (Lillie). If possible, it is important to demonstrate tumor calcifications on x-ray films of the skull, which, as a rule, makes the diagnosis easy.

Tumors of the pineal body and the midbrain

Tumors of the pineal body and midbrain, as parts of the brain that are in close proximity, are discussed in one section because their symptomatology is quite similar. By causing pressure on the roof of the midbrain, the quadrigeminate plate, tumors of the pineal body produce symptoms quite similar to those of neoplasms with the primary seat in the midbrain, especially in its cranial part. It also seems justified to discuss these tumors together from the standpoint of differential diagnosis. It is practically impossible to differentiate between tumors of the pineal body (Fig. 3-97) and the midbrain (Fig. 3-98) since, as a result of their close proximity, their symptoms are quite similar.

There are numerous references in the literature (Horrax and Bailey) to the fact that *pinealomas in boys* often produce *precocious puberty,* with premature development of the sex organs, the appearance of axillary and pubic hair, and breaking of the voice. We have a record of precocious puberty in only one patient with a pinealoma. In three other patients no sexual prematurity was noticeable. On the contrary, one boy showed a distinct dystrophia adiposogenitalis and one girl a marked obesity, especially of the hips and thighs. Three children in our series developed cerebellar symptoms such as trunk ataxia during the early stages that made it impossible for them to walk (effect of the tumor on the cerebellum or the spinocerebellar tract). The combination of cerebellar ataxia and aqueduct signs is called *Nothnagel's*

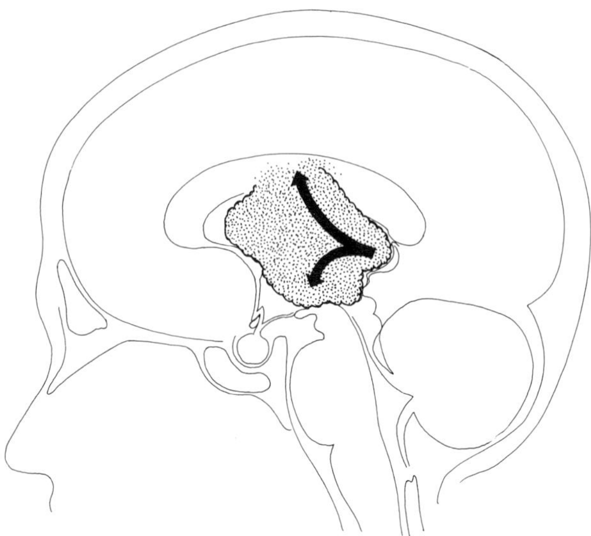

Fig. 3-97. Median section through the skull showing a pinealoma invading the third ventricle. Note the infiltration and compression of the roof of the midbrain and the narrowing of the Sylvian aqueduct.

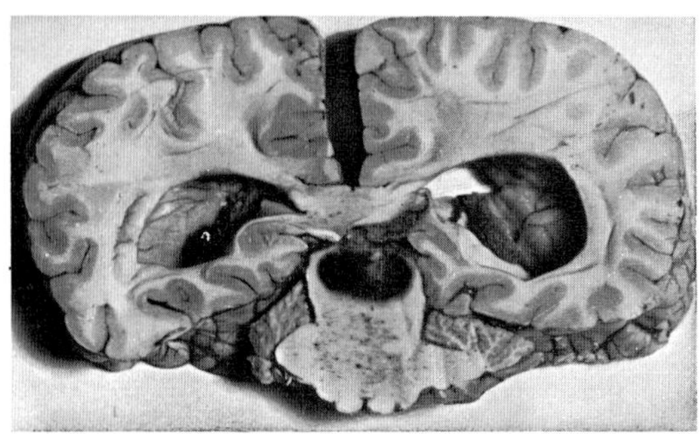

Fig. 3-98. Frontal section through the brain with a tumor (glioblastoma multiforme) in the region of the midbrain and aqueduct of Sylvius. An enormous internal hydrocephalus can be recognized by the enlarged lateral ventricles.

syndrome. Especially in children, the cerebellar symptoms in association with signs of increased intracranial pressure represent the first symptoms of tumors of the quadrigeminate plate (Tönnis). *Deafness* is another characteristic symptom resulting from involvement of the inferior colliculi. The acoustic impairment commences with the high tones. Occasionally pinealomas produce signs indicating impairment of the neighboring hypothalamus. They take the form of drowsiness, polyuria, polydipsia, and elevation of the body temperature. This is particularly the case if the tumor originates from vestiges of the pineal gland in the third ventricle. The symptomatology then is very similar to that of tumors of the third ventricle.

The symptoms of *tumors of the midbrain* are practically identical with those of the pineal body, especially if they develop in the roof of the midbrain (that is, the region of the corpora quadrigemina) (Glaser; Globus). Affection of the red nucleus and the ipsilateral oculomotor nerve results in the well-known *Benedikt syndrome* (rhythmic tremor of the arm on the contralateral side and oculomotor paresis on the side of the lesion). If the tumor has an even more ventral position, it usually causes a *Weber syndrome* consisting of an ipsilateral oculomotor paresis with hemiplegia on the opposite side. Frequently the tumors involve both sides. Thus a bilateral Benedikt syndrome or Weber syndrome in the form of a tetraplegia is not rare. Deafness due to compression of the lateral lemniscus and a contralateral sensory disturbance caused by involvement of the median lemniscus are rather late signs of tumors of the midbrain.

In the topical diagnosis of tumors of the pineal body and midbrain, ocular signs are of decisive significance. This can be expressed with the keyword *Parinaud's syndrome* (Posner and Horrax; Barker; Devic, Paufique, Girard, and Guinet). It consists of *supranuclear disturbances of conjugate eye movements in the form of a vertical gaze palsy* in addition to *pupillary disturbances of the Argyll Robertson type* and *nuclear oculomotor pareses.* This symptomatology is clear if one remembers the location of the supranuclear vertical gaze centers, the pupillary centers, and the oculomotor nuclei in the region of the midbrain, especially its periaqueductal area. (See Fig. 1-11.)

Accordingly, the *subjective ocular symptoms* caused by these tumors consist of relatively early *double vision* which, as a quasi preliminary stage of the diplopia (p. 34), may be preceded by an undefined sensation of blurred vision. In our own series of ten patients (six with pinealomas and four with tumors of the midbrain) this diplopia with vertically separated images, besides early general symptoms of increased intracranial pressure (headache and vomiting), was

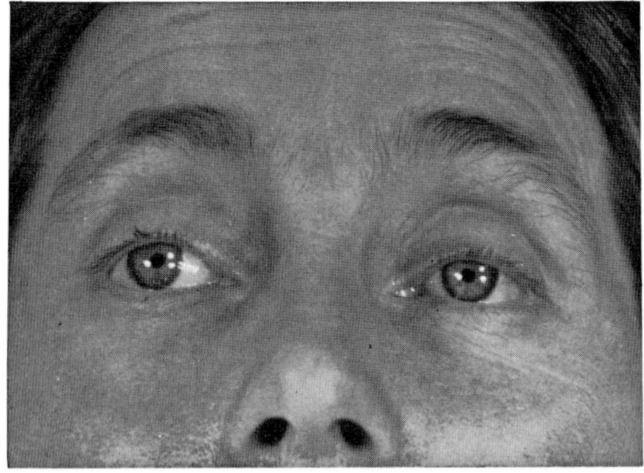

Fig. 3-99. Vertical gaze palsy in glioblastoma multiforme of the frontal lobe in a 37-year-old patient. Inability to perform schematic or following movements in a vertical direction, especially up. The photograph depicts a futile attempt by the patient to look up. This is recognizable by the raised upper lids. The frontal localization of the tumor suggests a remote effect on the region of the midbrain (vertical gaze centers) (p. 47).

one of the earliest and most striking symptoms and one that caused the patient to seek medical consultation. Other symptoms such as flickering, clouds, or blackouts must be interpreted as the results of papilledemas.

The most important and constant objective sign is a vertical gaze palsy. We have found it in all ten patients in our series and in every instance as a *gaze palsy of the upward movement* (Fig. 3-99). Only in one patient was the palsy of the upward movement combined with one of the downward movement. It is characteristic that neither spontaneous and command (schematic) movements nor pursuit movements, optically induced movements, and vestibular reflex movements can be executed simultaneously with both eyes in this one particular direction of gaze. In rare cases only the vestibular eye movements may remain. It can be elicited by means of the caloric nystagmus or the doll's head phenomenon. This is the so-called Roth-Bielschowsky type of a vertical gaze palsy. However, loss of all types of conjugate eye movements, that is, the schematic as well as the reflex types (occipital and vestibular fixation mechanism), is the rule. In the initial stages of a tumor of the pineal body or of the midbrain there may be only a vertical gaze paresis which, however, soon changes to complete paralysis: In combination with a vertical gaze palsy, there may be a consequent *downward or upward conjugate deviation of the eyes* according to the nature of the palsy, also either a *ptosis* or a *retraction of the upper lids* (Collier). Occasionally the very first sign of such a gaze paresis may be merely a vertical jerk (so to speak, gaze paretic) nystagmus in upward or, more rarely, downward gaze. Only a purely vertical gaze nystagmus is suggestive of a lesion of the midbrain, whereas a combination of horizontal and vertical nystagmus occurs in association with tumors of the vermis cerebelli (p. 266).

It has already been stressed in the section on gaze palsies (p. 47) that the supranuclear character of an impairment of the conjugate vertical eye movements (in other words, preservation of function of the individual nerves and their nuclei) can be proved by a positive Bell phenomenon (upward rotation of the eyes during lid closure or sleep). Such vertical gaze palsies must be due, as we know from newer publications (Balthasar; Nashold and Gills), to lesions in the *midbrain tegmentum*, especially its lateral periaqueductal segment. The earlier doctrine that vertical gaze palsy is produced by affections of the quadrigeminate plate(colliculi)can no longer be supported. Of great importance in this connection are the stereotactic experiments of Nashold and Gills, who were able to produce the ocular signs of Parinaud's syndrome by making lesions (unilateral) in the mesencephalic tegmentum (6 to 9 mm. from the midaqueductal plane). In most patients it is impossible to demonstrate a definite oculomotor paresis in addition to a vertical gaze palsy. Hence it must be assumed that the subjective diplopia is due to an unequal effect of the vertical gaze palsy on the two eyes (Duke-Elder; Bielschowsky; Böhringer and Koenig). With a more detailed test for ocular motility, such a supranuclear vertical gaze palsy with an unequal effect on the two eyes (a sort of "skew deviation," as described in midline lesions by Bielschowsky, Hoyt, and others) will not appear as a paresis of one extraocular muscle but as a persistent vertical divergence without essential change in the angle of strabismus in all fields of gaze (similar to a hyperphoria, a so-called latent vertical concomitant strabismus). However, one should not forget that true *nuclear oculomotor palsies or pareses* do occur with tumors of the pineal gland or, in particular, the midbrain. They may be responsible for some of the phenomena of diplopia. It is characteristic for nuclear oculomotor pareses that *only one isolated muscle or just some of them of either one eye or frequently both eyes may be involved.* Interestingly enough, we could demonstrate such nuclear oculomotor pareses in only four of the ten patients in our series. In about one fourth there was a paresis of the levator of the upper lid (mostly bilateral) causing a slight or moderate ptosis. The diplopia may also result from a sixth nerve paresis, which quite frequently occurs in association with these types of tumors, as they tend to cause early embarrassment of circulation of the cerebrospinal fluid and, consequently, increased intracranial pressure (two out of ten patients). We observed one instance of a *nuclear trochlear palsy in the very early stages of Parinaud's syndrome* caused by a pinealoma (Jaensch). Oculomotor palsy and a typical vertical gaze palsy, in addition to the trochlear nerve palsy, developed only some time later.

Part of Parinaud's syndrome in patients with tumors of the pineal body and midbrain also are characteristic *pupillary changes in the form of light rigidity of the pupil of the Argyll Robertson type.* We have observed such a pupillary disturbance (p. 28) in a more or less typical form in nine patients (50% according to Tönnis). Mostly it involved both sides and once only one side. Anisocoria was seen frequently, but corectopia (displaced pupils) seldom. The size of the pupils was rather larger than normal. The usual miosis of the typical Argyll Robertson pupil was quite rare. In contrast to the true Argyll Robertson phenomenon, the *light rigidity* of the pupils was *not always complete*. However, the reaction to light usually was quite markedly diminished. There may be signs of defective accommodation (Walsh and Hoyt) or accommodation spasms before disturbances of pupillary reaction appear. Another atypical mark of pupillary disturbance consisted of an *occasional additional impairment of the convergence reaction* (often combined with an actual *paralysis of convergence*), which is not true for the Argyll Robertson pupil (pseudo-Argyll Robertson pupil or incomplete general pupil rigidity according to Kestenbaum). We observed a paralysis of convergence in three of our ten patients. *Convergence spasm and nystagmus retractorius*—a jerking retraction of the eyes caused by clonic convergence spasms without simultaneous relaxation of the lateral recti— was found only once. We saw one case of true convergence spasm. (The intention to look upward causes a strong convergence movement of both eyes.)

By definition, Parinaud's syndrome consists only of a vertical gaze paresis, nuclear oculomotor pareses, and pupillary disturbances. If there are, in addition, tonic convergence spasms and a nystagmus retractorius (especially in an attempt to look up), the term *Sylvian aqueduct syndrome* is frequently used. We have seen these two signs, supposedly characteristic for the region of the Sylvian aqueduct, in association with pinealomas. Thus they are of no localizing value in the differentiation of Parinaud's and a Sylvian aqueduct syndrome. It would be more correct to consider the latter a rare and special form (although with additional symptoms) of the former.

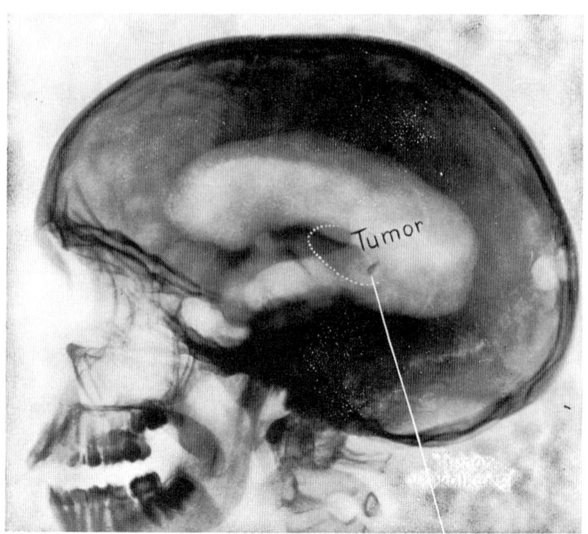

Fig. 3-100. Ventriculogram (lateral view) of the skull of a 15-year-old patient with a pinealoma. The localization of the tumor is indicated by the calcification. The tumor caused occlusion of the aqueduct, resulting in an enormous internal hydrocephalus. The dilated third ventricle extends toward the sella turcica. Clinical picture: typical vertical gaze palsy, pupillary disturbances of the Argyll Robertson type, and bilateral papilledema of 2 to 3 diopters elevation.

Because of the close proximity of these tumors to the Sylvian aqueduct (Yanagida), early blockage of the aqueduct and the development of an internal hydrocephalus (Fig. 3-100), with considerable increase in intracranial pressure, are possible. We have observed such an event with the occurrence of markedly prominent *papilledemas* in four of a total of ten patients. The others showed a perfectly normal eyeground. On the basis of our own experience, we cannot concur with the opinion of other authors (Walsh and Hoyt; Tönnis), according to which papilledemas occur almost invariably with such tumors. Tönnis found papilledemas in 80% of the patients in his series; one fifth of these already showed secondary optic atrophy. Twenty-five percent of all the patients with quadrigeminate plate tumors had visual disturbances; 15% already manifested amaurosis. We also were unable to record abnormal field changes in our patients, except occasional enlargement of the blind spot. Bitemporal or homonymous hemianopias due to interference with the visual pathways have been described.

In summary, the ocular symptoms are decisive for the diagnosis of tumors of the pineal gland and the midbrain. They can be defined as the so-called Parinaud syndrome, which consists of a vertical gaze palsy, pupillary disturbances of the true or pseudo-Argyll Robertson type, and nuclear oculomotor and trochlear palsies. Rarer additional phenomena are convergence paresis or convergence spasms and nystagmus retractorius. Occlusion of the Sylvian aqueduct may cause early papilledemas as well as a nonspecific sixth nerve paresis. Parinaud's syndrome is so extraordinarily characteristic that its presence alone is convincing proof of a lesion, most often a tumor in the region of the midbrain or the pineal gland.

Ectopic pinealomas may involve the third ventricle, the optic chiasm, and the optic nerves (Weber; Horrax and Wyatt). From such ectopic pineal tumors a chiasmal syndrome with primary optic atrophy, bitemporal hemianopia, diabetes insipidus, and hypopituitarism may result.

Only in rare instances will the remote effect of tumors located elsewhere cause Parinaud's syndrome. We have seen this a few times in patients with tumors of the corpus callosum and the frontal lobe, especially in association with signs of herniation and compression of the brain stem. In the interpretation of such disturbances of the vertical gaze utmost caution is indicated. Lack of attention (caused either by a psycho-organic syndrome or a disturbance of the sensorium) may lead to impairment of the conjugate upward eye movements. As a manifestation of an inhibited cooperation on the part of the patient, the vertical motors of the globe fail earlier than the lateral rotators. This is evident from the fact that even under normal conditions it is considerably more strenuous to look up than to any other direction. A disturbed sensorium (associated with tumors of the frontal lobe or the corpus callosum, general increase in intracranial pressure, or subarachnoidal hemorrhage) makes a "vertical gaze palsy" useless for localization. A lesion of the midbrain must be differentiated from a mere lack of attention with the help of other available signs (such as pupillary signs or extraocular muscle pareses).

INFRATENTORIAL TUMORS

Supratentorial tumors interfere primarily with the visual pathways and thus frequently produce visual field changes. For this reason the neuro-ophthalmologic symptomatology of supratentorial tumors is governed largely by the more or less characteristic visual field defects. Since direct interference with the visual tracts by infratentorial tumors is impossible, the visual field is of lessened significance in the ocular symptomatology. However, these tumors tend to affect the brain stem, the site of the nuclei of the cranial nerves, including those of the extraocular muscles and the medial longitudinal fascicle. *The picture of the ocular symptoms in infratentorial tumors thus is dominated by phenomena of motility disturbances* of the peripheral, nuclear, or supranuclear type. Common to both supratentorial and infratentorial tumors is an increase in intracranial pressure. A prematurely embarrassed circulation of the cerebrospinal fluid in patients with infratentorial tumors produces earlier and more pronounced formation of signs of increased intracranial pressure, including papilledemas.

Tumors of the cerebellum

The symptomatology of affections of the cerebellum can be expressed with the keywords *ataxia, asynergy, dysmetria, and hypotonia.* Depending on whether the lesion is in the hemispheres or the vermis, these signs may show somewhat different forms (Krayenbühl; de Martel and Guillaume; Stewart and Holmes). A clinical differentiation between the *syndrome*

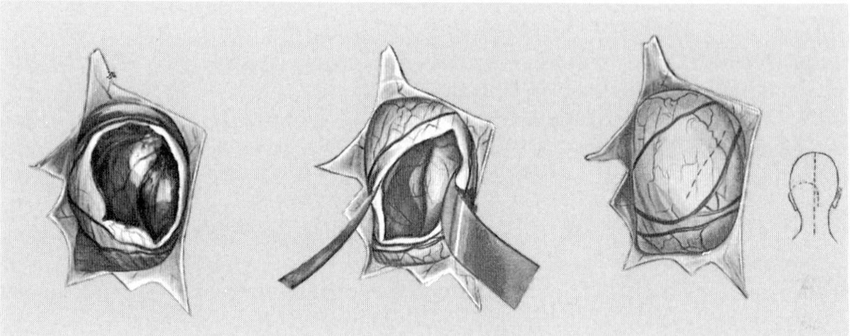

Fig. 3-101. Astrocytoma of the caudal and lateral parts of the left cerebellar hemisphere in a 9-year-old boy before and after surgical extirpation (left cerebellar exploration).

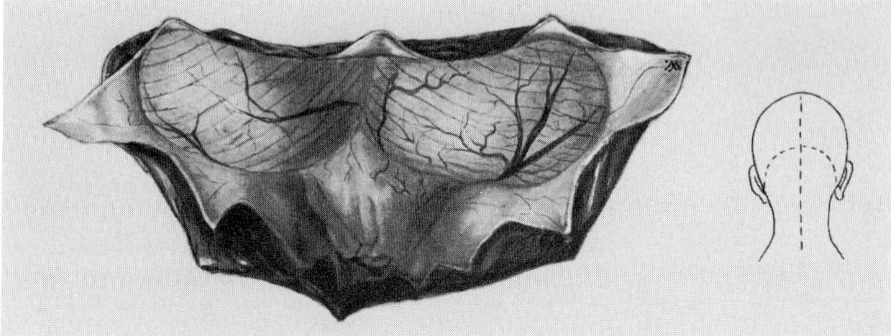

Fig. 3-102. Huge medulloblastoma of the cerebellar vermis filling the fourth ventricle in a 13-year-old patient. View after cerebellar exploration.

of the cerebellar hemispheres and *the syndrome of the cerebellar vermis* is possible. In children the syndrome of the cerebellar hemispheres is caused by an astrocytoma (Fig. 3-101). In adults it is usually caused by astrocytomas and hemangioblastomas. The syndrome of the cerebellar vermis in children is predominantly the result of a medulloblastoma (Fig. 3-102). In adolescents and adults it may be due to an astrocytoma, a hemangioblastoma, or other tumors. In addition to horizontal gaze nystagmus (to be discussed below), the characteristic signs of the syndrome of the cerebellar hemispheres are hypotonia (excessive range of movements on passive ductions) of the ipsilateral extremities with atactic and dysmetric movements and a distinct tendency to fall—frequently toward the side of the lesion. Ataxia and asynergy manifest themselves in the knee-heel test and in the test for diadochokinesis (rapidly executed pronation and supination movements of the hands or arms). Dysmetria as well as asynergy and intention tremor will be evident with the finger-nose test. The syndrome of the cerebellar vermis usually shows vertical gaze in addition to horizontal gaze nystagmus; ataxia and dysmetria of the extremities are insignificant, whereas ataxia of the gait and trunk are very pronounced. In children with medulloblastomas the symptoms of increased intracranial pressure (such as vomiting with or without headaches, double vision, and squint) frequently precede other signs. Marked loss of weight is characteristic in children with medulloblastomas. Only in the late stages will there be signs of herniation of the cerebellar tonsils through the foramen magnum in the form of stiffness of the neck, sensory disturbances and weakness of the extremities, and ultimately extensor spasms and respiratory paralysis. Less frequently there is an upward herniation of the cerebellum, expanded by a tumor through the tentorial opening, thus producing midbrain symptoms such as anisocoria, disturbances of the light and convergence reaction of the pupils, and vertical gaze palsies, especially upward. In children up to 14 years of age a tumor of the cerebellar vermis interfering with the circulation of the cerebrospinal fluid, and thus causing an internal hydrocephalus, will cause separation of the sutures of the skull and the characteristic "cracked-pot resonance" elicited by knocking on the head. Occasionally a chronic internal hydrocephalus associated with cerebellar tumors may be responsible for generalized epileptic seizures.

One should not forget that the symptomatology of cerebellar tumors is largely marked by symptoms of increased intracranial pressure. Especially with slowly growing tumors, signs attributable to the cerebellum are insignificant next to those caused by the increased intracranial pressure. In such cases, differentiation from supratentorial tumors often is not easy, especially since herniation of the cerebellum upward or downward may produce symptoms seen in all sorts of brain tumors. The sequence in which the symptoms appear is enormously important for proper evaluation. If cerebellar signs are present in the early stages before the onset of signs of increased pressure, a cerebellar tumor should be suspected. If the cerebellar signs appear late, they are not localizing.

As far as eye symptoms are concerned, it is unnecessary to differentiate between tumors of the vermis and the hemispheres of the cerebellum. There is no essential difference between these two types of localization and their effect on the visual apparatus.

In a little more than one half of the cerebellar tumors among the thirty adults and twenty-five children in our series, *ocular symptoms were present that were more or less directly related to increased intracranial pressure and the formation of papilledemas.* Patients frequently complain about flickering before the eyes, blackouts, and a peculiar quivering or regular transient amblyopic attacks (p. 102). Exceptionally frequent (45% according to Tönnis) are statements about a *diminution of vision* (usually bilateral). One patient stated that he saw his surroundings as through a "scraped-off" mirror. At times the loss of vision (usually bilateral) and a corresponding concentric constriction of the visual field (evidence of an incipient atrophy of a chronic papilledema, p. 115) is the *very first symptom that brings the patient to the physician and perhaps to the ophthalmologist.* The cerebellar signs are still absent or are present only in a very minor form. We have records of a few patients with cerebellar tumors who consulted an ophthalmologist first and in whom a diagnosis of bilateral papilledema led to referral to the neurosurgeon for further evaluation. The history reveals, in addition to loss of vision and obscurations, occasional double vision (30% according to Tönnis) and photophobia. The anatomic basis for the *diplopia* is a nonspecific, usually unilateral, sixth nerve paresis related to the increased intracranial pressure. Accordingly, the two images usually show horizontal separation. Diplopia and manifest strabismus (in addition to loss of vision, headache, and vomiting) are also among the first symptoms of a cerebellar tumor that bring the patient to his physician or ophthalmologist. The subjective ocular symptoms of patients with cerebellar tumors actually are absolutely nonspecific and should be regarded merely as the result of increased intracranial pressure.

One of the most frequent and perhaps earliest ocular signs of cerebellar tumors (especially in progressive cases) is *nystagmus.* We found it in practically all of the patients in our series. This *cerebellar nystagmus* is a jerk nystagmus (p. 55). It usually has the character of a central vestibular nystagmus, sometimes also of a symmetrical or asymmetrical gaze nystagmus or a vertical gaze nystagmus. The cerebellar nystagmus is predominantly horizontal in nature, rarely horizontal-rotary. It may be spontaneous with the eyes in the primary position or, more frequently, appear only during lateral movements, with the quick phase in the direction of gaze. In unilateral cerebellar lesions the amplitude varies with the direction of gaze. Despite some exceptions, it can be stated generally that *the direction of gaze associated with the coarser form of nystagmus corresponds to the side of the lesion. The nystagmus with cerebellar tumors is sometimes characteristically dissociated,* which means that the direction or amplitude of the oscillations is greater in one eye than in the other (Cogan). It is also said to vary with the position of the head. If the tumor has a mesial location (for instance, in the vermis), there may be a purely vertical gaze nystagmus either due to direct pressure upon the floor of the fourth ventricle and thus the posterior longitudinal bundle or due to a remote effect on the midbrain tegmentum. It also indicates a considerable anterior extension of the tumor from the posterior fossa anteriorly. This nystagmus is particularly noticeable in upward gaze. Among our own series, we have observed vertical gaze nystagmus (also usually in combination with horizontal nystagmus) almost exclusively in patients with tumors of the cerebellar vermis, especially in those with medulloblastoma (seven times) (Fig. 3-103). Twice it was noticed in patients with large tumors of the hemisphere that tended to expand toward the midline. It is still questionable whether nystagmus associated with cerebellar tumors represents a cerebellar sign or whether it is due to a remote effect on the brain stem, the pons, the posterior longitudinal bundle, or the vestibular nuclei.

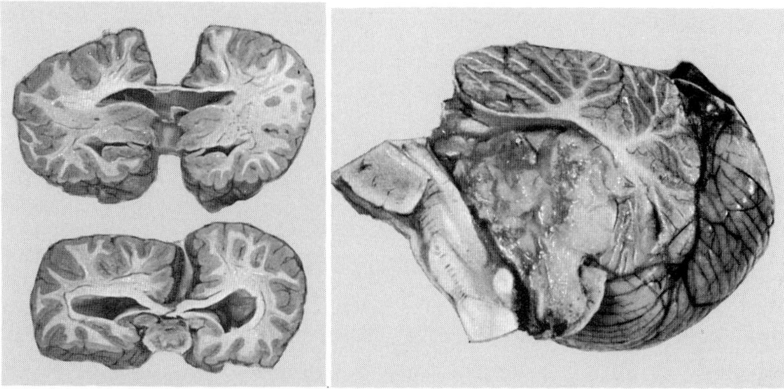

Fig. 3-103. Medulloblastoma of the cerebellum. Occlusion of the fourth ventricle and internal hydrocephalus (frontal section of the brain). (Specimen from the Pathologic-Anatomic Institute of the University of Zurich.)

Like the nystagmus of tumors of the pons and the cerebellopontine angle, cerebellar nystagmus belongs to the group of so-called *central nystagmus* (p. 54) (Kestenbaum). It may assume the character of a central vestibular nystagmus, a symmetrical or asymmetrical gaze nystagmus, or a vertical gaze nystagmus. The central vestibular nystagmus, a result of disturbed connections between the cerebellum and the vestibular nuclei, has a horizontal-rotary character. It shows first-, second-, or third-degree intensity. In contrast to peripheral vestibular nystagmus, it is permanent. As a rule, it is not associated with acoustic disturbances or dizziness. The symmetrical or asymmetrical gaze nystagmus is purely horizontal; it is missing in the primary position; and it becomes manifest only in gaze to the right or left and its intensity increases as the terminal positions are reached. It is caused by a remote effect of the cerebellar process on the brain stem structures. If vertical gaze nystagmus is an only sign, it is suggestive of a process in the midbrain tegmentum region. If there is a simultaneous horizontal nystagmus in the lateral positions of gaze, a vertical nystagmus is only one of the components of a severe symmetrical gaze nystagmus and has therefore no importance of its own. Undoubtedly these various components of a central nystagmus in cerebellar tumors merge, which makes their analysis difficult at times. For practical and clinical purposes it is sufficient to consider the criteria of cerebellar nystagmus previously outlined.

Some of the rarer phenomena of motility disturbances associated with cerebellar tumors are *horizontal gaze pareses* and *dissociated movements,* as seen in two patients in our series. Such manifestations must be regarded as the effect of a tumor on the pons or pressure on the floor of the fourth ventricle. (If the dissociation alternates, the lesion lies in both hemispheres.) One example of dissociated ocular movements is the "skew deviation," or the Hertwig-Magendie syndrome: one eye turns down and in and the other up and out. The dissociation remains unchanged in all directions of gaze. The side of the cerebellar lesion corresponds to the eye that is turned down.

With cerebrellar tumors one may sometimes observe ocular dysmetria, flutterlike oscillations, and disturbed pursuit movements in the sense of cogwheel eye movements. However, it is quite uncertain whether these signs are due to the primary cerebellar involvement or to a secondary affection of the brain stem.

Ocular dysmetria (Cogan), obviously an analogy to the dysmetria of the limbs, is characterized by an overshoot of the eyes and a subsequent correction by several oscillations when they have to change from one fixation point to another one. With unilateral lesions the overshoot is generally greatest on looking toward the side of the lesion.

Flutterlike oscillations, like dysmetria (according to Cogan), are typical of cerebellar lesions. These are intermittent to-and-fro movements of the eyes, lasting only seconds and induced mostly by changes in fixation.

Cogwheel eye movements are jerky, inaccurate pursuit movements, probably in relation to the cerebellar hypotonia (see discussion of gaze palsies).

Other ocular signs to be discussed are only nonspecific signs of an increased intracranial pressure. *Sixth nerve pareses* occur in about one fifth of the patients —interestingly, *mostly in unilateral,* rarely in bilateral form. These sixth nerve pareses are seen with particular frequency in association with tumors of the cerebellar vermis. Thus one cannot assume a relationship between the side of the tumor and a unilateral sixth nerve paresis. We maintain that, in case of cerebellar tumors, such pareses belong to the *general signs of an early increase in the intracranial pressure,* although a certain direct or indirect pressure effect on the brain stem and the abducent nuclei cannot be ruled out. Fifty of fifty-five patients with cerebellar tumors showed *bilateral papilledemas.* There was a particularly marked prominence, with numerous hemorrhages and exudates, in those with tumors of the vermis. It is significant that these papilledemas occur in a relatively early stage because of the well-known tendency of cerebellar tumors to compress the fourth ventricle and thereby to interfere with the circulation of the cerebrospinal fluid. In fact, general signs of increased intracranial pressure frequently first manifest cerebellar neoplasms. *Numerous papilledemas in cerebellar tumors, especially those of the vermis, already show distinct evidence of beginning atrophy at the first examination.* There is a grayish white discoloration of the disc, at first narrowing of the arteries and later of the veins, and together with loss of vision (progressing occasionally to blindness; 10% according to Tönnis), also a concentric constriction of the visual fields (p. 116 and Fig. 2-15).

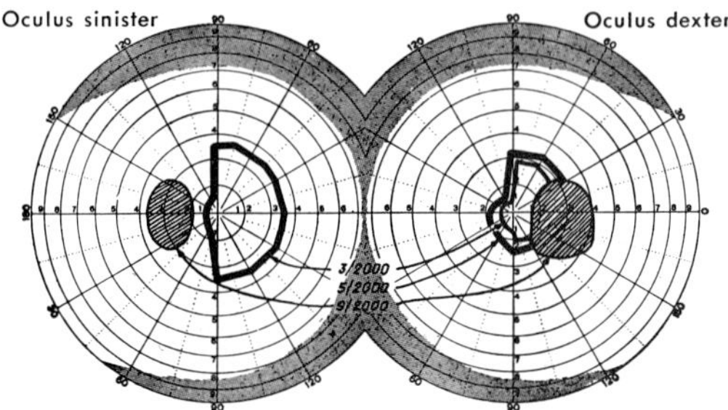

Fig. 3-104. Hemangioblastoma in the caudal section of the cerebellar vermis in a 38-year-old patient. Bilateral papilledema of 2 diopters prominence with marked venous congestion and numerous massive hemorrhages. Visual acuity, O.D. 1.0; O.S. 0.5. Amblyopic attacks! Enormous bilateral enlargement of the blind spot. Left homonymous hemianopia with sparing of the macula.

Except for an enlargement of the blind spot caused by the choked disc, the visual fields generally are not characteristic in patients with cerebellar tumors. There are references in the literature (Wagener and Cusick; Weinberger and Webster) pointing out the possibility of various field defects as the indirect result of an increased intracranial pressure or an internal hydrocephalus (Fig. 3-103). We have seen such *visual field changes very rarely.* A patient with a hemangioblastoma of the vermis showed an unquestionable *homonymous hemianopia* with sparing of the macula (Fig. 3-104). Another patient with hemangioma of the vermis showed a temporal hemianopia on one side and slight concentric constriction of the field on the other side. In a patient with an astrocytoma in the region of the vermis there was a superior bitemporal quadrant defect. *Bitemporal defects in cases of cerebellar tumors naturally must be regarded as a remote effect,* most likely *pressure on the chiasm* from above and behind *by a dilated third ventricle due to internal hydrocephalus* (p. 253). This also explains the occurrence of central scotomas (the macular fibers are situated at the posterior angle of the chiasm!) reported by a number of authors (Walsh and Hoyt and others).

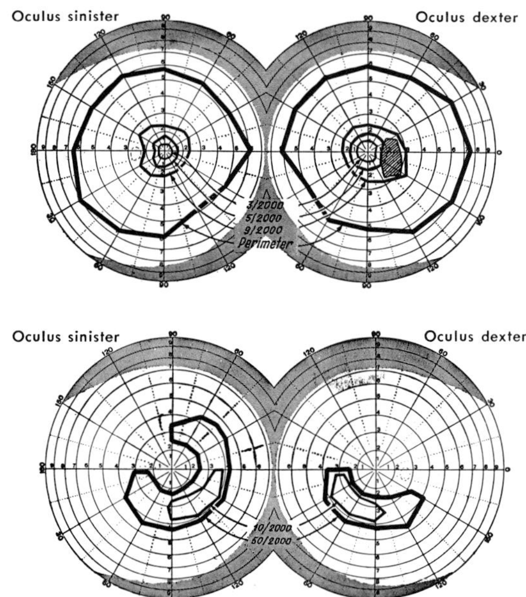

Fig. 3-105. Above, Astrocytoma in the lateral part of the left cerebellar hemisphere extending to the cerebellopontine angle in a 19-year-old patient. Beginning atrophy of bilateral papilledema. O.D. 3 diopters elevation; O.S. 4 diopters elevation. Visual acuity, O.U. 1.0. Concentric constriction of the central isopters with the suggestion of a beginning bitemporal hemianopia. Marked enlargement of the right blind spot. Below, Astrocytoma of the cerebellar vermis extending to the roof of the fourth ventricle in an 11-year-old girl. Beginning left papilledema. Normal right disc. Considerable internal hydrocephalus! Completely asymmetrical bitemporal hemianopia with loss of the right superior nasal quadrant and left kidney-shaped field remnants hugging the macula.

Cushing and Walker describe the occurrence of *binasal hemianopias* in patients with cerebellar tumors. They are of the opinion that a dilated third ventricle displaces the optic nerves laterally and presses them against the carotid arteries. It is difficult to find an explanation for homonymous field defects associated with cerebellar tumors except as a direct insult to one occipital lobe due to supratentorial expansion. At any rate, it can be stated that *a bitemporal, mostly asymmetrical hemianopia, combined with pronounced papilledemas, is suspicious of a tumor in the cerebellar or infratentorial region.* The significance of visual field changes in cerebellar tumors, however, is small considering their rare occurrence. On closer scrutiny, especially bitemporal visual field defects are much rarer in association with cerebellar neoplasms than is generally assumed (Fig. 3-105). Walsh and Hoyt observed homonymous defects more often than bitemporal defects.

An occasional observation of a *unilateral diminished corneal reflex* must be explained as a pressure effect on the trigeminal nerve in the region of the cerebellopontine angle. The facial and acoustic nerves usually are involved at a later stage.

Hemangioblastomas make up a significant number of tumors of the cerebellar hemispheres (Fig. 3-106). In our series of fifty-five patients we observed three instances of a simultaneous retinal angiomatosis, examples of the so-called *von Hippel–Lindau* disease (Craig, Wagener, and Kernohan; Danis; Martin; and others). A relationship between the occurrence of angiomas in the retina and the cerebellum is known to be loose: *retinal angiomas occur much more rarely and very infrequently in association with cerebellar hemangioblastomas.* Nevertheless, it is important to search carefully for angiomas of the retinal periphery (Traquair) in cases of similar cerebellar tumors.

Generally, the retinal angiomas tend to be located in the fundus periphery (seldom at the macula or the disc). They have a distinctly globular and quite prominent shape. Markedly dilated veins and arteries feed these angiomas, which can be distinguished only in the early stages. The hemangiomas appear as reddish globular structures in the fundus periphery. There may be a multiplicity of foci, predominantly in the lower parts of the retina. Frequently the angiomas are surrounded and eventually covered by a grayish white exudate (Fig. 3-107). Similar exudates accompany the enormously dilated and tortuous vessels. Massive

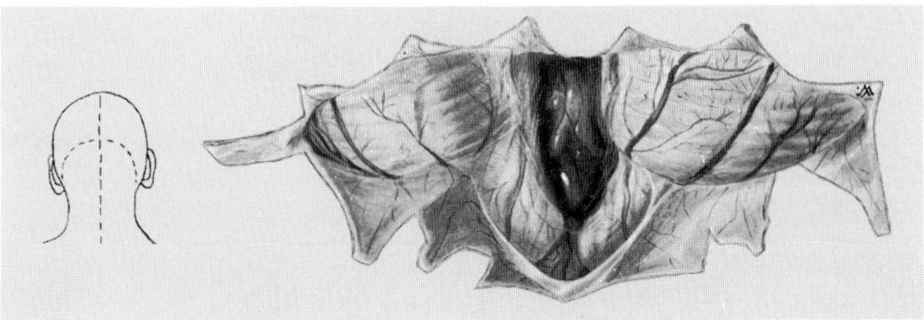

Fig. 3-106. Walnut-sized hemangioblastoma of the cerebellar vermis in a 57-year-old patient after bilateral cerebellar exploration.

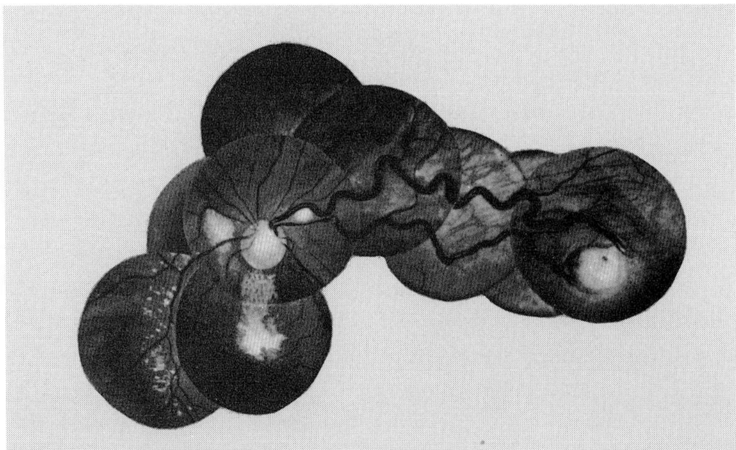

Fig. 3-107. Angiomatosis retinae in von Hippel–Lindau disease (with cerebellar hemangio-blastoma). Composite fundus photographs. The spherical retinal angioma in the right periphery of the fundus is covered almost completely by a grayish white exudate. The tumor is fed by markedly dilated and tortuous vessels (artery below, vein above). Retinopathy in the form of foci of fatty degeneration to the side and below the disc.

recurrent retinal hemorrhages are frequent. As a rule, there is a more or less distinct retinal detachment in the later stage of growth of these tumors, which tends to mask the original picture and make the diagnosis more difficult. An early diagnosis of the retinal angiomas is quite essential because it is frequently possible to destroy them with electro- or photo-coagulation if a retinal detachment has not yet developed. Such a procedure will prevent a secondary retinal detachment. In one of our patients, electrocoagulation of such a retinal angioma was successful in one eye. The patient completely lost the sight in the other eye as a result of a retinal detachment and a secondary complicated cataract.

The ocular symptoms associated with cerebellar tumors can be summarized as follows. Most of the signs are the result of an early increase in the intracranial pressure. Choked discs are almost always present and usually quite distinct. They cause flickering, foggy vision, and amblyopic attacks. Relatively early they develop atrophy, with loss of vision and concentric contraction of the visual fields. An outstanding ocular sign is nystagmus of the central vestibular type, which usually has a horizontal direction but may have a vertical component with tumors of the cerebellar vermis. As a rule, this nystagmus is coarser toward the tumor lesion than toward the opposite side. Nonspecific unilateral or bilateral sixth nerve pareses resulting from the increased intracranial pressure account for diplopia mentioned by the patient. Unilateral sixth nerve pareses are more frequent than the bilateral form. Visual field changes associated with cerebellar tumors are quite rare. They are not the result of a direct effect on the visual pathways but a remote effect via a dilated third ventricle and compression of the chiasm. Most common are unilateral temporal or bitemporal field defects. Occasional homonymous hemianopias are difficult to interpret. In rare cases a combination of a cerebellar and a retinal hemangioblastoma represents an

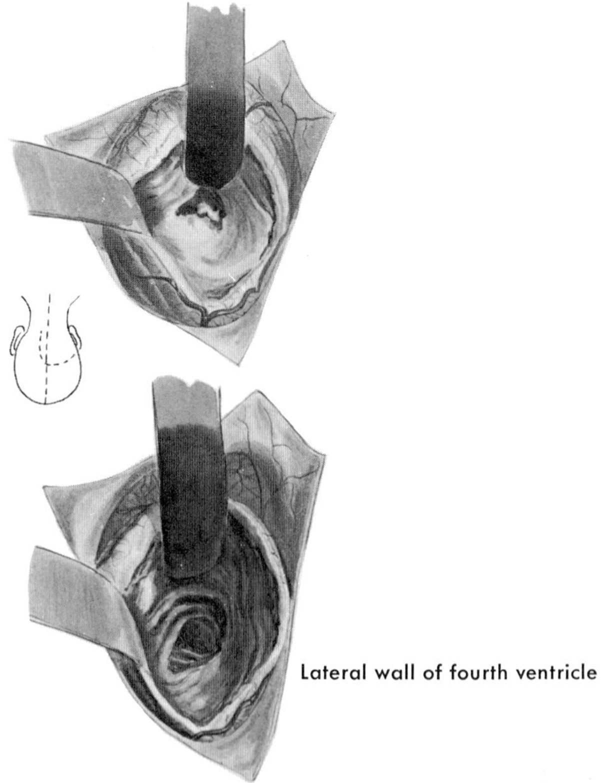

Fig. 3-108. Solid, partly cystic, walnut-sized papilloma of the plexus of the fourth ventricle in a medial and a caudal section of the left cerebellar hemisphere in a 57-year-old patient before and after extirpation via a left cerebellar exploration.

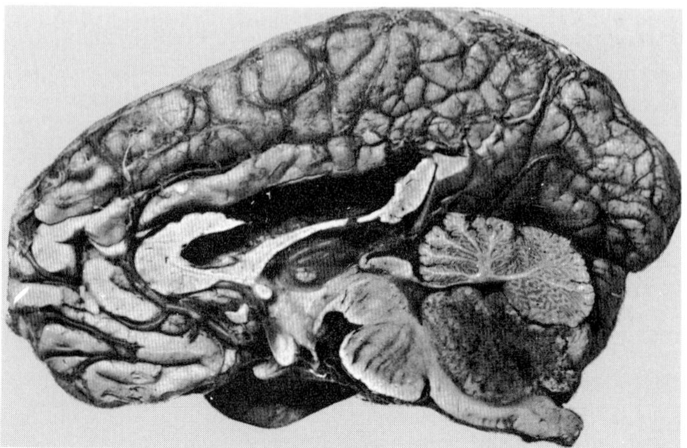

Fig. 3-109. Median section through the brain with ependymoma of the fourth ventricle in a 64-year-old patient. (Specimen from the Pathologic-Anatomic Institute of the University of Lausanne.)

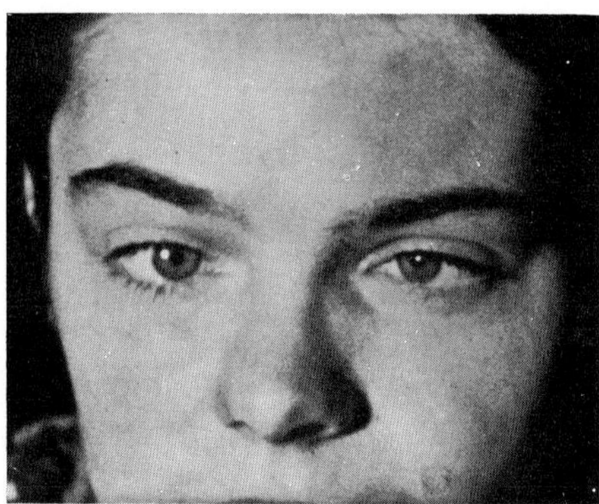

Fig. 3-110. Unclassified benign glioma originating from the floor of the fourth ventricle. Horizontal gaze palsy to both sides. Convergence spasm of the globes on command to look to the side or up.

example of von Hippel–Lindau disease. On the x-ray film, increased intracranial pressure manifests itself by demineralization and secondary enlargement of the structures of the sella turcica.

Tumors of the fourth ventricle

Actually, tumors of the cerebellar vermis cause symptoms very similar to those of neoplasms of the fourth ventricle (ependymomas, choroid plexus papillomas). The characteristic picture includes *early increase of intracranial pressure with bilateral choked discs, coarse horizontal gaze nystagmus in the lateral terminal positions,* and severe trunk ataxia. If the tumors originate from the floor of the fourth ventricle, there may be the additional *signs of disturbances of this area,* that is, facial palsy, a diminished corneal reflex on one side, or even a disturbed coordination of the ocular movements (Craig and Kernohan) (Figs. 3-108 and 3-109). In fourth ventricle tumors (also in third and lateral ventricle tumors) one may occasionally observe *Bruns' syndrome,* which is characterized by attacks of headache, vomiting, vertigo, and transient blindness produced by changes of position of the head. Between the attacks there is freedom of symptoms, but the head usually remains in a fixed position.

Our one patient showed very early disturbances of the extraocular muscles, which were said by his acquaintances to give him a "funny look." Detailed analysis of the ocular motility revealed a *severe gaze palsy to both sides.* Also, peculiarly, a pronounced *convergence spasm* occurred when the patient was asked to look up or to the side. We shall discuss these supranuclear motility disturbances in more detail in connection with tumors of the pons (p. 278), which produce similar gaze disorders. Our patient also showed very severe papilledemas of the chronic type. There were early episodes of amaurosis fugax as a form of amblyopic attacks (Fig. 3-110).

Tumors of the cerebellopontine angle

By far the greatest number of tumors of the cerebellopontine angle consists of *acoustic neuromas* (87% according to Tönnis), which originate from the neurilemma of the vestibular nerve near the internal porus acusticus (Fig. 3-111). The seat of these tumors produces characteristic symptoms that can be summarized as a *combination of signs of lesions of the fifth, sixth, seventh, and eighth cranial nerves with cerebellar symptoms* (Cushing; Edwards and Paterson; Lundberg). For a correct diagnosis it is of the utmost importance to note the chronologic development of the various signs. *One of the early symptoms is impairment of hearing,* such as tinnitus or other noises, combined in later stages with progressive loss of

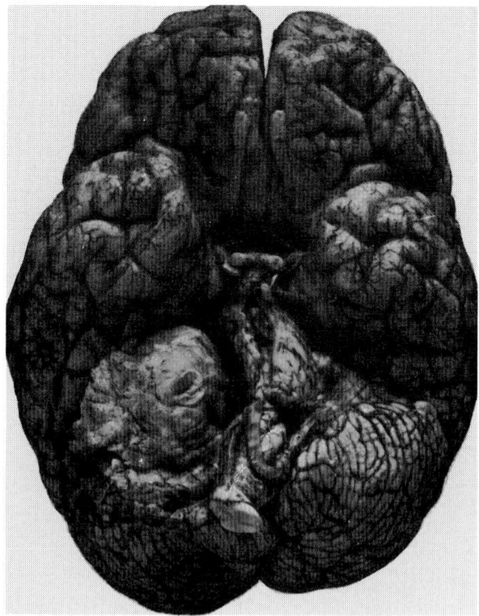

Fig. 3-111. Right acoustic neuroma with distinct displacement of the brain stem toward the left side. (Specimen from the Pathologic-Anatomic Institute of Basel.)

hearing. This loss of hearing leads to almost complete deafness. Although the tumor originates in the vestibular part of the acoustic nerve, there are usually no initial vestibular disturbances because of the slow progress of the tumor. As a rule, they are secondary to acoustic symptoms and take the form of vertigo (similar to that in Ménière's disease) and nystagmus. During the later stages of development of the tumor, an uncertain gait appears, which probably indicates damage to the cerebellum. Much later there is involvement of the other cranial nerves in the region of the cerebellopontine angle mentioned previously. The *facial nerve* is impaired quite frequently. It first shows signs of irritation in the form of spasms or twitching (for instance, blepharospasm). Later a peripheral facial palsy may be seen. *Trigeminal symptoms* manifest themselves as paresthesias (such as numbness or a dead feeling in parts of the face), later as slight neuralgias in the area of distribution of the trigeminal nerve. An important sequel of a trigeminal lesion is a diminished or abolished corneal reflex on the ipsilateral side, as will be described in greater detail below. There may be additional symptoms indicating impairment of the ninth, tenth, eleventh, and twelfth cranial nerves with a corresponding increase in size of the tumor. Predominant are disturbances of deglutition and articulation. As a rule, the appearance of lesions involving the cranial nerves is accompanied by *signs implicating the cerebellum,* such as unsteady gait, tendency to fall toward the tumor side, and other disturbances of coordination (p. 265). Apart from cerebellar signs in this stage, there are also *brain stem* symptoms such as hemiparesis (contralateral), hemianesthesia (contralateral), and Babinski's sign. The last stage is that of *increased intracranial pressure* due to hydrocephalus produced by closure of the Sylvian aqueduct. Headache, papilledema, and palsies of the sixth nerve are the common features of this stage.

As a rule, the acoustic neuroma is unilateral. Occasionally it may occur as a bilateral lesion and then is part of a systemic von Recklinghausen's disease (neurofibromatosis). In such cases one should search for neurofibromas of the skin or the characteristic "café au lait" spots. Besides acoustic neuromas, any expansive process in the angle between the petrous bone, tentorium, and pons may produce similar symptoms (for instance, meningiomas, dermoids, arachnoiditis of the pontocerebellar cistern, gliomas of the pons, or even cerebellar tumors).

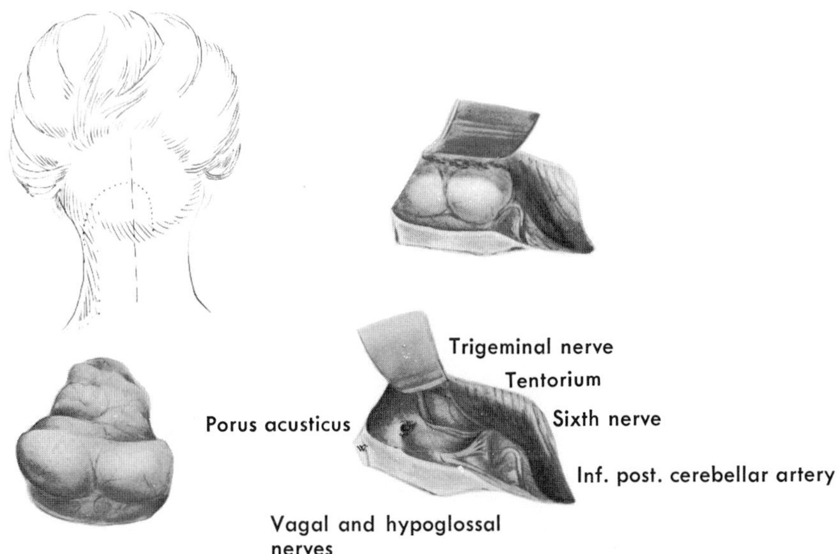

Fig. 3-112. Walnut-sized left acoustic neuroma in a 44-year-old patient before and after radical extirpation via a left cerebellar exploration and exposure of the left cerebellopontine angle.

Our own series includes thirty-one tumors of the cerebellopontine angle. Of these, twenty-five were acoustic neuromas (Fig. 3-112) (twenty-three unilateral, two bilateral) and six tumors of a different nature, such as meningioma, astrocytoma of the cerebellar hemisphere, medulloblastomas, and angiomas. Although it is evident from this discussion that the early diagnosis of acousticus tumors must be made on the basis of hearing disturbances, the *ocular symptomatology* has a certain significance (Best). It can be briefly summarized as *nystagmus, blepharospasm and facial nerve palsy* (seventh nerve), *disturbances of the corneal reflex* (fifth nerve), *and sixth nerve pareses.* Generally, signs of increased intracranial pressure are missing until the late stages. Consequently, *choked discs* are usually found rather late, and only in about one half of the patients.

In the later stages, among the most frequent *ocular symptoms* in patients with tumors of the cerebellopontine angle are loss of vision, blurred vision, flickering before the eyes, and occasional amblyopic attacks—all phenomena directly related to increased intracranial pressure. Next in frequency is *double vision.* It either may be more pronounced during violent attacks of headaches or may be noticeable only during these attacks. Because of its etiology (sixth nerve pareses), the diplopia is usually horizontal. It is interesting to note that *defects of the facial nerve* may also assume a subjective character. For instance, some patients stated that one eye had become larger and could not be closed so well; it showed frequent inflammations (lagophthalmic keratitis!). Some patients also noted an annoying blepharospastic twitching in one eye. The *trigeminal symptoms* involve less the eye itself than the sense of touch of the skin surrounding it, especially the forehead and cheek: a peculiar sensation of numb-ness, a feeling of swelling of the eye, and rarely a neuralgic pain in the area of distribution of the trigeminal nerve.

An early and most constant *ocular sign* is *nystagmus*. This is a *horizontal jerk nystagmus, usually coarser in its excursions with gaze toward the side of the tumor than toward the opposite side.* In some instances we have seen, in addition to this horizontal nystagmus, a rotary or, with eyes up, even a vertical component. The fact that nystagmus observed in association with tumors of the cerebello-pontine angle is persistent and does not disappear (as does vestibular nystagmus due to a lesion of the labyrinth or the vestibular nerve) suggests that it is less the result of a lesion of the vestibular nerve than a phenomenon of involvement of the brain stem or the cerebellar peduncles. In other words, this is a *central type* of nystagmus similar to the one seen in association with cerebellar tumors (p. 267). This is equally true for the frequent occurrence of a vertical component (that is, a remote effect on the midbrain tegmentum), especially if it appears as an isolated phenomenon and without horizontal nystagmus.

A visible involvement of the *ophthalmic branches of the facial nerve* mani-fested itself in only four of our thirty-one patients in the form of a distinctly widened lid fissure, although a general facial paresis could be demonstrated in almost all of the patients at an early date. In one patient the facial palsy was so pronounced that he was unable to close the eye on the involved side (in other words, a true lagophthalmos). This eye was the site of repeated inflammatory episodes in the form of a lagophthalmic keratitis. (See p. 15 and Fig. 1-3.)

One of the signs of trigeminal disturbance is a *diminished or abolished ipsilateral corneal reflex. It is known to become extinguished before sensory dis-turbances of corresponding skin areas can be demonstrated.* We have found this important ocular sign in a more or less distinct form among twenty-nine of our thirty-one patients. *It plays an important part in the diagnosis of tumors of the cerebellopontine angle.* The involvement of the corneal reflex may show varying gradations from slight diminution to complete anesthesia of the cornea. Often it can be demonstrated at first only in either the upper or lower quadrant (more fre-quently in the upper one). For testing of the corneal sensitivity see p. 21.

In a little more than one third of our patients a *unilateral sixth nerve paresis* on the side of the tumor could be demonstrated (due to the increased intracranial pressure rather than to direct tumor growth). This was the anatomic basis for the symptoms of double vision. In the initial stage these sixth nerve pareses often are minor and cannot be demonstrated by routine tests for motility dis-turbances. It is then imperative to use the red-green test to render the separation of the two images more distinct (p. 34). Gaze palsies (mostly the horizontal type) are extremely rare. They are caused by compression of the brain stem.

It has been mentioned previously that ocular signs of increased intracranial pressure such as *papilledemas* are present *in a little more than one half of the patients.* We observed all stages from slight blurring of the nasal and temporal disc margin to a fully developed choked disc of several diopters elevation. As in patients with cerebellar tumors, but much less prevalent, we found hemorrhages, white exudates, and frequently a beginning atrophy—a sign of transition into a chronic atrophic papilledema.

Visual field changes are extremely rare in patients with tumors of the cerebellopontine angle. Once we found concentric constriction of both fields resulting from chronic atrophic papilledemas and once a suggestion of homonymous hemianopia. Bitemporal or binasal defects as described in association with cerebellar tumors could never be demonstrated here. Due to their lateral position, these tumors are less likely to interfere with the circulation of the cerebrospinal fluid and to produce internal hydrocephalus with dilation of the third ventricle than are cerebellar tumors, especially neoplasms of the vermis with their mesial position.

Even for the ophthalmologist it may be of interest and importance to know that on x-ray films (Stenvers' projection) 70 to 80% of patients with acoustic neuromas show more or less pronounced *enlargement of the internal auditory canal*, progressing in later stages to erosion and even destruction of the whole petrous bone pyramid (Tönnis; Graf; Hitselberger).

The ocular symptoms of tumors of the cerebellopontine angle can be summarized as follows. The trigeminal and facial nerves are involved relatively early. A lesion of the trigeminal nerve manifests itself in the eye as a diminished or abolished corneal reflex on the ipsilateral side—one of the most frequent and most constant signs of tumors of the cerebellopontine angle. Less common is a facial palsy, especially widening of the lid fissure or actual inability to close the lids (that is, a lagophthalmos). One of the most important ocular signs is horizontal jerk nystagmus, usually coarser with gaze toward the tumor rather than to the opposite side. Sixth nerve pareses appear relatively late, with more extensive growth of the tumor. Ocular signs of increased intracranial pressure in the form of choked discs (nonspecific for this site) occur in a little more than one half of all patients.

Tumors of the pons and the medulla oblongata

Because of the extremely close topographic relationship between the pons and the medulla oblongata, it is hardly necessary to present tumors of these two structures as separate entities, especially since their neurologic symptoms are strikingly similar (Foerster; Horrax and Buckley). These symptoms can be divided into six groups, which ensue from the close proximity of various systems in the pons and the medulla oblongata (Kaufmann): (1) cranial nerves (fifth to twelfth), (2) their internuclear or supranuclear tracts, (3) the cerebellar system with its connections, (4) the pyramidal system, (5) the long sensory tracts, and (6) the important fact that a group of signs either are present in a bilateral form or some of them alternate between the two sides (for instance, impairment of the cranial nerves and horizontal gaze palsy on one side, lesions of the pyramidal or the long sensory tracts on the other side).

In summary, it can be stated that tumors of the pons and the medulla oblongata are distinguished by symptoms involving both sides as well as by a *combination of signs involving cranial nerves at the level of the tumor, with early signs of pyramidal and sensory tract disorders* (often together with ataxia, dysmetria, and horizontal and vertical nystagmus). It is also characteristic for tumors of this region that, in contrast to tumors of the midbrain, an increased intracranial pressure occurs relatively late (Fig. 3-113).

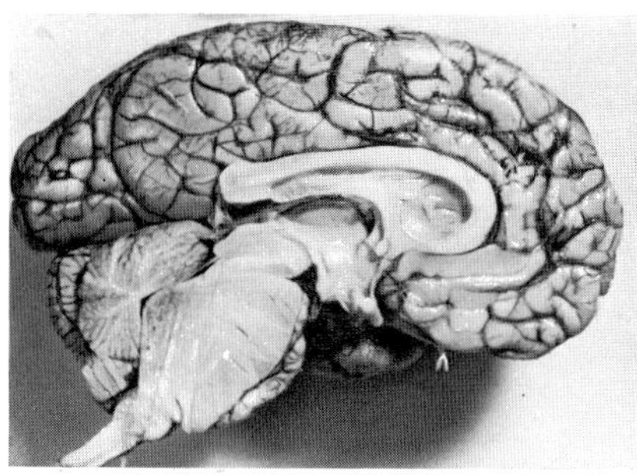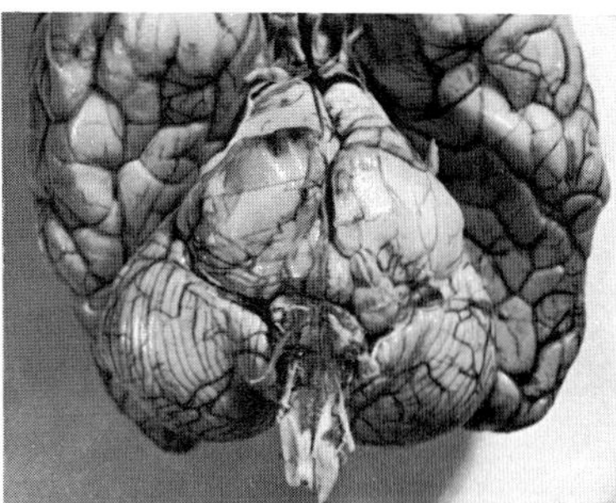

Fig. 3-113. Glioblastoma multiforme of the pons. Left, Median section through the brain. Right, Tumor seen from base of the skull. (Specimen from the Pathologic-Anatomic Institute of the University of Zurich.)

The ocular symptomatology of tumors of the pons and the medulla oblongata is mostly distinguished by *nuclear and supranuclear disturbances of the eye movements, especially of abduction and horizontal conjugate movements.* Thirteen of our cases were also reviewed by Kaufmann. In addition to vomiting, *disturbances of ocular motility are among the very first signs of a tumor of the pons or the medulla oblongata.* Next in frequency are disturbances of the equilibrium.

Quite frequently the history reveals an annoying and confusing horizontal *diplopia.* It is due to a nuclear, usually ipsilateral sixth nerve paresis (in seven of our thirteen patients). Only in three patients (and in the late stages) a paresis of isolated muscles supplied by the oculomotor nerve could be demonstrated in addition to a sixth nerve paresis.

Of greater localizing value than sixth nerve pareses are *horizontal gaze palsies,* which occurred in one half of our patients: three times in unilateral form (the globes could not be rotated past the midline) and three times in bilateral form (the globes remained in a primary parallel or convergent position) (Fig. 3-110). The anatomic basis is a *lesion in the center for conjugate lateral movements in the pons* (more exactly, the gray matter of the medulla somewhere dorsal to the pons and ventral to the fourth ventricle) *or in the adjacent medial longitudinal fascicle connecting the various oculomotor nuclei* (Fig. 1-11). It is characteristic for horizontal gaze palsies associated with lesions in the pons (Santha) that all the various forms of gaze movements (that is, schematic, pursuit, optically elicited, and vestibular movements) are involved. A unilateral horizontal gaze palsy is toward the side of the lesion. If there is a *compensatory deviation of the eyes, it is toward the opposite side.* We have observed one such case. Characteristically, compensatory conjugate deviation associated with pontine tumors is permanent in contrast to the deviation associated with cortical or subcortical lesions in the frontal optomotor centers of the hemispheres and their descending tracts (where we have a relatively rapid compensation—apparently from the contralateral side). Likewise, a horizontal gaze palsy caused by a lesion of the frontopontine tracts between the cortex and the pons is frequently marked by a turning of the head toward the side of the lesion. This phenomenon is missing in patients with pontine horizontal gaze palsies. Generally, in patients with tumors of the pons, *only the horizontal conjugate eye movements are paralyzed. Vertical gaze palsies are not part of this picture but are a distinct sign of a lesion in the midbrain* (p. 260). Only in one patient could we demonstrate a vertical gaze palsy in addition to a horizontal gaze lesion. Obviously, this was a sign of expansion of the tumor toward the region of the midbrain or, at least, of a remote effect.

Rarely, only one side of the pons is affected by a tumor: in such cases, sixth nerve palsy and homolateral peripheral facial palsy may be associated with a crossed hemiplegia (Millard-Gubler syndrome). If there is in addition a homolateral gaze palsy (to the side of the lesion), the *syndrome is that of Foville-Millard-Gubler.*

A lateral lesion of the pontine tegmentum may lead to the phenomenon of the pontine "skew deviation," which is characterized by a divergence of the eyes in the vertical plane with the eye on the side of the lesion lower than the other eye. Sometimes the lower eye shows intorsion, the higher one extorsion. Such a skew deviation in association with a pontine tumor must not be confused with an isolated palsy of a vertically acting single ocular muscle! Allerand reports paroxysmal skew deviation in a case of unilateral astrocytoma of the pons.

Hoyt describes still another motility disturbance in patients with metastatic tumors of the pons: *ocular bobbing,* characterized by jerky vertical conjugate movements of the eyes (downward movement more rapid than upward). Conjugate lateral gaze is abolished simultaneously with such bobbing of the eyes.

Next to horizontal gaze palsies, *central nystagmus* is an almost *constant sign* in patients with tumors of the pons. We found it recorded in twelve of thirteen cases of pontine tumors. Mostly *it is horizontal* (seven cases), frequently a combination of horizontal and vertical (four cases), and very rarely purely vertical (one case). In five patients, nystagmus manifested itself in combination with a horizontal gaze palsy. Most authors explain the vertical nystagmus as a pressure effect on the region of the quadrigeminate plate (respectively, on the midbrain tegmentum). The horizontal nystagmus, which is of the central type, must be interpreted as due to a lesion of the vestibular nuclei or their connections with the cerebellum or other brain stem structures (medial longitudinal fascicle or para-abducens reticular formation). Kaufmann is correct in stressing a relationship between horizontal gaze palsies and horizontal nystagmus. He considers them similar phenomena, *the nystagmus representing a quasi incomplete form of an impaired supranuclear gaze movement.* On this basis, such disturbances of the horizontal motility could be demonstrated in all thirteen patients in our series —eight times in bilateral form.

The *nystagmus* observed in association with tumors of the pons also belongs to the *central type* (p. 54). As mentioned previously, it is horizontal, or horizontal-vertical, with no rotary component. Characteristic for this "pontine nystagmus" is its asymmetry, which we could observe almost invariably among our patients (asymmetrical gaze nystagmus). With the eyes straight, there usually is no nystagmus. It increases in intensity as the eyes are turned toward the side of the lesion. If the eyes are turned toward the opposite side, they have to be moved for some distance before the nystagmus becomes manifest, and then only to a lesser extent. The asymmetrical gaze nystagmus is pathognomonic for an intrapontine lesion (with an involvement of the ipsilateral medial longitudinal fascicle).

An involvement of the *trigeminal nerve with impairment of the corneal reflex* could be observed in nine of our thirteen patients. The corneal reflex was either diminished or completely abolished. In one half of the patients this defect was unilateral and in the other half bilateral.

As mentioned already, *the late appearance of signs of increased intracranial pressure is characteristic for pontine tumors.* Papilledema was rare among our patients. We have seen it in only four of thirteen patients. It is important to state that three of these four patients had pontine tumors with a pronounced expansion into the fourth ventricle.

The ocular symptomatology of tumors of the pons and the medulla oblongata can be summarized as follows. Dominating the picture are the characteristic supranuclear horizontal gaze palsies. They must be regarded as an extraordinarily important localizing sign. In almost all instances there is a horizontal jerk nystagmus of the central type—either as an isolated sign or combined with a horizontal gaze palsy. Occasionally this nystagmus is horizontal-vertical. In addition to these supranuclear motility disturbances, we frequently find isolated, mostly unilateral, nuclear sixth nerve pareses. Pareses of the oculomotor nerve are seen rarely. A trigeminal lesion, which occurs frequently with tumors of this area, manifests itself in the form of a diminished or abolished corneal reflex. Choked discs are extremely rare in association with true pontine tumors. They appear late, and mostly if the tumor expands toward the fourth ventricle, causing embarrassment of the circulation of the cerebrospinal fluid.

CHAPTER FOUR *Relationship between type of tumor and ocular symptoms*

Our diagnostic endeavors are, on the one hand, to determine the *location* of a brain tumor and, on the other hand, to diagnose, if at all possible, the *type of tumor* before surgical intervention. On the basis of the general neurologic symptomatology and with the help of numerous methods of examination (pneumoencephalography, ventriculography, cerebral angiography, electroencephalography, neuroradiology, echo-encephalography, and others), this can be frequently accomplished in quite a satisfactory manner. It should be stated at the outset of this brief chapter that ocular symptoms occasionally may help to confirm the diagnosis of the type of tumor but that, as a rule, such symptoms alone can never establish such a diagnosis. This is particularly true for tumors of the hemispheres, but it is also true for those of the brain stem and the cerebellum. The situation is somewhat different for tumors of the sellar region and its surroundings. As discussed in the preceding chapter (pp. 156 to 270), it is actually possible in certain cases to determine the type of tumor (such as a pituitary adenoma, a craniopharyngioma, a meningioma of the sphenoid ridge, and others) from a more detailed analysis of the ocular symptoms. Even here, such a diagnosis cannot be accomplished with absolute certainty, in spite of our detailed and discriminate neuro-ophthalmologic diagnostic technique. The following statements will be limited to some general remarks that are not particularly aimed at diagnosing the type of tumor from the ocular symptoms but that merely stress the relationship between the type of tumor and the development of the ocular symptoms. Tumors with characteristic and localizing neuro-ophthalmologic symptoms (such as pituitary adenomas, craniopharyngiomas, gliomas of the chiasm, meningiomas of the sphenoid ridge, acoustic neuromas, etc.) will not be considered or discussed further.

One of the most frequent types of brain tumor is *glioblastoma multiforme* (Fig. 3-113). It is a highly malignant tumor noted for its rapid growth. It occurs in the hemispheres as well as in the basal ganglia and brain stem. The rapid and expansive growth of this tumor produces an alarming cerebral picture in a relatively short time. It is significant that the rapidly developing neurologic local symptoms are complicated by early psychic disturbances in the form of a psycho-organic syndrome. Sensory disturbances appear soon and progress from drowsiness to somnolence and coma. Statistical evaluation of our material shows that papilledemas are observed in about 50% of patients with glioblastoma, in contrast to about 65% of those with more slowly growing tumors (for instance, meningiomas). Although the difference is not significant, it seems that papilledemas in patients with rapidly growing tumors (such as, for instance, glioblastoma multiforme) occur somewhat less frequently than in those with slowly growing tumors. In the differential diagnosis of glioblastomas one must primarily consider subdural and epidural hematomas, brain abscesses, and cerebral metastases. They may produce a quite similar picture.

Resembling glioblastoma multiforme in malignancy are *medulloblastomas,* the prototype of childhood tumors, which are located predominantly in the cerebellar vermis (sometimes also in the cerebellar hemisphere and the pons). In the discussion of cerebellar tumors (p. 265) it has been mentioned that medulloblastomas have a striking tendency to invade the fourth ventricle and thus to cause an internal hydrocephalus. Thus symptoms of an increased intracranial pressure dominate the picture in the early stages of the disease: headaches, vomiting, the early appearance of pronounced papilledemas, and unilateral or bilateral sixth nerve palsies. In the latter case even a direct pressure effect on the nuclear region may be assumed, especially with an associated peripheral facial palsy. Occasionally medulloblastomas form metastases (for instance, in the spinal cord or the chiasm). They may then cause a chiasmal syndrome and eventually blindness. A chiasmal syndrome may also result from medulloblastomas, with their tendency to embarrass the circulation of the cerebrospinal fluid via an early dilation of the third ventricle with compression of the chiasm (p. 253 and Fig. 3-103).

The malignant and rapidly growing glioblastomas and medulloblastomas contrast with the relatively slowly progressing *astrocytomas,* the favorite localization sites of which are the hemispheres, the brain stem, and the cerebellum (also called cerebellar spongioblastoma). In accordance with the slow growth of astrocytomas of the hemispheres (Fig. 3-8), the neurologic symptoms dominate the picture. In contrast to glioblastomas, psychic changes are of rather secondary importance. It seems that, just as with meningiomas, papilledemas occur somewhat more frequently in patients with astrocytomas of the hemispheres than in patients with rapidly growing tumors (for instance, glioblastomas). The situation is, however, different with astrocytomas of the cerebellar hemispheres (Fig. 3-101). They cause rapid development of an internal hydrocephalus and, consequently, early increased intracranial pressure with pronounced papilledemas, which soon become atrophic. Not all, but most of our patients with cerebellar astrocytomas show papilledemas. *Oligodendrogliomas* are also slowly growing tumors that are localized especially in the hemispheres (frontal lobe), the basal ganglia, and in young patients in the thalamus. Apart from the local neurologic symptoms, they frequently produce epileptic attacks, which may be their only symptoms for a long time.

In the discussion of cerebellar tumors (p. 265) we mentioned the interesting relationship between cerebellar hemangioblastomas and retinal angiomas in *von Hippel–Lindau disease* (Fig. 3-107). In our series we have three patients with this combination of angiomatosis retinae and cerebellar hemangioblastoma. This relationship is quite loose. Nevertheless, one should keep in mind the possibility of a cerebral hemangioma in patients with cerebellar symptoms who show a retinal angioma. More rarely, *hemangioblastomas* occur also in the pons and the medulla oblongata.

Regarding pituitary adenomas, craniopharyngiomas, and acoustic neuromas, refer to the appropriate sections in Chapter 3 (pp. 189, 225, and 274). These tumors show a quite specific neuro-ophthalmologic symptomatology which, to some extent, is of localizing value.

Among the tumors with extracerebral development are the *meningiomas*, some of which (suprasellar meningiomas, meningiomas of the olfactory groove, meningiomas of the sphenoid ridge, meningiomas of the falx, parasagittal meningiomas) have also been discussed in Chapter 3 (pp. 187, 218, 225, and 231). These tumors are noted for their extraordinarily slow growth, which may extend over many years. Their clinical symptomatology shows a correspondingly slow and gradual progress. Signs of increased intracranial pressure as well as psychic changes and sensory disturbances may be absent for a long time. It is important for the general practitioner and the ophthalmologist to remember that meningiomas show hyperostoses, bone erosions, or actual tumor calcifications on a flat x-ray film of the skull. It has been stressed in Chapter 3 (p. 187) that certain meningiomas (for instance, those of the parasagittal region) may cause misleading visual field defects due to a remote effect that is still not properly understood (Klingler and Condrau). Possibly this results from the slow growth of these tumors or possibly from their extracerebral expansion, which does not involve actual brain substance but merely causes its displacement (Fig. 3-27).

Special forms of tumor that have to be considered frequently are cerebral *metastatic carcinoma* and, more rarely, cerebral *metastatic sarcoma* (Dandy; Livingston, Horrax, and Sachs; Scheid). Unless there are multiple, disseminated foci, the neurologic symptomatology of cerebral metastases does not differ from that of primary brain tumors. Frequently proof of the primary tumor is not possible. The following criteria suggest a metastasis to the brain: rapid development of cerebral symptoms, a disproportionate preponderance of psychic disturbances, a random appearance of new neurologic focal symptoms pointing to a site different from the original focus, a marked increase in the sedimentation rate, and general cachexia (Minkowski). However, in the differential diagnosis one must always consider a chronic subdural hematoma, a rapidly growing glioblastoma, and a brain abscess. Only proof of the primary tumor can make the diagnosis certain. Ocular signs of increased intracranial pressure (that is, papilledemas) are frequent but by no means characteristic. Nonspecific sixth nerve pareses caused by increased intracranial pressure seem to be

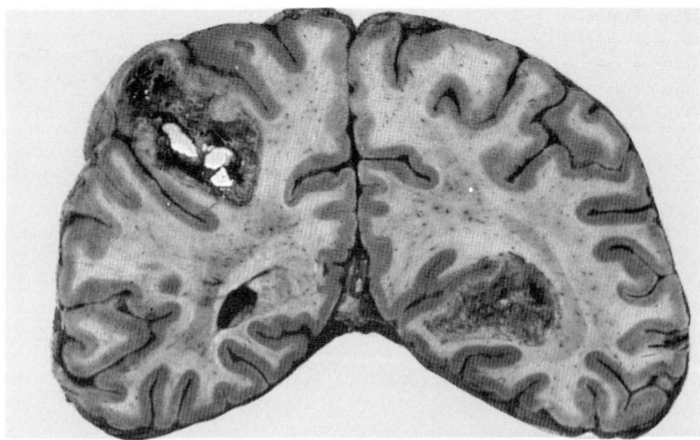

Fig. 4-1. Multiple metastases to the brain from small cell carcinoma of the left lung in a 50-year-old patient. (Specimen from the Pathologic-Anatomic Institute of the University of Lausanne.)

exceedingly rare in association with metastatic carcinomas of the brain (Minkowski). In men the primary tumor is mostly a carcinoma of the lung; in women, carcinoma of the breast. Much rarer are hypernephromas, melanomas, malignant tumors of the thyroid, and carcinomas of the prostate. According to Weber, cerebral metastases occur far more frequently in men than in women. Usually they are seated in the hemispheres immediately underneath the cortex (Fig. 4-1).

In comparison to these tumors, *teratomas, cholesteatomas, dermoid cysts, parasites,* and *granulomas* are less frequently located in the brain. Their symptomatology is not specific but depends largely on the site of the growth. Granulomas of the brain are almost always tuberculous, rarely syphilitic. The picture of a cerebral *tuberculoma* differs very little from other brain tumors. Febrile or subfebrile temperatures and tuberculous foci in other organs are of little diagnostic help. Brain tubercles usually are multiple but may occur in solitary form. They are said to occur mostly in the cerebellum but may occur in the hemispheres. In the latter location they rarely cause symptoms of intracranial pressure. Very rarely a solitary tubercle may develop in the chiasm or optic nerve. This is demonstrated once in our series, a patient with a tubercle in the extraorbital portion of the optic nerve causing ipsilateral amaurosis and contralateral temporal hemianopia (Fig. 3-88).

It should be worthwhile to discuss once more bilateral *papilledema and its relationship to the site and type of tumor* (see Table 3, p. 92). Tumors of the third and fourth ventricle as well as of the cerebellum most frequently cause papilledema, whereas tumors of the area of the sella produce papilledema least often. Papilledema is also relatively infrequently produced by tumors of the base of the skull, the pontine angle, and the brain stem. Among the tumors of the hemisphere, those of the occipital lobe produce papilledema most frequently, then follow those of the temporal lobe, and least frequently those

Table 5. Distribution of cases with primary optic atrophy by site and type of brain tumors (according to Tönnis)*

Type of tumor	Frontal	Temporal	Corpus callosum	Brain stem	Base	Pontine angle	Proximity of sella	Vessels	Total
Spongioblastoma	1			14			19		34
Oligodendroglioma	1	1					1		3
Astrocytoma		1		2					3
Ependymoma	1			3	1		3		8
Neuroma						3			3
Meningioma					35	1	26		62
Sarcoma					2				2
Pituitary adenoma							234		234
Cranopharyngioma							64		64
Epidermoid			1	1	1	1	5		9
Aneurysm								9	9
Metastases					1				1
Abscesses	1	1		1					3
Other tumors			2	4	7		1		14
Total	4	3	3	25	47	5	353	9	449 = 14.8%

*From a total of 3033 cases of intracranial tumors.

of the parietal lobe. Table 3 also shows the *difference in frequency of papilledema produced by various types of tumors.* Medulloblastoma most often produces papilledema, and this type of tumor is frequently found in the cerebellum. Then follow meningiomas, which also produce secondary atrophy, and astrocytomas and oligodendrogliomas. Rapidly growing glioblastomas produced papilledema in only 50% of the patients in our series (this differs markedly from the results of other authors, for instance, Tönnis who found a frequency of 80%). Neuromas occupy a position in the middle (75% according to Tönnis).

Approximately one sixth of all brain tumors produce primary optic atrophy. According to Table 5, which was calculated by Tönnis, we find that the tumors close to the sella and the tumors of the base of the skull most frequently produce such an atrophy. Pituitary adenomas, craniopharyngiomas, and spongioblastomas of the brain stem and hypothalamus are the types of tumors that most often produce primary optic atrophy. Tumors of the frontal and temporal lobe also have to be considered. Always, however, optic atrophy is the result of a direct compression of the optic nerve either by the tumor itself or by other tissues in the area of the basal cisterns that have been shifted.

Finally, we should consider those brain tumors that do not cause ocular involvement (9.9%). Again, we would like to refer to a table calculated by Tönnis (Table 6). Knowledge of such tumors and the site of such lesions is of greater importance than may be apparent on first glance. Such tumors are not diagnosed for a considerable period of time. These patients suffer from seizures, while symptoms of increased intracranial pressure occur much later. Among these tumors are especially neoplasms of the vessels and secondly tumors of the hemispheres, mostly of the frontal and the parietal lobes. Tumors of the cerebellum

Table 6. Cases without ocular involvement (according to Tönnis)[*]

	Number of cases
Cerebral tumors	170
Frontal	84
Temporal	20
Parietal	59
Occipital	3
Corpus callosum and lateral ventricles	4
Cerebellar tumors	7
Brain stem tumors	4
Tumors of base of brain	1
Pontine angle tumors	15
Tumors close to sella	37
Multiple tumors	
Vessels	67
Total	301 = 9.9%

[*]From a total of 3033 cases of intracranial tumors.

and the brain stem will only occasionally develop without affecting the visual system. The same is true, perhaps to a somewhat lesser extent, for tumors of the area of the sella and the pontine angles.

Following the discussion of the various types of tumors, it might be of interest to learn something about their relative distribution. Weber has compiled the following data based on 1645 cases at the Neurosurgical Clinic of the University of Zurich.

Gliomas and paragliomas		842
Angiogliomas	9	
Astroblastomas	65	
Astrocytomas (hemispheres)	139	
Astrocytomas (cerebellum, pons, midbrain)	67	
Ependymomas	26	
Glioblastomas multiforme	318	
Gliomas (unclassified)	92	
Medulloblastomas	70	
Oligodendrogliomas	28	
Pinealomas	10	
Spongioblastomas	15	
Sympathicoblastoma	1	
Gangliocytoneuromas	2	
Meningiomas		253
Metastases		135
Pituitary adenomas		95
Craniopharyngiomas		44
Neuromas (108 of acoustic nerve)		110
Hemangiomas (all cerebellar)		39
Papillomas		10
Teratomas, cholesteatomas, dermoid cysts		16
Chordomas and chondromyxomas		8
Other rare forms (reticuloma, colloid cyst, lymphoendothelioma)		3
Brain abscesses		60
Granulomas (20 tuberculous, 1 gumma, 6 nonspecific)		27
Parasites		3
Total		1645

 Pseudotumor cerebri, subdural hematomas, brain abscesses, and aneurysms

It was stated in the introduction to this book that subdural hematomas, brain abscesses, and aneurysms are not brain tumors in the true sense of the word. However, it would be a mistake to exclude them here because they play an important part in the differential diagnosis of brain tumors. Whereas the general neurologic symptoms are preeminent in patients with subdural hematomas and brain abscesses, the situation is quite different with some cerebral aneurysms. At times a topical diagnosis can be made merely on the basis of ocular signs. In this connection we would like to call attention also to the peculiar self-limited disease *pseudotumor cerebri* that sometimes simulates a cerebral neoplasm.

PSEUDOTUMOR CEREBRI

The term "pseudotumor cerebri" is used for those somewhat obscure cases in which the patients have bilateral papilledema and increased intracranial pressure but negative neurologic and general physical findings. Also characteristic are the normal-sized or small ventricles found on ventriculography (therefore no signs of obstructing hydrocephalus). A multitude of other names has been used for this syndrome, among which are "benign intracranial hypertension," "serous meningitis," "meningeal hydrops," and "otitic hydrocephalus." This disease occurs in young to middle-aged adults (especially in obese young women) and is characterized by headaches, sixth nerve palsy, papilledema, occasionally field defects, and rarely blindness. The affection lasts for weeks or sometimes for months and generally has a good prognosis as long as damage does not result from atrophy secondary to the papilledema. The pathogenesis of the syndrome is either a hypersecretion or an obstruction of the outflow of cerebrospinal fluid (combined with swelling of the brain). There are indications, at least in some of the patients, that a thrombosis of the sagittal or lateral sinuses is involved. Otitis media, especially in children, may be the preceding cause of such a thrombosis. The pseudotumor syndrome may be observed also at the time of menarche, during pregnancy, in Addison's disease (Walsh), after prolonged corticosteroid therapy, after vitamin A intoxication, and following the use of oral contraceptives (Walsh, Clark, Thompson, and Nicholsen) (Fig. 5-1).

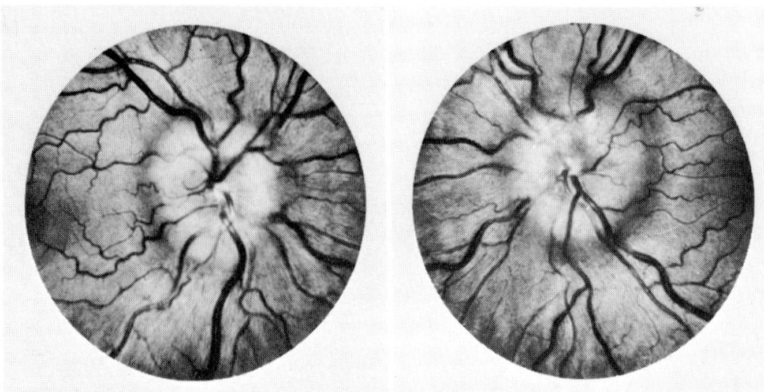

Fig. 5-1. Papilledema in a 35-year-old patient with cerebral pseudotumor caused by use of contraceptive pills. Distinct increase of diameter of the disc, blurred margins, mushroom-like prominence of about 2 diopters, and slight venous engorgement. Visual acuity, O.U. 1.0. Enlargement of both blind spots. No visual field defects. Increase of cerebrospinal fluid pressure up to 250 to 300 mm. H_2O.

SUBDURAL HEMATOMAS

Although the clinical picture of *chronic subdural hematoma* (Figs. 5-2 and 5-3) varies a great deal, certain general guiding principles can be stated. Following a more or less pronounced trauma to the head (which may not always be mentioned spontaneously in the history), there are attacks of headache, which increase in frequency and during later stages are associated with psychic disturbances and personality changes. These are memory disturbances, lassitude, apathy, loss of energy, later sensory disturbances and disorientation, and finally coma. The clinical picture may fluctuate. Improvements may alternate with relapses. There is a striking discrepancy between the severity of the general signs of increased intracranial pressure and the insignificance of focal symptoms. As a rule, the interval between trauma and onset of the alarming symptoms of the disease lasts from 4 to 8 weeks. In the differential diagnosis all types of brain tumors (in particular, glioblastoma multiforme and brain abscess), encephalitis, and cerebral arteriosclerosis must be considered.

In addition to traumatic chronic subdural hematoma (caused by the rupture of bridging veins traversing the subdural space from the cerebral cortex to the undersurface of the dura), there is also spontaneous subdural hematoma. It is due either to an *idiopathic internal hemorrhagic pachymeningitis* (capillary proliferation of a telangiectatic type) or to an *inflammatory internal hemorrhagic pachymeningitis.* The symptomatology of the traumatic and the spontaneous types of chronic subdural hematoma is alike, particularly since there are mixed or transitional forms.

Subdural hematoma of early childhood usually is the result of trauma, especially a severe birth trauma. Generalized epileptic attacks with unconsciousness, vomiting, irritability, restlessness, and stupor are outstanding signs. In addition to defective physical and mental development, there is an abnormal increase in the size of the head, with marked bulging of the large fontanel and suture ruptures. In about one fourth of the patients, ophthalmoscopic examination reveals diffuse retinal hemorrhages but rarely secondary optic atrophy or papilledema.

The following statements concerning neuro-ophthalmologic symptoms associated with *chronic subdural hematoma* are based on fifty cases from the Neurosurgical Clinic of the University of Zurich (Krayenbühl and Noto). Only eleven of the fifty patients showed *papilledemas,* which were in no way

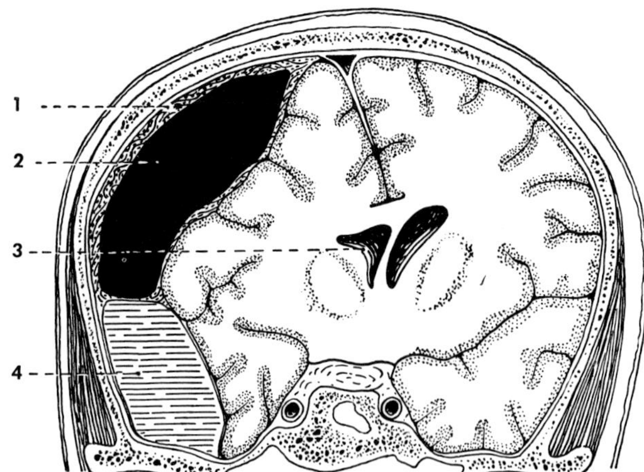

Fig. 5-2. Diagrammatic illustration (frontal section) of a chronic subdural hematoma and hydroma with compression of the involved part of the brain and displacement of the lateral ventricles. (After Dandy.)

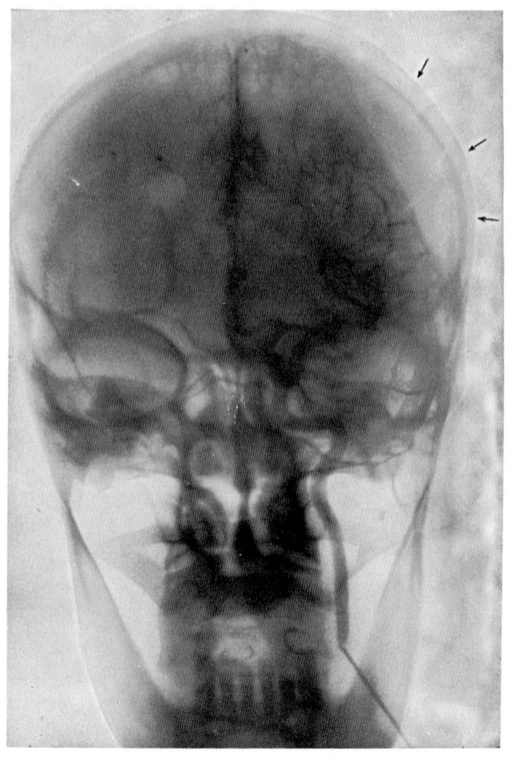

Fig. 5-3. Left cerebral arteriogram of a 36-year-old patient with bilateral chronic subdural hematoma of traumatic origin. Anteroposterior exposure. Massive displacement of the left Sylvian vessels away from the cranial vault. Arteriography renders the displaced surface of the hemisphere visible as an oblique line (cortical arteries and veins mark the surface of the brain). Contents of left hematoma, 100 ml.!

characteristic. The elevation may vary in degree from slight blurring to fully developed and markedly prominent choked discs. Occasionally the ipsilateral choked disc is more pronounced than the one on the opposite side. However, this sign is not dependable for purposes of localization (p. 90). It is interesting that even subdural hematomas, in certain localizations, can produce *visual field defects* (Govan and Walsh; Maltby). Among our own series we observed contralateral homonymous hemianopia in one patient with a parieto-occipital hematoma and quadrantanopia in two patients with the hematomas in the frontal and frontotemporal positions. However, such cases concern very large space-consuming lesions. The homonymous field defects may be explained on the basis of disorders of circulation within the posterior cerebral artery produced by the herniation of the hippocampal gyrus through the tentorium notch (see section on clivus ridge syndrome, p. 152).

Of some localizing significance are *pupillary changes,* most of all the *unilateral mydriasis and light rigidity which, as a rule, occur on the ipsilateral side* (de Quervain). This must be considered as a lesion of the peripheral oculomotor nerve, which is pressed against the ipsilateral clivus ridge by protruding portions of the brain. The damage to the oculomotor nerve forms part of the *clivus ridge syndrome* described by Fischer-Brügge (see p. 152 and Fig. 2-46). The value of a unilateral mydriasis and light rigidity is diminished by its relatively rare occurrence (in Krayenbühl's series of fifty patients there were only three cases). More frequent and thus of greater diagnostic value is a simple *anisocoria,* which occurs in about one half of the patients. The side of the wider pupil, which still reacts to light, usually corresponds to the side of the hematoma (75% according to McKissock, Richardson, and Bloom). If both pupils are rigid and small, a bilateral hematoma should be considered according to Krayenbühl. Obviously this miosis is a sign of irritation to the oculomotor nerve (that is, the very early stage of a clivus ridge syndrome). A paralysis of the trochlear nerve (one case in fifty patients) as well as the sixth nerve (six cases in fifty patients) apparently is based on the same mechanism. These extrinsic eye muscle pareses account for the diplopia observed in about one third of the patients. It has been previously mentioned that *the clivus ridge syndrome is by no means pathognomonic for a subdural hematoma—it occurs also with tumors.* Conjugate deviations of the eyes away from the lesion, ptosis on the side opposite to the hematoma (Govan and Walsh), nystagmus, and even palsy of upward gaze have been described as additional disturbances of ocular motility.

Acute subdural hematoma, which usually is traumatic in origin, can be distinguished from the chronic type by its decidedly more violent symptoms. Because of the usually severe cerebral trauma, the patient is either unconscious from the outset or becomes progressively more somnolent. Only rarely is an initial brief unconsciousness interrupted by a brief lucid interval. Sooner or later the unconsciousness, accompanied by vomiting and headaches, sets in again and progresses to a coma (with decrease of pulse rate, rapid respiration, and eventually Cheyne-Stokes breathing).

Pupillary disturbances associated with *traumatic acute subdural hematoma* (Kennedy and Wortis) are more frequent and more pronounced. Anisocoria is found in up to 70% of the patients. Mydriasis of one pupil is usually associated with light rigidity and is generally on the side of the hematoma. Bilateral light rigidity of the pupils is an unfavorable sign prognostically. It is due either to a bilateral clivus ridge reaction or to a lesion of the brain stem. In the latter case it may be combined with extensor spasms. A complete ophthalmoplegia is occasionally observed.

Similar ocular symptoms are seen in association with *epidural hematomas,* which usually are the sequel of a traumatic rupture of the middle meningeal artery. The trauma usually causes a skull fracture. After a brief lucid interval (with or without preceding unconsciousness immediately following the trauma), signs of rapidly increasing intracranial pressure become apparent which, without neurosurgical intervention, result in coma and eventual death. An ipsilateral mydriasis and light rigidity as signs of a clivus ridge syndrome do occur here too. Anisocoria is even more frequent than with acute subdural hematomas. Its localizing value is diminished by the fact that the mydriasis, which is usually ipsilateral to the hematoma, occasionally is on the opposite side in case of simultaneous edema of the brain on the other side. Pressure on the cavernous sinus may cause congestion of the ipsilateral globe as well as pareses or paralyses of isolated extraocular muscles. Retinal hemorrhages can occur if there is a simultaneous subarachnoidal hemorrhage (Woodhall, Devine, and Hart).

BRAIN ABSCESSES

During the *acute stage, signs of a systemic infection* (such as high fever, somnolence to coma, leukocytosis, increased sedimentation rate) and perhaps some focal signs dominate the picture. During the *chronic* stage the symptomatology is less specific and could be mistaken for that of a subdural hematoma or any brain tumor, especially glioblastoma multiforme. As a rule, the general condition of the patient is poor. Leukocytosis and a toxic count, an increased sedimentation rate, and bradycardia suggest an abscess. Blood-borne metastatic brain abscesses favor the hemispheres. Dissemination into the basal ganglia, the brain stem, or the cerebellum is rare. If the abscess is the result of an otitis media, it usually locates in the temporal or occipital lobe or in the cerebellum (Fig. 5-4).

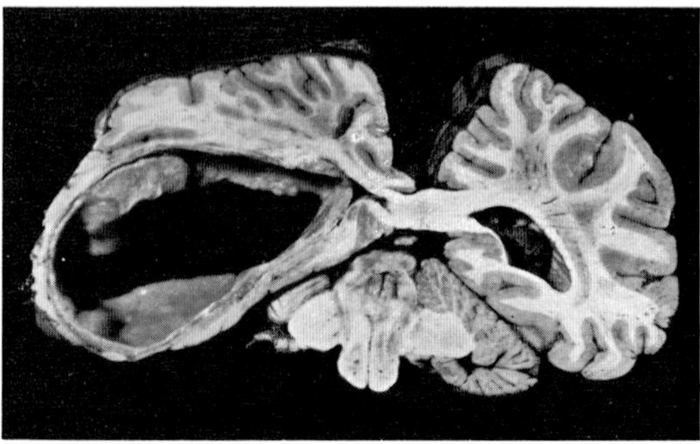

Fig. 5-4. Multiple brain abscesses in a patient with bronchiectases. Frontal section. A firm membranous wall surrounds the left large cavity of the abscess. (Specimen from the Pathologic-Anatomic Institute of the University of Zurich.)

The ocular symptomatology associated with brain abscesses varies according to the location of the abscess. In a frontal abscess, chemosis of the ipsilateral conjunctiva and exophthalmos with or without orbital infection are possible. Retrobulbar neuritis on the ipsilateral side and papilledema on the opposite side have been described several times in the literature as an atypical Foster Kennedy syndrome (Gros and Cazaban). Papilledema is particularly frequent and pronounced with an abscess in an occipital location. As a rule, it is not very characteristic (Lillie; Porto). There is no definite relationship between the size of the abscess and the severity of papilledema. Homonymous visual field defects are possible in an abscess with a temporal or occipital location. We have observed a patient with a chronic encapsulated abscess in the occipital lobe with complete homonymous hemianopia without sparing of the macula. The frequency of papilledema in patients with a brain process is quite similar to that in patients with brain tumors, depending more or less on the location of the abscess. Chronic brain abscesses produce papilledema more frequently than acute brain abscesses: Weber found choked discs in fifteen of twenty-three patients with chronic abscesses, but only in seven of twenty-six patients with acute abscesses. Papilledema usually is minimal or absent in patients with cerebellar abscesses (Colemann), because the gravity of the morbid abscess results in death or leads to surgical intervention before a true papilledema has a chance to develop. According to Colemann, there is no definite relationship between the side of the abscess and the extent of the swelling of the papilla.

It is frequently difficult to differentiate an acute brain abscess clinically from *a thrombophlebitis of the cortical veins or an intracranial sinus.* In the latter case there are frequently bilateral choked discs that are identical to those occurring in patients with a brain tumor or a brain abscess.

Ocular symptoms in patients with *subdural empyemas* and *abscesses* are not very characteristic and diagnostically are without much significance. They are quite similar to those associated with subdural hematoma (p. 290). What is more, subdural empyemas may originate from a metastatic infection of a chronic subdural hematoma (Fig. 5-5).

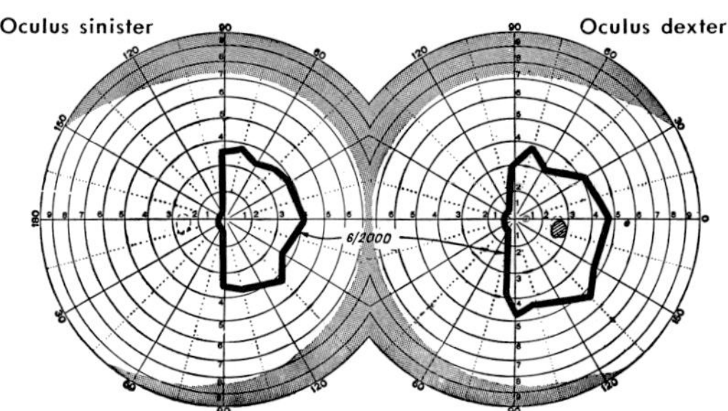

Fig. 5-5. Temporo-occipital subdural empyema on the right side of a 55-year-old patient. Hematogenous infected subdural hematoma. Left homonymous hemianopia with sparing of the macula. Visual acuity, O.D. 1.25; O.S. 0.9.

Extradural abscesses, originating most frequently from mastoiditis, may produce *Gradenigo's syndrome,* which is characterized by paresis of the homolateral sixth nerve, pain in the distribution of the fifth nerve (usually ophthalmic division), and sometimes ipsilateral corneal anesthesia (p. 21).

INTRACRANIAL ANEURYSMS

In certain tumor sites (for instance, the chiasmal region and the cavernous sinus), aneurysms always must be considered in the differential diagnosis in addition to true tumors. A large number of aneurysms are accompanied by typical ocular symptoms, which may be of localizing significance. Thus we feel it is important to discuss this form of "tumors" briefly. With due consideration to the extensive literature on the subject (Cushing and Bailey; Dandy; Jefferson; Walsh; and others), we plan to follow more or less the excellent monographs of Krayenbühl and Yasargil. At the same time, we want to consider some of our own cases.

In principle, two forms of aneurysms can be distinguished: (1) the *paralytic type* (increase in the size of the aneurysm and fine extravasations into the surrounding area result in pressure effects on certain cranial nerves and, con-

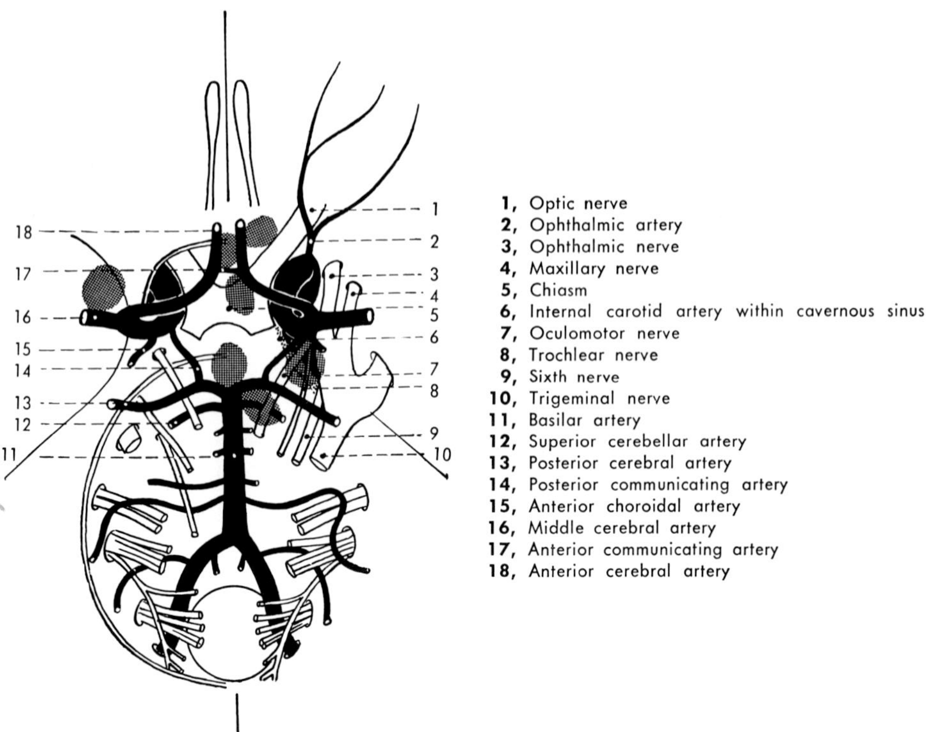

Fig. 5-6. Diagrammatic illustration of the base of the skull and position of the large cerebral arteries and cranial nerves. The shaded round areas indicate saccular cerebral aneurysms at typical areas of predilection for the arteries of the base of the brain (see pp. 297 to 302).

sequently, a neurologic syndrome similar to that of a brain tumor) and (2) the *apoplectic type* (as the name indicates, the apoplectic type is typified by the picture of a single or recurrent subarachnoid hemorrhage).

Paralytic type

It is the paralytic type of aneurysm which, due to the pressure effect on the cranial nerves (especially those of the extrinsic eye muscles), often gives rise to characteristic neuro-ophthalmologic syndromes that deserve consideration within the scope of our discussion (Fig. 5-6). The following groups of syndromes can be distinguished.

Aneurysms of the internal carotid artery within the cavernous sinus

In principle, the neurologic and especially the neuro-ophthalmologic signs of an aneurysm of the internal carotid artery within the cavernous sinus (also named intracavernous carotid aneurysm) correspond to those of tumors of the cavernous sinus, as described in detail in Chapter 3, p. 249. The more or less simultaneous involvement of the third, fourth, fifth, and sixth cranial nerves is characteristic for the *cavernous sinus syndrome.* According to Jefferson, posterior, middle, and anterior cavernous sinus syndromes can be distinguished depending on the impaired nerves (Fig. 3-91). The cavernous sinus syndrome due to an aneurysm sets in with *pain of the face (also behind the eye) or a sensation of numbness and is followed shortly by diplopia— with or without ptosis.* The pain usually is sharply demarcated in the area of distribution of the trigeminal nerve. In contrast to an ordinary trigeminal neuralgia, it is persistent rather than paroxysmal. In addition to these symptoms of irritation, it is always possible to demonstrate more or less distinct defects such as hypesthesias or hypalgesias in the area of distribution of the three branches of the trigeminal nerve. Of particular significance is the *corneal reflex: its decrease or loss may be the only trigeminal sign.* There is frequently a paresis of only the sixth nerve (Fig. 5-7) in association with the posterior cavernous sinus syndrome, whereas the oculomotor, trochlear, and sixth nerves may be involved simultaneously with the middle and anterior forms. *The oculomotor nerve is most frequently paralyzed.* Ptosis and internal ophthalmoplegia frequently are part of this characteristic picture (because of the simultaneous involvement of the oculosympathetic fibers surrounding the carotid artery, the pupil may be small, though fixed). A true exophthalmos is possible. Occasionally an aneurysm of unusual size and anterior extension may injure the ipsilateral optic nerve, causing atrophy with corresponding loss of vision and field defects. We have seen two such cases. During the further course of the disease, these signs of paralysis—after their sudden onset—may remain stationary or even may recede to a certain extent. This symptomatology actually is so typical that it almost permits the diagnosis. The final proof is the arteriographic demonstration of the aneurysm, which

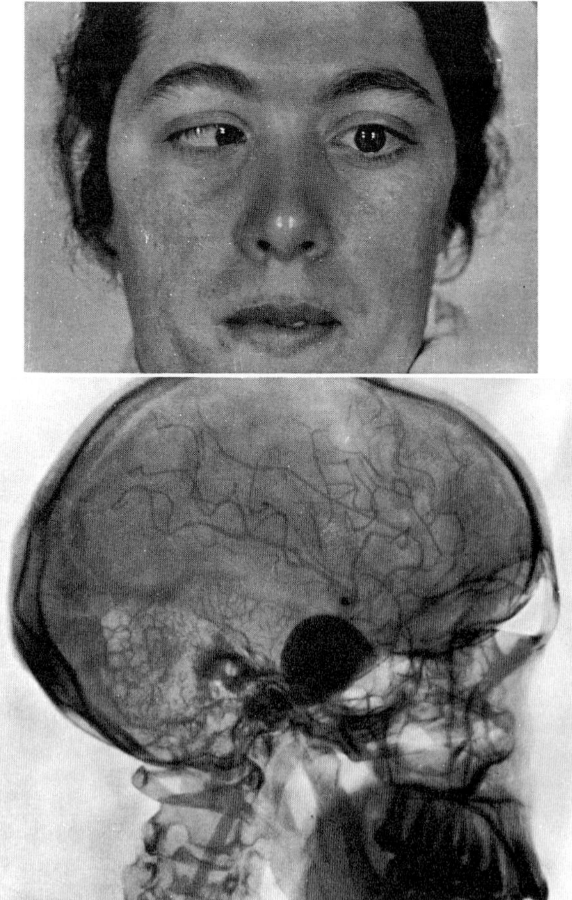

Fig. 5-7. Aneurysm of the left internal carotid artery within the cavernous sinus in a 21-year-old patient. Left sixth nerve paralysis. With the eyes turned left, the left globe cannot be rotated past the midline. Slight proptosis of the left eye. Distinct decrease of the left superior and inferior corneal reflexes. Below, Arteriogram (lateral view) after left cerebral arteriography. Enormous spherical aneurysm of the internal carotid artery within the cavernous sinus. (From Krayenbühl, H.: Schweiz. Arch. Neurol. Psychiat. 47:155, 1941.)

fails only in rare instances of complete thrombosis of the aneurysm. As previously mentioned, tumors in the region of the cavernous sinus always must be considered in the differential diagnosis.

Intrasellar aneurysms of the internal carotid artery

A break of an *intracavernous aneurysm of the internal carotid artery* (infraclinoid portion) through the inner wall of the cavernous sinus (with intrasellar expansion) may cause a more or less typical chiasmal syndrome. It results from direct compression of the chiasm by the aneurysm—usually more pronounced on one side, which causes a certain *asymmetry of the chiasmal*

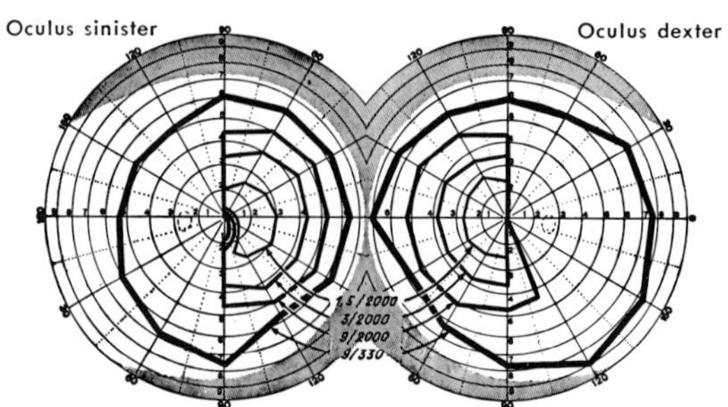

Fig. 5-8. Infraclinoid intrasellar aneurysm of the left internal carotid artery in a 36-year-old patient. Sudden onset of visual disturbances (at first on the left side, then on the right side) with formation of a slightly asymmetrical bitemporal hemianopia, involving especially the central isopters. Visual acuity, O.D. 0.3; O.S. 0.25. Temporal pallor of the left disc. Right disc normal. No x-ray changes evident in the sella. No endocrine disturbances. Aneurysm verified after surgical exploration of the chiasm. (From Krayenbühl, H.: Schweiz. Arch. Neurol. Psychiat. 47:155, 1941.)

syndrome (Jefferson). This is the reason why in such cases the bitemporal hemianopia is rather irregular and asymmetrical as compared with that in patients with pituitary tumors (p. 195 and Fig. 5-8). There are also reports of amaurosis of one eye with temporal hemianopia of the other eye. Frequently *the visual field defects vary in one and the same patient during different stages of the disease* (Klingler). Just as in patients with the chiasmal syndrome associated with pituitary tumors, there is a characteristic unilateral or bilateral loss of vision. With an aneurysm, however, it is greater in one eye. In patients with aneurysms it is particularly significant that the *visual disturbance often appears suddenly and is frequently associated with headaches. Sudden deterioration as well as improvement of vision is typical for an aneurysm.* In most patients, ophthalmoscopic examination reveals a bilateral primary optic atrophy with a sharp outline of the disc margin. As a rule, it is more distinct on one side.

The sudden appearance of a visual disturbance in the form of a chiasmal syndrome and associated with headaches or perhaps even with pareses or paralyses of extraocular muscles is highly suggestive of an intrasellar aneurysm. The absence of an enlarged sella on the x-ray film is important in differentiating an aneurysm from a pituitary tumor, although in the later stages destructive changes in the sella (even with enlargement) may occur. Arteriographic proof of an intrasellar aneurysm is not always possible. Definite proof is often obtained only on surgical exploration.

The other possible mechanism of a chiasmal syndrome is that of *suprasellar aneurysms* of the supraclinoid portion of the carotid artery. They originate from the anterior cerebral or the anterior communicating arteries (Fig. 5-9).

Aneurysms of the proximal segment of the internal carotid artery generally grow to a considerable size and may show extension toward the midline, with

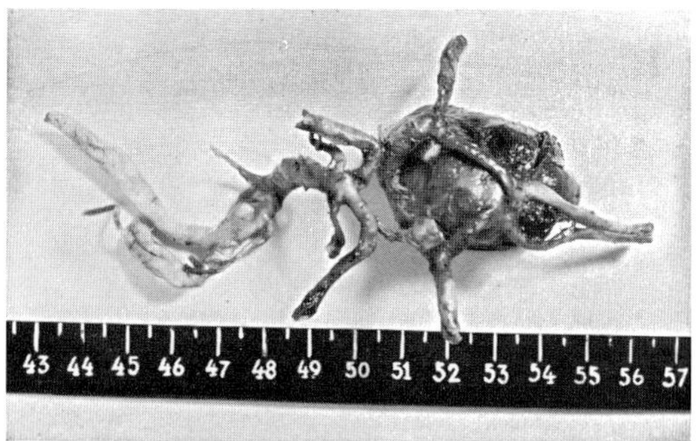

Fig. 5-9. Aneurysm of the anterior communicating artery. (Specimen from the Pathologic-Anatomic Institute of the University of Zurich.)

displacement of the chiasm laterally and upward. Therefore this form of aneurysm may also produce chiasmal syndromes. Ipsilateral blindness and temporal field defects in the contralateral eye are the most common findings.

Aneurysms at the internal carotid bifurcation, if large enough and if they extend anteriorly and medially, may involve the chiasm and produce the symptomatology of a suprasellar tumor with bitemporal field defects (Krayenbühl). Posterior and medial extension of such an aneurysm leads to compression of the optic tract, and posterior and inferior extension may involve the oculomotor nerve.

Intracranial aneurysms of the ophthalmic artery

These aneurysms, also termed supraclinoid or carotid-ophthalmic aneurysms, originate from the junction of the ophthalmic artery with the intracranial internal carotid artery. They are rare (Krayenbühl and Yasargil found only 7 such aneurysms among 290 intracranial aneurysms). Because of their site in the immediate neighborhood of the optic nerve, these aneurysms produce signs of optic nerve compression that may mimic the symptoms of retrobulbar neuritis. The usual symptomatology of intracranial aneurysm of the ophthalmic artery is a scotoma in one eye that progresses to a loss of the nasal field on the involved side. Later there are an upper temporal depression of the contralateral field and finally blindness on the side of the aneurysm. The temporal field defect in the contralateral eye is caused by the compression of the lateral side of the chiasm by the upward growing aneurysm. Erosion of the anterior clinoid and of the intracranial portion of the optic canal may be visible on x-ray films.

Aneurysms of the anterior cerebral and anterior communicating arteries

Although the majority of aneurysms of the anterior cerebral and anterior communicating arteries belong to the apoplectic type that leads to subarachnoidal hemorrhages, a paralytic form may occur in this location. Like the

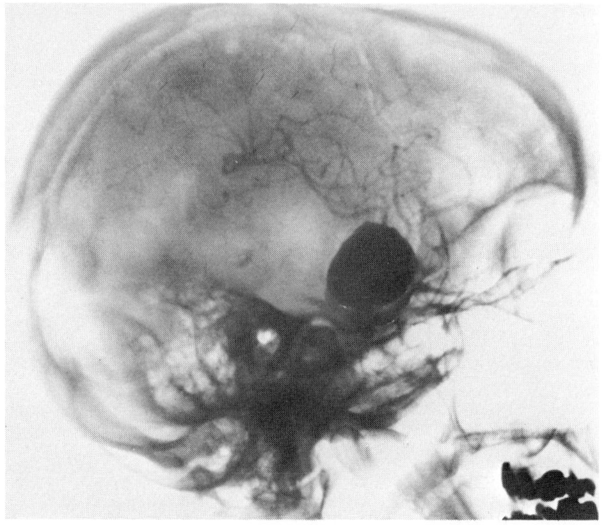

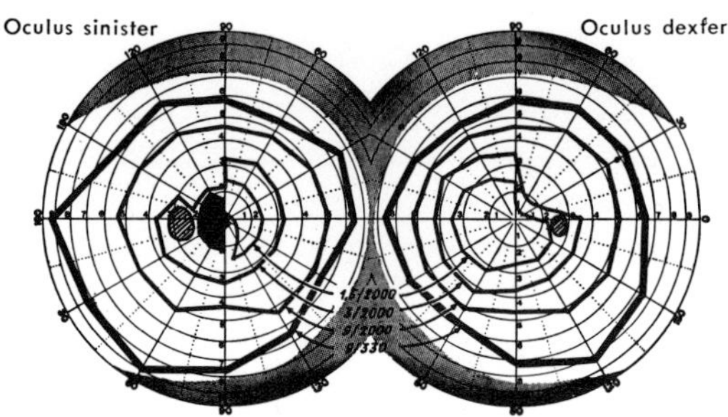

Fig. 5-10. Suprasellar aneurysm of the left anterior cerebral and anterior communicating arteries in a 46-year-old patient. Above, Arteriogram (lateral view) after left cerebral angiography. Walnut-sized saccular aneurysm of the anterior communicating artery at the sella entrance (giant aneurysm). Below, Bitemporal superior quadrantanopia with paracentral temporal scotoma of the hemianopic type of the left field. Visual acuity, O.D. 1.0; O.S. 0.05. Both discs unchanged. X-ray film of the skull normal. No endocrine disturbances. (From Krayenbühl, H.: Schweiz. Arch. Neurol. Psychiat. 47:155, 1941.)

intrasellar intracavernous aneurysm, this suprasellar form is marked by *a more or less atypical chiasmal syndrome. Sudden onset of bitemporal field defects with a certain preponderance on one side and fluctuations or actual improvement of impaired vision* are of diagnostic significance for such an aneurysm. According to Jefferson, onset of the visual field defects in both inferior temporal quadrants is suggestive of a suprasellar aneurysm, provided the other signs suggest an aneurysm. Sometimes the visual symptoms consist only in a relatively *acute loss of vision in one eye* preceding the subarachnoidal hemorrhage. In contrast to the intrasellar intracavernous aneurysm, the cranial nerves (except the olfactory nerve) remain intact (Fig. 5-10).

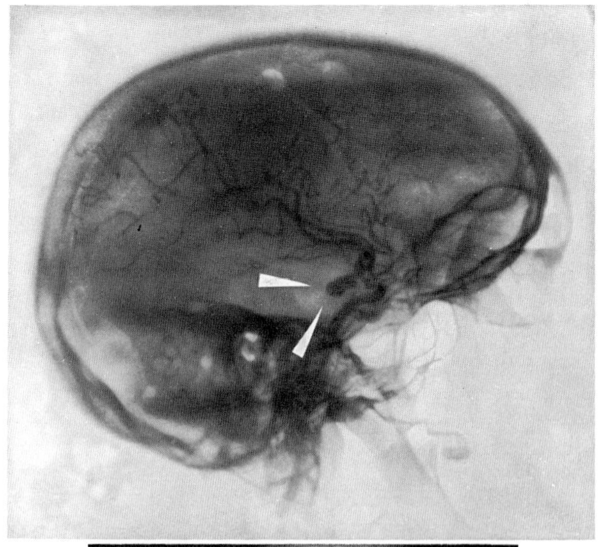

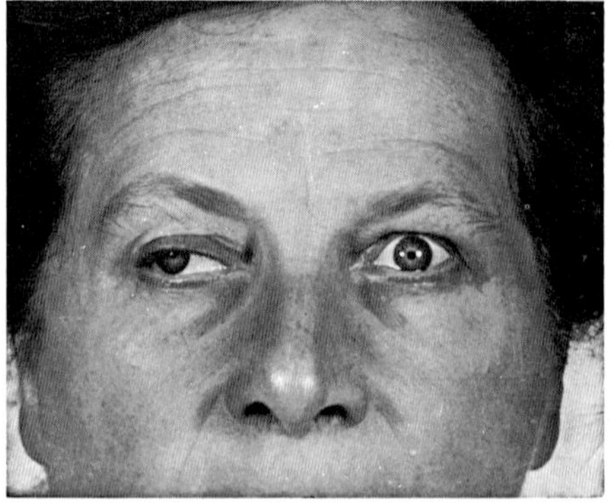

Fig. 5-11. Saccular aneurysm of the right carotid–posterior communicating artery junction in a 48-year-old patient. Above, Arteriogram (lateral view) after right cerebral angiography. Irregularly shaped lentil-sized aneurysm of the posterior communicating artery. Normal configuration of the vascular tree of the middle cerebral artery. Anterior and posterior cerebral arteries not filled. Below, Corresponding clinical picture. Right oculomotor paresis with partial ptosis. Right globe in divergent position.

Aneurysms of the carotid–posterior communicating artery junction

These aneurysms favor a location very close to or directly at the point where the posterior communicating artery branches off the internal carotid artery (Alpers and Schlezinger) (Fig. 5-11). Characteristic symptoms and signs are *sudden appearance of unilateral headache, pain in the forehead or the eye, and the simultaneous or slightly delayed appearance of a partial or complete oculomotor palsy with or without mydriasis and paresis of accommodation* (Jefferson). Usually the pupil is rigid to light and mydriatic. The intense unilateral headaches at the onset of the disease are localized in the forehead or the eye and must be regarded as a trigeminal disturbance in the area of distribution of the ophthalmic nerve. Objective sensibility disturbances can be demonstrated very rarely—mostly as a slight decrease of the ipsilateral superior corneal reflex. This form can be differentiated from an aneurysm of the internal carotid artery within the cavernous sinus by the minimal impairment of the trigeminal nerve and its limitation to the oculomotor nerve.

A sudden unilateral oculomotor paralysis (ptosis, exotropia, dilated, fixed pupil) associated with violent ipsilateral headaches, pain in the forehead, or pain in the eye should always arouse suspicion of an aneurysm of the carotid–posterior communicating artery junction. This is particularly true since an isolated peripheral oculomotor palsy (Fig. 1-8) is actually rarely associated with a tumor at the base of the skull—because of its expansion, it would involve other structures, especially other cranial nerves. Hoyt is right in this connection, emphasizing that in an isolated oculomotor palsy due to aneurysm the pupil is always involved and that "pupillary sparing" excludes the aneurysmatic etiology. In rare instances a peripheral oculomotor paralysis may be due to an aneurysm of the posterior part of the posterior communicating artery. It should be considered if the oculomotor palsy is not accompanied by trigeminal pain but by severe pain in the neck. It should be stressed here that *frequently the picture of the so-called ophthalmoplegic migraine may be simulated by such an aneurysm* (Sjöqvist; Frankel; Patrikios).

Aneurysms of the posterior cerebral artery

According to Jefferson, a thrombosis of the aneurysm and the posterior cerebral artery producing an ischemic infarct of the occipital lobe or a compression of the optic tract by the aneurysm may cause a *homonymous hemianopia.* Apart from homonymous field defects, *Weber's syndrome* (third nerve palsy and contralateral hemiplegia), similar to acute occlusive vascular disease, may occur as a focal sign before the rupture of the aneurysm (Jamieson; Hanafee and Jannetta). This type of aneurysm is extremely rare.

Aneurysms of the basilar and vertebral arteries

The clinical diagnosis of an aneurysm of the basilar artery is difficult. Frequently there are no symptoms. Occasionally there is the syndrome of an acute or chronic bulbar paralysis. Krayenbühl found a bilateral sixth nerve paresis in a patient with a vertebral aneurysm. The oculomotor nuclei are in the area immediately adjacent to the bifurcation of the basilar artery. Thus bilateral oculomotor pareses are possible. Frequently this aneurysm is of the apoplectic type and may cause massive subarachnoidal hemorrhages. After these have been absorbed, bilateral paralyses of extrinsic eye muscles may manifest themselves. Large aneurysms (basilar and vertebral) may produce pyramidal signs, gaze palsy, nystagmus (vertical and rotary), and cerebellar or cerebellopontine angle signs.

Apoplectic type (subarachnoidal hemorrhage)

The symptomatology of the apoplectic type of aneurysm is determined by an *acute subarachnoidal hemorrhage.* The picture is extremely characteristic. The onset is marked by *violent frontal or occipital headaches and vomiting, followed in about one third of the patients by unconsciousness* lasting several hours or days. In addition to the severe meningeal symptoms, there are usually bilateral pyramidal signs. Lumbar puncture reveals a characteristic *hemorrhagic cerebrospinal fluid.* Actually, any aneurysm (saccular or arteriovenous) outside the cavernous sinus can cause such an acute subarachnoidal hemorrhage.

What has been said about subarachnoidal hemorrhage is applicable also to *intracerebral hemorrhages,* which can also be produced by rupture of saccular or arteriovenous aneurysms. The blood escaping from these aneurysms may either flow into the subarachnoid space or into the substance of the brain itself.

It is of interest from the neuro-ophthalmologic point of view that a subarachnoidal or intracerebral hemorrhage may cause grave ocular changes, although in a relatively small number of patients (Ballantyne; Paton; Cardarello; and others). Papilledema (Griffith, Jeffers, and Fry), repeatedly mentioned in the literature as occurring in patients with subarachnoidal hemorrhages, is rather rare (about 20%) and is seen mostly in the late stages of the disease and usually is moderate in degree (Fig. 5-12). Much more important are *intraocular hemorrhages,* which may result in considerable impairment of the visual acuity. At first, such hemorrhages are either *retinal* (either punctate or forming concentric circles around the disc) or *preretinal* (with a typical horizontal level), between the retina and hyaloid membrane. The preretinal hemorrhages after some days frequently tend to spread into the *vitreous* (Drews and

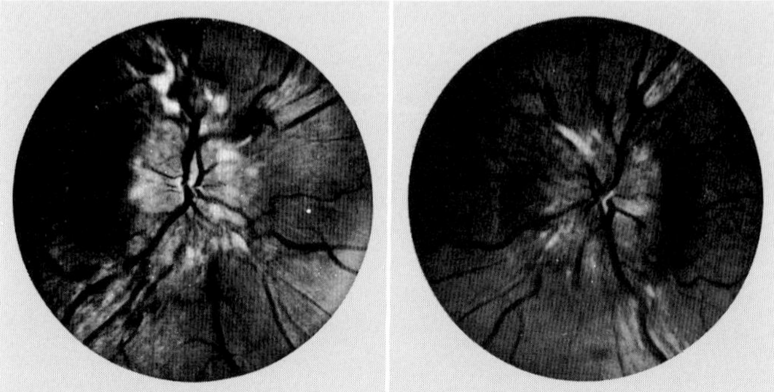

Fig. 5-12. Bilateral papilledema after subarachnoidal hemorrhage from an aneurysm in a 26-year-old patient. Enlargement, edematous swelling, and elevation (2.5 diopters) of both papillae with hemorrhages and white spots. Bilateral venous congestion. Bending of the vessels at the disc margins. Radial folding of the retina adjacent to the disc. The left illustration corresponds to the right eye; the right illustration, to the left eye.

Minckler; Wagener and Foster). In such cases more or less extensive vitreous hemorrhages make an ophthalmoscopic examination impossible. Loss of vision following a subarachnoidal hemorrhage may be caused by such vitreous hemorrhages. In a small number of patients who show no evidence of vitreous or retinal hemorrhages, the loss of vision must be interpreted as being due to acute circulatory disturbances (ischemic or hemorrhagic) in the region of the optic nerve or chiasm resulting from the subarachnoidal hemorrhage. The retinal peripapillary hemorrhages may be unilateral or bilateral. The preretinal hemorrhages, with their tendency to spread into the vitreous, are mostly bilateral. Retinal and vitreous hemorrhages occur also with severe intracranial hemorrhages from any cause.

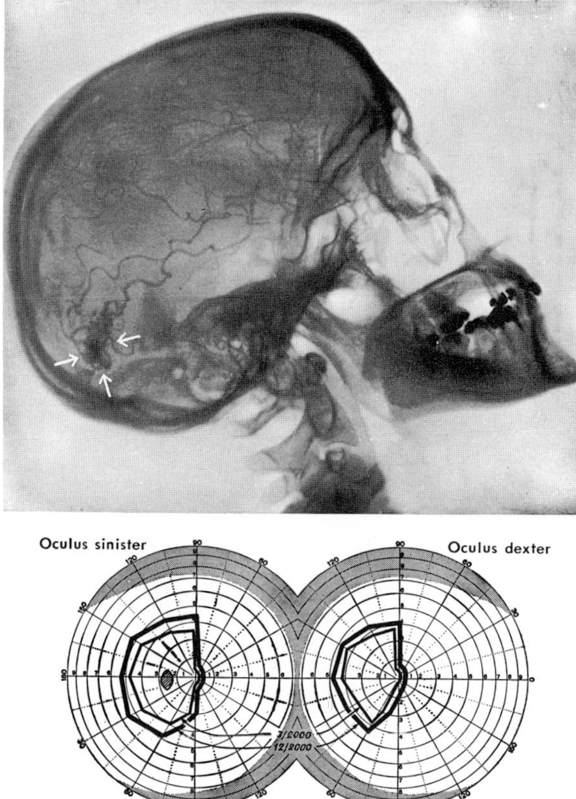

Fig. 5-13. Arteriovenous aneurysm in the left occipital region of a 32-year-old patient. Above, Arteriogram (lateral view) after left cerebral angiography. Two strong branches of the posterior cerebral artery lead to the arteriovenous vessel convolution (arrows) in the lateral and middle region of the left occipital lobe. Below, Right homonymous hemianopia with sparing of the macula. Visual acuity, O.U. 0.5. Patient had unformed visual hallucinations in the form of fire balls!

The pathogenesis of these retinal and preretinal hemorrhages is still quite controversial. It is generally assumed (MacDonald) that the cause of these hemorrhages is more or less the same as that of the hemorrhages in papilledema (that is, an obstruction of the central retinal vein at the site where it pierces the sheath of the optic nerve after emerging from the latter). A block of the venous drainage causes the characteristic hemorrhages in the area of distribution of the central retinal vein (that is, the retina and, secondarily, the preretinal space). At any rate, it is an accepted fact that in subarachnoidal hemorrhages the extravasations, which may reach the sheaths of the optic nerve, can never enter the intraocular space because the intervaginal space of the nerve forms a cul-de-sac-like termination near the bulbus wall. As a result of his investigations, Ballantyne comes to the conclusion that obstruction of the central retinal vein cannot be blamed solely for the retinal and preretinal hemorrhages. He assumes an additional embarrassment of the venous drainage channels of the globe and orbit in their entirety, the cause of which is a sudden increase in the intracranial pressure resulting from the subarachnoidal hemorrhage.

Arteriovenous aneurysms

The arteriovenous aneurysm, too, may assume the form of the apoplectic type (that is, a subarachnoidal or intracerebral hemorrhage). The general neurologic symptomatology of the nonapoplectic type depends on the location of the aneurysm. Frequently its seat is a hemisphere (Dandy; Weber; Olivecrona and Ladenheim), the location being pial, pial and dural, or extracranial.

Eighty-five percent of the cerebral arteriovenous aneurysms are supratentorial (Krayenbühl and Yasargil). Subarachnoidal hemorrhage is one of the initial symptoms. Headache, epileptic seizures, and intellectual and psychic disorders are frequent symptoms. Transient attacks of monocular blindness, unformed photopsias, or transient homonymous hemianopias are occasional initial

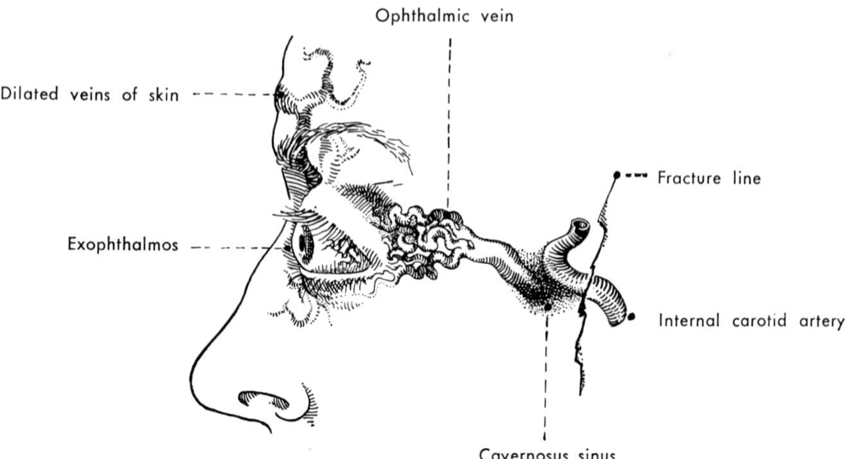

Fig. 5-14. Diagrammatic illustration of the development of venous stasis and exophthalmos after traumatic arteriovenous aneurysm of the internal carotid artery within the cavernous sinus (carotid-cavernous fistula). A rupture in the internal carotid artery following a fracture of the base of the skull leads to a shunt with the cavernous sinus and, consequently, an impairment (that is, a stasis) in the area of drainage of the ophthalmic vein. The pulse wave of the internal carotid artery is transmitted to the globe: pulsating exophthalmos. (After Dandy.)

Table 7. Symptomatology of intracranial aneurysms[*]

Location	Ocular symptoms	Trigeminal signs	Other signs
Infraclinoid aneurysms			
1. Aneurysm of internal carotid within cavernous sinus	Unilateral ophthalmoplegia, especially of third nerve (rarely optic atrophy with loss of vision and field defects)	Intense pain in area of first, second, and third branches of trigeminal nerve; hypesthesia and hypalgesia in same area (sensation of numbness)	Violent headache (generally sudden onset); destructions in region of lesser sphenoid wing and clinoid processes; enlargement of superior orbital fissure
(a) Posterior syndrome	Mostly only paresis of sixth nerve	Involvement of all sensory and motor fibers of trigeminal nerve	
(b) Middle syndrome	Complete ophthalmoplegia	Involvement of first and second branch	
(c) Anterior syndrome	Paresis of third nerve or complete ophthalmoplegia	Involvement of first branch	
2. Intrasellar aneurysm of internal carotid artery within cavernous sinus	Chiasmal syndrome with asymmetrical bitemporal hemianopia; sudden onset; fluctuations of visual field defects; occasionally ophthalmoplegia	Negative	More or less pronounced headaches; no roentgenologic evidence of enlargement of sella
Supraclinoid aneurysms			
3. Aneurysm of anterior cerebral and anterior communicating arteries	More or less atypical chiasmal syndrome (bitemporal inferior quadrantanopia); unilateral or bilateral loss of vision, subject to fluctuations; possible optic atrophy	Negative	With exception of olfactory nerve, cranial nerves remain intact; aneurysm frequently apoplectic in type
4. Aneurysm of carotid–posterior communicating artery junction	Partial or complete oculomotor palsy with mydriasis and disturbance of accommodation	Negative except pain in forehead and eye; neuralgic pain of ophthalmic branch	Sudden onset of unilateral headache, pain in forehead or eye ("ophthalmoplegic migraine"); with severe pain in neck, aneurysm located in posterior part of posterior communicating artery

[*]Modified after Franceschetti.

Table 7. Symptomatology of intracranial aneurysms—cont'd

Location	Ocular symptoms	Trigeminal signs	Other signs
5. Aneurysm of posterior cerebral artery (very rare)	Possibly homonymous hemianopia (optic tract or radiation!)	Negative	Possibly Weber's syndrome
6. Aneurysm of basilar and vertebral artery (rare)	Possible bilateral abducens or oculomotor pareses		Frequently no symptoms whatsoever; occasionally bulbar, pyramidal, or cerebellar signs; this aneurysm frequently belongs to apoplectic type

ocular signs. *The most common ophthalmologic complication of supratentorial arteriovenous aneurysm is homonymous hemianopia* (sometimes bilateral, causing cerebral blindness). Here we would like to call attention to our observation of a patient with homonymous hemianopia with an occipital arteriovenous aneurysm (Fig. 5-13).

Intracranial arteriovenous malformations occasionally may be associated with *retinal arteriovenous aneurysms* (Fig. 2-42) and arteriovenous malformations in the orbit, the optic nerve, the maxilla, the pterygoid fossa, and the mandible *(Wyburn-Mason syndrome)*. The cerebral arteriovenous aneurysms in this syndrome generally involve the basofrontal area, the Sylvian fissure, or the posterior fossa, including the midbrain.

The arteriovenous aneurysm of the internal carotid artery within the cavernous sinus, also called carotid-cavernous fistula (traumatic or spontaneous) (Fig. 5-14), *is especially interesting from the neuro-ophthalmologic point of view.* In contrast to the saccular aneurysm of the cavernous sinus (intracavernous aneurysm), this arteriovenous aneurysm does not result in a cavernous sinus syndrome (Dandy; Walsh; Kaeser).

Our own series includes thirteen cases of arteriovenous aneurysms of the internal carotid artery within the cavernous sinus, nine of them proved by angiography. Their symptomatology can be summarized as follows. The most important sign is *exophthalmos* (Fig. 5-15), which usually occurs unilaterally (on the side of the fistula). Contralateral exophthalmos may follow in about one third of the cases days or weeks later. The exophthalmos may be pulsating (three of thirteen cases). The pulsation is synchronous with the radial pulse (Martin and Mabon; Sugar and Meyer). A characteristic *bruit synchronous with the pulse* can be heard with the stethoscope in the frontotemporal region or directly over the eye. Often this noise is noted by the patient himself and may disturb his sleep. In case of pronounced exophthalmos there may be additional motility disturbances—mostly sixth nerve pareses, more rarely oculomotor pareses (lesions of

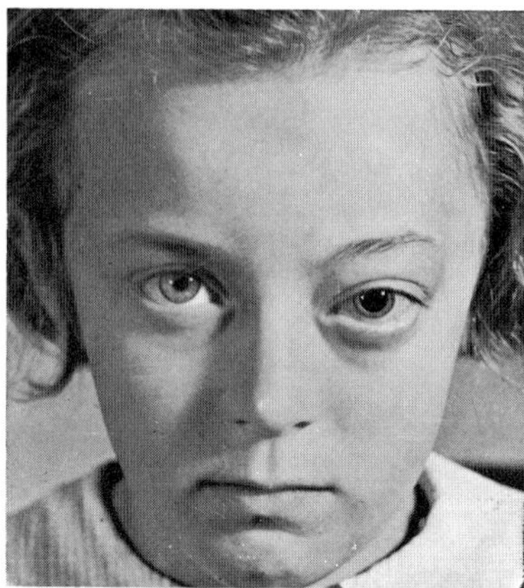

Fig. 5-15. Traumatic arteriovenous aneurysm of the left internal carotid artery within the cavernous sinus (carotid-cavernous fistula) of an 8-year-old girl. Left exophthalmos measuring 4 mm. with minimal pulsation of the corneal apex. Distinct congestion of the conjunctival and episcleral veins. Marked prominence of the veins in the left upper lid and left temporal region. Pulse-synchronous bruit with maximum over left eye and left temple. Compared with right side, left retinal veins are distinctly congested and tortuous. Visual acuity, O.U. 1.0. Trauma—a small steel rib of an umbrella penetrated the left orbit.

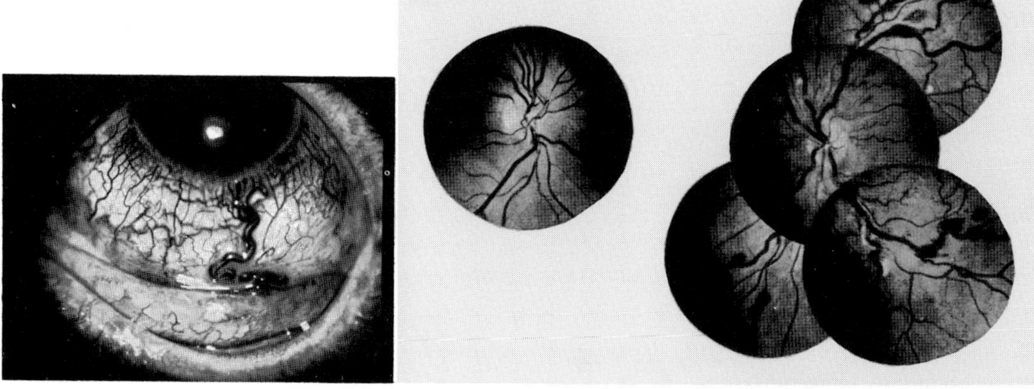

Fig. 5-16. Traumatic arteriovenous aneurysm of the internal carotid artery within the cavernous sinus (carotid-cavernous fistula) of a 43-year-old patient. Above, Appearance of the left exophthalmic bulbus with enormous venous congestion of the conjunctival as well as, in particular, the episcleral veins. Large tortuous vessel in 5:30 o'clock position. Below, Fundus photographs demonstrating the enormous congestion of the left retinal veins with patches of retinal hemorrhages. O.S., Beginning papilledema. Right fundus normal. Visual acuity, O.D. 1.0.; O.S. 0.25.

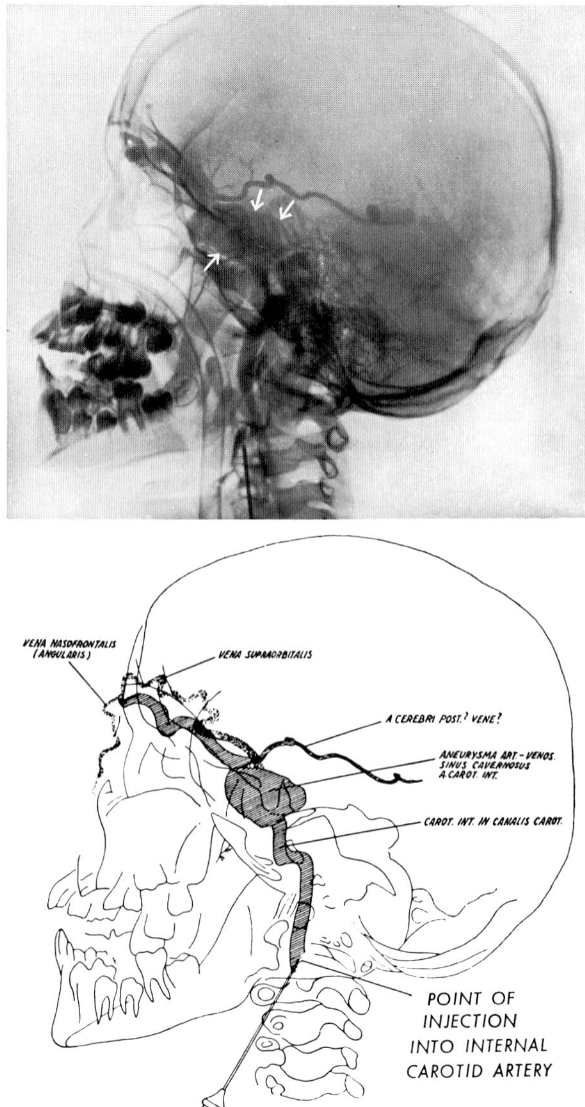

Fig. 5-17. Arteriogram (lateral view) after left internal carotid arteriography. Large arteriovenous aneurysm (arrows) of the left internal carotid artery within the cavernous sinus (carotid-cavernous fistula). Retrograde congestion of the orbital veins. Arteriogram of same patient shown in Fig. 5-16. Below, Diagrammatic sketch illustrating the vascular areas filled by the contrast dye.

the nerve within the cavernous sinus). Signs of the latter are an occasional ptosis of the upper lid and mydriatic pupils fixed to light. The arteriovenous shunt causes *venous stasis,* which manifests itself in dilated conjunctival veins (with chemosis), dilated veins of the lids (with lid swelling), dilated scleral veins, and especially dilated retinal veins with retinal hemorrhages (Fig. 5-16). Ophthalmoscopically, there is frequently slight choking of the disc, which changes to optic atrophy with blurred disc margins (two out of thirteen cases) at a later stage. There are no visual field changes or loss of vision in the initial stages. They develop at a later stage (impaired retinal circulation, secondary glaucoma), an important fact in the differential diagnosis of intraorbital and perhaps intrasellar aneurysms. The neuro-ophthalmologic syndrome of an arteriovenous shunt between the internal carotid artery and the cavernous sinus is so characteristic that it permits a clinical diagnosis. Definite proof, however, can be obtained only by means of cerebral angiography (Fig. 5-17). *It is most important to differentiate this arteriovenous aneurysm from tumors of the orbit or a possible meningioma of the sphenoid wing. Next in importance are tumors at the base of the skull, which may create a superior orbital fissure syndrome.* Although exophthalmos and motility disturbances may occur in association with tumors as well as with arteriovenous aneurysms, the extensive venous congestion, the pulsation of the exophthalmos, and the bruit synchronous with the pulse are missing in patients with tumors. Their presence will aid in drawing the right conclusion.

Finally, it must be mentioned that pulsating exophthalmos may be produced also by aneurysms of the ophthalmic artery or by arteriovenous malformations within the orbit. In the differential diagnosis the orbital varix (intermittent exophthalmos syndrome) must also be considered. *For correct diagnosis of carotid–cavernous sinus fistulas there is no doubt that bilateral carotid angiography* (including orbital angiography) *supplemented by subtraction technique represents the method of choice.*

The purpose of this book has been to present the ocular symptoms seen in association with brain tumors. The book is based on the 2500 cases of brain tumors on record at the Neurosurgical Clinic at the University of Zurich. Personal experience in the evaluation of these cases is emphasized.

The special methods of the neuro-ophthalmologic examination in patients suspected of having a brain tumor receive consideration in Chapter 1. The first section of that chapter is devoted to the *history and subjective symptoms* as reported by the patients. These symptoms consist of visual field disturbances, disturbances of primitive and higher visual functions, photopsias, hallucinations, mind blindness, cortical blindness, alexia, agraphia, disturbances of motility (especially diplopia), pain, photophobia, and others. The *objective examination* includes the external aspect of the globe (lid fissure, exophthalmos, pupils), corneal sensitivity, pupils and pupillary reactions, motility (extraocular palsies, gaze palsies, nystagmus), fundus, visual acuity, color sense, dark adaptation, and finally visual fields (gross methods, perimeter, tangent screen, general pathology of visual fields). The special technique and the importance of the neuro-ophthalmologic examination justify this detailed and comprehensive presentation.

Chapter 2 concerns itself with the *general symptoms of increased intracranial pressure* in patients with brain tumors. The ocular signs consist of *papilledema, extraocular pareses, clivus ridge syndrome,* and rarely *exophthalmos.* First the frequency of papilledemas is discussed. Then the ophthalmoscopic appearance of fully developed papilledema is described in detail. The subjective symptoms associated with it, especially amblyopic attacks, are considered. The ways of diagnosing incipient papilledema, including the relationship of the blood pressure in the retinal vessels, are discussed. There is a separate section on unilateral papilledema and its differential diagnosis as well as chronic atrophic papilledema with its grave significance for vision and the visual fields. The pathologic anatomy of the disc and pathogenesis of papilledema are summarized. The *differential diagnosis of papilledema,* with its great importance in the neuro-ophthalmology of brain tumors and especially its significance in *papillitis, the edema of malignant hypertension,* and *various congenital anomalies* (drusen, pseudoneuritis, pseudochoked disc) are considered. In addition to papilledema, increased intracranial pressure causes nonspecific *extraocular muscle pareses,* especially a unilateral sixth nerve paresis. Another result of increased intracranial pressure is the *clivus ridge syndrome*—a compression of the oculomotor nerve by protruding brain substance at the tentorial notch results in unilateral mydriasis and rigidity to light. In later stages there may be ptosis and paralysis of the extrinsic muscles supplied by the oculomotor nerve. Exophthalmos caused by a general increase in the intracranial pressure is mostly bilateral. A unilateral exophthalmos should be interpreted rather as a focal sign, as for instance, invasion of the orbit by the tumor.

In Chapter 3 the local symptoms of brain tumors are presented from a topographic point of view. In order to present a true scope of the ocular signs in relation to other neurologic signs, a brief summary of general neurologic syndromes for each location under discussion is included. The brain tumors are divided into supratentorial and infratentorial groups. The *supratentorial group* includes tumors of the hemispheres (frontal, temporal, parietal, and occipital lobes), the pituitary region, the anterior and middle fossae, the third ventricle, the thalamus, the basal ganglia, the midbrain with the Sylvian aqueduct, and the pineal body. The infratentorial group involves cerebellar tumors (cerebellar spheres and vermis), the cerebellopontine angle, the pons, and the medulla oblongata. The fourth ventricle is considered in detail in the discussion of tumors of the cerebellar vermis. No special summary of the ocular symptomatology of the various tumor sites is given since a brief summary of these symptoms is presented following each topographic description.

In Chapter 4 the *relationship between the type of tumor and the ocular symptoms* is discussed. Some tumor types produce a very characteristic ocular symptomatology (for instance, pituitary adenomas, craniopharyngiomas, chiasmal gliomas, and others). In addition, there are numerous tumor types with a neuro-ophthalmologic symptomatology that does not depend on their histologic structure but merely on their localization. There is a discussion of the characteristics of glioblastoma multiforme, astrocytoma, oligodendroglioma, medulloblastoma, hemangiomas, and meningiomas. A special section deals with metastatic tumors to the brain—a special type of brain tumors. If there are specific ocular signs, they are always considered in connection with these special types of tumors.

Chapter 5 considers tumorlike changes—an important group from the standpoint of differential diagnosis that always must be considered in the discussion of ocular symptomatology. Included are *pseudotumor cerebri, subdural hematomas, brain abscesses,* and, in the broadest sense of the word, *aneurysms.* Subdural hematomas are notable especially because of unilateral pupillary disturbances (mydriasis, possibly light rigidity, as in the clivus ridge syndrome). The ocular symptomatology of brain abscesses does not differ essentially from that of other tumors. Of important localizing value, however, are the ocular signs of various forms of cerebral aneurysms—due to their close vicinity to cranial nerves, they result in characteristic syndromes. We mention especially aneurysms of the internal carotid artery within the cavernous sinus (cavernous sinus syndrome), intrasellar and suprasellar aneurysms (chiasmal syndrome), and aneurysms of the carotid–posterior communicating artery junction (peripheral oculomotor paralysis). In addition to the paralytic type of aneurysm, there is an apoplectic form, which is characterized by an *acute subarachnoidal hemorrhage.* The ocular changes associated with such subarachnoidal hemorrhages consist of retinal and especially vitreous hemorrhages.

The importance of ocular symptoms in brain tumors is underlined by the fact that a little more than 50% of patients with brain tumors show ocular signs in one form or another. These signs are not merely insignificant accompanying manifestations but frequently important initial symptoms. Frequently they point to a general increase in the intracranial pressure. Very often they are extremely valuable as localizing symptoms.

Literature

Abbreviations of journals used in the bibliography

Acta Med. Scand.	Acta medica scandinavica, Stockholm
Acta Neurochir.	Acta neurochirurgica, Vienna
Acta Neurol. Scand.	Acta neurologica scandinavica, Kobenhavn
Acta Ophthal.	Acta ophthalmologica, Copenhagen
Acta Soc. Ophthal. Jap.	Acta Societatis ophthalmologica Japonica, Tokyo
Advances Ophthal.	Advances in Ophthalmology, Basel
Allg. Z. Psychiat.	Allgemeine Zeitschrift für Psychiatrie und ihre Grenzgebiete, Berlin
Amer. J. Ophthal.	American Journal of Ophthalmology, Chicago
Amer. J. Roentgen.	American Journal of Roentgenology, Radium therapy and Nuclear Medicine, Springfield, Ill.
Ann. Oculist.	Annales d'oculistique, Paris
Ann. Ottal.	Annali di ottalmologia, Parma
Ann. Roy. Coll. Surg. Eng.	Annals of the Royal College of Surgeons of England, London
Annee Ther. Clin. Ophtal.	Année thérapeutique et clinique en ophtalmologie, Marseille
Arch. Klin. Chir.	Archiv für klinische Chirurgie, Berlin
Arch. Augenheilk.	Archiv für Augenheilkunde, Munich
Arch. Dis. Child.	Archives of Disease in Childhood, London
Arch. Neurol. Psychiat.	Archives of Neurology and Psychiatry, Chicago
Arch. Ophtal.	Archives d'ophtalmologie, Paris
Arch. Ophthal.	Archives of Ophthalmology, Chicago
Arch. Otolaryng.	Archives of Otolaryngology, Chicago
Arch. Path.	Archives of Pathology, Chicago
Arch. Oftal. B. Air.	Archivos de oftalmologia de Buenos Aires, Buenos Aires
Arch. Soc. Oftal. Hispano-Amer.	Archivos de la Sociedad oftalmológica hispano-americana, Madrid
Atti Soc. Oftal. Ital.	Atti della Società oftalmologica italiana, Rome
Ber. Deutsch. Ophth. Ges.	Bericht; deutsche ophthalmologische Gesellschaft, Heidelberg
Boll. Oculist.	Bollettino d'oculistica, Bologna
Brain	Brain; Journal of Neurology, London
Brit. J. Ophthal.	British Journal of Ophthalmology, London
Brit. J. Surg.	British Journal of Surgery, Bristol
Bull. Neurol. Inst. N. Y.	Bulletin of the Neurological Institute of New York, New York
Bull. N. Y. Acad. Med.	Bulletin of the New York Academy of Medicine, New York
Bull. Soc. Belg. Ophtal.	Bulletin de la Société belge d'ophtalmologie, Brussels
Bull. Soc. Ophtal. Franc.	Bulletin des Sociétés d'ophtalmologie de France, Paris
Bull. Soc. Franc. Ophtal.	Bulletins et mémoires de la Société française d'ophtalmologie, Paris
Circulation	Circulation; Journal of the American Heart Association, New York
Confin. Neurol.	Confinia neurologica, Basel
Deutsch. Med. Wschr.	Deutsche medizinische Wochenschrift, Stuttgart
Deutsch. Z. Nervenheilk.	Deutsche Zeitschrift für Nervenheilkunde, Berlin

Docum. Ophthal.	Documenta ophthalmologica, The Haag
Encephale	L'Encéphale, journal de neurologie, de psychiatrie et de médecine psychosomatique (Supplément: L'Hygiène mentale), Paris
Ergebn. Physiol.	Ergebnisse der Physiologie, Biologischen Chemie und experimentellen Pharmakologie, Berlin
Eye Ear Nose Throat Monthly	Eye, Ear, Nose and Throat Monthly, Chicago
Folia Ophthal. Jap.	Folia ophthalmologica Japonica, Osaka
Fortschr. Roentgenstr.	Fortschritte auf dem Gebiete der Roentgenstrahlen und der Nuklearmedizin, Stuttgart
G. Ital. Oftal.	Giornale italiano di oftalmologia, Florence
Graefe. Arch. Ophthal.	Albrecht von Graefe's Archiv für Ophthalmologie, Berlin
Helv. Med. Acta	Helvetica medica acta, Basel
Helv. Paediat. Acta	Helvetica paediatrica acta, Basel
Helv. Physiol. Pharmacol. Acta	Helvetica physiologica et pharmacologica acta, Basel
Int. Arch. Allerg.	International Archives of Allergy and Applied Immunology, Basel
Int. Ophthal. Clin.	International Ophthalmology Clinics, Boston
J.A.M.A.	The Journal of the American Medical Association, Chicago
J. Med. Lyon	Journal de médecine de Lyon
J. Nerv. Ment. Dis.	Journal of Nervous and Mental Diseases, Baltimore
J. Neurosurg.	Journal of Neurosurgery, Springfield, Ill.
Klin. Mbl. Augenheilk.	Klinische Monatsblätter für Augenheilkunde, Stuttgart
Lancet	Lancet, London
Lyon Med.	Lyon médical, Lyon
Mayo Clin. Proc.	Mayo Clinic Proceedings, Rochester, Minn.
Minerva Oftal.	Minerva oftalmologica, Torino
Mschr. Psychiat. Neurol.	Monatsschrift für Psychiatrie und Neurologie, Basel
Neurochirurgie	Neuro-chirurgie, Paris
Neurology	Neurology, Minneapolis
Nord. Ophthal. Tidsskr.	Nordisk ophthalmologisk tidsskrift, Copenhagen
Ophthal. Lit.	Ophthalmic Literature, London
Ophthalmologica	Ophthalmologica, Basel
Praxis	Praxis, Bern
Proc. All-India Ophthal. Soc.	Proceedings of the All-India Ophthalmological Society, Madras
Proc. Roy. Soc. Med.	Proceedings of the Royal Society of Medicine, London
Radiol. Med.	La radiologia medica, Turin
Radiology	Radiology, Syracuse, N. Y.
Rass. Ital. Ottal.	Rassegna italiana di ottalmologia, Turin
Rev. Méd. Suisse Rom.	Revue médicale de la Suisse Romande, Lausanne
Rev. Otoneuroophtal.	Revue d'oto-neuro-ophtalmologie, Paris
Rev. Esp. Otoneurooftal.	Revista española de oto-neuro-oftalmologia y neurocirurgia, Valencia
Riv. Otoneurooftal.	Rivista oto-neuro-oftalmologica, Bologna
Schweiz. Arch. Neurol. Psychiat.	Schweizer Archiv für Neurologie und Psychiatrie, Zurich
Schweiz. Med. Wchr.	Schweizerische medizinische Wochenschrift, Basel
Sist. Nerv.	Sistema nervoso, Milan
Surg. Gynec. Obstet.	Surgery, Gynecology and Obstetrics, Chicago
Trans. Amer. Acad. Ophthal. Otolaryng.	Transactions of the American Academy of Ophthalmology and Otolaryngology, Rochester, Minn.
Trans. Amer. Ophthal. Soc.	Transactions of the American Ophthalmological Society, New York
Trans. Ophthal. Soc. U. K.	Transactions of the Ophthalmological Society of the United Kingdom, London
Tunisie Med.	Tunisie médicale, Tunis
Wien. Klin. Wschr.	Wiener klinische Wochenschrift, Vienna
Zbl. Chir.	Zentralblatt für Chirurgie, Leipzig
Zbl. Neurochir.	Zentralblatt für Neurochirurgie, Leipzig
Zbl. Ges. Ophthal.	Zentralblatt für die gesamte Ophthalmologie und ihre Grenzgebiete, Berlin
Z. Augenheilk.	Zeitschrift für Augenheilkunde, Berlin

Handbooks, textbooks, and monographs

Adler, F. H.: Textbook of ophthalmology, Philadelphia, 1950, W. B. Saunders Co.

Adrogue, E.: Neurologia ocular, Buenos Aires, 1942, El Ateneo.

Altenburger, H.: Die raumbeengenden Krankheiten des Schädelinnern. In Bergmann and Staehelin, editors: Handbuch der inneren Medizin, Berlin, 1939, Julius Springer, vol. 5, pt. 1.

Amsler, M., Brückner, A., Franceschetti, A., Goldmann, H., and Streiff, E. B.: Lehrbuch der Augenheilkunde, Basel, 1954, S. Karger AG.

Bailey, P.: Die Hirngeschwülste (deutsche Übertragung), Stuttgart, 1951, Ferdinand Enke.

Bailey, P., Buchmann, D. N. P., and Bucy, P. C.: Intracranial tumors of infancy and childhood, Chicago, 1939, University of Chicago Press.

Bailliart, P.: La circulation rétinienne à l'état normal et à l'état pathologique, Paris, 1923, Gaston Doin & Cie.

Bailliart, P., Coutela, C., Redslob, E., and Velter, E.: Traité d'ophtalmologie, Paris, 1939, Masson & Cie.

Bärtschi-Rochaix, W.: Neurologische Diagnostik, Basel, 1952, E. Reinhardt.

Behr, C.: Auge und Zentralnervensystem, Zbl. Ges. Ophthal. 45:1, 1940.

Bender, M. B.: Ophthalmoneurology. In Progress in neurology and psychiatry, New York, 1946, Grune & Stratton, Inc.

Berens, C.: The eye and its diseases, Philadelphia, 1949, W. B. Saunders Co.

Berens, C., and Zuckermann, J.: Diagnostic examination of the eye, Philadelphia, 1946, J. B. Lippincott Co.

Bergstrand, H., Olivecrona, H., and Tönnis, W.: Gefässmissbildungen und Gefässgeschwülste des Gehirns, Leipzig, 1936, Georg Thieme.

Best, F.: Die Augenveränderungen bei den organischen nicht entzündlichen Erkrankungen des Zentralnervensystems. In Schieck, F., and Brückner, A., editors: Kurzes Handbuch der Ophthalmologie, Berlin, 1931, Julius Springer, vol. 6.

Bing, R.: Allgemeine Anatomie, Physiologie, Pathologie und Symptomatologie des Gehirnes. In Bergmann and Staehelin: Handbuch der inneren Medizin, Berlin, 1939, Julius Springer, vol. 5, p. 33.

Bing, R.: Kompendium der topischen Gehirn- und Rückenmarkdiagnostik, Basel, 1945, Benno Schwabe & Co.

Bing, R.: Lehrbuch der Nervenkrankheiten, Basel, 1945, Benno Schwabe & Co.

Bing, R., and Brückner, R.: Gehirn und Auge, Grundriss der Neuro-Ophthalmologie, Basel, 1954, Benno Schwabe & Co.

Bodechtel, G.: Differentialdiagnose neurologischer Krankheitsbilder, Stuttgart, 1958, Georg Thieme.

Bonamour, G., Brégeat, P., Bonnet, M., and Juge, P.: La papille optique, Paris, 1968, Masson & Cie.

Brégeat, P.: L'oedème papillaire, Paris, 1956, Masson & Cie.

Bumke, O., and Förster, O.: Handbuch der Neurologie, Berlin, 1936, Julius Springer, vol. 14.

Cattaneo, D.: Oftalmoangioscopia, Bologna, 1947, L. Cappelli.

Cogan, D. G.: Neurology of the visual system, Springfield, Ill., 1968, Charles C Thomas, Publisher.

Cushing, H.: Intrakranielle Tumoren, Berlin, 1935, Julius Springer.

Cushing, H., and Bailey, P. B.: Tumors arising from the blood vessels of the brain: angiomatous malformations and hemangioblastomas, Springfield, Ill., 1928, Charles C Thomas, Publisher.

Cushing, H., and Eisenhardt, L.: Meningiomas: their classification, regional behavior, life, history, surgical end results, Springfield, Ill., 1938, Charles C Thomas, Publisher.

Dandy, W. E.: Intracranial arterial aneurysms, Ithaca, N. Y., 1944, Comstock Publishing Co., Inc.

Dandy, W. E.: Surgery of the brain, Hagerstown, Md., 1945, W. F. Prior Co., Inc.

Dubois-Poulsen, A.: Le champ visuel, Paris, 1952, Masson & Cie.

Duke-Elder, S.: Textbook of ophthalmology, St. Louis, 1938, 1945, 1949, The C. V. Mosby Co., vols. 2 to 4.

Ewald, G.: Lehrbuch der Neurologie und Psychiatrie, Munich, 1948, Urban & Schwarzenberg.

Focosi, M.: Le paralisi dei muscoli oculomotori estrinseci, Rome, 1948, Abruzzini.

Fulton, J. F.: Physiology of the nervous system, London, 1943, Oxford University Press.

Gagel, O.: Einführung in die Neurologie, Berlin, 1949, Julius Springer.

Gerlach, J.: Grundriss der Neurochirurgie, Stuttgart, 1967, J. F. Steinkopf.

Gerlach, J., and Simon, G.: Erkennung, Differentialdiagnose und Behandlung der Geschwülste und Entzündungen der Schädelknochen. In Olivecrona, H., and Tönnis, W., editors: Handbuch der Neurochirurgie, Berlin, 1962, Julius Springer, pp. 211-366.

Guillaumat, L., Morax, P. V., and Offret, G.: Neuro-ophtalmoıogie, Paris, 1959, Masson & Cie.

Guillot, P., Saraux, H., and Sedan, R.: L'exploration neuroradiologique en ophtalmologie, Paris, 1966, Masson & Cie.

Harrington, D. O.: Visual fields. A textbook and atlas of clinical perimetry, St. Louis, 1964, The C. V. Mosby Co.

Hess, R.: Elektroencephalographische Studien bei Hirntumoren, Stuttgart, 1958, Georg Thieme.

Hughes, B.: The visual fields, Oxford, 1954, Blackwell Scientific Publications.

Hughes, B.: The visual fields: a study of the applications of quantitative perimetry to the anatomy and pathology of the visual pathways, Springfield, Ill., 1954, Charles C Thomas, Publisher.

Jentzer, A.: Klinische Neurochirurgie. In Brunner et al., editors: Lehrbuch der Chirurgie, Basel, 1949, Benno Schwabe & Co., pp. 659-778.

Kehrer, F.: Die Allgemeinerscheinungen der Hirngeschwülste, Leipzig, 1931, Georg Thieme.

Kershner, C. M.: Blood supply of the visual pathway, Boston, 1943, Meador Publishing Co.

Kestenbaum, A.: Clinical methods of neuro-ophthalmologic examination, London, 1947, William Heinemann.

Krayenbühl, H.: Allgemeine hirnchirurgische Diagnostik und Therapie. In Brunner et al., editors: Lehrbuch der Chirurgie, Basel, 1949, Benno Schwabe & Co., pp. 605-658.

Krayenbühl, H., and Richter, W.: Die cerebrale Angiographie, Leipzig, 1952, Georg Thieme.

Krayenbühl, H., and Yasargil, G. M.: Die zerebrale Angiographie, Stuttgart, 1965, Georg Thieme.

Krieg, W. J. S.: Functional neuroanatomy, Philadelphia, 1942, The Blakiston Co.

Kuntz, A.: A textbook of neuroanatomy, Philadelphia, 1942, Lea & Febiger.

Kyrieleis, W.: Klinik der Augensymptome bei Nervenkrankheiten, Berlin, 1954, Walter de Gruyter.

Laubenthal, F.: Leitfaden der Neurologie, Leipzig, 1948, Georg Thieme.

Lauber, H.: Das Gesichtsfeld, Berlin, 1944, Julius Springer.

Lhermitte, J.: Les hallucinations, Paris, 1951, Gaston Doin & Cie.

Lindenberg, R.: Neuropathology involving the lateral geniculate bodies, the optic radiation and the calcarine cortex. In Smith, J. L., editor: Neuro-ophthalmology, St. Louis, 1965, The C. V. Mosby Co., vol. 2.

Lyle, J. D.: Neuro-ophthalmology, Springfield, Ill., 1945, Charles C Thomas, Publisher.

McLean, A. J.: Intracranial tumors. In Bumke and Foerster, editors: In Handbuch der Neurologie, Berlin, 1936, Julius Springer, vol. 14.

Malbran, J.: Campo visual normal y pathologico, Buenos Aires, 1936, El Ateneo.

Mifka, P.: Die Augensymptomatik bei der frischen Schädelhirn-Verletzung, Berlin, 1968, Walter de Gruyter.

von Monakow, C.: Die Lokalisation im Grosshirn und der Abbau der Funktion durch kortikale Herde, Munich, 1914, J. F. Bergmann.

Moniz, E.: Diagnostic des tumeurs cérébrales et épreuve de l'encéphalographie artérielle, Paris, 1931, Masson & Cie.

Moniz, E.: Die zerebrale Arteriographie und Phlebographie, Berlin, 1940, Julius Springer.

Netter, F. H.: The Ciba collection of medical illustrations. Nervous system, Summit, N. J., 1953, Ciba Pharmaceutical Products, Inc., vol. 1.

Nielsen, J. M.: A textbook of clinical neurology, New York, 1946, Paul B. Hoeber, Inc.

Olivecrona, H.: Chirurgische Behandlung der Gehirngeschwülste. In von Fedor, K., editor: Die spezielle Chirurgie der Gehirnkrankheiten, Stuttgart, 1941, Ferdinand Enke, vol. 3.

Olivecrona, H., and Tönnis, W.: Handbuch der Neurochirurgie, Berlin, 1954, 1955, etc., Julius Springer.

Ostertag, B.: Anatomie und Pathologie der raumfordernden Prozesse des Schädelbinnenraumes. Neue Deutsche Chirurgie, Stuttgart, 1941, Ferdinand Enke, vol. 50.

Ottonello, P., and Vassura, G. W.: Neuro-oftalmologia, Bologna, 1959, L. Capelli.

Perria, L.: Sintomi oculari e tumori endocranici, Sist. Nerv. 3:9-17, 1949.

Polyak, S.: The retina, Chicago, 1941, University of Chicago Press.

Rasmussen, A. T.: The principal nervous pathways, New York, 1941, The Macmillan Co.

Rea, R. L.: Neuro-ophthalmology. St. Louis, 1941, The C. V. Mosby Co.

Rucker, W. C.: The interpretation of visual fields, American Academy of Ophthalmology and Otolaryngology, ed. 3, Omaha, 1957, Douglas Printing Co.

Sachs, E.: Diagnosis and treatment of brain tumors, St. Louis, 1940, The C. V. Mosby Co.

Sachs, E.: Brain tumors and care of the

neurosurgical patient, St. Louis, 1949, The C. V. Mosby Co.

Scheid, W.: Lehrbuch der Neurologie, Stuttgart, 1966, Georg Thieme.

Schieck, F., and Brückner, A.: Kurzes Handbuch der Ophthalmologie, Berlin, 1930-1932, Julius Springer.

Smith, J. L., editor: Neuro-ophthalmology, St. Louis, 1965, 1967, 1968, The C. V. Mosby Co., vols. 2 to 4.

Spalteholz, W.: Handatlas der Anatomie des Menschen, Leipzig, 1933, S. Hir el, vol. 3.

Spiegel, E. A., and Sommer, I.: Ophthalmo- und Oto-Neurologie, Berlin, 1931, Julius Springer.

Spiegel, E. A., and Sommer, I.: Neurology of the eye, ear, nose and throat, New York, 1944, Grune & Stratton, Inc.

Streiff, E. B.: Krankheiten des Nervensystems. In Amsler, M., et al., editors: Lehrbuch der Augenheilkunde, Basel, 1954, S. Karger AG, p. 829.

Töndury, G.: Angewandte und topographische Anatomie, Zürich, 1949, Fretz & Wasmuth.

Tönnis, W.: Die Chirurgie des Gehirns und seiner Häute. In von Kirschner-Nordmann, editor: Die Chirurgie, Munich, 1948, Urban & Schwarzenberg.

Tönnis, W.: Diagnostik der intrakraniellen Geschwülste. In Olivecrona, H., and Tönnis, W., editors: Handbuch der Neurochirurgie, Berlin, 1962, Julius Springer, vol. 4, pt. 3, pp. 1-579.

Traquair, H. M.: An introduction to clinical perimetry, London, 1949, Henry Kimpton.

Troncoso, M. U.: Internal disease of the eye: an atlas of ophthalmoscopy, Philadelphia, 1942, F. A. Davis, Co.

Tschermack-Seysenegg, A.: Einführung in die physiologische Optik, Berlin, 1947, Julius Springer.

Uhthoff, W.: Die Augenveränderungen bei den Erkrankungen des Gehirnes. In Graefe-Saemisch: Handbuch der gesamten Augenheikunde, Leipzig, 1915, Wilhelm Engelmann, vol. 11, pp. 1143-1375.

Walsh, F. B., and Hoyt, W. F.: Clinical neuro-ophthalmology, Baltimore, 1969, The Williams & Wilkins Co.

Wilbrand, H., and Saenger, A.: Handbuch der Neurologie des Auges, Munich, 1900-1922, 1927, J. F. Bergmann.

Wilson, S. A.: Neurology, Baltimore, 1940, The Williams & Wilkins Co.

Wolff, E.: The anatomy of the eye and orbit, London, 1933, H. K. Lewis & Co., Ltd.

Zülch, K. J.: Die Hirngeschwülste, Leipzig, 1951, Johann Ambrosius Barth.

Zülch, K. J.: Biologie und Pathologie der Hirngeschwülste. In Olivecrona, H., and Tönnis, W., editors: Handbuch der Neurochirurgie, Berlin, 1956, Julius Springer, vol. 3.

CHAPTER ONE *Methods of neuro-ophthalmologic examination in patients with suspected brain tumors*

History and subjective symptoms

Amsler, M., and Huber, A.: Allgemeine Symptomatologie. In Amsler, M., et al., editors: Lehrbuch der Augenheilkunde, Basel, 1954, S. Karger AG, pp. 53-67.

Bay, E.: Analyse eines Falles von Seelenblindheit, Deutsch. Z. Nervenheilk. **168:** 1-23, 1952.

Bender, M. B., and Furlow, L. T.: Phenomenon of visual extinction of homonymous fields and psychologic principles involved, Arch. Neurol. Psychiat. **53:**29-33, 1945.

Bender, M. B., and Kahn, R. L.: After-imagery in defective fields of vision, J. Neurol. Neurosurg. Psychiat. **12:**196-204, 1949.

Bender, M. B., and Kanzer, M. G.: Metamorphopsia and other psychovisual disturbances in a patient with tumor of the brain, Arch. Neurol. Psychiat. **45:**481-485, 1941.

Bender, M. B., and Savitsky, N.: Micropsia and teleopsia limited to the temporal fields of vision, Arch. Ophthal. **29:**904-908, 1943.

Best, F.: Über optische Agnosie, Klin. Mbl. Augenheilk. **116:**14-18, 1950.

van Bogaert, L.: Sur les hallucinations visuelles au cours des affections organiques du cerveau (Contributions à l'étude du syndrome des hallucinations lilliputiennes), Encephale **21:**657-679, 1926.

Brégeat, Klein, Thiébaut, and Bouniol: Rev. Otoneuroophtal. **11:**238, 1947.

Busch, E., and Moller, H. U.: Veränderungen im zentralen Sehen bei neurochirurgischen Erkrankungen, Ugesk. Laeg., pp. 1174-1175, 1940.

Clark, W. E. L.: Visual centres of brain and their connexions, Physiol. Rev. **22:** 205-232, 1942.

Critchley, M.: The problem of awareness or non-awareness of hemianopic field defects, Trans. Ophthal. Soc. U. K. **69**:95, 1949.

Critchley, M.: Metamorphopsia of central origin, Trans. Ophthal. Soc. U. K. **69**: 111-121, 1949.

Critchley, M.: Types of visual perseveration: "Paliopsia" and "illusory visual spread," Brain **74**:267-299, 1951.

Cushing, H.: Trans. Amer. Neurol. Ass., p. 374, 1921.

Duensing, F.: Beitrag zur Frage der optischen Agnosie, Arch. Psychiat. **188**:131-161, 1952.

Duke-Elder, S.: Textbook of ophthalmology, St. Louis, 1949, The C. V. Mosby Co., vol. 4, pp. 3169, 3639-3642, 3643, 3654-3665, 3670-3681, and 3693.

Ethelberg, S., and Jensen, V. A.: Obscurations and further time-related paroxysmal disorders in intracranial tumors, Arch. Neurol. Psychiat. **68**:130, 1952.

Fischer-Brügge, K.: Das Klivuskanten-Syndrom, Acta Neurochir. **1**:36, 1951.

Förster, O.: Über Rindenblindheit, Arch. Ophthal. **36**:94-108, 1890.

Fuchs, W.: Z. Angew. Psychol. **84**:67, 1920.

Goldstein, K., and Gelb, A.: Psychologische Analysen hirnpathologischer Fälle auf Grund der Untersuchungen Hirnverletzter, Leipzig, 1920, Johann Ambrosius Barth.

Hecaen, H., de Ajuriaguerra, J., and Massonnet, J.: Les troubles visuo-constructifs par lésion pariéto-occipitale droite, Encephale **40**:122-178, 1951.

Henschen, K.: Klinische und anatomische Beiträge zur Pathologie des Gehirns, Upsala, 1890-1922.

Hess, W. R.: Die graphische Darstellung von Bewegungsstörungen der Augen mit Beispieltafeln zur Diagnose von Augenmuskellähmungen, Arch. Augenheilk. **70**: 1911.

Horrax, G.: Visual hallucinations as a cerebral localizing phenomenon, with special reference to their occurrence in tumors of the temporal lobes, Arch. Neurol. Psychiat. **10**:532, 1923.

Horrax, G., and Putnam, T. J.: Distortion of the visual fields in cases of brain tumor; the field defects and hallucinations produced by tumors of the occipital lobe, Brain **55**:499-523, 1932.

Krayenbühl, H.: Primary tumors of the root of the fifth cranial nerve, Brain **59**: 337, 1936.

Lebensohn, J. E., and Bellows, J.: The nature of photophobia, Arch. Ophthal. **12**:380-390, 1934.

Lhermitte, J.: Les fondements anatomophysiologiques de certaines hallucinations visuelles, Confin. Neurol. **9**:43-57, 1949.

Lhermitte, J., and Ajuriaguerra, J.: Hallucinations visuelles et lésions de l'appareil visuel, Ann. Med.-Physiol. **94**:321-351, 1936.

Lhermitte, J., and Ajuriaguerra, J.: Psychopathologie de la vision, Paris, 1942, Masson & Cie.

Lippmann, O.: Paralysis of divergence due to cerebellar tumor, Arch. Ophthal. **31**: 299, 1944.

Lisch, K.: Cerebrale Metamorphopsie, Graefe. Arch. Ophthal. **141**:554-558, 1940.

Lotmar, F.: Zur Kenntnis der herdanatomischen Grundlagen leichterer optischagnostischer Störungen, Schweiz. Arch. Neurol. Psychiat. **42**:299, 1938.

Lunn, V.: Über mangelnde Wahrnehmung der eigenen Blindheit (Anton's symptom). Eine Übersicht und klinische Studie, Acta Psychiat. Neurol. **16**:191-242, 1941.

Lyle, T. K.: Eye symptoms and signs caused by intra-cranial lesions, Ann Roy. Coll. Surg. Eng. **7**:316, 1950.

McKendree, C. A., and Doshay, L. J.: Visual disturbances of obscure etiology, produced by focal intracranial lesions implicating optic nerve; study of 6 cases of optic nerve compression with proven lesions, Bull. Neurol. Inst. N. Y. **5**:223-246, 1936.

Magitot, A.: Photophobie, Ann. Oculist. **174**: 817-832, 1937.

Masson, C. B.: The disturbances in vision and in visual fields after ventriculography, Bull. Neurol. Inst. N. Y. **3**:190-209, 1933.

Mooney, A. J., Carey, P., Ryan, M., and Bofin, P.: Parasagittal parieto-occipital meningioma with visual hallucinations, Amer. J. Ophthal. **59**:197-205, 1965.

Moretti, G.: Su alcun rilievi a proposito della cosidetta sindromo di Anton, Riv. Otoneurooftal. **38**:605-609, 1963.

de Morsier, G.: Les hallucinations, Rev. Otoneuroophtal. **16**:241-352, 1938.

de Morsier, G., and Fellmann, H.: Disorders of body sensation caused by traumatic encephalopathy; some remarks on the pathogenesis of visual hallucinations (les troubles du schéma corporel dans l'encéphalopathie traumatique), Schweiz. Arch. Neurol. **70**:42-47, 1952.

Nielsen, J. M., and FitzGibbon, J. P.: Agnosia, apraxia, aphasia: their value in cerebral localization, Bull. Los Angeles Neurol. Soc., 1936.

Nielsen, J. M., and von Hagen, K. O.: Three cases of mind blindness (visual agnosia): one due to softening in occipital lobes (autopsy), one due to anterior poliomyelitis (non-fatal), one due to drugs (transient), J. Nerv. Ment. Dis. **84**:386-398, 1936.

Orlando, R., and Arndt, M.: La agnosia optica, Neuropsiquiatria **2**:36-63, 1951.

Paillas, J. E., Alliez, J., and Tamalet, J.: Sur la valeur localisatrice des hallucinations visuelles paroxystiques, Ann. Med.-Psychol. 2:473, 1949.

Parkinson, D., Rucker, C. W., and McCraig, K. W.: Visual hallucinations associated with tumors of occipital lobe, Arch. Neurol. Psychiat. 68:66, 1952.

Poppelreuter, W.: Die psychischen Schädigungen durch Kopfschuss im Kriege 1914-16: Die Störungen der niederen und höheren Sehleistungen durch Verletzung des Occipitalhirns, Leipzig, 1917, Voss.

Quensel, F.: Die Erkrankungen der höheren optischen Zentren. In Schieck, F., and Brückner, A., editors: Kurzes Handbuch der Ophthalmologie, Berlin, 1931, Julius Springer, vol. 6, p. 324.

Renard, G.: Les hémianopsies: recherche et valeur diagnostique, Presse Med. 58:174-177, 1950.

Riese, W.: Craniopharyngiome chez une femme âgée de 57 ans; hallucinations visuelles et auditives; deuxième note sur la genèse des hallucinations survenant chez les malades atteints de lésions cérébrales, Rev. Neurol. 82:137-139, 1950.

Riley, H. A., Yaskin, J. C., Riggs, M. E., and Torney, A. S.: Bilateral blindness due to lesions in both occipital lobes, New York J. Med. 43:1619, 1943.

Sanford, H. S., and Blair, H. L.: Visual disturbances associated with tumors of the temporal lobe, Arch. Neurol. Psychiat. 42:21, 1939.

Savitsky, N., and Madonick, M. J.: Arch. Neurol. Psychiat. 53:135, 1945; 55:232, 1946.

Seitelberger, F.: Über zerebrale Metamorphopsie, Wien. Med. Wschr. 102:980-983, 1952.

Souter, W. C.: Visual disturbances of central origin. In Berens, C., editor: The eye and its diseases, Philadelphia, 1949, W. B. Saunders Co.

Tarachow, S.: Clinical value of hallucinations in localizing brain tumors, Amer. J. Psychiat. 97:1434-1442, 1941.

Thiébaut, F., Guillaumat, L., and Brégeat, P.: L'hémianopsie relative, Bull. Soc. Franc. Ophtal. 60:73-80, 1947.

Triska, H. I.: Amaurose und Tumor Cerebri, Wien, Med. Wschr. 103:323-326, 1953.

Tyler, H. R.: Cerebral disorders of vision. In Smith, J. L., editor: Neuro-ophthalmology, St. Louis, 1968, The C. V. Mosby Co., vol. 4, pp. 266-281.

Vogel, P.: Die Bedeutung der Anamnese für die Diagnostik der Hirntumoren, Deutsch. Med. Wschr. 2:1277-1281, 1938.

Walsh, F. B., and Hoyt, W. F.: Aphasia, apraxia, agnosia, alexia. In Clinical neuro-ophthalmology, Baltimore, 1969, The Williams & Wilkins Co., pp. 98-119.

Weinberger, L. M., and Grant, F. C.: Visual hallucinations and their neuro-optical correlates (review), Arch. Ophthal. 23:166-199, 1940.

Wernicke, E.: Der aphasische Symptomenkomplex, Breslau, 1874.

Wilbrand, H.: Seelenblindheit als Herderscheinung, Wiesbaden, 1887.

Wilbrand, H.: Die hemianopischen Gesichtsfeldformen und das optische Wahrnehmungszentrum, Wiesbaden, 1890.

External aspect of the eyes

Cushing, H.: The meningiomas, Brain 45:282, 1922.

Elsberg, C. A., Hare, C. C., and Dyke, C. H.: Unilateral exophthalmos in intracranial tumors with special reference to its occurrence in meningiomas, Surg. Gynec. Obstet. 55:681, 1932.

Uhthoff, W.: Die Augenveränderungen bei den Erkrankungen des Gehirns. In Graefe-Saemisch: Handbuch der gesamten Augenheilkunde, Leipzig, 1915, Wilhelm Engelmann, vol. 11, p. 1143.

Corneal sensitivity

Boberg-Ans, J.: Corneal sensitivity with special reference to clinical methods of examination (in Danish with an English summary), M.D. Thesis, Cophenhagen, 1952.

Duke-Elder, S.: Textbook of ophthalmology, St. Louis, 1938, The C. V. Mosby Co., vol. 2, p. 1155.

von Frey, M.: Physiologische Sensibilitätsprüfungen, Verh. Deutsch. Ges. Inn. Med., pp. 19, 74, 1925.

von Frey, M., and Strughold, H.: Weitere Untersuchungen über das Verhalten von Hornhaut und Bindehaut des menschlichen Auges gegen Berührungsreize, Z. Biol. 84:321, 1926.

Pannabecker, C. L.: Keratitis neuroparalytica: corneal lesions following operations for trigeminal neuralgia, Arch. Ophthal. 32:456-463, 1944.

Skrzypczak, J.: Der Kornealreflex und seine Bedeutung für die topische Hirntumordiagnostik, Klin. Mbl. Augenheilk. 152:465-475, 1968.

Weve, H.: Prüfung der Hornhautsensibilität. In Amsler, M., et al., editors: Lehrbuch der Augenheilkunde, Basel, 1954, S. Karger AG, p. 34.

Pupils and pupillary reactions

Adie, W.: Complete and incomplete forms of the benign disorder characterized by tonic pupils and absent tendon reflexes, Brit. J. Ophthal. 16:449-461, 1932.

Adie, W.: Disorder characterized by tonic pupil and absent tendon reflexes, Brit. J. Ophthal. 16:449, 1932.

Adie, W.: Tonic pupils and absent tendon reflexes: a benign disorder sui generis: its complete and incomplete forms, Brain 55: 98-113, 1932.

Adler, F. H., and Scheie, H. G.: The site of the disturbance in tonic pupils, Trans. Amer. Ophthal. Soc. 38:183, 1940.

Argyll Robertson, D.: Four cases of spinal miosis; with remarks on the action of light on the pupil, Edinburgh Med. J. 15:487, 1869.

Barrios, R. R.: Diagnosis of localization of Claude Bernard-Horner's syndrome, Arch. Oftal. B. Air. 18:629, 1943.

Bing, R., and Franceschetti, A.: Die Pupille. In Schieck, F., and Brückner, A., editors: Kurzes Handbuch der Ophthalmologie, Berlin, 1931, Julius Springer, vol. 6, p. 80.

Bolsi, D.: Rilievi clinici e nosologici sulla sindrome di Adie, Riv. Otoneurooftal. 27: 361-366, 1952.

Bourbon, O. P.: An improved pupillometer, Amer. J. Ophthal. 25:1107, 1942.

Bürki, E.: Über Pupillotonie, Klin. Mbl. Augenheilk. 99:145, 1937.

Bürki, E.: Zur differentialdiagnostischen Bedeutung der reflexorischen Pupillenstarre, Schweiz. Med. Wschr. 34:774, 1937.

Demmler, P.: Pupillenstörungen bei Hirntumoren und ihre Beziehungen zu anderen Schädigungen des Sehorgans, Arch. Psychiat. 107:701-710, 1938.

Dressler, M., and Wagner, H.: Das Syndrom von Adie, Schweiz. Med. Wschr. 22: 1937.

Dressler, M., and Wagner, H.: Über das Adiesche Syndrom, Schweiz. Arch. Neurol. Psychiat. 39:246; 40:50, 1937.

Druault-Toufesco: Sur ce que peut apprendre l'examen des pupilles, Bull. Soc. Franc. Ophtal. 65:62-66, 1952.

Duke-Elder, S.: Textbook of ophthalmology, St. Louis, 1938, 1949, The C. V. Mosby Co., vols. 2 and 4, pp. 3733-3805.

Dynes, J. B.: Adie's syndrome: its recognition and importance, J.A.M.A. 119:1493, 1942.

Edinger, L.: Vorlesungen über den Bau der nervösen Zentralorgane des Menschen der Tiere, Leipzig, 1911, F. W. C. Vogel.

Fanta, H., and Reisner, H.: Argyll Robertson-sches Syndrom bei suprasellärem Tumor, Klin. Mbl. Augenheilk. 121:63-68, 1952.

Faure-Beaulieu, M., Christophe, J., and Isorni, P.: Pupillary areflexia and Parinaud's syndrome, Rev. Neurol. 73:29, 1941.

Fischer-Brügge, K.: Das Klivuskanten-Syndrom, Acta Neurochir. 2:36-68, 1951.

Foerster, O., and Gagel, O.: Z. Ges. Neurol. Psychiat. 88:1, 1932.

Foerster, O., Gagel, O., and Mahoney, W.: Über die Anatomie und Pathologie der Pupillarinnervation, Verh. Deutsch. Ges. Inn. Med. 48:386, 1936.

Gunn, R. Marcus: Retrobulbar neuritis, Lancet 2:412, 1904.

Harms, H.: Entwicklungsmöglichkeiten der Perimetrie, Graefe. Arch. Ophthal. 150:28, 1950.

Harms, H.: Hemianopische Pupillenstarre, Klin. Mbl. Augenheilk. 118:133-147, 1951.

Hartmann, E.: Les pupilles dans les traumatismes et les tumeurs cérébrales, Rev. Neurol. 69:646-656, 1938.

Hess, C.: Untersuchungen zur Physiologie und Pathologie des Pupillenspieles, Arch. Augenheilk. 60:327-389, 1908.

Horner, F.: Über eine Form von Ptosis, Klin. Mbl. Augenheilk. 7:193-198, 1869.

Inciardi, J. A.: A routine for diagnosing abnormal pupils, Dis. Eye Ear Nose Throat 2:219, 1942.

Ingvar, S.: The pathogenesis of the Argyll Robertson phenomenon, Bull. Johns Hopkins Hosp. 43:363, 396, 1928.

Jackson, E.: A simple pupillometer, Amer. J. Ophthal. 25:871, 1942.

Jackson, Hughlings: Lancet 1:11, 1894.

Jaensch, P. A.: Pupille, Handbuch der Neurologie, Berlin, 1936, Julius Springer. vol. 4.

Jaensch, P. A.: Grenzgebiet der Ophthalmologie und Neurologie; Pupille, 1933-1937, Fortschr. Neurol. Psychiat. 10:366-384, 1938.

Jaffe, N. S.: Localization of lesions causing Horner's syndrome, Arch. Ophthal. 44: 710-728, 1950.

Jefferson, G.: The tentorial-pressure cone, Arch. Neurol. Psychiat. 40:857, 1938.

Kearns, T. P.: The neuro-ophthalmologic examination. In Smith, J. L., editor: Neuroophthalmology, Springfield, Ill., 1964, Charles C Thomas, Publisher, vol. 1.

Kehrer, F.: Die Kuppelungen von Pupillenstörungen mit Aufhebung der Sehnervenreflexe (Adie-Syndrom, Pupillotonie, Pseudotabes, konstitutionelle Areflexie.), Leipzig, 1937, Georg Thieme.

Kennedy, F., Wortis, H., Reichard, J. D., and Fair, B. B.: Adie's syndrome, Arch. Ophthal. 19:68-80, 1938.

Kerr, F. W. L.: The pupil: functional anatomy and clinical correlation. In Smith, J. L., editor: Neuro-ophthalmology, St. Louis, 1968, The C. V. Mosby Co., vol. 4, p. 49.

Krayenbühl, H., and Noto, G.: Das intrakranielle subdurale Hämatom, Bern, 1949, Hans Huber.

Langdon, H. M.: Three cases of Argyll Robertson pupil, apparently non-luetic, Amer. J. Ophthal. 23:331, 1940.

Leathart, P. W.: The tabetic pupil, Brit. J. Ophthal. 25:111, 1941.

Leathart, P. W.: The tonic pupil syndrome, Brit. J. Ophthal. 28:60, 1942.

Lowenstein, O.: Pupillography: its significance in clinical neurology, Arch. Neurol. Psychiat. 44:227, 1940.

Lowenstein, O.: Clinical pupillary symptoms in lesions of the optic nerve: optic chiasm and optic tract, Arch. Ophthal. 52:385, 1954.

Lowenstein, O., and Friedman, E. D.: Adie's syndrome (pupillotonie, pseudotabes), Arch. Ophthal. 28:1042-1068, 1942.

Lowenstein, O., and Friedman, E. D.: Pupillographic studies: present state of pupillography; its method and diagnostic significance, Arch. Ophthal. 27:969-993, 1942.

Lowenstein, O., Murphy, S. B., and Loewenfeld, I. E.: Functional evaluation of the pupillary light reflex pathways: experimental pupillographic studies in cats, Arch. Ophthal. 49:656-670, 1953.

McKinney, J. M., and Frocht, M.: Adie's syndrome: a nonluetic disease simulating tabes dorsalis, Amer. J. Med. Sci. 199:546, 1940.

Marin-Amat, M.: Valor clinico de la anisocoria, Medicina (Madrid) 20:129-148, 1952.

Nathan, P. W., and Turner, J. W. A.: Argyll Robertson's pupil: efferent pathway for contraction, Brain 65:343, 1942.

Pagliarani, N.: On the localizing value of Foerster's pharmacodynamic test in lesions of the cervical sympathetic, Proceedings of the Sixteenth International Congress on Ophthalmology, London, 1950, vol. 1, pp. 375-380.

de Quervain, F.: Die starre Pupillenerweiterung in der Diagnostik der Schädel- und Hirntraumen, Schweiz. Med. Wschr. 1:75, 1935.

Rook, J. T.: Adie's syndrome, Arch. Ophthal. 29:936, 1943.

Rosen, E.: Adie's syndrome, Arch. Ophthal. 30:553, 1943.

Schachter, M.: Pupillary disturbances in the course of cerebral tumors, Confin. Neurol. 5:298, 1943.

Scharfetter, F.: Mydriasis und Lichtstarre, Schweiz. Arch. Neurol. Psychiat. 96:386-392, 1965.

Scheie, H. G.: Site of disturbance in Adie's syndrome, Arch. Ophthal. 24:225-237, 1940.

Schwab, R. S.: Differential diagnosis between Argyll Robertson and Adie's myotonic pupil, Amer. J. Ophthal. 23:456, 1940.

Spiegel, E. A.: Argyll Robertson pupil, Urol. Cutan. Rev. 45:428-432, 1941.

Stern, H. J.: A simple method for the early diagnosis of abnormalities of the pupillary reaction, Brit. J. Ophthal. 28:275, 1944.

Tönnis, W.: Kopfverletzungen. In Tönnis W., et al., editors: Taschenbücher d. Truppenarztes Munich, 1938, J. F. Lehmann Verlag, vol. 2.

Turner, E. A.: Pupillary inequalities in man, Brain 68:98, 1945.

Walsh, F. B., and Hoyt, W. F.: The pupil in neurologic diagnosis. In Clinical neuro-ophthalmology, Baltimore, 1969, The Williams & Wilkins Co.

Wernicke, C.: Hemianopische Pupillen-Reaktion, Virchows Arch. Path. Anat. 56:397, 1872.

Westphal, A.: Weiterer Beitrag zur Pathologie der Pupille, Neurol. Centralbl. 32:517, 1913.

Weve, H.: Vorrichtung zur Untersuchung der hemianopischen Pupillarreaktion, Klin. Mbl. Augenheilk. 61:140, 1918.

Weve, H.: Untersuchung der Pupille. In Amsler, M., et al., editors: Lehrbuch der Augenheilkunde, Basel, 1954, S. Karger AG, p. 10.

Wilson, S. A. K.: Some problems in neurology. I. The Argyll Robertson pupil, J. Neurol. Psychopath. 2:1-25, 1921.

Wilson, S. A. K., and Gerstle, M.: The Argyll Robertson sign in mesencephalic tumors, Arch. Neurol. Psychiat. 22:9-18, 1929.

Wormser, P.: Die Reaktion der Pupille auf Pharmaka nach Unterbrechung der sympathischen Pupillenbahn, Confin. Neurol. 8:5, 249, 1947-1948.

Motility of the eyes (extraocular muscle palsies, palsies of conjugate eye movements, nystagmus)

Alpers, B. J.: Partial paralysis of upward gaze (incomplete Parinaud syndrome), Confin. Neurol. 15:1-12, 1942.

Aragones-Olle, J. M., and Obach-Tuca, J.: Contribucion al estudio electronistagmografico, con registro simultaneo de la actividad de los musculos del cuello, durante las pruebas vestibulares caloricas en los tumores endocraeneales, Rev. Esp. Otoneurooftal. 25:378-397, 1966.

Arganaraz, R.: Nystagmus, conjugate deviations and paralyses of associated movements in cerebral diseases, Arch. Oftal. B. Air. 17:113-123, 1942.

Arganaraz, R.: Study of opticokinetic nystagmus as a means of diagnosis, Arch. Oftal. B. Air. **17**:323-329, 1942.

Bárány, R.: Zur Klinik und Theorie des Eisenbahnnystagmus, Arch. Augenheilk. **88**: 139, 1921.

Behr, C.: Die Erkrankungen der Augennerven. In Schieck, F., and Brückner, A., editors: Kurzes Handbuch der Ophthalmologie, Berlin, 1931, Julius Springer, vol. 6, p. 156.

Behr, C.: Der Nystagmus retractorius (Koerber). In Schieck, F., and Brückner, A., editors: Kurzes Handbuch der Ophthalmologie, Berlin, 1931, Julius Springer, vol. 6, p. 182.

Bender, M. B., and Savitsky, N.: Paralysis of divergence, Arch. Ophthal. **23**:1046-1051, 1940.

Beres, C., and MacAlpine, P. T.: Motor anomalies of the eye. In Berens, C., editor: The eye and its diseases, Philadelphia, 1949, W. B. Saunders Co.

Best, F.: Localization of cerebral lesions producing disturbances of homolateral eye movements, Arch. Ophthal. **144**:25-40, 1941.

Bielschowsky, A.: Lectures on motor anomalies of eyes; paralysis of conjugate movements of eyes, Arch. Ophthal. **13**:569-583, 1935.

Bielschowsky, A.: Disturbances of vertical motor muscles, Acta Ophthal. **16**:235-270, 1938.

Bielschowsky, A.: Lectures on motor anomalies; oculomotor nerve paralysis and ophthalmoplegias, Amer. J. Ophthal. **22**: 484-498, 1939.

Bielschowsky, A.: Lectures on motor anomalies; paralysis of individual eye muscles: abducens-nerve paralysis, Amer. J. Ophthal. **22**:357-367, 1939.

Bielschowsky, A.: Lectures on motor anomalies; supranuclear paralyses, Amer. J. Ophthal. **22**:603-613, 1939.

Björk, A., and Kugelberg, E.: Motor unit activity in the human extraocular muscles, Electroenceph. Clin. Neurophysiol. **5**:271, 1953.

Blodi, F. C., and Van Allen, M.: Electromyography of extraocular muscles in fusional movements, Amer. J. Ophthal. **44**: 136-144, 1957.

van Bogaert, L.: Contribution anatomoclinique à l'étude du nystagmus optique, Rev. Otoneurooculist. **5**:793, 1927.

Böhringer, H. R., and Koenig, F.: Die diagnostische Bedeutung der vertikalen Blicklähmung, Ophthalmologica **125**:357, 1953.

Breinin, G. M.: The electrophysiology of extraocular muscle, Toronto, 1962, University of Toronto Press.

Breinin, G. M.: Research in strabismus. In Vision and its disorders, Bethesda, 1967, National Institute of Neurological Diseases and Blindness, Monograph No. 4, pp. 50-64.

Cairns, H.: Peripheral ocular palsies from neuro-surgical point of view, Trans. Ophthal. Soc. U. K. **58**:464-482, 1938.

Cawthorne, T., Dix, M. R., and Hood, J. D.: Recent advances in electronystagmographic technique with special reference to its value in clinical diagnosis. In Smith, J. L., editor: Neuro-ophthalmology, St. Louis, 1968, The C. V. Mosby Co., vol. 4, p. 81-109.

Chamlin, M., and Davidoff, L. M.: Divergence paralysis with increased intracranial pressure, Arch. Ophthal. **46**:145-147, 1951.

Chandy, J., and Isaiah, P.: Ophthalmoplegia —a clinical study, Indian J. Surg. **15**:268-274, 1953.

Cogan, D. G.: Neurology of ocular muscles, Springfield, 1956, Charles C Thomas, Publisher.

Cogan, D. H., and Loeb, D. R.: Optokinetic response and intracranial lesions, Arch. Neurol. Psychiat. **61**:183, 1949.

Coppez, H., and Buys, E.: Graphic records of nystagmus, Ophthalmoscope, 1909.

Coppez, L.: Nystagmus, Ophthalmologica **104**:102-120, 1942.

Coppez, L.: Review of literature on nystagmus for 1939, Ophthalmologica **105**:107-119, 1943.

Cords, R.: Die Bedeutung des optischmotorischen Nystagmus für die neurologische Diagnostik, Deutsch. Z. Nervenheilk. **84**:125, 1925.

Cords, R.: Optomotorisches Feld und optomotorische Bahn, Graefe. Arch. Ophthal. **117**:58, 1926.

Cranmer, R.: Nystagmus related to lesions of the central vestibular apparatus and the cerebellum, Ann. Otol. **60**:186-196, 1951.

Cushing, H.: Strangulation of the nervi abducentes by lateral branches of the basilar artery in cases of brain tumor, with an explanation of some obscure palsies on the basis of arterial obstruction, Brain **33**:204-235, 1910-1911.

DeJong, R. N.: Nystagmus: appraisal and classification, Arch. Neurol. Psychiat. **55**: 43-56, 1946.

De Recondo, J.: Les mouvements conjugés oculaires et leurs diverses modalités d'atteinte, Paris, 1967, Masson & Cie.

Druckman, R., Ellis, P., Kleinfeld, J., and Waldman, M.: Seesaw nystagmus, Arch. Ophthal. **76**:668-675, 1966.

Duke-Elder, S.: Textbook of ophthalmology, St. Louis, 1949, The C. V. Mosby, Co., vol.

4, pp. 4031-4046, 4149-4150, and 4199-4239.

Enoksson, P.: Internuclear ophthalmoplegia and paralysis of horizontal gaze, Acta Ophthal. 43:697-707, 1965.

Esslen, E., and Papst, W.: Die Bedeutung der Elektromyographie für die Analyse von Motilitäts-störungen der Augen, Bibl. Ophthal. 57: 1961.

Fink, W. H.: Etiologic considerations of vertical muscle defects, Amer. J. Ophthal. 36:1427, 1551, 1953.

Fox, J. C.: Disorders of optic nystagmus due to cerebral tumors, Arch. Neurol. Psychiat. 28:1007-1029, 1932.

Fox, J. C., and Holmes, G.: Optic nystagmus tumors, Brain 49:333, 1926.

Franceschetti, A.: Le diagnostic des paralysies des différents muscles oculaires, Bull. Soc. Ophtal. Franc. 5:471-498, 1936.

Franceschetti, A., and Blum, J. D.: Motilitätsstörungen des Auges. In Amsler, M., et al., editors: Lehrbuch der Augenheilkunde, Basel, 1954, S. Karger AG.

Franceschetti, A., Monnier, M., and Dieterle, P.: Electronystagmography in the analysis of congenital nystagmus, Trans. Ophthal. Soc. U. K. 72:515, 1952.

Freeman, W.: Paralysis of associated lateral movements of the eyes, Arch. Neurol. Psychiat. 7:454-487, 1922.

Frenzel, H.: Spontan- und Provokations-Nystagmus als Krankheitssymptom, Berlin, 1955, Julius Springer.

Gormann, W. F., and Brock, S.: Nystagmus: its mechanism and significance, Amer. J. Med. Sci. 220:225-233, 1950.

von Graefe, A.: Symptomlehre der Augenmuskellähmungen, 1867.

Green, W. R., Hackett, E. R., and Schlesinger, N. S.: Neuro-ophthalmologic evaluation of oculomotor nerve paralysis, Arch. Ophthal. 72:154-167, 1964.

Hagedoorn, A.: A new diagnostic motility scheme, Amer. J. Ophthal. 25:726-728, 1942.

Hallpike, C. S., Hood, J. D., and Trinder, E.: Some observations on the technical and clinical problems of electronystagmography, Confin. Neurol. 20:232, 1960.

Harms, H.: Examination in paralysis of ocular muscles, Arch. Ophthal. 144:129-149, 1941.

Hart, C. W. J.: The role of nystagmography in clinical diagnoses, Arch. Otolaryng. 84: 631-633, 1966.

Henderson, J. W., and Crosby, E. C.: An experimental study of optokinetic responses, Arch. Ophthal. 47:43-54, 1952.

Hess, W. R.: Die graphische Darstellung von Bewegungsstörungen der Augen mit

Beispieltafeln zur Diagnose von Augenmuskellähmungen, Arch. Augenheilk. 70: 10, 1912.

Holmes, G.: Palsies of the conjugate ocular movements, Brit. J. Ophthal. 5:241-250, 1921.

Holmes, G., and Horrax, G.: Disturbances of spatial orientation and visual attention with loss of stereoscopic vision, Arch. Neurol. Psychiat. 1:385, 1919.

Huber, A.: Die peripheren Augenmuskellähmungen, Ber. Deutsch. Ophth. Ges. 67: 25-46, 1965.

Huber, A.: Topographische und ätiologische Analyse von Augenmuskellähmungen im Elektromyogramm, Ophthalmologica 149: 359-374, 1965.

Huber, A., and Esslen, E.: Diagnostic des myopathies oculaires à l'aide de l'electromyographie, Bull. Soc. Franc. Ophtal. 80: 460-472, 1967.

Huber, A., and Lehner, F. H.: Zur Elektromyographie der Augenmuskeln, Ophthalmologica 131:238-247, 1956.

Huber, A., and Wiesendanger, M.: L'analyse des paralysies oculaires par l'électromyographie, Bull. Soc. Franc. Ophtal. 75: 438-448, 1962.

Jaensch, P. A.: Störungen der Augenbewegungen bei Erkrankungen des Zentralnervensystems, Fortschr. Neurol. Psychiat. 9: 114-130, 1937.

Jayle, G. E. A.: A propos des paralysies d'origine supra-nucléaire; essai de classification, Bull. Soc. Ophtal. Franc. 50:144-152, 1938.

Jung, R., and Kornhuber, H. H.: Results of electronystagmography in man; the value of optokinetic, vestibular and spontaneous nystagmus for neurologic diagnosis and research. In Bender, M. B., editor: The oculomotor system, New York, 1964, Harper & Row, Publishers.

Kestenbaum, A.: Zur Klinik des optokinetischen Nystagmus, Graefe. Arch. Ophthal. 124:113, 1930.

Kestenbaum, A.: Blickbewegungen und Blicklähmungen, Confin. Neurol. 2:121, 1939.

Kestenbaum, A.: Topical diagnosis of disturbed oculomotor motility, Amer. J. Ophthal. 29:94-95, 1946.

Knapp, E.: Kommt Spontan-Nystagmus bei Gesunden vor? HNO. Beih. Hals-Nasen-Ohren Heilk. 2:17-19, 1950.

Lancaster, W.: Fifty years' experience in ocular motility, Amer. J. Ophthal. 24:485, 619, 741, 1941.

Leiva, A. V.: Paralisis de los nervios oculomotores, estudio de 40 casos, Arch. Soc. Cubana Oftal. 3:104-122, 1953.

Lenz, H.: Das Verhalten des optokinetischen

Nystagmus bei einigen Fällen von Lappen-resektionen, Nervenarzt **14**:124-126, 1941.

Leydhecker, W.: Divergenzlähmung, Klin. Mbl. Augenheilk. **123**:83-86, 1953.

Ling, W., and Gay, A. J.: Optokinetic nystagmus: a proposed pathway and its clinical application. In Smith, J. L., editor: Neuro-ophthalmology, St. Louis, 1968, The C. V. Mosby Co., vol. 4, pp. 117-123.

Lippmann, O.: Paralysis of divergence due to cerebellar tumor, Arch. Ophthal. **31**: 299, 1944.

Lutz, A.: Über die Bahnen der Blickwendung und deren Dissoziierung, Klin. Mbl. Augenheilk. **70**:213-235, 1923.

Lyle, D. J.: Divergence insufficiency, Arch. Ophthal. **52**:858, 1954.

Lyle, D. J., and Mayfield, F. H.: Retraction nystagmus, Amer. J. Ophthal. **37**:177, 1954.

Maddox, E. E.: Test and studies of the ocular muscles, Philadelphia, 1907, Keystone Publishing Co.

Magni, S.: Il segno di Kestenbaum nelle atrofie ottiche, Atti Soc. Oftal. Ital. **12**: 254-255, 1950.

Malbrán, J.: Anomalies of the vertical movements of the eyes, Arch. Oftal. B. Air. **15**:65-102, 1940.

Monnier, M., and Hufschmidt, H. J.: Das Elektro-Oculogramm (EOG) und Elektro-nystagmogramm (ENG) beim Menschen, Helv. Physiol. Pharmacol. Acta **9**:348-366, 1951.

Nicolai, H.: Weitere Erfahrungen auf dem Gebiet der objektiven Sehschärfenbestimmung mit optokinetischem Nystagmus, Klin. Mbl. Augenheilk. **124**:81, 1954.

Norden, von G., and Preziosi, T. J.: Eye movement recordings in neurological disorders, Arch. Ophthal. **76**:162-171, 1966.

Nordmann, J.: Considérations sur les paralysies oculo-motrices verticales, Rev. Otoneuroophtal. **24**:152-154, 1952.

O'Brien, F. H., and Bender, M. B.: Localizing value of vertical nystagmus, Arch. Neurol. Psychiat. **54**:378-380, 1945.

Ohm, J.: Registriervorrichtung für waagrechte Augenbewegungen, Z. Augenheilk. **36**: 198, 1916.

Ohm, J.: Die klinische Bedeutung des optischen Drehnystagmus, Klin. Mbl. Augenheilk. **68**:323, 1922.

Ohm, J.: Der optische Drehnystagmus bei Augen- und Allgemeinleiden, Graefe. Arch. Ophthal. **114**:169-191, 1924.

Ohm, J.: Zur Augenzitternkunde, Graefe. Arch. Ophthal. **117**:174, 1926, pt. 3.

Ohm, J.: Zur Augenzitternkunde; über den Einfluss der gleichseitigen Halbblindheit auf den optokinetischen Nystagmus, Graefe. Arch. Ophthal. **135**:200-219, 1936.

Olsen, C. W., and Millitzer, M. M.: Parinaud's syndrome; case with autopsy, Bull. Los Ángeles Neurol. Soc. **5**:224-226, 1940.

Parinaud, M.: Paralysie de la convergence, paralysie de la divergence, Ann. Oculist. **95**:205, 1886.

Parinaud, M.: Paralysie des mouvements associés des yeux, Ann. Oculist. **175**:47-66, 1938.

Rintelen, F.: Zur Kenntnis der Blickparesen, Ophthalmologica **114**:325-331, 1947.

Robbins, A. R.: Divergence paralysis with autopsy report, Amer. J. Ophthal. **24**:556-557, 1941.

Roelofs, C. O.: Optokinetic nystagmus, Docum. Ophthal. **7-8**:579, 1954.

Roper, K. L.: Paralysis of convergence, Arch. Ophthal. **25**:336-353, 1941.

Roussel, F.: Die Diagnostik der vertikalen Doppelbilder: (a) Koordinometer nach Hess-Lees; (b) Untersuchung nach Franceschetti, Bull. Soc. Belg. Ophtal. **101**:439-444, 1952.

Rucker, C. W.: Nystagmus, Amer. J. Ophthal. **2**:250, 1953.

Sager, O., and Voiculescu, V.: Etudes expérimentales sur les voies cortico-oculogyres de verticalité, Folia. Psychiat. Neurol. Neurochir. Neerl. **53**:394-407, 1950.

Salleras, A., Carrillo, A., and Amezuà, L.: Divergencia vertical de la mirada (Fenomeno de Hertwig-Magendie), Arch. Oftal. B. Air. **25**:317-321, 1950.

Savitsky, N., and Madonick, M. J.: Arch. Neurol. Psychiat. **53**:135, 1945; **55**:232, 1946.

Scala, N. P., and Spiegel, E. A.: Mechanism of optokinetic nystagmus, Trans. Amer. Acad. Ophthal. **43**:277-303, 1938.

Schifferli, P.: Etude par enregistrement photographique de la motricité oculaire dans l'exploration, dans la reconnaissance et dans la représentation visuelles, Basel, 1953, S. Karger AG.

Schuster, H.: Zur Pathologie der vertikalen Blicklähmung, Münchner Med. Wschr. **16**: 497, 1921.

Sédan, J., and Sédan-Bauby, S.: Les diplopies de provocation, leurs premiers résultats en neuro-clinique oculaire, Rev. Otoneuroophtal. **19**:257-276, 1947.

Sédan, J., and Séden-Bauby, S.: Principe, but, mécanisme, technique, variétés et résultats de la provocation diplopique, Rev. Otoneuroophtal. **21**:4, 200, 1949.

Sen Gupta, M.: Upward gaze palsy with convergence paralysis (Parinaud's syndrome) with a case report, Proc. All-India Ophthal. Soc. **12**:187-201, 1951.

Spaeth, E. B.: Nystagmus: its diagnostic

326 Literature

significance, Arch. Ophthal. 44:549-560, 1950.

Spiller, W. G.: Corticonuclear tracts for associated ocular movements, Arch. Neurol. Psychiat. 28:251, 1932.

Stadlin, W.: Hémianopsie et nystagmus opto-cinétique, Ophthalmologica 118:383, 1949.

Stenvers, H. W.: On the optic (opto-kinetic, opto-motorial) nystagmus, Acta Otolaryng. 8:545, 1925.

Stenvers, H. W.: Die Wichtigkeit der opto-motorischen Reaktionen für die klinische Diagnostik, Z. Ges. Neurol. Psychiat. 152: 197-207, 1935.

Stenvers, H. W.: Über die klinische Bedeutung des optischen Nystagmus für die cerebrale Diagnostik, Schweiz. Arch. Neurol. 14:279, 1949.

Strauss, H.: Hirnlokalisatorische Bedeutung des einseitigen Ausfalls des optokinetischen Nystagmus, Z. Ges. Neurol. Psychiat. 143: 427, 1933.

Tamler, E., and Jampolsky, A.: Electromyographic study following movements of the eye between tertiary positions, Arch. Ophthal. 62:804-809, 1959.

Teuber, H. L., and Bender, M. B.: Neuro-ophthalmology: the oculomotor system, Progr. Neurol. Psychiat. 6:148-178, 1951.

Torok, N., Guillemin, V., Jr., and Barnothy, J. M.: Photoelectric nystagmography, Ann. Otol. 60:917-926, 1951.

Walsh, F. B.: Certain abnormalities of ocular movements: their importance in general and neurologic diagnosis, Bull. N. Y. Acad. Med. 19:253-272, 1943.

Walsh, F. B., and Hoyt, W. F.: The ocular motor system. In Clinical neuro-ophthalmology, Baltimore, 1969, The Williams & Wilkins Co., pp. 130-347.

Walsh, F. B., and Hoyt, W. F.: Disorders of muscles. In Clinical neuro-ophthalmology, Baltimore, 1969, The Williams & Wilkins Co., pp. 1242-1311.

Weekers, R., and Roussel, F.: La mesure de la fréquence de fusion en clinique, Docum. Ophthal. 2:130, 1948.

Weve, H.: Untersuchung der Bewegungsstörungen. In Amsler, M., et al., editors: Lehrbuch der Augenheilkunde, Basel, 1954, S. Karger AG, p. 47.

White, J. W.: What is the minimum routine examination of muscles? Arch. Ophthal. 24:112-131, 1941.

Wilson, Kinnier: Cited in Rea, R. L.: Neuro-ophthalmology, 1941, The C. V. Mosby Co., p. 158.

Fundus

Amalric, R., Bessou, P., and Aubry, J. P.: Angioscopie et angiographie rétiniennes à la fluorescéine en oto-neuro-ophtalmologie, Soc. Otoneuroophtal. de Toulouse, 1966.

Asenjo, A., and Contreras, M.: Optic atrophy in neuro-surgery (in Portuguese), J. Med. Porto. 20:479-485, 1952.

Bailliart, P.: La circulation rétinienne à l'état normal et à l'état pathologique, Paris, 1923, Gaston Doin & Cie.

Barut, C.: Valeur séméiologique de la mesure de la pression de l'artère centrale de la rétine, Conférences Lyonnaises d'Ophtalmologie, 1968, p. 93.

Bedavanija, A.: Dynamometer-Eichkurve mit Hilfe der Applanationstonometrie, Med. Dissertation, Bonn, 1960.

Bettelheim, H.: Vergleichende ophthalmodynamometrische und ophthalmodynamographische Untersuchungen bei obliterierenden Prozessen der Karotiden, Klin. Mbl. Augenheilk. 146:801-819, 1965.

Bettelheim, H.: Experience with ophthalmodynamography in the diagnosis of carotid occlusion, Amer. J. Ophthal. 64:689, 1967.

Bonamour, G., and Wertheimer, J.: Valeur sémiologique de l'atrophie optique unilatérale, Ann. Oculist. 183:199-215, 1950.

Brückner, A.: Ophthalmoskopie. In Schieck, F., and Brückner, A., editors: Kurzes Handbuch der Ophthalmologie, Berlin, 1931, Julius Springer, vol. 2, p. 862.

Charamis, J., Katsourakis, N., and Mandras, G.: The study of the cerebroretinal circulation by intravenous fluorescein injection, Amer. J. Ophthal. 61:1078-1080, 1966.

Duke-Elder, S.: Textbook of ophthalmology, St. Louis, 1938, The C. V. Mosby Co., vol. 2, p. 1159.

Ferrer, O.: Serial fluorescein fundus photography of retinal circulation. A description of technique, Amer. J. Ophthal. 60:587-591, 1965.

Gay, A. J.: Clinical ophthalmodynamometry, Int. Ophthal. Clin. 7:729-744, 1967.

Hager, H.: Objektive elektrische Dynamometrie mit Hilfe des Bulbus-Orbita-Pulses, XVII. Concilium Ophthalmologicum, Belgica, 1958.

Hardy, H.: Ophthalmoscopy. In Berens, C., editor: The eye and its diseases, Philadelphia, 1949, W. B. Saunders Co., p. 174.

Hartmann, E., and Guillaumat, L.: Aspect du fond d'oeil dans les tumeurs intra-crâniennes: étude statistique, Ann. Oculist. 175:717-737, 1938.

Hollenhorst, R. W.: Ophthalmodynamometry in the diagnosis of intracerebral hypotension, Mayo Clin. Proc. 38:532, 1963.

Horrax, G.: Importance of optic nerve atrophy in diagnosis of favorable brain tumors, S. Clin. N. Amer. 21:903-912, 1941.

Jaensch, P. A.: Atrophia nervi optici; Bericht über die Jahre 1933-1938, Fortschr. Neurol. Psychiat. **12**:83-100, 1940.

Justice, J., and Sever, R. J.: Technique of fluorescein fundus photography. In Smith, J. L., editor: Neuro-ophthalmology, St. Louis, 1965, The C. V. Mosby Co., vol. 2, pp. 82-83.

Jütte, A., and Lemke, L.: Intravitalfärbung am Augenhintergrund mit Fluoreszein-Natrium, Bücherei des Augenarztes, no. 49, Stuttgart, 1968, Ferdinand Enke.

Matsui, J., Koh, K., and Tashiro, T.: Studies on the fluorescence fundus photography. II. Clinical significances of the serial fluorescence fundus photography in some retinal vascular lesions, Acta Soc. Ophthal. Jap. **70**:613-619, 1966. Cited in Zbl. Augenheilk. **97**:430, 1967.

Niesel, P.: Ophthalmodynamometrie, Ophthalmologica **158**:342-352, 1969.

Norton, E. W. D.: Angiography of ocular fundus. In Smith, J. L., editor: Neuro-ophthalmology, St. Louis, 1965, The C. V. Mosby Co., vol. 2, pp. 62-81.

Novotny, H., and Alvis, D. L.: A method of photographing fluorescene in circulating blood in the human retina, Circulation **24**:82-86, 1961.

Raverdino, E.: Sull'importanza dei sintomi oftalmoscopici prima e dopo gli interventi per tumori endocranici, Riv. Otoneurooftal. **15**:532-543, 1938.

Rintelen, F.: Zur diagnostischen Bedeutung der Dynamometrie, Klin. Mbl. Augenheilk. **100**:469, 1938.

Salmon, M. L., and Gay, A. J.: Ophthalmodynamography, Int. Ophthal. Clin. **7**:744, 1967.

Smith, J. L., editor: Neuro-ophthalmology, Springfield, Ill., 1964, Charles C Thomas, Publisher, vol. 1.

Streiff, B., and Monnier, M.: Der retinale Blutdruck im gesunden und kranken Organismus, Berlin, 1946, Julius Springer.

Vogt, A.: Die Nervenfaserzeichnung der menschlichen Netzhaut im rotfreien Licht, Klin. Mbl. Augenheilk. **66**:718, 1921.

Walsh, F. B., and Hoyt, W. F.: Optic atrophy. In Clinical neuro-ophthalmology, Baltimore, 1969, The Williams & Wilkins Co., p. 631.

Weigelin, E., and Lobstein, A.: Ophthalmodynamometry, New York, 1963, Hepner.

Wessing, A.: Fluoreszenzangiographie der Retina, Stuttgart, 1968, Georg Thieme.

Weve, H.: Ophthalmoskopie. In Amsler, M., et al., editors: Lehrbuch der Augenheilkunde, Basel, 1954, S. Karger AG, p. 11.

Witmer, R.: Differential diagnostische Aspekte bei retinalen Gefäss-störungen und Papillenödem, Ophthalmologica **156**:313-321, 1968.

Woltman, H. W.: The ophthalmoscope in neurologic diagnosis, Mayo Clin. Proc. **26**:226-231, 1951.

Visual acuity

Birkhäuser, R.: Scalae typographicae, Basel, A. E. Birkhäuser.

Brückner, A.: Die Untersuchung der Sehschärfe. In Schieck, F., and Brückner, A., editors: Kurzes Handbuch der Ophthalmologie, Berlin, 1931, Julius Springer, vol. 2, p. 921.

Busch, E., and Moller, H. U.: Ophthalmological symptoms in intracranial tumors with special reference to visual acuity, Acta Ophthal. **16**:453-454, 1938.

Duke-Elder, S.: Textbook of ophthalmology, St. Louis, 1938, The C. V. Mosby Co., vol. 2, p. 1194.

Elsberg, C. A., and Spotnitz, H.: Value of quantitative visual test for localization of supratentorial tumors of brain; preliminary report, Bull. Neurol. Inst. N. Y. **6**:411-420, 1937.

Goldmann, H.: Objektive Sehschärfenbestimmung, Ophthalmologica **105**:240-252, 1943.

Günther, G.: Objektive Sehschärfenbestimmung, Samml. Zwangl. Abh. aus d. Gebiete d. Augenheilk., Halle a.d.S., 1950.

Holmes, G.: Disturbances of vision by cerebral lesions, Brit. J. Ophthal. **2**:353-384, 1918.

Jaeger, E.: See under Brückner, A.

Keil, F. C.: Visual acuity. In Berens, C., editor: The eye and its diseases, Philadelphia, 1949, W. B. Saunders Co.

Leinfelder, P. J.: Unilateral loss of vision in neurological disease, Amer. J. Ophthal. **22**:1337-1342, 1939.

Lemoine, P., and Valois, G.: Les optotypes. In Bailliart, P., et al., editors: Traité d'ophtalmologie, Paris, 1939, Masson & Cie, vol. 2, p. 937.

Ohm, J.: Objektive Prüfung der Sehleistungen mit Hilfe der optokinetischen Augenbewegungen, Stuttgart, 1953, Ferdinand Enke.

Schumann, W. P.: The objective determination of visual acuity on the basis of the optokinetic nystagmus, Amer. J. Optom. **29**:575-583, 1952.

Snellen, H.: See under Brückner, A.

Weve, H.: Bestimmung der Sehschärfe. In Amsler, M., et al., editors: Lehrbuch der Augenheilkunde, Basel, 1954, S. Karger AG, p. 35.

Color sensation

Chance, B.: The color sense and its derangements. In Berens, C., editor: The eye and its diseases, Philadelphia, 1949, W. B. Saunders Co., p. 196.

Duke-Elder, S.: Textbook of ophthalmology, St. Louis, 1938, The C. V. Mosby Co., vol. 2, p. 1206.

Gallagher, J. R., Gallagher, C. D., and Sloane, A. E.: A critical evaluation of pseudo-isochromatic plates and suggestions for testing color vision, Yale J. Biol. Med. 15: 79-98, 1942.

Gallagher, J. R., Gallagher, C. D., and Sloane, A. E.: A brief method of testing color vision with pseudo-isochromatic plates, Amer. J. Ophthal. 26:178, 1943.

Hardy, L. H., and Rand, G.: Tests for the detection and analysis of color-blindness. I. The Ishihara test, an evaluation, J. Opt. Soc. Amer. 35:268, 1945.

Hardy, L. H., and Rand, G.: Tests for the detection and analysis of color-blindness. II. The Ishihara test, comparison of editions, J. Opt. Soc. Amer. 35:350, 1945.

Hardy, L. H., and Rand, G.: Tests for the detection and analysis of color-blindness. III. The Rabkin test, J. Opt. Soc. Amer. 35:481, 1945.

Hardy, L. H., and Rand, G.: Color vision and recent developments in color vision, Arch. Ophthal. 36:603, 1946.

Hardy, L. H., Rand, G., and Rittler, M. C.: Color vision and recent developments in color vision testing, Arch. Ophthal. 35: 603, 1946.

Helmbold, R.: Der Farbensinn. In Schieck, F., and Brückner, A., editors: Kurzes Handbuch der Ophthalmologie, Berlin, 1931, Julius Springer, vol. 2, p. 295.

Hertel, E.: Stilling's Pseudo-isochromatische Tafeln, Leipzig, 1929, Georg Thieme.

Ishihara, S.: Tests for colour-blindness, London, 1944, H. K. Lewis & Co., Ltd.

Loken, R. D.: The color-meter: a quantitative color vision test, Amer. J. Physiol. 55: 563, 1942.

Martin, L. C.: A standardized colour-vision testing lantern: transport type, Brit. J. Ophthal. 27:255, 1943.

Murray, E.: Color blindness: current tests and the scientific charting of cases, Psychol. Bull. 39:165, 1942.

Murray, E.: Evolution of color vision tests, J. Opt. Soc. Amer. 33:316-334, 1943.

Murray, E.: Color vision tests. In Glasser, O., editor: Handbook of medical physics, Chicago, 1944, Year Book Publishers, Inc.

Pickford, R. W.: The Ishihara test for colour blindness, Nature 153:656, 1944.

Polack, A.: Anomalies du sens chromatique. In Bailliart, P., et al., editors: Traité d' ophtalmologie, Paris, 1939, Masson & Cie, vol. 3, p. 339.

Schwichtenberg, A. H.: Review of color vision with some practical suggestions for medical examiners, Arch. Ophthal. 27:887, 1942.

Shoemaker, R. E.: The pseudoisochromatic plate test of color vision: practical application, Arch. Ophthal. 29:909, 1943.

Sloan, L. L.: The use of pseudo-isochromatic charts in detecting central scotomas due to lesions in the conducting pathways, Amer. J. Ophthal. 25:1352, 1942.

Stilling, J.: Über das Sehen der Farbenblinden, Berlin, 1883.

Traquair, H. M.: An introduction to clinical perimetry, London, 1949, Henry Kimpton.

Weve, H.: Prüfung des Farbensinnes. In Amsler, M., et al., editors: Lehrbuch der Augenheilkunde, Basel, 1954, S. Karger AG, p. 41.

Wilbrand, H.: Ophthalmologischer Beitrag zur Diagnose der Gehirnkrankheiten, Wiesbaden, 1884.

Dark adaptation

Behr, C.: Das Verhalten und die diagnostische Bedeutung der Dunkeladaptation bei den verschiedenen Erkrankungen des Sehnervenstammes, Klin. Mbl. Augenheilk. 5:193, 1915.

Bourdier, F.: Anomalies du sens lumineux. In Bailliart et al., editors: Traité d'ophthalmologie, Paris, 1939, Masson & Cie, vol. 3, p. 305.

Comberg, W.: Lichtsinn. In Schieck, F., and Brückner, A., editors: Kurzes Handbuch der Ophthalmologie, Berlin, 1931, Julius Springer, vol. 2, p. 172.

Della Casa, F.: Ein Adaptometer für den praktischen Arzt, Ophthalmologica 106: 143, 1943.

Duke-Elder, S.: Textbook of ophthalmology, St. Louis, 1939, The C. V. Mosby Co., vol. 2, p. 1208.

Gasteiger, H.: Über Störungen der Dunkeladaptation bei Sehnervenkrankheiten, Klin. Mbl. Augenheilk. 78:827, 1927.

Goldmann, H.: Un nouvel adaptomètre automatique, Bull. Soc. Ophtal. Franc. 63: 4, 1950.

Hamburger, F. A.: Das Sehen in der Dämmerung, Berlin, 1949, Julius Springer.

Hartmann, N. B.: Testing night vision, Brit. Med. J. 2:347, 1941.

Holmes, W. J.: Night vision, Arch. Ophthal. 30:367, 1943.

Mandelbaum, J.: Dark adaptation, Arch. Ophthal. 26:203, 1941.

Nagel, W. A.: Zwei Apparate für die augenärztliche Funktionsprüfung: Adaptometer und kleines Spektralphotometer, Z. Augenheilk. 17:201, 1907.

Rutgers, G. E.: Die Dunkeladaptation bei der

Opticusatrophie, Klin. Mbl. Augenheilk. 71:449, 1934.

von Studnitz, G.: Physiologie des Sehens und retinale Primärprozesse, Akad. Ver.-Ges. Leipzig, 1940.

Verplanck, W. S.: Night vision, the terminal thresholds. In Berens, C., editor: The eye and its diseases, Philadelphia, 1949, W. B. Saunders Co., p. 203.

Weve, H.: Adaptationsprüfung. In Amsler, M., et al., editors: Lehrbuch der Augenheilkunde, Basel, 1954, S. Karger AG, p. 42.

Yudkin, S.: New dark adaptation tester, Brit. J. Ophthal. 25:231, 1941.

Visual field

Allen, T. D., and Carmann, H. F.: Homonymous hemianopic paracentral scotoma, Arch. Ophthal. 20:846-849, 1938.

Amsler, M.: L'examen qualitatif de la fonction maculaire, Ophthalmologica 114:248, 1947.

Amsler, M.: Quantitative and qualitative vision, Trans. Ophthal. Soc. U. K. 69:397, 1949.

Aulhorn, E.: Perimetrie. In Sautter, H., editor: Entwicklung und Fortschritt der Augenheilkunde, Stuttgart, 1963, Ferdinand Enke, pp. 701-706.

Bair, H. L., and Harley, R. D.: Midline notching in the normal field of vision, Amer. J. Ophthal. 23:183, 1940.

Bender, M. B., and Wechsler, I. S.: Irregular and multiple homonymous visual field defects, Arch. Ophthal. 28:904-912, 1942.

Berkeley, W. L., and Bussey, F. R.: Altitudinal hemianopia, Amer. J. Ophthal. 33:593-600, 1950.

Bessière, E., Rougier-Houssin, J., and Verin, P.: Contribution à la séméiologie précoce des scotomes centraux, Arch. Ophtal. 27:359-386, 1967.

Biemond, A.: La projection des quadrants rétiniens dans la radiation optique étudiée dans un cas d'hémianopsie bilatérale, Folia Psychiat. Neurol. Neurochir. Neerl. 53:159-164, 1950.

Bjerrum, J.: About a supplementary examination of the visual field and about the field in glaucoma, Nord. Ophthal. Tidskr., pp. 2, 3, 11, 23, 27, 49, 1889.

Bjerrum, J.: Ein Zusatz zur gewöhnlichen Gesichtsfelduntersuchung und über das Gesichtsfeld bei Glaukom, Verh. 10. Int. Med. Kongr. Berlin 4:66, 1890.

Bonnet, P.: Valeur sémiologique de l'hémianopsie, J. Med. Lyon 33:587-624, 1952.

Bourbon, O. P.: Selection of the test object in perimetry, Amer. J. Ophthal. 23:1260, 1940.

Brückner, A.: Die Untersuchung des Gesichsfeldes. In Schieck, F., and Brückner, A., editors: Kurzes Handbuch der Ophthalmologie, Berlin, 1931, Julius Springer, vol. 2, p. 931.

Buffat, J. D.: L'hémianopsie binasale en neurochirurgie, Rev. Med. Suisse Rom. 70:174-183, 1950.

Chamlin, M.: Visual field studies in neurosurgical problems, Eye Ear Nose Throat Monthly 31:415-422, 1952.

Chamlin, M., and Davidoff, L.: The 1/2000 field in chiasmal interference, Arch. Ophthal. 44:53, 1950.

Chamlin, M., and Davidoff, L. M.: Papilledema: its differential diagnosis, with special reference to minimal testing to the blind spot at two meters, Arch. Neurol. Psychiat. 68:213, 1952.

Comberg, U., and Lommatssch, P.: Zur Bewertung des Zentralskotomes, Klin. Mbl. Augenheilk. 142:336-347, 1963.

Cuendet, J. F., and Dufour, R.: Appréciation quantitative du champ visuel à l'aide de la planimétrie, Confin. Neurol. 14:143, 1954.

Cushing, H.: Distortion of the visual fields in cases of brain tumor: the field defects produced by temporal lobe lesions, Brain 44:341, 1921.

Cushing, H.: The field defects produced by temporal lobe lesions, Brain 44:341, 1921.

Davis, L.: The blind spots in patients with intracranial tumors, J.A.M.A. 92:794, 1929.

D'Orio, R., and Scarinci, A.: L'Hemianopsia binasale, Riv. Otoneurooftal. 34:617-659, 1959.

Dubois-Poulsen, A.: Périmétrie, campimétrie, scotométrie. In Bailliart, P., et al., editors: Traité d'ophtalmologie, Paris, 1939, Masson & Cie, vol. 2, p. 1011.

Dubois-Poulsen, A.: Variations de la sensibilité et interactions nerveuses dans le champ visuel, Ann. Oculist. 185:1025-1052, 1952.

Dufour, R.: Hémianopsie relative et ataxie, Confin. Neurol. 9:413, 1949.

Duke-Elder, S.: Textbook of ophthalmology, St. Louis, 1938, The C. V. Mosby Co., vol. 2, p. 1214.

Ecker, A. D., and Anthony, E. W.: Exanopic central scotoma, a pitfall in diagnosis, J. Neurosurg. 2:47, 1945.

Elsberg, C. A., and Spotnitz, H.: Sense of vision; Reciprocal relation of area and light intensity and its significance for localization of tumors by functional visual test, Bull. Neurol. Inst. N. Y. 6:243-252, 1937.

Enoksson, P.: Internuclear ophthalmoplegia, Acta Ophthal. 43:697, 1965.

Evans, J. N.: Classic characteristics of defects of visual field, Arch. Ophthal. **22**:410-431, 1933.

Evans, J. N.: An introduction to clinical scotometry, London, 1938, Oxford University Press.

Falconer, M. A.: Visual field changes and the value of quantitative perimetry in compression of the optic chiasm and the optic nerve, Trans. Ophthal. Soc. U. K. **3**:8, 1949.

Fanta, H.: Das Gesichtsfeld für Bewegung und Weiss bei intrakraniellen Prozessen, Graefe. Arch. Ophthal. **161**:492-501, 1960.

Ferree, C. E., and Rand, G.: An illuminated perimeter with campimeter features, Amer. J. Ophthal. **5**:455, 1922.

Ferree, C. E., and Rand, G.: Methods for increasing the diagnostic sensitivity of perimetry and scotometry with the form fields stimulus, Amer. J. Ophthal. **13**:2, 1930.

Foerster, R.: Das Perimeter, Sitzungsber. Ophthal. Ges. Heidelberg, 1869, p. 411.

François, J.: L'hemianopsie binasale, Ophthalmologica **113**:321-343, 1947.

Friedmann, A. I.: Serial analysis of changes in visual field defects employing a new instrument, to determine the activity of disease involving the visual pathways, Ophthalmologica **152**:1-12, 1966.

Goldmann, H.: Grundlagen exakter Perimetrie, Ophthalmologica **109**:57, 1945.

Goldmann, H.: Ein selbstregistrierendes Projektionskugelperimeter, Ophthalmologica **109**:71, 1945.

Goldmann, H.: Demonstration unseres neuen Kugelperimeters samt theoretischen und klinischen Bemerkungen über Perimetrie, Ophthalmologica **3**:187, 1946.

Goldmann, H.: La périmétrie en oto-neuro-ophtalmologie, Confin. Neurol. **14**:102, 1954.

Guillaumat, L., and Robin, A.: Etude statistique du champ visuel dans les affections neurochirurgicales non traumatiques, Bull. Soc. Franc. Ophtal. **65**:5-16, 1952.

Harms, H.: Entwicklungsmöglichkeiten der Perimetrie, Graefe. Arch. Ophthal. **150**:28, 1950.

Harms, H.: Die praktische Bedeutung quantitativer Perimetrie, Klin. Mbl. Augenheilk. **121**:683, 1952.

Harms, H.: Quantitative Perimetrie bei sellanahen Tumoren, Ophthalmologica **127**:255, 1954.

Harms, H.: Quantitative Perimetrie bei sellanahen Tumoren, Ophthalmologica **127**:255-261, 1954.

Harms, H.: Diagnostische Bedeutung der Gesichtsfelduntersuchung. In Rohrschneider, W., editor: Augenheilkunde in Klinik und Praxis, Stuttgart, 1958, Ferdinand Enke, pp. 22-49.

Harms, H.: Leitsymptom "Zentralskotom." In Sautter, H., editor: Entwicklung und Fortschritt in der Augenheilkunde, Stuttgart, 1963, Ferdinand Enke, pp. 516-539.

Harms, H., and Raabe, M.: Besondere perimetrische Methoden bei ophthalmo-neurologischen Erkrankungen, Oesterreichische Ophthalmologie Gesellschaft, Vienna, 1960, Brüder Hollinek.

Harrington, D. O.: Localizing value of incongruity in defects in the visual fields, Arch. Ophthal. **21**:453, 1939.

Harrington, D. O.: The visual fields, St. Louis, 1964, The C. V. Mosby Co.

Head, H.: Aphasia and kindred disorders of speech, London, 1926, Cambridge University Press.

Henschen, S. E.: Zur Anatomie der Sehbahn und des Sehzentrums, Graefe. Arch. Ophthal. **117**:403-418, 1926.

Holmes, G.: Disturbances of vision by cerebral lesions, Brit. J. Ophthal. **2**:353-384, 1918.

Holmes, G., and Lister, W. T.: Disturbances of vision from cerebral lesions with special reference to the cortical representation of the macula, Brain **39**:34-73, 1916.

Huber, A.: Zur homonymen Hemianopsie nach occipitaler Lobektomie, Schweiz. Med. Wschr. **80**:1227, 1950.

Hughes, E. B. C.: Selected cases illustrating the value of quantitative perimetry in neurosurgical diagnosis, Trans. Ophthal. Soc. U. K. **63**:143-147, 1944.

Hylkema, B. S.: Klinische Anwendung der Bestimmung der Verschmelzungsfrequenz, Graefe. Arch. Ophthal. **146**:110, 241, 1943.

Igersheimer, J.: Zur Pathologie der Sehbahn: Über Hemianopsie, Graefe. Arch. Ophthal. **97**:105, 1918.

Igersheimer, J.: Binasal hemianopia, Arch. Ophthal. **38**:248, 1947.

Johnson, T. H.: Homonymous hemianopia: some practical points in its interpretation with report of 49 cases in which lesion in brain was verified, Trans. Amer. Ophthal. Soc. **33**:90-113, 1935; Arch. Ophthal. **15**:604-616, 1936.

Kennedy, F.: Retrobulbar neuritis as an exact diagnostic sign of certain tumors and abscesses in the frontal lobe, Amer. J. Med. Sci. **142**:355, 1911.

Kennedy, R. J.: Value of perimetry in brain lesions, Cleveland Clin. Quart. **6**:290-303, 1939.

Kestenbaum, A.: Einfache Methode der groben Gesichtsfeldprüfung, Wien. Med. Wschr. **46**:2533, 1925.

Kestenbaum, A.: Wertung der neuen topischen Diagnostik der Hemianopsie, Proceedings of the Fifteenth International Congress on Ophthalmology, Cairo, 1937, vol. 4, pp. 120-123.

Klingler, M., and Condrau, G.: Lokalisatorisch irreführende Gesichtsfeldsymptome bei Hirntumoren, Ophthalmologica **120**:5, 270, 1950.

Kluyskens, J., and Titeca, J.: Examen électro-encéphalographique du champ visuel, Ophthalmologica **126**:3, 129, 1953.

Krainer, L.: Zur Anatomie und Pathologie der Sehbahn und der Sehrinde, Deutsch. Z. Nervenheilk. **141**:177-190, 1936.

Kravitz, D.: Studies of visual fields in cases of verified tumor of brain, Arch. Ophthal. **20**:437-470, 1938.

Krayenbühl, H.: Die Bedeutung der Gesichtsfeldbestimmung in der Diagnostik von Tumoren des Schläfen- und Hinterhauptlappens, Schweiz. Med. Wschr. **69**:1028-1032, 1939.

Kreiger, H. P.: Visual function in perimetrically blind fields, Arch. Neurol. Psychiat. **65**:72, 1951.

Kronfeld, P. C.: The central visual pathway, Arch. Ophthal. **2**:709-732, 1929.

Kronfeld, P. C.: The temporal half-moon, Trans. Amer. Ophthal. Soc. **30**:431, 1932.

McGavic, J. S.: Visual field defects due to head injuries, Surg. Gynec. Obstet. **84**:823-827, 1947.

Meyer, A.: The connections of the occipital lobes and the present status of the cerebral visual affections, Trans. Ass. Amer. Physicians **22**:7, 1907.

Miles, P. W.: Flicker fusion fields, Amer. J. Ophthal. **33**:769, 1069, 1950.

Miles, P. W.: Testing visual fields by flicker fusion, Arch. Neurol. Psychiat. **55**:39-47, 1951.

Minkowski, M.: Zur Kenntnis der cerebralen Sehbahnen, Schweiz. Med. Wschr. **69**:990-995, 1939.

Mooney, A. J., and McConnell, A. A.: Visual scotomata with intracranial lesions affecting the optic nerve, J. Neurol. Neurosurg. Psychiat. **12**:205-218, 1949.

Parsons, O. A., Chandler, P. J., Teed, R. W., and Haase, G. R.: Comparison of flicker perimetry and standard visual fields in brain-damaged patients, Acta Neurol. Scand. **42**:207-212, 1966.

Pasino, L.: Dati comparativi ottenuti col perimetro di Maggiore e col perimetro di Goldmann, Rass. Ital. Ottal. **22**:318-329, 1953.

Penido Burnier, F.: Consideration on campimetry and perimetry. Paper read before the Medical Association of the Penido Burnier Institute, 1941.

Peter, L. C.: The principles and practice of perimetry, New York, 1923, Lea & Febiger.

Posner, A.: Perimetry: principles and methods, Eye Ear Nose Throat Monthly **31**:378-379, 387, 1952.

Raiford, M. B.: Binasal hemianopia, Amer. J. Ophthal. **32**:99, 1949.

Roenne, H.: Zur Theorie und Technik der Bjerrumschen Gesichtsfelduntersuchung, Arch. Augenheilk. **78**:4, 1915.

Roenne, H.: The different types of defects of the fields of vision, J.A.M.A. **89**:1860-1865, 1927.

Roenne, H.: The focal diagnostic of the visual path, Acta Ophthal. **16**:446-453, 1938.

Rosen, L.: Contribution à l'étude des hémianopsies: symptomatologie et étiologie de 97 cas examinés à la Clinique ophthalmologiques de Bâle, entre 1925 et 1940, Confin. Neurol. **4**:271, 1942.

Rucker, C. W.: Bitemporal defects in visual fields resulting from developmental anomalies of optic disks, Arch. Ophthal. **35**:546-554, 1946.

Sanford, H. S., and Bair, H. L.: Visual disturbances associated with tumors of the temporal lobes, Arch. Neurol. Psychiat. **42**:21, 1939.

Schmidt, T.: Über die Gesichtsfelduntersuchung am Goldmann-Perimeter, Klin. Mbl. Augenheilk. **126**:209, 1955.

Shenkin, H. A., and Leopold, J. H.: Localizing value of temporal crescent defects in visual fields, Arch. Neurol. Psychiat. **54**:97-101, 1945.

Stenvers, H. W.: Localisation directe et localisation par image dans les champs visuels périphériques, Rev. Otoneuroophtal. **23**:6-14, 1951.

Straub, W.: Über die binasale Hemianopsie, Acta Ophthal. **30**:229-252, 1952.

Streiff, E. B.: Traumatismes des voies optiques, Rev. Otoneuroophtal. **26**:356-390, 1951.

Suarez-Villafranca, M. R.: Hemianopsia binasal de localización chiasmatica, Arch. Soc. Oftal. Hispano-Amer. **11**:252-256, 1951.

Suda, K.: Interesting ocular symptoms and quantitative perimetry in brain tumors, Folia Ophthal. Jap. **16**:818-821, 1965.

Thiébaut, F., Guillaumat, L., and Brégeat, P.: L'hémianopsie relative, Bull. Soc. Franc. Ophtal. **60**:73, 1947.

Traquair, H. M.: Perimetry and the visual pathway (Mackenzie memorial lecture), Glasgow Med. J. **133**:105-118, 1940.

Traquair, H. M.: An introduction to clinical perimetry, London, 1949, Henry Kimpton.

Traquair, H. M.: Peripheral vision and perimetry. In Berens, C., editor: The eye and its diseases, Philadelphia, 1949, W. B. Saunders Co., p. 210.

Verrey, F.: Pseudo-hémianopsie inférieure post-hémorragique, Ophthalmologica **125**: 351-356, 1953.

Walker, C. B.: Quantitative perimetry: practical devices and errors, Arch. Ophthal. **46**: 537-561, 1917.

Walsh, F. B.: Selected visual field studies. In Smith, J. L., editor: Neuro-ophthalmology, St. Louis, 1965, The C. V. Mosby Co., pp. 84-108.

Walsh, F. B., and Ford, F. R.: Central scotomas: their importance of topical diagnosis, Arch. Ophthal. **24**:500-534, 1940.

Walsh, F. B., and Hoyt, W. F.: Topical diagnosis of lesions in the visual pathways with particular reference to the visual fields. In Clinical neuro-ophthalmology, Baltimore, 1969, The Williams & Wilkins Co., p. 60.

Weekers, R., and Roussel, F.: Introduction à l'étude de la fréquence de fusion en clinique, Ophthalmologica **112**:305-316, 1946.

Westby, R. K.: Central, paracentral scotoma and intracranial tumour, Acta Ophthal. **41**: 749-756, 1963.

Weve, H.: Untersuchung des Gesichtsfeldes. In Amsler, M., et al., editors: Lehrbuch der Augenheilkunde, 1954, S. Karger AG, p. 43.

Wiesli, P.: Eine Methode zur Frühdiagnose der bitemporalen Hemianopsie bei Hypophysentumoren, Schweiz. Med. Wschr. **19**: 479, 1928.

Wilbrand, H.: Die hemianopischen Gesichtsfeldformen, Munich, 1890, J. F. Bergmann.

Williamson, W. P.: After-image perimetry: rapid method of obtaining visual fields, Arch. Ophthal. **33**:40, 1945.

Yanagida, N.: Ophthalmic symptoms of brain tumor, particularly of the visual field, Acta Soc. Ophthal. Jap. **55**:836-842, 1951.

Zuckerman, J.: Perimetry, Philadelphia, 1954, J. B. Lippincott Co.

CHAPTER TWO *General symptoms of increased intracranial pressure in patients with brain tumors*

Papilledema*

Adrogué, E., and Insausti, T.: Etiology of papilledema, Arch. Oftal. B. Air. **17**:285, 1942.

Amsler, M.: Neuritis nervi optici (Neuritis optica). In Amsler, M., et al., editors: Lehrbuch der Augenheilkunde, Basel, 1954, S. Karger AG, p. 689.

Amsler, M.: Papillitis. In Amsler, M., et al., editors: Lehrbuch der Augenheilkunde, Basel, 1954, S. Karger AG, p. 690.

Amsler, M.: Stauungspapille. In Amsler, M., et al., editors: Lehrbuch der Augenheilkunde, Basel, 1954, S. Karger AG, p. 686.

Anastasopoulos, G.: Klinische Untersuchungen an Hirntumoren zur Frage der Entstehung der Stauungspapille, Basel, 1937, S. Karger AG.

Anderson, W. A.: Medullated nerve fibers, Trans. Ophthal. Soc. U. K. **62**:343, 1942.

Babel, J.: La pathogénie de la stase papillaire, Praxis **50**:3, 1948.

Bailliart, P.: La circulation rétinienne à l'état normal et pathologique, Paris, 1923, Gaston Doin & Cie.

Bailliart, P.: La circulation rétinienne, Docum. Ophthal. **7/8**:357, 1954.

Baumann, M.: Zur Differentialdiagnose zwischen Stauungspapille und Papillitis, Graefe. Arch. Ophthal. **134**:189-191, 1935.

Bedell, A. J.: Ophthalmoscopical differentiation between papilledema and papillitis, Southern Med. J. **30**:37-44, 1937.

Bedell, A. J.: Papilledema without increased intracranial pressure, Amer. J. Ophthal. **25**: 685, 1942.

Behr, C.: Z. Augenheilk. **71**:275, 1930.

Behr, C.: Neue anatomische Befunde bei Stauungspapille: ein weiterer Beitrag zu ihrer Pathogenese, Graefe. Arch. Ophthal. **137**:1-60, 1937.

Behr, C.: Zur klinischen Diagnose der Stauungspapille, Nervenarzt **10**:337-340, 1937.

Bettaies, A.: Cécité par hypertension intracrânienne, Tunisie Med. **4**:227-233, 1966.

Biemond, A.: Poliomyelitis anterior acuta mit Stauungspapillen, Psychiat. Quart. **46**:424-426, 1942.

---

*A few references dealing with the blood pressure in the retinal vessels in papilledema that were cited in the text are not listed here (Coppez, Rasvan, Magitot, Pereyra, Spinelli, Gallois, Bauwens, Winthers, Ascher, Rossano, Gauddisart, Suvina, Serr, Baurmann, Marchesani). They can be found in detail in the monograph by Streiff, B., and Monnier, M.: Der retinale Blutdruck im gesunden und kranken Organismus, Vienna, 1946, Julius Springer.

Bietti, G. B.: Considerazioni sulla comparsa della papilla da stasi negli occhi miopi, Riv. Otoneurooftal. **15**:47, 1938.

Bing, R.: Stauungspapille, Arch. Neurol. Psychiat. **39**:49-71, 1937.

Bing, R.: Zur diagnostischen Bewertung intrakranieller Drucksteigerung, Schweiz. Med. Wschr. **15**:407, 1942.

Blum, J.: Disparition d'une stase papillaire hypertensive pseudo-tumorale, Rev. Otoneuroophtal. **19**:61-62, 1947.

Bollack, J., and Delthil, S.: Maladies de la papille. In Baillert, P., et al., editors: Traité d'ophtalmologie, Paris, 1939, Masson & Cie, vol. 5, p. 673.

Bonamour, G., Brégeat, P., Bonnet, M., and Juge, P.: La papille optique, Paris, 1968, Masson & Cie.

Bondi: Prag. Med. Wschr. **26**:311, 1901.

Bonnet, M.: Les tumeurs de la papille, Conf. Lyon. Ophtal. **86**:1-27, 1966.

Braun, W.: Über familiäres Vorkommen von Drusen der Papille, Klin. Mbl. Augenheilk. **94**:734-738, 1935.

Brégeat, P.: L'œdème papillaire, Paris, 1956, Masson & Cie.

Brégeat, P., David, M., and Fischgold, H.: Anévrysme cirsoïde de la rétine et du cerveau; procédé d'exploration, Bull. Soc. Franc. Ophtal. **65**:77-85, 1952.

Brégeat, P., David, M., and Lelièvre, A.: Einige Betrachtungen zum falschen Papillenödem, Rev. Neurol. **87**:545-549, 1952.

Brückner, A.: Dynamometrie. In Schieck, F., and Brückner, A., editors: Kurzes Handbuch der Ophthalmologie, Berlin, 1931, Julius Springer, vol. 2, p. 915.

Butterfield, D. L.: Types and locations of brain tumors and other space-displacing masses within the cranial cavity occurring without choked disc, New York J. Med. **465**:1935.

Bynke, H. G.: Rapid variations of the prominent optic disc. IV. Outline of the pulse curve, Acta Ophthal. **40**:171-180, 1962.

Bynke, H. G.: On early diagnosis and prevention of secondary optic atrophy in papilloedema, Acta Ophthal. **44**:801-813, 1966.

Cameron, J.: Marked papilloedema in pulmonary emphysema, Brit. J. Ophthal. **17**:167, 1933.

Casanovas Carnicer, J.: Estasis papilar, Arch. Soc. Oftal. Hispano-Amer. **8**:221-299, 1948.

Chalmers, J. W., and Walsh, F. B.: Hyaline bodies in the optic disks, Brain **74**:95, 1951.

Chamlin, H., and Davidoff, L. M.: Drusen of optic nerve simulating papilledema, J. Neurosurg. **7**:70-78, 1950.

Chamlin, H., and Davidoff, L. M.: Drusen of the optic nervehead: ophthalmoscopic and histopathologic study, Amer. J. Ophthal. **35**:1599-1605, 1952.

Chamlin, H., and Davidoff, L. M.: Papilledema—its differential diagnosis, with special reference to minimal testing of the blind spot at two meters, Arch. Neurol. Psychiat. **68**:213-232, 1952.

Champion de Crespigny, C. T.: Papilloedema, Australia Med. J. **2**:911-914, 1937.

Cone, W., and MacMillan, J. A.: The optic nerve and papilla: cytology and cellular pathology of the nervous system, New York, 1932, Paul B. Hoeber, Inc., vol. 2, pp. 837-901.

Cordes, F. C.: Congenital and acquired anomalies of the optic disk, Arch. Ophthal. **23**:1063, 1940.

Cordes, F. C., and Aiken, S. D.: Papilledema (choked disk) and papillitis (optic neuritis); their differential diagnosis, J. Nerv. Ment. Dis. **99**:576-582, 1944.

Cushing, H., and Bordley, J.: Observations on experimentally induced choked disk, Bull. Johns Hopkins Hosp. **20**:95-101, 1909.

Dandy, W. E.: Papilledema without intracranial pressure (optic neuritis), Ann. Roy. Coll. Surg. **110**:161-168, 1939.

Davis, L.: The blind spots in patients with intracranial tumors, J.A.M.A. **92**:794, 1929.

Desvignes, P.: Les faux aspects de stase papilláire, Bull. Soc. Franc. Ophtal. **66**:343, 1953.

Dill, J. L., and Crowe, S. J.: Thrombosis of the sigmoid or lateral sinus: reports of thirty cases, Arch. Surg. **28**:705-722, 1934.

Doggart, J. H.: On diagnosing papilledema, Trans. Ophthal. Soc. U. K. **62**:141, 1942.

Drew, J. H., and Grant, F. C.: Polycythemia as neurosurgical problem; review with report of 2 cases, Arch. Neurol. Psychiat. **54**:25, 1945.

Duke-Elder, S.: Textbook of ophthalmology, St. Louis, 1945, The C. V. Mosby Co., vol. 3, pp. 2944-2967, 3038.

Dunphy, E. B.: Unilateral papilledema, Amer. J. Ophthal. **50**:1084-1087, 1960.

Dupont, F.: Pseudo-névrite et excavation physiologique de la papille, Bull. Soc. Ophtal. Franc. **3**:373-377, 1952.

Dupuy-Dutemps, L.: Pathogénie de la stase papillaire dans les affections intracrâniennes, Thèse, Paris, 1900, Steinheil.

Ectors, L., and Bégaux-van Boven, C.: Glaucoma and papilledema, Ophthalmologica **180**:113-120, 1944.

Ethelberg, S., and Jensen, V. A.: Obscurations and further time-related paroxysmal disorders in intracranial tumors, Arch. Neurol. Psychiat. **68**:130, 1952.

Euzière, J., Pagès, P., Lafon, R., Cabanettes, S., and Labauge, R.: Hyperostose du canalicule optique extériorisée par une stase papillaire unilatérale et symptomatique d'une hyperostose frontale interne, Rev. Otoneuroophtal. 24:331-334, 1952.

Finke, J.: Ophthalmodynamographie. In Neurologie und Psychiatrie, Berlin, 1966, Springer Verlag.

Fischer, F.: Zur Frage sog. permanenter Stauungspapille, Klin. Mbl. Augenheilk. 85: 672-673, 1930.

Freusberg, O.: Ergebnisse der Dynamometrie bei Hirntumoren, Klin. Mbl. Augenheilk. 113:279-280, 1948.

Freusberg, O., and Moellmann, D.: A oftalmodinamometria en casos de tumores cerebrais, Rev. Brasil. Oftal. 11:275, 1953.

Fry, W. E.: The pathology of papilledema: an examination of 40 eyes with special reference to compression of the central vein of the retina, Amer. J. Ophthal. 14:874-883, 1931.

Fry, W. E., and DeLong, P.: Tumor formation at disk; report of case, Arch. Ophthal. 30:417, 1943.

Gallais, P.: Pseudo-stase papillaire avec corps hyalins, Rev. Otoneuroophtal. 24:369-372, 1952.

Ghersi, J. A., and Lara, F. D.: Contribución al conocimiento de la patogenia del edema de la papila, Rev. Asoc. Med. Argent. 64: 15-18, 1950.

Ghirardi, L., and Maione, M.: Considerazioni sull'arteriografia dell'oftalmica con speciale riguardo alla sua visualizzazione nell'ipertensione endocranica, Riv. Otoneurooftal. 26:426-437, 1951.

Gibbs, F. A.: Intracranial tumor with unequal choked disk: relationship between the side of greater choking and the position of the tumor, Arch. Neurol. Psychiat. 27:828-835, 1932.

Girard, P., Devic, M., and de Gevigney, D.: Neuro-papillite œdémateuse aiguë et discopathie cervicale. Récupération de la vision et disparition de l'œdème par élongation du cou. Le syndrome "névrite optique et acroparesthésie," J. Med. Lyon 183:161-168, 1950.

Glees, M.: Arteriovenöses Aneurysma des Augenhintergrundes und der gleichseitigen Grosshirnhemisphäre, Klin. Mbl. Augenheilk. 124:457, 1954.

Glees, M.: Ueber Anomalien der Papillengefässe bei intrakraniellen Angiomen, Klin. Mbl. Augenheilk. 130:403, 1957.

Good, P.: Choked discs in head encephalopathy, Amer. J. Ophthal. 24:794-797, 1941.

Gowers: The ophthalmoscope in medicine, 1904.

Griffith, J. Q., Jr., Fry, W. E., and McGuinness, A. C.: Experimental and clinical studies in hydrocephalus, with special reference to occurrence of papilledema, Amer. J. Ophthal. 23:245, 1940.

Griffith, J. Q., Jr., Fry, W. E., and Roberts, E.: Studies of criteria for classification of arterial hypertension; increased intracranial pressure and papilledema, Amer. Heart J. 21:94, 97, 1941.

Griffith, J. Q., Jr., Jeffers, W., and Fry, W. E.: Papilledema associated with subarachnoid haemorrhage: experimental and clinical study, Arch. Intern. Med. 61:880, 1938.

Gross, A. G.: The papilledema of toxic hydrocephalus, Trans. Ophthal. Soc. U. K. 68:181, 1948.

Gullstrand, A.: Das vereinfachte grosse Gullstrandsche Ophthalmoskop, Klin. Mbl. Augenheilk. 67:118, 1921.

Gunn, R. Marcus: Retrobulbar neuritis, Lancet, p. 412, 1904.

Hager, H.: Die Ophthalmo-Dynamographie als Methode zur Beurteilung des Gehirnkreislaufes, Klin. Mbl. Augenheilk. 142:827-846, 1963.

Hammes, E. M., Jr.: Papilledema in optic neuritis and tumor of the brain, Med. Clin. N. Amer. 28:957, 1944.

Harms: Ber. Ophthal. Ges. Heidelberg 33: 253, 1906.

Hegner, H.: Über Stauungspapille bei Blutkrankheiten, Klin. Mbl. Augenheilk. 50: 119, 1912.

Heinz, K.: Eine Methode zur Vermessung der Stauungspapille, Med. Rundschau 1: 1947.

Herzberger, E.: Papilledema due to unknown cause, Proc. Staff Meet. Beilinson Hosp. 1: 51-52, 1953.

van Heuven, J. A.: Papilloedema, Trans. Ophthal. Soc. U. K. 58:549-560, 1938.

Holmes, G.: Prognosis in papilloedema, Brit. J. Ophthal. 21:337-342, 1937.

Holmes, G.: Tr. Ophth. Soc. U. K. 57:3, 1937.

Howell, C.: The craniostenoses, Amer. J. Ophthal. 37:359, 1954.

Hoyt, W. F., and Beeston, D.: The ocular fundus in neurologic disease, St. Louis, 1966, The C. V. Mosby Co.

Huber, A.: Einseitige Stauungspapille, Pseudoneuritis und Pseudopapillenödem, Ophthalmologica 123:262, 1952.

Huber, A.: Papillenoedem bei Leukämien, Ophthalmologica 141:290-300, 1961.

Igersheimer, J.: Beiträge zur Pathologie und Pathogenese der Stauungspapille, Folia Ophthal. Orientalia 2:1-10, 1935.

Irisch, C. W.: Sinus thrombosis: longitudinal

sinus thrombosis, Ann. Otol. **47**:402-410, 1938.

Jackson, H. R.: L. O. H. Rep. **7**:573, 1873.

Jaensch, P. A.: Graefe. Arch. Ophthal. **122**:618, 1929.

Jaensch, P. A.: Stauungspapille (Bericht über die Jahre 1931-1935), Fortschr. Neurol. Psychiat. **8**:387-398, 1936.

Jensen: Chorioiditis juxtapapillaris, Graefe. Arch. Ophthal. **69**:41, 1908.

Kafer, J. P.: Sobre un sinal oftalmoscopico de hipertensao intracraniana, Selecoes Cientificas Med. Fra. **3**:53-55, 1948.

Keith, N. M., Rucker, C. W., and Parkhill, E. M.: Recession of retinal papilledema during terminal stage of malignant hypertension; report of case, Arch. Ophthal. **26**:240-246, 1941.

Kennedy, F.: Retrobulbar neuritis as an exact sign of certain tumors and abscesses in the frontal lobe, Amer. J. Med. Sci. **142**:355, 1911.

Koch, F. L. P.: Ophthalmodynamometry, Arch. Ophthal. **34**:234-247, 1945.

Koziak, P. H.: Craniostenosis, Amer. J. Ophthal. **37**:380, 1954.

Kravitz, D., and Lloyd, R. S.: Dilated and tortuous retinal vessels: report of a case of congenital arteriovenous communication, Arch. Ophthal. **14**:591-598, 1935.

Krayenbühl, H.: Allgemeine hirnchirurgische Diagnostik und Therapie. In Brunner, et al., editors: Lehrbuch der Chirurgie, Basel, 1949, Benno Schwabe & Co., pp. 605-658.

Kwaskowski, A.: Diagnostische Schwierigkeiten bei der Stauungspapille, Klin. Oczna **23**:63-74, 1953.

Laje Weskamp, R.: Evolución y pronostico del edema papilar, Arch. Oftal. B. Air. **14**:303-340, 1939.

Lambert, R. K., and Weiss, H.: Optic pseudoneuritis and pseudopapilledema, Arch. Neurol. Psychiat. **30**:580, 1933.

Larsson, L., and Nord, B.: Klinisk bild av retrobulbärneurit vid intrakraniella tumörer, Nord. Med. **34**:1059-1065, 1947.

Lasco, F., and Arseni, C.: La pseudonévrite optique postérieure comme manifestation rare des néoformations intracrâniennes, Rev. Otoneuroophtal. **32**:385-391, 1960.

Lauber, H.: Die Entstehung der Stauungspapille, Wien. Klin. Wschr. **47**:1547, 1934.

Leber: In Graefe-Saemisch: Handbuch der gesamten Augenheilkunde, 1877, vol. 5, p. 778.

Leimgruber, M.: Erbforschungen über die Drusen der Sehnervenpapille, Graefe. Arch. Ophthal. **136**:364-376, 1936.

Leinfelder, P. J.: Choked disks and low intrathecal pressure occurring in brain tumor, Amer. J. Ophthal. **26**:1294, 1943.

Leinfelder, P. J., and Paul, W. D.: Papilledema in general diseases, Arch. Ophthal. **28**:983-987, 1942.

Levatin, P.: Increased intracranial pressure without papilledema, Arch. Ophthal. **58**:683-688, 1957.

Lloberas-Camino, L., Ribas-Clotet, F., and Rodríguez-Arias, B.: Sobre el diagnóstico diferencial del éstasis papilar de origin tumoral, Rev. Esp. Otoneurooftal. **11**:266-270, 1952.

Lobstein, A.: L'ophtalmodynamométrie clinique, Bull. Soc. Ophtal. Franc. **11**:690-692, 1959.

Love, J. G., Wagner, H. P., and Woltmann, H. W.: Tumours of the spinal cord associated with choking of the optic disk, Arch. Neurol. Psychiat. **66**:171, 1951.

Lyle, T. K.: Some pitfalls in the diagnosis of plerocephalic oedema, Trans. Ophthal. Soc. U. K. **73**:87, 1953.

Magitot, A. P.: How to know the blood pressure in the vessels of the retina, Amer. J. Ophthal. **5**:777, 1922.

Mansuy, L., Rougier, J., Aimard, G., and Thierry, A.: Expérience neuro-chirurgicale de dix-sept cas d'oedème papillaire de stase de cause rare ou inconnue; avenir éloigné de ces malades, Lyon Med. **49**:1389-1404, 1967.

Marchesani, O.: Schwierigkeiten der Diagnose der Stauungspapille bei Myopie, Arch. Psychiat. **95**:447, 1931.

Martinez Moreno, J., and Cuevas Cancino, D.: Diferenciación clinica entre papiledema y papilitis, Boll. Hosp. Oftal. Mexico **5**:195-200, 1952.

Matavulj, N.: Stase papillaire unilatérale d'origine intra-crânienne, Bull. Soc. Franc. Ophtal. **63**:19-25, 1950.

Morone, G.: Sur les facteurs oculaires qui peuvent empêcher le développement d'une stase papillaire par hypertension endocrânienne, avec considération particulière de l'influence de la myopie, Rev. Otoneuroophtal. **21**:1, 80, 1949.

de Morsier, G., Monnier, M., and Streiff, B.: Rev. Neurol. Belge. **6**:702, 1939.

Morton, A. S., and Parsons, J. H.: Hyaline bodies (Drusenbildungen) at the optic disk, Trans. Ophthal. Soc. U. K. **23**:135-153, 1903.

Nano, H. M.: Ensayo de una clasificación practica de los aspectos oftalmoscopicamente edematosos de papila, Arch. Oftal. B. Air. **27**:49-53, 1952.

Nottbeck, B.: Ein Beitrag zur kongenitalen Pseudoneuritis optica (Scheinneuritis), Graefe. Arch. Ophthal. **44**:31, 1897.

Ohashi: Acta Soc. Ophthal. Jap. **36**:11, 1932.

Parnitzke, K. H.: Zur Bedeutung der Stauungspapille in der Hirntumordiagnostik, Deutsch. Gesundh. **8**:89-91, 1953.

Paton, L.: Papilledema and optic neuritis; a retrospect, Trans. Sect. Ophthal. Amer. Med. Assoc. pp. 98-119, 1935.

Paton, L.: Brain **32**:65, 1909.

Paton, L., and Holmes, G.. The pathology of papilloedema; a histological study of 60 eyes, Brain **33**:389-432, 1911.

Paufique, L., and Bonamour, G.: Aspects et diagnostic ophthalmoscopiques de la papille œdémateuse, J. Med. Lyon **658**: 433-444, 1947.

Pennybacker, J.: Papilledema due to intracranial venous obstruction, Trans. Ophthal. Soc. U. K. **63**:333-339, 1944.

Petrohelos, M. A., and Henderson, J. W.: Ocular findings of intracranial tumor: study of 358 cases, Amer. J. Ophthal. **34**:1387, 1951.

Pigassou, R.: Contribution à l'étude de la pression veneuse rétinienne, Thèse, Toulouse, 1947.

Primrose, J.: Mechanism of production of papilloedema, Brit. J. Ophthal. **48**:19-29, 1964.

Raimondo, N.: La macchia cieca "enorme" senza papilledema, Riv. Otoneurooftal. **35**:288-297, 1960.

Redslob, E.: Sémiologie critique de la stase papillaire, Rev. Otoneuroophtal. **22**:1-18, 1950.

Reese, A. B.: Relation of drusen of the optic nerve to tuberous sclerosis, Arch. Ophthal. **24**:187, 1940.

Rehwald, E.: Die differentialdiagnostiche Bewertung der Stauungspapille vom Standpunkt des Neurologen, Med. Klin. **35**: 1108-1110, 1939.

Rentz, S.: Aneurysma racemosum retinale, Arch. Augenheilk. **95**:84-91, 1924.

Richter, H.: Über hochgradige Stauungspapille bei bereits organisierter Sinusthrombose, HNO. Beih. Hals-Heilk. **2**:394-395, 1951.

Rintelen, F.: Zur diagnostischen Bedeutung der Dynamometrie, Klin. Mbl. Augenheilk. **100**:469-471, 1938.

Rintelen, F.: Über Stauungspapille und ihre Entstehung, Schweiz. Med. Wschr. p. 575, 1946.

Riser, Calmettes, Garipuy, Pigassou, and Pigassou: Tension veineuse rétinienne et hypertension crânienne, Rev. Otoneuroophtal. **21**:1, 1949.

Riser, Couadau, and Gayral: Tension artérielle rétinienne et pression intra-crânienne mesurée par ponction sous-occipitale, Rev. Otoneuroophtal. **19**:371, 1947.

Riser, Géraud, Caizergues, and Dreyfus: Hypotension crânienne avec œdèma papillaire et signes méningés, Rev. Otoneuroophtal. **21**:448, 1949.

Roberts, W. L., and Nielsen, R. F.: Uveoparotid fever with bilateral papilledema, Amer. J. Ophthal. **28**:1252, 1945.

Rollin, A.: Mesure de la pression dans les vaisseaux rétiniens. In Bailliart, P., et al., editors: Traité d'ophtalmologie, Paris, 1939, Masson & Cie, vol. 2, p. 1057.

Rucker, C. W.: Defects in visual fields produced by hyaline bodies in optic disks, Arch. Ophthal. **32**:56-59, 1944.

Rucker, C. W.: Defects in the visual fields resulting from increased intracranial pressure, New York J. Med. **49**:2417-2421, 1949.

Sadoughi, G.: La stase papillaire dans les hypotensions ventriculaires, Bull. Soc. Ophtal. Franc. **1**:17-23, 1950.

Samuels, B.: Histopathology of papilledema, Amer. J. Ophthal. **21**:1242-1258, 1938.

Samuels, B.: Drusen of the optic papilla; a clinical and pathologic study, Arch. Ophthal. **25**:412-423, 1941.

Sanders, M. D., and Fytche, T. J.: Fluorescein-angiography in the diagnosis of drusen of the optic disc, Trans. Ophthal. Soc. U. K. **87**:457-468, 1967.

Satanowsky de Neumann, P., and Brodsky, M.: Edemas de papila sin sintomas de hipertensión endocraneana, Arch. Oftal. B. Air. **28**:128-133, 1953.

Scheie, H. G.: Evaluation of ophthalmoscopic changes of hypertension and arterial sclerosis, Arch. Ophthal. **49**:117, 1953.

Schieck, F.: Über die Entstehungsart der Stauungspapille, Verh. Physikal.-Med. Ges. Würzburg **51**:121, 1926.

Schieck, F.: Beiträge zur Frage der Entstehung der Stauungspapille, Graefe. Arch. Ophthal. **138**:48-54, 1937.

Schieck, F.: Die ophthalmoskopische Diagnose der Stauungspapille, Graefe. Arch. Ophthal. **137**:203-215, 1937.

Schieck, F.: Das Wesen der Stauungspapille, Bücherei d. Augenarztes H. 12, Stuttgart, 1942, Ferdinand Enke.

Schlezinger, N. S., Waldmann, J., and Alpers, B. J.: Drusen of optic nerve simulating cerebral tumor, Arch. Ophthal. **31**:509-516, 1944.

Schupfer, F.: Diagnostische Schwierigkeiten und klinisches Bild der Myelinfasern in Fällen von Stauung oder Entzündung der Sehnervenpapille, Boll. Oculist. **19**:647-656, 1940.

de Schweinitz, G. E.: The relation of cerebral decompression to the relief of the ocular manifestation of increased intracranial tension, Ann. Ophthal. **20**:271-289, 1911.

Scott, G. I.: Optic disc oedema, Trans. Ophthal. Soc. U. K. 87:733-753, 1967.

Sédan, J.: La diplopie de provocation au service de l'expert, Rev. Otoneuroophtal. 24:351-352, 1952.

Selinger, E.: Choked disk and papillitis: differential diagnosis by the protein content of the aqueous, Arch. Neurol. Psychiat. 33:360-367, 1935.

Serr, H.: Die Stauungspapille in ihrer allgemein klinischen und ophthalmologischen Bedeutung, Graefe. Arch. Ophthal. 135:431-450, 1936.

Simpson, T.: Papilloedema in emphysema, Brit. Med. J. 2:639, 1948.

Smith, E. G.: Unilateral papilledema; its significance and pathologic physiology, Arch. Ophthal. 21:856-878, 1939.

Smith, J. L.: Ophthalmodynamometric techniques. In Smith, J. L., editor: Neuro-ophthalmology, Springfield, 1964, Charles C Thomas, Publisher, vol. 1.

Sobanski, J.: Der Wert dynanometrischer Untersuchung für die Erklärung der Entstehung der Stauungspapille, Graefe. Arch. Ophthal. 137:84, 1937.

Stokes, W. H.: Racemose arteriovenous aneurysm of the retina (aneurysma racemosum arteriovenosum retinae), Arch. Ophthal. 11:956, 1934.

Stopford, J. S. B.: Increased intracranial pressure, Brain 51:485-507, 1928.

Stough, J. T.: Choking of the optic disc in diseases other than tumor of the brain, Arch. Ophthal. 8:821-830, 1932.

Streiff, B.: Zwei Fälle von Allgemeinleiden mit Stauungspapille infolge Hirntumors, Klin. Mbl. Augenheilk. 106:246-247, 1941.

Streiff, B.: L'ophthalmodynamométrie de Bailliart, sa valeur et sa précision, Docum. Ophthal. 7/8:27, 1954.

Streiff, B., and Monnier, M.: Der retinale Blutdruck im gesunden und kranken Organismus, Berlin, 1946, Julius Springer.

Taylor, R. D., Corcoran, A. C., and Page, I. H.: Increased cerebrospinal fluid pressure and papilledema in malignant hypertension, Arch. Intern. Med. 93:818, 1954.

Thiébaut, F., Phillippidès, D., Rohmer, F., and Mengus, M.: Stase papillaire unilatérale, Rev. Otoneuroophtal. 25:303-304, 1953.

Tönnis, W.: Pathophysiologie und Klinik der intracraniellen Drucksteigerung. In Olivecrona, H., and Tönnis, W., editors: Handbuch der Neurochirurgie, Berlin, 1959, Julius Springer, vol. 1.

Tönnis, W., and Krenkel, W.: Grosshirngeschwülste ohne Stauungspapille, Acta Neurochir. 5:458-487, 1957.

Traquair, H. M.: An introduction to clinical perimetry, London, 1949, Henry Kimpton.

Uhthoff, W.: Trans. Ophthal. Soc. U. K. 34:47, 1914.

Uhthoff, W.: Zur Pseudoneuritis optica, Deutsch. Z. Nervenheilk. 50:258, 1913.

Uhthoff, W.: Die Augenveräderungen bei den Erkrankungen des Gehirnes. In Graefe-Saemisch: Handbuch der gesamten Augenheilkunde, Leipzig, 1915, Wilhelm Engelmann, vol. 11, p. 1143.

Vissi, F., and Dobrina, D.: Papilla da stasi da probabile distonia vasomotoria, Riv. Otoneurooftal 25:437-452, 1950.

Voisin, J.: Les signes oculaires de l'hypertension intra-crânienne, Presse Med. 58:421, 1950.

Volhard, F.: Wesen und Behandlung des roten und blassen Hochdruckes. In von Thiel, R., editor: Gegenwartsprobleme der Augenheilkunde, Leipzig, 1937, Georg Thieme.

Voutres, J.: Un cas d'œdème papillaire ayant la valeur d'un signe de localisation de tumeur intracrânienne, Bull. Soc. Ophtal. Franc. 1:169-171, 1949.

van Wagenen, W. P.: The incidence of intracranial tumors without "choked disk" in one years series of cases, Amer. J. Med. Sci. 176:746, 1928.

Wagener, H. P.: Drusen (hyaline bodies) of the optic disk, Amer. J. Med. Sci. 210:262-268, 1945.

Wagener, H. P.: Pseudotumor cerebri, Amer. J. Med. Sci. 227:214, 1954.

Wagener, H. P., and Keith, N. M.: Diffuse arteriolar disease and hypertension, Proceedings of the Fifteenth International Congress of Ophthalmology, Cairo, 1937, vol. 1, pp. 1-86.

Walsh, F. B.: Ocular signs of thrombosis of the intracranial venous sinuses Arch. Ophthal. 17:46-65, 1937.

Walsh, T. J., Garden, J., and Gallagher, B.: Relationship of retinal venous pulse to intracranial pressure. In Smith, J. L., editor: Neuro-ophthalmology, St. Louis, 1968, The C. V. Mosby Co., vol. 4, pp. 288-292.

Walsh, F. B., and Hoyt, W. F.: Optic neuritis. In Clinical neuro-ophthalmology, Baltimore, 1969, The Williams & Wilkins Co., p. 607.

Walsh, F. B., and Hoyt, W. F.: Papilledema. In Clinical neuro-ophthalmology, Baltimore, 1969, The Williams & Wilkins Co., p. 567.

Watkins, C. H., Wagener, H. P., and Brown, R. W.: Cerebral symptoms accompanied by choked optic disks in types of blood dyscrasia, Amer. J. Ophthal. 24:1374, 1941.

Weber, G.: Hirnabszesse im Rahmen infektiöser intrakranieller Erkrankungen, Stuttgart, 1956, Georg Thieme.

Weed, L. H.: Experimental studies in intracranial pressure. The intracranial pressure in health and diseases, Baltimore, 1929, The Williams & Wilkins Co.

Weigelin, E.: Zur Genese der Stauungspapille, Ber. Deutsch. Ophth. Ges. **56:** 181, 1950.

Weigelin, E.: Beurteilung des intrakraniellen Kreislaufes mit Hilfe der Netzhautarteriendruckmessung, Docum. Ophthal. **7/8:**183, 1954.

Weigelin, E., and Müller, H. K.: Über die praktische Bedeutung der Blutdruckmessung an der Zentralarterie der Netzhaut, Docum. Ophthal. **5/6:**357-402, 1951.

Weigelin, E., and Lobstein, A.: Ophthalmodynamometrie, Basel, 1962, S. Karger A G.

Weigelin, E., Kazuo, I., and Halder, M.: Fortschritte auf dem Gebiet der Blutdruckmessung am Auge, Fortschr. Augenheilk. **15:**44-84, 1964.

Weiman, C. G., McDowell, F. H., and Plum, F.: Papilledema in poliomyelitis, Arch. Neurol. Psychiat. **66:**722, 1951.

Weve, H.: Ophthalmodynamometrie. In Amsler, M., et al., editors: Lehbuch der Augenheilkunde, Basel, 1954, S. Karger AG, p. 30.

Williamson-Noble, F. A.: Venous pulsation, Trans. Ophthal. Soc. U. K. **72:**317-326, 1952.

Witmer, R.: Differentialdiagnostische Aspekte bei retinalen Gefässtörungen und Papillenoedem, Ophthalmologica **156:**313-321, 1968.

Wohlwill, F.: Die Affektionen der grossen venösen Blutleiter der Dura. In Schieck, F., and Brückner, A., editors: Kurzes Handbuch der Ophthalmologie, Berlin, 1931, Julius Springer, vol. 6, pp. 17-19.

Wyburn-Mason, R.: Arteriovenous aneurysm of midbrain and retina, facial naevi and mental changes, Brain **66:**163, 1943.

Yegen, V.: The importance of papilloedema in the localization and diagnosis of intracranial tumours (in Turkish), Otonorooftal. **7:**96-116, 1952.

Pareses of extraocular muscles*

Cushing, H.: Strangulation of the nervi abducentes by lateral branches of the basilary artery in cases of brain tumor, Brain **33:** 204-235, 1910-1911.

Van Allen, M. W.: Transient recurring paralysis of ocular abduction. A syndrome

---

*For additional literature see references for motility of the eyes in Chapter 1.

of intracranial hypertension, Arch. Neurol. **17:**81-88, 1967.

Zielinski, H. W.: Paresen der äusseren Augenmuskeln bei intrakraniellen raumfordernden Prozessen, ein Ueberblick über die Beobachtungen an über 3000 Fällen, Zbl. Neurochir. **19:**235-251, 1959.

Clivus ridge syndrome

Fischer-Brügge, K.: Das Klivuskanten-Syndrom, Acta Neurochir. **11:**36, 1951.

Hoyt, W. F.: Vascular lesions of the visual cortex with brain herniation through the tentorial incisura. Neuro-ophthalmologic considerations, Trans. Pacif. Coast. Otoophthal. Soc. **41:**301-327, 1960.

Jefferson, G.: The tentorial pressure cone, Arch. Neurol. Psychiat. **40:**857-876, 1938.

Lyle, D. J.: Eye symptoms produced by tentorial herniation from increased intracranial pressure, Proceedings of the Seventeenth International Congress on Ophthalmology, New York, 1954.

McKenzie, K. G.: Extradural hemorrhage, Brit. J. Surg. **26:**336, 1938.

Scharfetter, F.: Mydriasis und Lichtstarre, Schweiz. Arch. Neurol. Psychiat. **96:**386-392, 1965.

Schwarz, G. A., and Rosner, A. A.: Displacement and herniation of the hippocampal gyrus through the incisura tentorii: clinicopathologic study, Arch. Neurol. Psychiat. **46:**297, 1941.

Welte, E.: Zur formalen Genese der traumatischen Mydriasis. Oculomotorius-wurzelschädigung durch einseitiges Vorquellen des Uncus hippocampi, Zbl. Neurochir. **6:**217, 1943.

Exophthalmos

Brain, W. R.: Neurological and general medical causes of exophthalmos, Trans. Ophthal. Soc. U. K. **58:**27-32, 1938.

Cohn, H.: Messung der Prominenz der Augen mittels eines neuen Instruments, des Exophthalmometers, Klin. Mbl. Augenheilk. **5:**339, 314, 1905.

Crawford, J. S.: Proptosis in children, Trans. Canad. Ophthal. Soc. **5:**81-93, 1952.

Dixon, G. J.: Unilateral exophthalmos (proptosis), causation and differential diagnosis, Brain **64:**73-89, 1941.

Elsberg, C. A., Hare, C. C., and Dyke, C. H.: Unilateral exophthalmos in intracranial tumors with special reference to its occurrence in meningiomata, Surg. Gynec. Obstet. **55:**681, 1932.

Godtfredsen, E.: Exophthalmos due to malignant tumours in the paranasal sinuses, Acta Ophthal. **25:**295, 1947.

Hertel, E.: Ein einfaches Exophthalmomometer, Graefe. Arch. Ophthal. 60:171, 1905.

Knudtzon, K.: Exophthalmos and primary intracranial tumours (In Danish with an English summary), M.D. Thesis, Copenhagen, 1952, Danish Scientific Press.

Lazorthes, G., and Geraud, J.: L'exophtalmie unilatérale neurochirurgicale; analyse de 17 observations, Rev. Neurol. 83:373-379, 1950.

Morin, G., Tuset, J., Bouchacourt, A., Force, L., and Beauchamp, P.: Exophthalmie par tumeur de la fosse cérébrale moyenne, Rev. Otoneuroophtal. 33:270-275, 1961.

Skydsgaard, H.: Exophthalmus coincident to intracranial tumors, Acta Ophthal. 16:474-480, 1938.

Thurel, R.: Exophtalmie par distension de la corne temporo-sphénoïdale du ventricule latéral et refoulement de la paroi externe de l'orbite, Rev. Otoneuroophtal. 23:286, 1951.

Weve, H.: Exophthalmometrie und Orbitotonometrie. In Amsler, M., et al., editors: Lehrbuch der Augenheilkunde, Basel, 1954, S. Karger AG, p. 8.

Yasargil, G. M.: Die Röntgendiagnostik des Exophthalmus unilateralis, Basel, 1957, S. Karger AG.

CHAPTER THREE *Local symptoms of brain tumors*

Reviews of local symptoms in brain tumors

Abott, W. D.: Ocular symptoms in the diagnosis of tumor of the brain, Arch. Ophthal. 6:244, 1931.

Bardram, M., and Möller, H. U.: Diagnosis of tumours in the anterior and middle cranial fossa, Acta Ophthal. 30:65-96, 1952.

Behr, C.: Die Erkrankungen der Sehbahn vom Chiasma aufwärts. In Schieck, F., and Brückner, A., editors: Kurzes Handbuch der Ophthalmologie, Berlin, 1931, Julius Springer, vol. 6, p. 245.

Benda, C.: Die topische Diagnostik der Hirntumoren, Mschr. Psychiat. Neurol. 93:332-354, 1936.

Best, F.: Die Augenstörungen bei Hirntumoren. In Schieck, F., and Brückner, A., editors: Kurzes Handbuch der Ophthalmologie, Berlin, 1931, Julius Springer, vol. 6, p. 563.

Bonnet, P.: Les signes ophtalmologiques des tumeurs cérébrales, J. Med. Lyon 37:511-530, 1956.

Bonnet, P.: Les signes ophtalmologiques des tumeurs cérébrales, Rev. Med. Moyen-Orient 1:1-32, 1956.

Brégeat, P., and David, M.: Hémianopsie latérale homonyme due à la compression de la bandelette par un athérome carotidien, Bull. Soc. Franc. Ophtal. 63:60-63, 1950.

Bucy, P. C.: Early recognition of tumors of brain, Dis. Nerv. Syst. 1:356-362, 1940.

Cairns, H.: Accessory methods of diagnosis in intracranial tumor and allied diseases, Trans. Med. Soc. London 58:50-74, 1935.

Cairns, H.: Ergebnisse der Behandlung der intrakraniellen Tumoren, Schweiz. Med. Wschr. 44:1043, 1937.

Collier, J.: The false localizing signs of intracranial tumour, Brain 27:490-508, 1904.

Coughlin, W. T.: Subtentorial tumors, Southern Med. J. 30:665-674, 1937.

Cushing, H., and Walker, C. B.: Distortion of the visual fields in cases of brain tumor; binasal hemianopia, Arch. Ophthal. 41:559, 1912.

Dandy, W. E.: Brain tumors, general diagnosis and treatment. In Lewis, Dean, editor: Lewis' practice of surgery, Hagerstown, Md., 1932, W. F. Prior Co., Inc., vol. 12, pp. 443-674.

David, M.: Evolution et valeur des examens complémentaires dans le diagnostic des tumeurs intracrâniennes. Essai synthétique, Neurochirurgie 13:181-205, 1967.

Dubois-Poulsen, A., editor: Le champ visuel, Paris, 1952, Masson & Cie.

Duke-Elder, S.: Textbook of ophthalmology, St. Louis, 1949, The C. V. Mosby Co., vol. 4, p. 3579.

Evans, J. N.: An introduction to clinical scotometry, London, 1938, Oxford University Press.

Förster, O.: Über die Wechselbeziehungen von Herdsymptomen und Allgemeinsymptomen bei Hirntumor, Verh. Deutsch. Ges. Inn. Med., pp. 458-485, 1938.

Gjessing, H. G. A.: Importance of eye symptoms in diagnosis and localization of brain tumors, Trans. Ophthal. Soc. U. K. 54:581-602, 1934.

Globus, J. H., and Silverstone, S. M.: Diagnostic value of defects in the visual fields and other ocular disturbances associated with supratentorial tumors of the brain, Arch. Ophthal. 14:325, 1935.

Gorton, L. W.: Eye findings as aid in diagnosing and localization of brain tumors, New Orleans Med. Surg. J. 90:315-318, 1937.

Guillaumat, L., and Robin, A.: Statistique des modifications du champ visuel dans les affections neurochirurgicales non traumatiques. In Dubois-Poulsen, A., editor: Le champ visuel, Paris, 1952, Masson & Cie.

Guillaumat, L., Morax, P. V., and Offret, G.: Neuro-ophtalmologie, Paris, 1959, Masson & Cie.

Hartmann, E.: Ocular symptoms in diseases of brain, spinal cord, and meninges. In Berens, C., editor: The eye and its diseases, Philadelphia, 1949, W. B. Saunders Co., p. 824.

Hartmann, E., and Guillaumat, L.: Aspect du fond d'œil dans les tumeurs intra-crâniennes. Etude statistique, Ann. Oculist. 175:717-737, 1938.

Hobbs, H. E.: Visual defect as an early sign of intracranial tumor, Eye Ear Nose Throat Monthly 46:602-610, 1967.

Hollwich, F.: Irrtümer beim Nachweis von Hirntumoren, Ber. Deutsch. Ophth. Ges. 59:36-39, 1955.

Horrax, G., and Putnam, T. J.: Distortions of the visual fields in cases of brain tumour, Brain 55:499, 1932.

Huber, A.: Die ophthalmologische Symptomatologie der Hirntumoren. Allgemeinsymptome und supratentorielle Tumoren, Ophthalmologica 125:287-319, 1953.

Huber, A.: Eye symptoms in brain tumors, Highlights Ophthal. 7:3-98, 1964.

Jelsma, F.: Intracranial tumors, Kentucky Med. J. 39:203-208, 1941.

Jentzer, A.: Les tumeurs cérébrales, Helv. Med. Acta 5:6, 720, 1938.

Junius, P.: Diagnostik der Hirntumoren mit besonderer Berücksichtigung der Erscheinungen am Sehorgan, Zbl. Ges. Ophthal. 26:5, 273, 1932.

Kennedy, F.: Symptomatology and diagnosis of lesions of the brain, with special reference to disturbances of vision, hearing, taste, smell and speech, Trans. Amer. Acad. Ophthal. 30:8-25, 1925.

Kraus, H.: Gutartige basale Tumoren der vorderen und mittleren Schädelgrube, Wien. Med. Wschr. 100:156, 1950.

Krayenbühl, H.: Hilfsmethoden der Diagnostik raumbeschränkender intrakranieller Erkrankungen, Schweiz. Med. Wschr. 5:89, 1937.

Krayenbühl, H.: Die Symptomatologie der Tumoren der hinteren Schädelgrube, Schweiz. Med. Wschr. 31:901, 1938.

Krayenbühl, H., and Schmid, A. E.: Zur Lokalisation intrakranieller orbitaler Dermoide, Ophthalmologica 106:251, 1943.

Krayenbühl, H., and Weber, G.: Diagnostik

und Grundzüge der Therapie der Hirntumoren im Kindesalter, Helv. Paediat. Acta 2:2, 115, 1947.

Lauber, H.: Das Gesichtsfeld, Munich, 1944, Bergmann & Springer.

Lobeck, E.: Über die Bedeutung des Augenbefundes für die Diagnostik und Therapie von Hirntumoren (nebst Bemerkungen über die Entstehungsweise von Zentralskotomen bei Tumoren der vorderen Schädelgrube), Graefe. Arch. Ophthal. 140:599-628, 1939.

Lyle, D. J.: Symposium on manifestations of disease affecting cerebral nerves supplying eye, ear, nose and throat due to involvement in diseases of the central nervous system, Trans. Amer. Acad. Ophthal. 41:49-68, 1936.

McCulloch, R. J. P.: Eye signs in intracranial disease, Canad. Med. Ass. J. 42:236, 1940.

Marburg, O.: Some remarks on tumors of the brain in childhood, Trans. Amer. Neurol. Ass. 67:35-40, 1942.

Miller, R. H., McCraig, W., and Kernohan, J. W.: Supratentorial tumors among children, Arch. Neurol. Psychiat. 68:797-814, 1952.

Moersch, F. P., McCraig, W., and Kernohan, J. W.: Tumors of the brain in aged persons, Arch. Neurol. Psychiat. 45:235, 1941.

Moller, H. U.: Zur neuro-ophthalmologischen Diagnostik (in Danish), Nord. Med. 14:1405-1409, 1942.

Monbrun, A.: Affections des voies optiques rétro-chiasmatiques et de l'écorce visuelle. In Bailliart, P., et al., editors: Traité d'ophtalmologie, Paris, 1939, Masson & Cie, vol. 6, p. 903.

Mooney, A. J.: Lesions of visual pathway and their relation to neuro-surgery (Mary Louise Prentice Montgomery Memorial Lecture), Irish J. Med. Sci., pp. 315-327, 1938.

Müller, R., and Wohlfahrt, G.: Intracranial teratomas and teratoid tumours, Acta Psychiat. Neurol. 22:69-95, 1947.

Pandit, Y. K. C.: Intracranial space occupying lesions, J. All-India Ophthal. Soc. 13:1-34, 1965.

Payne, F.: Neuro-ophthalmology, Arch. Ophthal. 50:644-666, 1953.

Perria, L.: Sintomi oculari e tumori endocranici, Sist. Nerv. 1:9-17, 1949.

Petrohelos, M. A., and Henderson, J. W.: The ocular findings of intracranial tumour, Trans. Amer. Acad. Ophthal. 55:89-98, 1950.

Petrohelos, M. A., and Henderson, J. W.: Ocular findings of intracranial tumor; study of 358 cases, Amer. J. Ophthal. 34:1387, 1951.

Raaf, J.: Perimetric diagnosis of intracranial tumors, Trans. Pacif. Coast Otoophthal. Soc. 27:131-144, 1942.

Rand, C. W., and Wagenen, R. J.: Brain tumors in childhood; review of 38 cases, J. Pediat. 6:322-339, 1935.

Remky, H.: Ophthalmoneurologie. Zur Diagnose von Ort und Art intrakranieller Prozesse (insbesondere Tumoren), Almanach Augenheilk. München, pp. 107-115, 1960.

Reymond, A.: Classification anatomo-pathologique des tumeurs cérébrales, Ophthalmologica 125:204-230, 1953.

Rintelen, F.: Die ophthalmoneurologischen Symptome und Syndrome bei infratentoriellen Hirntumoren, Ophthalmologica 125: 320-340, 1953.

Rintelen, F.: Gesichtsfelddefekte bei Hirntumoren, Ophthalmologica 158:451-468, 1969.

Rucker, C. W.: Neuro-ophthalmology, Arch. Ophthal. 44:733, 1950.

Ruedemann, A. D.: Differential diagnosis of brain lesions; comparison by visual fields, encephalography and ventriculography, Trans. Sect. Ophthal. Amer. Med. Ass., pp. 32-37, 1940.

Schreck, E.: Ophthalmologische Diagnostik bei Hirntumoren, Regensb. Jb. Arztl. Fortbild. 6:294-308, 1958.

Stender, A.: Über die Frühdiagnose von Hirntumoren, Deutsch. Med. Wschr. 4: 210-213, 1953.

Tönnis, W.: Augensymptome bei 3033 Hirngeschwülsten, Ber. Deutsch. Ophth. Ges. 59:6-27, 1955.

Tönnis, W.: Diagnostik der intrakraniellen Geschwülste. In Olivecrona, H., and Tönnis, W., editors: Handbuch der Neurochirurgie, Berlin, 1962, Julius Springer, vol. 4, pt. 3, pp. 1-579.

Traquair, H. M.: An introduction to clinical perimetry, London, 1949, Henry Kimptom.

Walsh, F. B., and Hoyt, W. F.: Orbital, ocular and intracranial tumors. In Clinical neuro-ophthalmology, Baltimore, 1969, The Williams & Wilkins Co., pp. 1927-2330.

Weber, G.: Zur Diagnose und Prognose intrakranieller Tumoren, Ophthalmologica 125: 231-286, 1953.

Yanagida, N.: Ophthalmic symptoms of brain tumour, particularly of the visual field, Acta Soc. Ophthal. Jap. 55:836, 1951.

Yanagida, N.: Ocular symptoms of various intracranial tumours (in Japanese), Acta Soc. Ophthal. Jap. 57:1377-1381, 1392-1398, 1464-1477, 1953.

Tumors of the frontal lobe

Bailey, P.: Syndrom des Frontallappens. Die Hirngeschwülste, Stuttgart, 1951, Ferdinand Enke, p. 214.

Calmettes, L., Deodati, F., and Bec, P.: Hémianopsie horizontale et tumeur frontale, Rev. Otoneuroophtal. 36:175-178, 1964.

Croll, M.: Frontal lobe tumors and their ocular manifestations, Amer. J. Ophthal. 59:206-211, 1965.

David, M., and Sourdille, G.: Sur la valeur localisatrice des signes dits de compression directe due nerf optique, en particulier du syndrome de Foster-Kennedy, dans les tumeurs cérébrales, J. Ophtal. 3:215, 1942.

Dilenge, D., David, M., and Fischgold, H.: Artère ophtalmique et tumeurs frontales intracrâniennes, Neurochirurgie 8:379-384, 1962.

Duke-Elder, S.: Textbook of ophthalmology, St. Louis, 1949, The C. V. Mosby Co., vol. 4, p. 3523.

Ecker, A. D.: Concentric contraction of visual fields associated with tumor of the frontal lobe, Mayo Clin. Proc. 12:679, 1937.

François, J., and Neetens, A.: Le syndrome de Foster Kennedy et son étiologie, Ann. Oculist. 188:219-253, 1955.

Fridenberg, P.: The "Foster Kennedy" syndrome, Arch. Ophthal. 26:288, 1941.

Galvez Montes, J., and de Frederico Antras, A.: Sindrome de Foster-Kennedy invertido, Arch. Soc. Oftal. Hispano-Amer. 21:225-233, 1961.

Glees, M.: Dem Foster Kennedyschen Syndrom ähnliche Veränderungen der Sehnerven durch Arteriosklerose, Klin. Mbl. Augenheilk. 100:865-873, 1938.

Halpern, L.: Frontalhirnsyndrome, Mschr. Psychiat. Neurol. 101:4, 239, 1939.

Holmes, G., and Fox, J. C.: Optic nystagmus and its value in the localization of cerebral lesions, Brain 49:333, 1926.

Hyland, H. H., and Botterell, E. H.: Frontal lobe tumors; clinical and physiological study, Canad. Med. Ass. J. 37:530-540, 1937.

Jenkner, F. L., and Kutschera, E.: Frontal lobes and vision, Confin. Neurol. 25:63-78, 1965.

Kennedy, F.: Retrobulbar neuritis as an exact diagnostic sign of certain tumors and abscesses in the frontal lobe, Amer. J. Med. Sci. 142:355, 1911.

Kennedy, F.: A further note on the diagnostic value of retrobulbar neuritis in expanding lesions of the frontal lobes, J.A.M.A. 67:1360, 1916.

Kestenbaum, A.: Zur Klinik des optokinetischen Nystagmus, Graefe. Arch. Ophthal. 124:339, 1930.

Lillie, W. I.: Ocular phenomena produced by basal lesions of the frontal lobe, J.A.M.A. 89:2099-2103, 1927.

Pittrich, H.: Stirnhirngeschwülste, Arch. Psychiat. 113:1-60, 1941.

de Recondo, J.: Les mouvements conjugués oculaires et leurs diverses modalités d'atteinte, Paris, 1967, Masson & Cie.

Silberpfennig, J.: Contributions to problem of eye movements; disturbances of ocular movements with pseudohemianopia in frontal lobe tumors, Confin. Neurol. 4:1-13, 1941.

Skrzypczak, J.: Zur Bedeutung des Foster-Kennedy Syndroms, Klin. Mbl. Augenheilk. 150:504-509, 1967.

Tönnis, W.: Geschwülste des Stirnhirns. Diagnostik der intrakraniellen Geschwülste. In Olivecrona, H., and Tönnis, W., editors: Handbuch der Neurochirurgie, Berlin, 1962, Julius Springer, vol. 4, p. 369.

Voris, H. C., Adson, A. W., and Moersch, F. P.: Tumors of the frontal lobe: clinical observations in series verified microscopically, J.A.M.A. 104:93-99, 1935.

Walsh, F. B., and Hoyt, W. F.: Tumors of the frontal lobe. In Clinical neuro-ophthalmology, Baltimore, 1969, The Williams & Wilkins Co., p. 2166.

Tumors of the temporal lobe

Bailey, P.: Syndrom des Temporallappens. Die Hirngeschwülste, Stuttgart, 1951, Ferdinand Enke, p. 246.

Barcia Goyanes, J. J.: Anatomie du lobe temporal, Rev. Otoneuroophtal. 22:71-102, 1950.

Bender, M. B., and Savitsky, N.: Micropsia and teleopsia limited to the temporal fields of vision, Arch. Ophthal. 29:904, 1943.

de Crinis, M.: Zur Symptomatologie der Schläfenlappentumoren, Z. Ges. Neurol. Psychiat. 174:169-193, 1942.

Cushing, H.: Distortion of the visual fields in cases of brain tumor: the field defects produced by temporal lobe lesions, Brain 44:341-396, 1922.

Duke-Elder, S.: Textbook of ophthalmology, St. Louis, 1949, The C. V. Mosby Co., vol. 4, p. 3601.

Fischer-Brügge, K.: Das Klivuskantensyndrom, Acta Neurochir. 2:36, 1951.

Fonte Barcena, A.: Visual disturbances in tumors of temporal lobe, An. Soc. Mex. Oftal. Otorinolaring. 19:1-16, 1944.

Game, J.: Temporal lobe tumors, Med. J. Aust. 1:366-368, 1953.

Géraud, Lazorthes, Caizergues, and Ribaut:

Atteinte du réflexe cornéen par méningiome temporal, Rev. Otoneuroophtal. 23:169-170, 1951.

Gordon, A.: Visuel field defects as a deciding diagnostic factor in lesions of temporal lobe simulating cerebellar involvement, Trans. Amer. Neurol. Ass. 60:212-215, 1934.

Guillot, P., and Casanovas, J.: Le lobe temporal en O. N. O.: séminologie ophtalmique, Rev. Otoneuroophtal. 22:2, 219, 1950.

Harrington, D. O.: Localizing value of incongruity in defects in the visual fields, Arch. Ophthal. 21:453, 1939.

Heppner, F., and Vogler, E.. Symptomatologie und Diagnostik der Schläfenlappentumoren, Wien. Klin. Wschr. 64:969-976, 1952.

Horrax, G.: Visual hallucinations as a cerebral localizing phenomenon: with special reference to their occurrence in tumors of the temporal lobe, Arch. Neurol. Psychiat. 10:532-547, 1923.

Insausti, T., and Salleras, A.: Sintomatologia oftalmologica de los tumores temporales, Neuropsiquiatria 3:143-157, 1952.

Kennedy, F.: The symptomatology of temporo-sphenoidal tumors, Arch. Intern. Med. 8:317, 1911.

Koch, G.: Oculomotoriuswurzelschädigung durch einseitige Quellung des Uncus gyri hippocampi, Deutsch. Z. Nervenheilk. 159:417, 1948.

Kolodny, A.: The symptomatology of tumours of the temporal lobe, Brain 51:385-417, 1928.

Kravitz, D.: The value of quadrant defects in the localization of temporal lobe tumors, Amer. J. Ophthal. 14:781-785, 1931.

Krayenbühl, H.: Erkennung und Behandlung von Gehirngeschwülsten, Nachrichtenbl. Schweiz. Pflegerinnenschule, Zurich, 1938.

Krayenbühl, H.: Die Bedeutung der Gesichtsfeldbestimmung in der Diagnostik von Tumoren des Schläfen- und Hinterhauptlappens, Schweiz. Med. Wschr. 69:43, 197, 1939.

Matavulj, N.: Syndrome ophtalmologique dans les tumeurs temporales, Rev. Otoneuroophtal. 22:8, 1950.

Meyer, A.: The temporal lobe detour of the radiations, etc., Amer. Neurol. Ass. 1911; Trans. Ass. Amer. Physicians 22:7, 1907.

Obrador Alcalde, S.: Physiologie du lobe temporal, Rev. Otoneuroophtal. 22:103-120, 1950.

Paillas, J. E., and Tamalet, J.: Les glioblastomes temporaux, Marseille Chir. 2:461-465, 1950.

Paillas, J. E., Boudouresques, J., and Tamalet, J.: Hallucinations visuelles d'origine temporale à propos de 15 observations anatomo-cliniques, Rev. Neurol. 81:154-156, 1949.

Strobos, R. R. J.: Tumors of the temporal lobe, Neurology 3:752-760, 1953.

Taveras, J. M.: Roentgenologic aspects of the diagnosis of temporal tumors. In Smith, J. L., editor: Neuro-ophthalmology, St. Louis, 1968, The C. V. Mosby Co., vol. 4, pp. 215-229.

Thiébaut, F.: Hémianopsie relative et lobe temporal, Rev. Otoneuroophtal. 23:236-238, 1951.

Thurel, R.: Paralysie du moteur oculaire commun par engagement du lobe temporal, Rev. Otoneuroophtal. 23:285-286, 1951.

Tönnis, W.: Geschwülste des Schläfenlappens. Diagnostik der intrakraniellen Geschwülste. In Olivecrona, H., and Tönnis, W., editors: Handbuch der Neurochirurgie, Berlin, 1962, Julius Springer, vol. 4, p. 410.

Traquair, H. M.: The course of the geniculolocalcarine visual path in relation to the temporal lobe, Brit. J. Ophthal. 6:251-259, 1922.

Walker, A. E., and Walsh, F. B.: Visual disturbances in temporal lobectomized patients. In Smith, J. L., editor: Neuroophthalmology, St. Louis, 1968, The C. V. Mosby Co., vol. 4, pp. 230-248.

Walsh, F. B., and Hoyt, W. F.: Temporal lobe tumors. In Clinical neuro-ophthalmology, Baltimore, 1969, The Williams & Wilkins Co., p. 2178.

Weinberger, L. M., and Grant, F. C.: Visual hallucinations and their neuro-optical correlates, Arch. Ophthal. 23:166-199, 1940.

Tumors of the parietal lobe

Bailey, P.: Syndrom des Parietallappens. Die Hirngeschwülste, Stuttgart, 1951, Ferdinand Enke, p. 230.

Cogan, D. G.: Hemianopia and associated symptoms due to parietotemporal lobe lesions, Amer. J. Ophthal. 50:1056-1066, 1960.

David, M., and Hecaen, H.: Sur certains troubles de la latéralité du regard dans les lésions pariétales s'accompagnant de troubles de la somatognosie, Bull. Soc. Ophtal. Paris 1:103-105, 1947.

Duke-Elder, S.: Textbook of ophthalmology, St. Louis, 1949, The C. V. Mosby Co., vol. 4, pp. 3603, 3666.

Offret, G.: Modern concepts of physiology and psychophysiology of parietooccipital lobes of brain, Arch. Ophthal. 6:145-172, 1946.

Penfield, W., and Rasmussen, T.: The cerebral cortex of man, New York, 1950, The Macmillan Co.

Pouliquen, P., and Lobstein, A.: Champs visuels atypiques au cours de l'évolution d'une tumeur pariéto-temporale, Rev. Otoneuroophtal. 24:158, 1952.

Tönnis, W.: Geschwülste des Scheitellappens. Diagnostik der intrakraniellen Geschwülste. In Olivecrona, H., and Tönnis, W., editors: Handbuch der Neurochirurgie, Berlin, 1962, Julius Springer, vol. 4, p. 406.

Walsh, F. B., and Hoyt, W. F.: Parietal lobe tumor. In Clinical neuro-ophthalmology, Baltimore, 1969, The Williams & Wilkins Co., p. 2203.

Tumors of the occipital lobe

Allen, I. M.: Clinical study of tumors involving the occipital lobe, Brain 53:194, 1930.

Bailey, P.: Syndrom des Occipitallappens. Die Hirngeschwülste, Stuttgart, 1951, Ferdinand Enke, p. 250.

Barkan, O., and Boyle, S. F.: Paracentral homonymous hemianopic scotoma, Arch. Ophthal. 14:956-959, 1930.

Le Beau, J., Wolinetz, E., and Rosier, M.: Phénomène de persévération des images visuelles dans un cas de méningiome occipital droit, Rev. Neurol. 86:692-695, 1952.

Bender, M. B., and Strauss, I.: Defects in visual field of one eye only in patients with a lesion of one optic radiation, Arch. Ophthal. 17:765-787, 1937.

Brégeat, P., Klein, M., Thiébaut, F., and Bouniol: Hémimacropsie homonyme droite et tumeur occipitale gauche, Rev. Otoneuroophtal. 19:238-240, 1947.

Clark, W. E. L.: The visual centers of the brain and their connections, Physiol. Rev. 22:205, 1942.

Clark, W. E. L.: The anatomy of cortical vision, Trans. Ophthal. Soc. U. K. 62:229, 1942.

David, H., Hecaen, R., Angelergues, R., and Magis, C.: Les tumeurs occipitales, Neurochirurgie 1:177-191, 1955.

Dubois-Poulsen, A., Magis, C., de Ajuriaguerra, J., and Hecaen, R.: Les conséquences visuelles de la lobectomie occipitale chez l'homme, Ann. Oculist. 185: 4, 305, 1952.

Duke-Elder, S.: Textbook of ophthalmology, St. Louis, 1949, The C. V. Mosby Co., vol. 4, p. 3603.

Edmund, J.: Visual disturbances associated with gliomas of the temporal and occipital lobe, Acta Psychiat. Neurol. Scand. 29: 291, 1954.

Evans, J. N., and Browder, J.: Problem of split macula: study of visual fields, Arch. Ophthal. 31:43-53, 1944.

Frazier, C. H., and Waggoner, R. W.: Tumors of the occipital lobe: a review of 40 cases, Arch. Neurol. Psychiat. 22:1096-1099, 1929.

Halstead, W. C., Walker, A. E., and Bucy, P. C.: Sparing and nonsparing of "macular" vision associated with occipital lobectomy in man, Arch. Ophthal. 24:948-966, 1940.

Holmes, G. A.: Contribution to the cortical representation of vision, Brain 54:470, 1931.

Holmes, G. A., and Listei, W. T.: Disturbances of vision from cerebral lesions with special reference to the cortical representation of the macula, Brain 39:34-73, 1916.

Horrax, G., and Putnam, T. J.: Distortion of the visual fields in cases of brain tumor: the field defects and hallucinations produced by tumors of the occipital lobe, Brain 55:499, 1932.

Huber, A.: Zur homonymen Hemianopsie nach occipitaler Lobektomie, Schweiz. Med. Wschr. 80:1227, 1950.

Klüver, H.: Visual functions after removal of the occipital lobes, J. Psychol. 11:23, 1941.

Krayenbühl, H.: Die Bedeutung der Gesichtsfeldbestimmung in der Diagnostik von Tumoren des Schläfen- und Hinterhauptlappens, Schweiz. Med. Wschr. 69:43, 197, 1939.

Levenson, D. S., and Smith, J. L.: Optokinetic nystagmus and occipital lesions, Amer. J. Ophthal. 61:753-762, 1966.

Parkinson, D., and Kernohan, J. W.: Tumours of the occipital lobe; J. Neurosurg. 7:555, 1950.

Parkinson, D., and McCraig, K. W.: Tumours of the brain, occipital lobe; their signs and symptoms, Canad. Med. Ass. J. 64:111-113, 1951.

Parkinson, D., Rucker, C. W., and McCraig, K. W.: Visual hallucinations associated with tumors or occipital lobe, Arch. Neurol. Psychiat. 68:66-68, 1952.

Penfield, W., Evans, J. P., and MacMillan, J. A.: Visual pathways in man; with particular reference to macular representation, Arch. Neurol. Psychiat. 33:816-834, 1935.

Porsaa, K.: Central visual field after occipital lobectomy, Acta Ophthal. 22:243-250, 1944.

Putnam, T. J., and Liebmann, S.: Cortical representation of the macula lutea, with special reference to the theory of bilateral representation, Arch. Ophthal. 28:415, 1943.

Scarlett, H. W., and Ingham, S. D.: Visual defects caused by occipital lobe lesions; report of 13 cases, Arch. Neurol. Psychiat. 8:225-246, 1922.

Stadlin, W.: Hémianopsie et nystagmus optocinétique, Ophthalmologica 118:383, 1949.

Tönnis, W.: Geschwülste des Occipitallappens. Diagnostik der intrakraniellen Geschwülste. In Olivecrona, H., and Tönnis, W., editors: Handbuch der Neurochirurgie, Berlin, 1962, Julius Springer, vol. 4, p. 417.

Walsh, F. B., and Hoyt, W. F.: Occipital lobe tumors, In Clinical neuro-ophthalmology, Baltimore, 1969, The Williams & Wilkins Co., p. 2195.

Weinberger, L. M., and Grant, F. C.: Visual hallucinations and their neuro-optical correlates (review), Arch. Ophthal. 23:166-199, 1940.

Tumors of the corpus callosum

Bailey, P.: Syndrom des Balkens. Die Hirngeschwülste, Stuttgart, 1951, Ferdinand Enke, p. 254.

Walsh, F. B., and Hoyt, W. F.: Corpus callosum tumor. In Clinical neuro-ophthalmology, Baltimore, 1969, The Williams & Wilkins Co., p. 2261.

Parasagittal meningiomas and meningiomas of falx

Bailey, P., and Bucy, P. C.: The origin and nature of meningeal tumors, Amer. J. Cancer 1:15-54, 1931.

Courville, C. B., and Abbot, K. H.: On classification of meningiomas; survey of 99 cases in light of existing schemes, Bull. Los Angeles Neurol. Soc. 6:21-31, 1941.

Custodis, E.: Meningioma der Falx und lokalisatorisch irreführendes Foster Kennedysches Syndrom, Klin. Mbl. Augenheilk. 101:823-830, 1938.

Gassel, M. M.: False localizing signs. A review of the concept and analysis of the occurrence in 250 cases of intracranial meningioma, Arch. Neurol. 4:526-554, 1961.

Klingler, M., and Condrau, G.: Lokalisatorisch irreführende Gesichtsfeldsymptome bei Hirntumoren, Ophthalmologica 120:270, 1950.

Krayenbühl, H.: Klinik und Therapie des intrakraniellen Meningeoms, Schweiz Med. Wschr. 78:841, 1948.

Newell, F. W., and Beaman, T. C.: Ocular signs of meningioma, Amer. J. Ophthal. 45:30-40, 1958.

Olivecrona, H.: Die parasagittalen Meningiome, Leipzig, 1934, Georg Thieme.

Olivecrona, H.: The parasagittal meningiomas, J. Neurosurg. 4:327, 1947.

Rucker, C. W., and Kearns, T. P.: Mistaken diagnoses in some cases of meningioma. Clinics in perimetry no. 5, Amer. J. Ophthal. 51:15-19, 1961.

Tönnis, W.: Meningiome. Diagnostik der intrakraniellen Geschwülste. In Olivecrona, H., and Tönnis, W., editors: Handbuch der Neurochirurgie, Berlin, 1962, Julius Springer, vol. 4, p. 39.

Vincent, C., Hartmann, E., and Le Beau, J.: Atrophie optique primitive par méningiome à distance de la région opto-chiasmatique, Zbl. Neurochir. 3:145, 1938.

Walsh, F. B., and Hoyt, W. F.: Parasagittal and falx meningiomas. In Clinical neuro-ophthalmology, Baltimore, 1969, The Williams & Wilkins Co., pp. 2269-2271.

Pituitary adenomas

Adler, F. H., Austin, G., and Grant, F. C.: Localizing value of visual fields in patients with early chiasmal lesions, Arch. Ophthal. 40:579-600, 1948.

Amat, M., and Enciso,, M. M.: Los trastornos visuales como primera manifestación subjectiva de los tumores de la hipofisis, Arch. Soc. Oftal. Hispano-Amer. 7: 815-817, 1947.

Amsler, M.: Krankheiten des Chiasma. In Amsler, M., et al., editors: Lehrbuch der Augenheilkunde, Basel, 1954, S. Karger AG, p. 703.

Amsler, M.: Opticusatrophie. In Amsler, M., et al., editors: Lehrbuch der Augenheilkunde, Basel, 1954, S. Karger AG, p. 649.

Arganaraz, R.: Heteronymous hemianopia and intracranial tumors in the neighborhood of the chiasma, Arch. Oftal. B. Air. 16:349, 1941.

Asbury, T.: Unilateral scotoma as the presenting sign of pituitary tumor, Amer. J. Ophthal. 59:510-512, 1965.

Aulhorn, E.: Perimetrie. In Sautter, H., editor: Entwicklung und Fortschritte in der Augenheilkunde, Stuttgart, 1963, Ferdinand Enke.

Bailey, P.: Das Hypophysensyndrom. Die Hirngeschwülste, Stuttgart, 1951, Ferdinand Enke, p. 73.

Bailey, P.: Syndrom des Chiasma opticum. Die Hirngeschwülste, Stuttgart, 1951, Ferdinand Enke, p. 270.

Bakay, L.: The Results of 300 pituitary adenoma operations (Prof. Herbert Olivecrona's series), J. Neurosurg. 7:240-255, 1950.

Bardram, M. T.: Oculomotor paresis and nonparetic diplopia in pituitary adenomata, Acta Ophthal. 27:225-258, 1949.

Bergland, R., and Ray, B. S.: The arterial supply of the human optic chiasm, J. Neurosurg. 31:327-334, 1969.

Best, F.: Die Augensymptome bei Hypophysenerkrankungen. In Schieck, F., and Brückner, A., editors: Kurzes Handbuch der Ophthalmologie, Berlin, 1931, Julius Springer, vol. 6, p. 596.

Biemond, A.: Nervus opticus und chiasma, Ophthalmologica 102:287-302, 1941.

Cairns, H.: Peripheral ocular palsies from the neuro-surgical point of view, Trans. Ophthal. Soc. U. K. 58:464, 1938.

Chamlin, M., and Davidoff, L. M.: The 1/ 2000 field in chiasmal interference Arch. Ophthal. 44:53-70, 1950.

Chamlin, M., and Davidoff, L. M.: Symposium on pituitary tumors. II. Ophthalmologic criteria in diagnosis and management of pituitary tumors, J. Neurosurg. 19: 9-18, 1962.

Chamlin, N., Davidoff, L. M., and Feiring, E. H.: Ophthalmologic changes produced by pituitary tumors, Amer. J. Ophthal. 40: 353-368, 1955.

Chiasserini, F.: Diagnostic et thérapeutique de quelques affections sellaires, Schweiz. Med. Wschr., p. 425, 1939.

Cullen, J. F.: Pituitary and parapituitary tumors. Value of perimetry in diagnosis, Brit. J. Ophthal. 48:590-596, 1964.

Cushing, H.: The pituitary body and its disorders, Philadelphia, 1912, J. B. Lippincott Co.

Cushing, H., and Walker, C. B.: Chiasmal lesions with especial reference to bitemporal hemianopia, Brain 37:341-400, 1915.

Dandy, W. E.: Hypophyseal tumors. In Lewis, Dean, editor: Lewis' practice of surgery, Hagerstown, Md., 1937, W. F. Prior Co., Inc., vol. 12, pp. 556-605.

Davidoff, L. M.: The anamnesis and symptomatology in one hundred cases of acromegaly, Endocrinology 10:461-482, 1926.

Desvignes, P.: Le syndrome de compression directe du nerf optique intracrânien, Ann. Oculist. 174:289-308, 1937.

Duke-Elder, S.: Textbook of ophthalmology, St. Louis, 1949, The C. V. Mosby Co., vol. 4, pp. 3473 and 3488.

Franceschetti, A., and Werner, A.: Syndrome chiasmatique du à un abcès intrasellaire chronique, Rev. Otoneuroophtal. 29: 177-182, 1957.

Franceschetti, A., Babel, J., Doret, M., and Werner, A.: A propos du syndrome chiasmatique, la fréquence de son origine non-hypophysaire; la nécessité d'un diagnostic précoce avec la névrite optique rétrobulbaire, Confin. Neurol. 16:206-207, 1956.

François, J., Hoffmann, G., and de Brabandere, J.: Le diagnostic visuel des opérations pour tumeurs hypophysaires, Ann. Oculist. 193:993-1033, 1960.

Franklin, C. R.: Visual studies in pituitary adenoma, Bull. Neurol. Inst. New York 5: 180-198, 1936.

Fujie, Y.: Ocular symptoms of tumors in the pituitary region. Course of visual field in the pituitary adenoma, Jap. J. Ophthal. 21: 1053-1060, 1967.

Gartner, S.: Ocular pathology in the chiasmal

syndrome, Amer. J. Ophthal. 34:593-596, 1951.

Gil Espinosa, M.: Über nicht hypophysäre Chiasmasyndrome, Bibliotheca Ophthalmologica, fasc. 32, Basel, 1946, S. Karger AG.

Glees, M., and Zielinksi, H.: Ueber bitemporale zentrale und parazentrale Skotome bei verschiedenen krankhaften Prozessen im Sellabereich, Klin. Mbl. Augenheilk. 129: 145-160, 1956.

di Gugliemo, G.: Diagnosi e terapia delle sindromi sellari e parasellari, Arch. Ed. Atti Soc. Ital. Chir. 44:334-413, 1938.

Guillaumat, L.: Le champ visuel dans les affections chiasmatiques, Annee Ther. Clin. Ophtal. 11:7-190, 1960.

Hagedoorn, A.: The chiasmal syndrome and retrobulbar neuritis in pregnancy, Amer. J. Ophthal. 20:690, 1937.

Harms, H.: Quantitative Perimetrie bei Sellanahen Tumoren, Ophthalmologica 127: 255-261, 1954.

Harms, H.: Leitsymptom "Zentralskotom." In Sautter, H., editor: Entwicklung und Fortschritte in der Augenheilkunde, Stuttgart, 1963, Ferdinand Enke, pp. 516-539.

Hartmann, E., and David, M.: Affections du Chiasma. In Bailliart, P., et al., editors: Traité d'ophthalmologie, Paris, 1939, Masson & Cie, vol. 6, p. 877.

Henderson, W. R.: The pituitary adenomas, Brit. J. Surg. 26:811-911, 1939.

Hirsch, O.: Die Diagnose und Therapie der Hypophysen-Tumoren, Proceedings of the Fifteenth International Congress on Ophthalmology, Cario, 1937, vol. 1, pp. 119-123.

Hirsch, O.: Die Bedeutung des Augenhintergrundes für die Diagnose eines Hypophysentumors, Mschr. Psychiat. Neurol. 117: 236-240, 1949.

Hobbs, H. E.: Visual field defects in chiasmal lesions, Trans. Canad. Ophthal. Soc. 8:111-121, 1956.

Hoyt, W. F., and Luis, O.: The primate chiasm: details of visual fiber organization studied by silver impregnation techniques, Arch. Ophthal. 70:69-85, 1963.

Jeandelize, P., and Drouet, P. L.: L'œil et l'hypophyse, Proceedings of the Fifteenth International Congress on Ophthalmology, Cairo, 1937, vol. 1, pp. 227-400.

Jefferson, G.: Extrasellar extension of pituitary adenomas; President's Address, Proc. Roy. Soc. Med. 3:433-458, 1940.

Karbacher, P., and Schinz, H. R.: Augensymptome bei Patienten mit Hypophysentumoren aus der Zürcher Augenklinik und dem Zürcher Röntgeninstitut von 1924-1939, Klin. Mbl. Augenheilk. 103:541-557, 1939.

Kearns, T. P., Salassa, E. M., Kernohan, J. W., and MacCarty, C. S.: Ocular manifestations of pituitary tumor in Cushing's syndrome, Arch. Ophthal. 62:242-247, 1959.

Kestenbaum, A.: Clinical methods of neuro-ophthalmologic examination, London, 1947, William Heinemann, Ltd., p. 81.

Knapp, P.: Diagnostische und therapeutische Fragen bei Tumoren der Chiasmagegend, Klin. Mbl. Augenheilk. 105:401, 1940.

Kosic, H.: Die Tumoren der Gegend der Sella turcica, Arch. Klin. Chir. 201:89-108, 1941.

Kraft, H.: Die Bewertung der Augenmuskelstörungen für die Differential-diagnose der Chiasmasyndrome, Ber. Deutsch. Ophth. Ges. 67:378-380, 1965.

Kravitz, D.: Visual field interpretations in chiasmal lesions, Amer. J. Ophthal. 31: 415, 1948.

Lillie, W. I.: Ocular phenomena in acromegaly, Amer. J. Ophthal. 8:32-39, 1925.

Lindenberg, R.: Neuropathology of the optic chiasma and adnexa. In Smith, J. L., editor: Neuro-ophthalmology, Springfield, Ill., 1963, Charles C Thomas, Publisher, vol. 1, pp. 385-425.

Loepp, W.: Gegenseitige Auswertung der Augen- und Röntgensymptome bei der Tumordiagnostik im Sellabereich, Berlin, 1936, S. Karger AG.

Lombardi, G.: Radiology in neuro-ophthalmology, Baltimore, 1967, The Williams & Wilkins Co.

Lyle, T. K.: Carcinoma of a parapituitary residue, Trans. Ophthal. Soc. U. K. 69: 285-291, 1949.

Lyle, T. K., and Clover, P.: Ocular symptoms and signs in pituitary tumours, Trans. Roy. Soc. Med. 54:611-619, 1961.

Maddox, E. E.: Symposium on nystagmus, Proc. Roy. Soc. Med. 1:29-41, 1914.

Malagon Castro, V.: La exoftalmia hipofisaria, Med. Cir. Bogota 16:255-273, 1952.

Mark, V. H., Smith, J. L., and Kjellberg, R. D.: Suprasellar epidermoid tumor. A case report with the presenting, complaint of see-saw nystagmus, Neurology 10:81-83, 1960.

Matavulj, N., and Guillaumat, L.: Quelques cas d'affections de la région chiasmatique à périmétrie atypique, Bull. Soc. Ophtal. Franc. 7:716-728, 1951.

Meadows, S. P.: Pituitary tumors, Ann. Roy. Coll. Surg. Eng. 9:224, 1951.

Meadows, S. P.: Unusual clinical features and modes of presentation in pituitary adenoma, including pituitary apoplexy. In Smith, J. L., editor: Neuro-ophthalmology, St. Louis, 1968, The C. V. Mosby Co., vol. 4, pp. 178-189.

Montaldi, M., and Zingirian, M.: La perimetria mesopica, cinetica e statica nelle affezione produttive delle'ipofisi, Riv. Otoneurooftal. 38:15-36, 1963.

Mooney, A., J.: Perimetry and angiography in the diagnosis of lesions in the pituitary region, Trans. Ophthal. Soc. U. K. 72: 49, 1952.

Moreu, A.: Considerations on the ophthalmic diagnosis of hypophyseal tumors, Arch. Soc. Oftal. Hispano-Amer. 1:181, 1942.

Mundinger, F., Riechert, T., and Reisert, P. M.: Hypophysentumoren, Hypophysektomie, Stuttgart, 1967, Georg Thieme.

Nover, A.: Augensymptome bei sellanahen Tumoren, Deutsch. Med. Wschr. 88:2139-2142, 1963.

Obrador, A. S.: Tumores y procesos inflamatorios de la region quiasmatica-hipofisaria, Arch. Soc. Oftal. Hispano-Amer. 25:305-332, 1965.

Okonek, G.: Das Syndrom der Sehnervenkreuzung, Klin. Mbl. Augenheilk. 116:113, 1950.

Paufique, L., and Etienne, R.: Le tuberculome du chiasma, Bull. Soc. Franc. Ophtal. 65: 97-109, 1952.

Prskavec, F.: Über einen Fall von Sehnervenatrophie durch Getässdruck bei Hypophysentumor, Klin. Med. 2:1024-1028, 1947.

Ray, B. S.: Surgical lesions of optic nerves and chiasm in infants and children. In Smith, J. L., editor: Neuro-ophthalmology, St. Louis, 1967, The C. V. Mosby Co., pp. 77-99.

Remky, H.: Augensymptome bei Tumoren im Bereich des Türkensattels, Klin. Mbl. Augenheilk. 150:436, 1967.

Rönne, H.: On non-hypophyseal affections of the chiasma, Acta Ophthal. 6:332-343, 1928.

Rougerie, M.: Paralysie unilatérale de l'accommodation, signe rélévateur d'une tumeur de l'hypophyse, Bull. Soc. Ophtal. Franc. 64:724-728, 1964.

Rucker, C. W.: Chiasmal lesions. In The interpretation of visual fields, ed. 3, 1957, American Academy of Ophthalmology, pp. 22-35.

Rucker, C. W.: Tumors of the region of the optic chiasm, Fourth Pan-Amer. Cong. Oftal. 1:178-183, 1952.

Schneider, J. A.: Sellabrücke und Konstitution, Leipzig, 1939, Georg Thieme.

Siebeck, R., and Schiefer, W.: Zur Symptomatologie und Therapie sellanaher Prozesse, Klin. Mbl. Augenheilk. 142:194-212, 1963.

Smith, J. L., and Marl, V. H.: See-saw nystagmus with suprasellar epidermoid tumor, Arch. Ophthal. 62:280-283, 1959.

Terrien, F.: Les troubles visuels d'origine hypophysaire, Progr. Med., pp. 639-644, 1937.

Timmerman, J. M. E. N.: The diagnosis of tumors in the region of the optic chiasma, Ophthalmologica 152:530-536, 1966.

Tiwisina, T., and Haar, H.: Die Geschwülste der Chiasmagegend im cerebralen Angiogramm, Nervenarzt 24:58-63, 1953.

Tönnis, W.: Geschwülste im Bereich des Chiasma opticum (Differentialdiagnose, Indikation zur Operation, Therapie, Prognose), Zeitfragen Augenheilk., Berlin, 1938.

Tönnis, W.: Anzeigestellung zur operativen Behandlung der Geschwülste im Bereiche des Türkensattels, Klin. Mbl. Augenheilk. 114:1, 1949.

Tönnis, W.: Das chromophobe Adenom. Diagnostik der intrakraniellen Geschwülste. In Olivecrona, H., and Tönnis, W., editors: Handbuch der Neurochirurgie, Berlin, 1962, Julius Springer, vol. 4, p. 165.

Tönnis, W.: Das eosinophile Adenom. Diagnostik der intrakraniellen Geschwülste. In Olivecrona, H., and Tönnis, W., editors: Handbuch der Neurochirurgie, Berlin, 1962, Julius Springer, vol. 4, p. 155.

Tönnis, W., and Rausch, F.: Sellaveränderungen bei gesteigertem Schädelinnendruck, Deutsch. Z. Nervenheilk. 171:351-369, 1954.

Tönnis, W., Friedmann, G., and Albrecht, H.: Zur röntgenologischen Differentialdiagnose der Hypophysenadenome, unter Berücksichtigung der primären und sekundären Sellaveränderungen, Fortschr. Roentgenstr. 87:678-686, 1957.

Trumble, H. C.: Pituitary tumours: observations on large tumours which have spread widely beyond the confines of the sella turcica, Brit. J. Surg. 39:7, 1951.

Velter, E.: Le diagnostic des affections de la région chiasmatique et sellaire, Bull. Soc. Ophtal. Paris, pp. 216-234, 1936.

Vercesi, G., and Gatti, E.: La paralisi del nervo oculomotore comune quale segno di esordino clinico di un adenoma ipofisario. Tre osservazioni personali, Riv. Otoneurooftal. 41:17-34, 1966.

Vogt, A.: Die Ophthalmoskopie im rotfreien Licht. In Graefe-Saemisch: Handbuch der gesamten Augenheilkunde, Leipgiz, 1925, Wilhelm Engelmann, vol. 3.

Walker, A. E.: The neurosurgical evaluation of the chiasmal syndromes, Amer. J. Ophthal. 54:563-581, 1962.

Walker, C. B., and Cushing, H.: Studies of optic nerve atrophy in association with chiasmal lesions, Arch. Ophthal. 45:407-437, 1916.

Walsh, F. B.: Bilateral total ophthalmoplegia

with adenoma of the pituitary gland, Arch. Ophthal. 42:646-654, 1949.

Walsh, F. B.: Concerning the optic chiasma. Selected pathologic involvement and clinical problems. The DeSchweinitz lecture, Amer. J. Ophthal. 50:1031-1047, 1960.

Walsh, F. B., and Hoyt, W. F.: Pituitary adenomas. In Clinical neuro-ophthalmology, Baltimore, 1969, The Williams & Wilkins Co., p. 2130.

Weber, G.: Symptomatologie und Chirurgie der sellären und suprasellären Tumoren, Ophthalmologica 149:326-342, 1965.

Weinberger, L. M., Adler, F. H., and Grant, F. C.: Primary pituitary adenoma and the syndrome of the cavernous sinus, Arch. Ophthal. 24:1197, 1940.

Weinstein, P.: Data concerning the ophthalmological diagnosis of hypophyseal tumors, Ophthalmologica 125:169-171, 1953.

Weskamp, C.: Contribución al diagnostico differencial de los tumores hipofisarios, Arch. Oftal. B. Air. 28:173-181, 1953.

Weyland, R. D., McCraig, W., and Rucker, C. W.: Ungewöhnliche Krankheitsprozesse im Bereich des Chiasma, Mayo Clin. Proc. 27:505-511, 1952.

White, J. C., and Warren, S.: Unusual size and extension of a pituitary adenoma, J. Neurosurg. 2:126, 1945.

Wijngaarden, G. K.: Optic nerve and chiasma. Review of the literature 1951-1961, Advances Ophthal. 15:1-43, 1964.

Williamson-Noble, F. A.: Die oculären Folgen gewisser Läsionen des Chiasmas, Trans. Ophthal. Soc. U. K. 59:627-682, 1939.

Wybar, K.: Visual field defects in chiasmal lesions, Trans. Amer. Med. Ass. Sect. Ophthal. 112:16-39, 1963.

Yanagida, N.: Ocular symptoms of pituitary and pontine tumours (in Japanese), Acta Soc. Ophthal. Jap. 57:1293-1305, 1953.

Craniopharyngiomas

Armenise, B., Regli, F., and Del Vivo, R.: I craniofaringiomi: sintomatologia oftalmologica, Ann. Ottal. 87:652-662, 1961.

Bailey, P.: Das Hypothalamussyndrom. Die Hirngeschwülste, Stuttgart, 1951, Ferdinand Enke, p. 105.

Bianchi, G., Rosselet, E., and Buffat, J. D.: Cecité par craniopharyngiome et récupération fonctionelle après opération, Confin. Neurol. 17:155-162, 1957.

Bingas, B., and Wolter, M.: Augensymptome beim Kraniopharyngiom, Klin. Mbl. Augenheilk. 149:275-276, 1966.

van Bogaert, L.: Le diagnostic des tumeurs suprasellaires et en particulier des tumeurs de la poche pharyngienne de Rathke, Rev. Otoneuroophtal. 7:645, 1929.

Brégeat, P.: Le champ visuel des cholestéatomes suprasellaires. Quelques particularités, Bull. Soc. Ophtal. Franc. 3:446-448, 1949.

Calmettes, L., and Déodati, F.: Les craniopharyngiomes. Symptomatologie ophtalmologique, Rev. Otoneuroophtal. 27:99-105, 1955.

Dandy, W. E.: Hypophyseal duct tumors. In Lewis, Dean, editor: Lewis' practice of surgery, Hagerstown, Md., 1936, W. F. Prior Co., Inc., vol. 12, pp. 598-605.

Dott, N. M., and Bailey, P.: A consideration of the hypophysial adenomata, Brit. J. Surg. 13:314-366, 1925.

Duke-Elder, S.: Textbook of ophthalmology, St. Louis, 1949, The C. V. Mosby Co., vol. 4, p. 3511.

Erdheim, J.: Ueber Hypophysenganggeschwülste und Hirncholesteatome, S.-B. Akad. Wiss. Wien. Math.-Nat. Klin. 113: 537, 1904.

Franceschetti, A., and Blum, J. D.: Névrite rétrobulbaire aiguë transitoire dans un cas de tumeur cérébrale, Confin. Neurol. 11: 302, 1951.

Frazier, C. H., and Alpers, B. J.: Adamantinoma of the craniopharyngeal duct, Arch. Neurol. Psychiat. 26:905-965, 1931.

Freimann, G.: Sudden blindness and optic atrophy with craniopharyngeal pouch tumour, Amer. J. Ophthal. 10:579, 1927.

Globus, J. H., and Gang, K. M.: Craniopharyngioma and suprasellar adamantinoma, J. Mt. Sinai Hosp. 12:220-276, 1945.

Hirsch, O., and Hamlin, H.: Symptomatology and treatment of the hypophyseal duct tumors (craniopharyngiomas), Confin. Neurol. 19:153-219, 1959.

Irsigler, F. J.: Considérations cliniques et chirurgicales sur les cranio-pharyngiomes et les kystes suprasellaires, Neurochirurgie 8:242-252, 1962.

Lombardi, G.: Radiology in neuro-ophthalmology, Baltimore, 1967, The Williams & Wilkins Co.

Love, J. G., and Marshall, T. M.: Craniopharyngiomas (pituitary adamantinomata), Surg. Gynec. Obstet. 90:591-601, 1950.

McCritchley, D., and Ironside, R. N.: The pituitary adamantinomata, Brain 40:437-481, 1926.

Müller, R., and Wohlfahrt, G.: Craniopharyngiomas, Acta Med. Scand. 138:121, 1950.

Palomares-Petit, F., and Ley-Garcia, A.: Evolucion de la sintomatologia oftalmoneurologica en los craneofaringiomas, Arch. Soc. Oftal. Hispano.-Amer. 22:741-774, 1962.

Rintelen, F., and Leuenberger, A.: Zur Differentialdiagnose und Therapie des Kraniopharyngeoms, Schweiz. Med. Wschr. 87: 1189-1190, 1957.

Schulze, A.: Zur Differentialdiagnose der suprasellären Tumoren, Klin. Mbl. Augenheilk. **136**:166-185, 1960.

Seidenari, R.: Kraniopharyngeom bei Erwachsenen mit Stauungspapille, Osp. Magg. Milano Rev. Otol. Ecc. **18**:293-299, 1941.

Thébaut, F.: Klinik und Histologie der Kraniopharyngeome, Wien. Klin. Wschr. **59**:409-413, 1947.

Tönnis, W.: Kraniopharyngiome. Diagnostik der intrakraniellen Geschwülste. In Olivecrona, H., and Tönnis, W., editors: Handbuch der Neurochirurgie, Berlin, 1962, Julius Springer, vol. 4, p. 186.

Wagener, H. P., and Love, J. G.: Studies of the fields of vision in cases of Rathke's pouch tumors, Trans. Amer. Ophthal. Soc. **40**:305, 1942.

Wagener, H. P., and Love, J. G.: Fields of vision in cases of tumor of Rathke's pouch, Arch. Ophthal. **29**:873, 1943.

Walsh, F. B., and Hoyt, W. F.: Craniopharyngioma. In Clinical neuro-ophthalmology, Baltimore, 1969, The Williams & Wilkins Co., p. 2157.

Meningiomas of the tuberculum sellae

Alfonso, G. F., and Marini, G.: I segni oculari da meningiomi del tubercolo della sella. Osservazioni di 38 casi, Boll. Oculist. **63**:524, 1964.

Bailey, P.: Syndrom des Tuberculum sellae. Dei Hirngeschwülste, Stuttgart, 1951, Ferdinand Enke, p. 143.

Celotti, H., and Frera, C.: La sinomatologica oculare dei meningiomi del tubercolo della sella, Ann. Ottal. Clin. Oculist. **85**:1-13, 1959.

Coulonjou, R., Le Bozec, E., and Salaun, A.: Une erreur de diagnostic instructive: méningiome suprasellaire pris pour une arachnoïdite opto-chiasmatique, Rev. Otoneuroophtal. **22**:563-571, 1950.

Cushing, H.: The chiasmal syndrome of primary optic atrophy and bitemporal field defects in adults with a normal sella turcica, Arch. Ophthal. **3**:505-551, 704-735, 1930.

Cushing, H., and Eisenhardt, L.: Meningiomas arising from the tuberculum sellae, Arch. Ophthal. **1**:1-41, 168-205, 1929.

Cushing, H., and Eisenhardt, L.: Meningiomas, Springfield, Ill., 1938, Charles C Thomas, Publisher.

Desvignes, Brun, and Bounes: Sur un cas de cholestéatome suprasellaire, Bull. Soc. Ophtal. Paris **3**:382-385, 1947.

Duke-Elder, S.: Textbook of ophthalmology, St. Louis, 1949, The C. V. Mosby Co., vol. 4, p. 3516.

Forster, H. W., and Bouzarth, W. F.: Meningioma of the tuberculum sellae as a cause of optic atrophy, Amer. J. Ophthal. **64**:908-910, 1967.

Grant, F. C., and Hedges, T. R., Jr.: Ocular findings in meningiomas of the tuberculum sellae, Arch. Ophthal. **56**:163-170, 1956.

Guillaumat, L.: Les méningiomes suprasellaires, Paris, 1937, Gaston Doin & Cie.

Gunn, R. M.: Retrobulbar neuritis, Lancet, p. 412, 1904.

Harms, H.: Leitsymptom "Zentralskotom." In Entwicklung und Fortschritte in der Augenheilkunde, Stuttgart, 1963, Ferdinand Enke, p. 306.

Hartmann, E.: Cas atypique de méningiome suprasellaire avec dilatation unilatérale du canal optique, Bull. Soc. Ophtal. Paris **3**:100, 1948.

Hartmann, E., and Guillaumat, L.: Symptomes oculaires des méningiomes suprasellaires, Ann. Oculist. **174**:1-39, 1937.

van der Hoeve, J.: Diagnosis of suprasellar tumors, Ophthalmologica **99**:258, 1940.

Jane, J. A., and McKissock, W.: Importance of failing vision in early diagnosis of suprasellar meningiomas, Brit. Med. J. **2**:5-7, 1962.

Joy, H. H.: Suprasellar meningioma; report of an atypical case, Amer. J. Ophthal. **35**:1139-1146, 1952.

Kuwajima, J.: Problem cases of intracranial tumor simulating retrobulbar neuritis, Jap. J. Ophthal. **20**:1405-1411, 1966.

Lasco, F., and Arseni, C.: La pseudonévrite optique postérieure comme manifestations rare des néoformations intracrâniennes, Rev. Otoneuroophtal. **32**:386-391, 1960.

Mathewson, W. R.: Meningioma of tuberculum sellae with bitemporal hemianopia, Brit. J. Ophthal. **30**:92-102, 1946.

Mundinger, F., and Riechert, T.: Hypophysentumoren, Hypophysektomie, Stuttgart, 1967, Georg Thieme.

Philippidès, D., Montrieul, B., und Lévy, R.: A propos de trois cas de méningiome du tubercule de la selle, Rev. Otoneuroophtal. **25**:16-22, 1953.

Reuter, A., and Lamprecht, J.: Zur Klinik der suprasellären Tumoren, Deutsch. Med. Wschr. **2**:1033-1037, 1937.

Schlezinger, N. S., Alpers, B. J., and Weiss, B. P.: Suprasellar meningiomas associated with scotomatous field defects, Arch. Ophthal. **35**:620-642, 1946.

Walsh, F. B., and Hoyt, W. F.: Meningioma of the tuberculum sellae. In Clinical neuro-ophthalmology, Baltimore, 1969, The Williams & Wilkins Co., p. 2263.

Meningiomas of the olfactory groove

Bailey, P.: Syndrom der Olfaktoriusrinne. Die Hirngeschwülste, Stuttgart, 1951, Ferdinand Enke, p. 150.

Bonamour, G., and Wertheimer, I.: Valeur sémiologique de l'atrophie optique unilatérale, Ann. Oculist. 183:199, 1950.

Cushing, H.: Lancet 1:1329, 1927.

Cushing, H., and Eisenhardt, L.: Meningiomas, Springfield, Ill., 1938, Charles C Thomas, Publisher.

David, M., and Askenasy, H.: Méningiomes olfactifs, Rev. Neurol. 68:489, 1937.

David, M., and Sourdille, G.: Sur la valeur localisatrice des signes dits de compression directe du nerf optique, en particulier du syndrome de Foster Kennedy, dans les tumeurs cérébrales, J. Ophtal. 3:215, 1942.

Dilenge, D., and Fischgold, H.: L'angiographie orbitaire, Paris, 1968, Masson & Cie, p. 302.

Duke-Elder, S.: Textbook of ophthalmology, St. Louis, 1949, The C. V. Mosby Co., vol. 4, p. 3529.

Hagedoorn, A.: The chiasmal syndrome and retrobulbar neuritis in pregnancy, Amer. J. Ophthal. 20:690, 1937.

Hartmann, E., David, M., and Desvignes, P.: Les symptômes oculaires dans les méningiomes olfactifs, Ann. Oculist. 174: 505-527, 1937.

van der Hoeve, J.: Diagnosis of tumors of the suprasellar region with special reference to the Foster Kennedy syndrome, Ophthalmologica 99:258-264, 1940.

Lukasiewicz, W.: Diagnose, Differentialdiagnose und Therapie des Olfactoriusmeningeoms, Dissertation, Zurich, 1948.

Mylius, K.: Über das Meningeom der Olfactoriusrinne, eine für den Augenarzt besonders wichtige Geschwulst, Z. Augenheilk. 82:257, 1934.

Walsh, F. B., and Hoyt, W. F.: Olfactory groove tumors. In Clinical neuro-ophthalmology, Baltimore, 1969, The Williams & Wilkins Co., p. 2175.

Meningiomas of the sphenoid ridge

Bailey, P.: Syndrom des Keilbeinflügels. Die Hirngeschwülste, Stuttgart, 1951, Ferdinand Enke, p. 145.

Cushing, H., and Eisenhardt, L.: Meningiomas, Springfield, Ill., 1938, Charles C Thomas, Publisher.

David, M., and Hartmann, E.: Les symptômes oculaires dans les méningiomes de la petite aile du sphénoïde, Ann. Oculist. 172:177, 1935.

Doret, M.: Un cas d'hémicraniose—méningiome en plaque—avec importantes altérations du fond de l'œil, Rev. Otoneuroophtal. 22:460-466, 1950.

Duke-Elder, S.: Textbook of ophthalmology, St. Louis, 1949, The C. V. Mosby Co., vol. 4, p. 3531.

Elsberg, C. A., and Dyke, C. G.: Meningiomas attached to the mesial part of the sphenoidal ridge, Arch. Ophthal. 12:644, 1934.

Elsberg, C. A., Hare, C. C., and Dyke, C. H.: Unilateral exophthalmos in intracranial tumors with special reference to its occurrence in meningiomata, Surg. Gynec. Obstet. 55:681, 1932.

Fischer, E.: Paraselläre Geschwülste, Zbl. Chir., p. 2893, 1938.

Henderson, W. R.: Anterior basal meningiomas, Brit. J. Surg. 26:124-165, 1938.

Kearns, T. P., and Wagener, H. P.: Ophthalmologic diagnosis of meningiomas of the sphenoidal ridge, Amer. J. Med. Sci. 226: 221-228, 1953.

Kennedy, F.: The symptomatology of temporosphenoidal tumors, Arch. Intern. Med. 8:317-350, 1911.

Kennedy, F.: Retrobulbar neuritis as an exact diagnostic sign of certain tumors and abscesses in the frontal lobe, Amer. J. Med. Sci. 142:355, 1911.

Kraft, H.: Zur unterschiedlichen Symptomatologie der Keilbeinflügelmeningiome, Klin. Mbl. Augenheilk. 144:944, 1964.

Mortada, A.: Misinterpretation of sphenoidal ridge meningiomata, Brit. J. Ophthal. 51: 829-838, 1967.

Mylius, K.: Über das Meningeom des Keilbeinflügels, Klin. Mbl. Augenheilk. 113: 105-110, 1948.

Smith, J. W.: Meningioma producing unilateral exophthalmos; syndrome of tumor of pterional plaque arising from outer third of sphenoid ridge, Arch. Ophthal. 22:540-549, 1939.

Thorkildsen, A.: Meningioma of the sphenoid wing with proptosis (in Norwegian), Nord. Med. 38:1027-1028, 1948.

Tönnis, W.: Meningiome des kleinen Keilbeinflügel. Diagnostik der intrakraniellen Geschwülste. In Olivecrona, H., and Tönnis, W., editors: Handbuch der Neurochirurgie, Berlin, 1962, Julius Springer, p. 50.

Tönnis W., and Schürmann, K.: Meningiome der Keilbeinflügel, Zbl. Neurochir. 11:1-13, 1951.

Truemann, R. H.: Syndrome of meningeal fibroblastoma arising from the lesser wing of the sphenoid bone; analysis from an ophthalmologic standpoint, Amer. J. Ophthal. 30:1585, 1947; Arch. Ophthal. 38:566, 1947.

Uihlein, A., and Weyand, R. D.: Meningiomas of anterior clinoid process as a cause of unilateral loss of vision: surgical consideration, Arch. Ophthal. 49:261-270, 1953.

Walsh, F. B., and Hoyt, W. F.: Meningiomas of the sphenoid ridge. In Clinical neuro-ophthalmology, Baltimore, 1969, The Williams & Wilkins Co., p. 2266.

Aneurysms*

Jefferson, G.: Compression of the chiasma, optic nerves and optic tracts by intracranial aneurysms, Brain **60**:444, 1937.

Thomas, H., Tridon, P., Laxenaire, M., and Montaut, J.: Anévrisme simulant une tumeur hypophysaire avec hypopituitarism, Rev. Otoneuroophtal. **36**:128-135, 1964.

Indirect compression of the chiasm

Klingler, M., and Condrau, G.: Lokalisatorisch irreführende Gesichtsfeldsymptome bei Hirntumoren, Ophthalmologica **120**:270, 1950.

Hughes, E. B. C.: Some observations on visual fields in hydrocephalus, J. Neurol. Neurosurg. Psychiat. **9**:30, 1946.

Mooney, A. J., and McConnel, A. A.: Visual scotomata with intracranial lesions affecting the optic nerve, J. Neurol. Neurosurg. Psychiat. **12**:205, 1949.

Chiasmal arachnoiditis

Bollack, J., David, M., and Puech, P.: Les arachnoïdites optochiasmatiques, Paris, 1937, Masson & Cie.

Hartmann, E.: Optochiasmic arachnoiditis, Arch. Ophthal. **33**:68, 1945.

Kravitz, D.: Arachnoiditis, Arch. Ophthal. **37**:199-210, 1947.

Meyer, F. W.: Zur Symptomatologie der Arachnoiditis opticochiasmatica, Klin. Mbl. Augenheilk. **107**:274-280, 1941.

Vail, D.: Optochiasmic arachnoiditis; importance of mixed type of atrophy of optic nerve as diagnostic sign, Arch. Ophthal. **20**:384-394, 1938.

Gliomas of the chiasm;
tumors of the optic nerve

Amsler, M.: Tumoren des Sehnerven. In Amsler, M., et al., editors: Lehrbuch der Augenheilkunde, Basel, 1954, S. Karger AG, p. 701.

Brégeat, P.: Les gliomes du chiasma, Thèse, Paris, 1942.

Buffat, J. D.: Tumeurs du nerf optique et neurochirurgie, Confin. Neurol. **9**:411, 1949.

Bürki, E.: Über den primären Sehnerventumor und seine Beziehungen zur Recklinghausenschen Neurofibromatose. Bibliotheca Ophthalmologica, fasc. 30, Basel, 1944, S. Karger AG.

Bürki, E.: Über den klinischen West der

Röntgenaufnahme des Canalis opticus, Ophthalmologica **123**:243-248, 1952.

Christensen, E., and Ry Andersen, S.: Primary tumours of the optic nerve and chiasma, Acta Psychiat. Neurol. **27**:5-16, 1952.

Chutorian, A. M., Schwartz, J. F., Evans, R. A., and Carter, S.: Optic gliomas in children, Neurology **14**:83-95, 1964.

Cohen, I.: Tumours of the intracranial portion of the optic nerve, J. Mt. Sinai Hosp. **17**:738-745, 1951.

Daum, S., and Guillaumat, L.: Tumeurs de la gaine méningée du nerf optique, Rev. Otoneuroophtal. **21**:18, 1949.

Davis, F. A.: Primary tumors of optic nerve (phenomenon of Recklinghausen's disease); clinical and pathologic study with report of five cases and review of literature, Arch. Ophthal. **23**:735-957, 1940.

Dresner, E., and Montgomery, D. A. D.: Primary optic atrophy in von Recklinghausen's disease (multiple neurofibromatosis), Quart. J. Med. **18**:93-104, 1949.

Dufour, R., and Cometta, F.: Aspects radiologiques de l'orbite en ONO, Confin. Neurol. **10**:201, 1950.

Epstein, B. S., and Kulick, M.: Technique for optic foramen roentgenography, Radiology **42**:186-187, 1944.

Floris, V., and Castorina, G.: I gliomi del nervo ottico; Varietà endocranica, Riv. Neurol. **22**:799-821, 1952.

Foerster, O., and Gagel, O.: Ein Fall von sog. Gliom des Nervus opticus—Spongioblastoma multiforme ganglioides, Z. Ges. Neurol. Psychiat. **136**:335-366, 1931.

François, J., and Rabaey, M.: Tumeurs primitives du nerf optique, Acta Ophthal. **30**:203-221, 1952.

Goldmann, H., and Grunthal, E.: Über einen Tumor des Sehnerven und seiner Leptomeningen bei Recklinghausenscher Krankheit, Ophthalmologica **102**:79-92, 1941.

Greenway, R.: The optic foramina, X-ray Techn. **25**:31-35, 1953.

Grote, W.: Zur Klinik und Behandlung der Tumoren des Nervus opticus und des Chiasmas, Klin. Mbl. Augenheilk. **144**:841-855, 1964.

van der Hoeve, J.: Röntgenphotographie des Foramen Opticum bei Geschwülsten und Erkrankungen des Sehnerven, Graefe. Arch. Ophthal. **115**:355, 1925.

van der Hoeve, J.: Eye symptoms in phakamatoses (the Doyne Memorial Lecture), Trans. Ophthal. Soc. U. K. **52**:380-401, 1932.

Javett, S. N., and Samuel, E.: Tumor of the optic chiasma, Arch. Dis. Childhood **22**:248-250, 1947.

---

*For additional literature see references for intracranial aneurysms in Chapter 5.

Ingvar, A.: Primary tumors of optic nerve and their relation to Recklinghausen's disease, Acta Paediat. 32:262, 1945.

Katzin, H. M.: Glioma of optic nerve, J. Mt. Sinai Hosp. 11:332, 1945.

Kessel, F. K.: Chiasmagliom, Nervenarzt 10:414-415, 1937.

Löhlein, W., and Tönnis, W.: Die operative Behandlung der das Foramen opticum überschreitenden Sehnervengeschwülste, Graefe. Arch. Ophthal. 149:318-354, 1949.

Love, J. G., and Rucker, C. W.: Meningioma of sheath of optic nerve, Arch. Ophthal. 23:377, 1940.

McCraig, W., and Gogela, L. J.: Intraorbital meningiomas, Amer. J. Ophthal. 32:1663-1680, 1949.

McCraig, W., and Gogela, L. J.: Meningioma of the optic foramen as a cause of slowly progressive blindness, J. Neurosurg. 7:44-48, 1950.

McFarland, P. E., and Eisenbeiss, J.: Glioma of the optic nerve, Amer. J. Ophthal. 33:463-466, 1950.

Mannheimer, M.: Glioma of optic nerve, Amer. J. Ophthal. 29:323, 1946.

Marinkovic, A.: Imagine stratigrafica dei canali ottici, Radiol. Med. 37:987-997, 1951.

Marshall, D.: Glioma of the optic nerve as a manifestation of von Recklinghausen's disease, Amer. J. Ophthal. 37:15, 1954.

Martin, P., and Cushing. H.: Primary gliomas of the chiasm and optic nerves in their intracranial portion, Arch. Ophthal. 52:209-241, 1923.

Meadows, S. P.: Optic nerve compressions and its differential diagnosis, Proc. Roy. Soc. Med. 42:1017, 1949.

Minton, J.: Glioma of optic chiasma, Proc. Roy. Soc. Med. 38:594, 1945.

Nordmann, J.: Les tumeurs du nerf optique. In Bailliart, P., et al., editors: Traité d'ophthalmologie, Paris, 1939, Masson & Cie, vol. 6, p. 385.

Okonek, G.: Das Syndrom der Sehnervenkreuzung, Klin. Mbl. Augenheilk. 116:113-130, 1950.

Paillas, J. E., Bonnal, J., Sedan, R., Berard-Badier, M., and Combalbert, A.: Tumeurs du chiasme, Neurochirurgie 7:278-297, 1961.

Penfield, W., and Young, A. W.: Nature of von Recklinghausen's disease and tumors associated with it, Arch. Neurol. Psychiat. 23:320-344, 1930; Trans. Amer. Neurol. Ass. 55:319-343, 1929.

Pereira Gomes, J.: Tumors of the optic nerve; study of 13 cases in Brazil: relation to Recklinghausen's disease, Amer. J. Ophthal. 24:1144-1169, 1941.

Ray, B. S.: Surgical lesions of optic nerves and chiasm in infants and children, In Smith, J. L., editor: Neuro-ophthalmology, St. Louis, 1967, The C. V. Mosby Co.

Rettelbach, E., and Schutzbach: Über Sehnerventumoren, ihre Beziehungen zur Neurofibromatosis Recklinghausen und ihr klinisches Krankheitsbild, Graefe. Arch. Ophthal. 145:179-241, 1942.

Schoen, D.: Übersichtsaufnahme beider Foramina nervi optici, Fortschr. Roentgenstr. 78:349-350, 1953.

Schwarz, K.: Zur Klinik und Anatomie der Geschwülste des Sehnerven, Graefe. Arch. Ophthal. 135:247-264, 1936.

Tita, C.: Primary glioma of optic nerve, Boll. Oculist. 18:271, 1939.

Tönnis, W.: Die operative Behandlung der das Foramen opticum überschreitenden Geschwülste des N. opticus, Acta Neurochir. 1:52-71, 1950.

Udvarhelyi, G. B., Khodadoust, A. A., and Walsh, F. B.: Gliomas of the optic nerve and chiasm in children, Clin. Neurosurg. 13:204-237, 1966.

Wagener, H. P.: Gliomas of the optic nerve, Amer. J. Med. Sci. 237:238-261, 1959.

Wagner, F.: Beitrag, zur Frage der primären Opticusgliome, Klin. Mbl. Augenheilk. 103:606, 1939.

Walsh, F. B.: The ocular signs of tumors involving the anterior visual pathway, Amer. J. Ophthal. 42:347-377, 1956.

Walsh, F. B., and Hedges, T. R., Jr.: Optic nerve sheath haemorrhage, Trans. Amer. Acad. Ophthal. 54:29-48, 1950.

Walsh, F. B., and Hoyt, W. F.: Primary tumors of the optic nerve and chiasm. In Clinical neuro-ophthalmology, Baltimore, 1969, The Williams & Wilkins Co., p. 2075.

Wilson, J. M., and Farmer, W. D.: Glioma of optic nerve, Arch. Ophthal. 23:605, 1940.

Wilson, J. M., and Farmer, W. D.: Spongioblastoma of the optic nerve: a critical review; report of 2 cases with autopsy observation in one, Arch. Ophthal. 23:605-618, 1940.

Tumors of the cavernous sinus

Aron-Rosa, D.: Utilité de la phlébographie orbitaire dans l'étude anatomique et pathologique du sinus caverneux, Bull. Soc. Ophtal. Franc. 66:912-916, 1966.

Fairclough, W. A.: Drainage in infected cavernous sinus thrombosis, Aust. New Zeal. J. Surg. 16:193-196, 1947.

Foix, C.: Syndrome de la paroi externe du sinus caverneux, Rev. Neurol. 37/38:827, 1922.

Franceschetti, Jentzer, Junet, and Dorex: Deux cas d'anévrisme artério-veineux de la carotide interne dans le sinus caverneux, Rev. Otoneuroophtal. 19:53-54, 1947.

Godtfredsen, E.: Ophthalmologic and neurologic symptoms of malignant nasopharyngeal Tumors, Acta Path. Microbiol. Scand. (supp.) 55: 1944.

Huber, A.: Les anévrismes de l'orbite et du sinus caverneux, Bull. Soc. Franc. Ophtal. 64:374, 1951.

Huber, A.: Über das Syndrom des Sinus cavernosus, Ophthalmologica 121:118, 1951.

Jefferson, G.: On the saccular aneurysms of the internal carotid artery in the cavernous sinus, Brit. J. Surg. 26:267, 1938-1939.

Jefferson, G.: Concerning injuries, aneurysms and tumours involving the cavernous sinus, Trans. Ophthal. Soc. U. K. 73:117-152, 1953.

Kaeser, E. H.: Das arteriovenöse Aneurysma im Sinus cavernosus, Ophthalmologica 126: 237-282, 1953.

Krayenbühl, H.: Das Hirnaneurysma, Schweiz. Arch. Neurol. Psychiat. 47:155, 1941.

McNeal, W. J., Frisbee, F. C., and Blevine, A.: Thrombophlebitis of the cavernous sinus; review of reported recoveries with special reference to thrombophlebitis of staphylococcic origin, Arch. Ophthal. 29: 231-257, 1943.

Weber, G.: Hirnabszesse im Rahmen infektiöser intrakranieller Erkrankungen, Stuttgart, 1956, Georg Thieme.

Weinberger, L. M., Adler, F. H., and Grant, F. C.: Primary pituitary adenoma and the syndrome of the cavernous sinus, Arch. Ophthal. 24:1197, 1940.

Tumors of the third ventricle

Dandy, W. E.: Benign tumors of the third ventricle of the brain: diagnosis and treatment, Springfield, Ill, 1933, Charles C Thomas, Publisher.

Duke-Elder, S.: Textbook of ophthalmology, St. Louis, 1949, The C. V. Mosby Co., vol. 4, p. 3525.

François, M. J.: La pathogénie de l'hémianopsie binasale, Bull. Soc. Belg. Ophtal. 81: 1945.

François, M. J.: L'hémianopsie binasale, Ophthalmologica 113:321, 1947.

Hughes, E. B. C.: Some observations on visual fields in hydrocephalus, J. Neurol. Neurosurg. Psychiat. 9:30, 1946.

Jefferson, G., and Jackson, H.: Tumors of lateral and third ventricles, Proc. Roy. Soc. Med. 32:1105-1137, 1939.

Jentzer, A.: Les tumeurs du troisième ventricule, Schweiz. Arch. Neurol. Psychiat. 14: 2, 1939.

Newton, T. H.: Neuroradiologic evaluation of lesions affecting the third ventricle and adjacent structures. In Smith, J. L., editor: Neuro-ophthalmology, St. Louis, 1967, The C. V. Mosby Co.

Oldberg, E., and Eisenhardt, L.: Neurological diagnosis of tumors of third ventricle, Trans. Amer. Neurol. Ass. 64:33-36, 1938.

Oldberg, E., and Eisenhardt, L.: Die neurologische Diagnose der Geschwülste des dritten Ventrikels, Nervenarzt, p. 615, 1939.

Pecker, J., Ferrand, B., and Javalet, A.: Tumeurs du troisième ventricule, Neurochirurgie 12:7-136, 1966.

Penfield, W.: Diencephalic autonomic epilepsy, Arch. Neurol. Psychiat. 22:358-375, 1929.

Penfield, W., and Erickson, T.: Epilepsy and cerebral localization, Springfield, Ill., 1941, Charles C Thomas, Publisher.

Rand, R. W., and Lemmen, L. J.: Tumours of the posterior portion of the third ventricle, J. Neurosurg. 10:1-18, 1953.

Tönnis, W.: Tumoren im dritten Ventrikel. Diagnostik der intrakraniellen Geschwülste. In Olivecrona, H., and Tönnis, W., editors: Handbuch der Neurochirurgie, Berlin, 1962, Julius Springer, vol. 4, p. 429.

Wagener, H. P., and Cusick, P. L.: Chiasmal syndromes produced by lesions in posterior fossa, Arch. Ophthal. 18:887-891, 1937.

Walsh, F. B., and Hoyt, W. F.: Third ventricle tumor. In Clinical neuro-ophthalmology, Baltimore, 1969, The Williams & Wilkins Co., p. 2248.

Tumors of the thalamus

Bailey, P.: Thalamussyndrom. Die Hirngeschwülste, Stuttgart, 1951, Ferdinand Enke, p. 228.

Knapp, P., and Schwarzmann, A.: Contraction in visual fields caused by diencephalic lesion, Ophthalmologica 111:270-278, 1946.

McKissock, W., and Paine, K. W. E.: Primary tumors of the thalamus, Brain 81: 41-63, 1958.

Névin, S.: Thalamic hypertrophy or gliomatosis of the optic thalamus, J. Neurol. Neurosurg. Psychiat. 1:342, 1938.

Ranson, S. W., and Magoun, H. W.: The hypothalamus, Ergebn. Physiol. 41:56-163, 1939.

Tönnis, W.: Tumoren des Thalamus. Diagnostik der intrakraniellen Geschwülste. In Olivecrona, H., and Tönnis, W., editors: Handbuch der Neurochirurgie, Berlin, 1962, Julius Springer, vol. 4, p. 433.

Walsh, F. B., and Hoyt, W. F.: Tumors of the thalamus. In Clinical neuro-ophthalmology, Baltimore, 1969, The Williams & Wilkins Co., p. 2221.

Weinberger, L. M., and Grant, F. C.: Precocious puberty and tumors of the hypothalamus, Arch. Intern. Med. 67:762, 1941.

Tumors of the basal ganglia; tumors of the pineal body and the midbrain

Argyll Robertson, D., Four cases of spinal myosis; with remarks on the action of light on the pupil, Edinburgh Med. J. 15:487, 1869.

Bailey, P.: Syndrom der Epiphysengegend. Die Hirngeschwülste, Stuttgart, 1951, Ferdinand Enke, p. 282.

Balthasar, K.: Gliomas of the quadrigeminal plate and eye movements, Ophthalmologica 155:249-270, 1968.

Barker, L. F.: Parinaud's syndrome, Amer. J. Ophthal. 18:827-832, 1935.

Best, F.: Erkrankungen der Vierhügel, die Zirbeldrüsengeschwülste. In Schieck, F., and Brückner, A., editors: Kurzes Handbuch der Ophthalmologie, Berlin, 1931, Julius Springer, vol. 6, p. 497, 499.

Bielschowsky, A.: Lectures on motor anomalies: paralysis of conjugate movements of eyes, Arch. Ophthal. 13:569, 1935.

Böhringer, H. R., and Koenig, F.: Die diagnostische Bedeutung der vertikalen Blicklähmung, Ophthalmologica 125:357, 1935.

Calmettes, L., and Deodati, F.: Nystagmus retractorius, Annee Ther. Clin. Ophtal. 15:71-91, 1964.

Collier, J.: Nuclear ophthalmoplegia. With special reference to the retraction of lids and ptosis and to lesions of posterior commissure, Brain 50:488-498, 1927.

Devic, A., Paufique, L., Girard, P., and Guinet, P.: Syndrome of Parinaud: anatomico-clinical study, Ann. Oculist. 178:199, 1945.

Foerster, O., Gagel, O., and Mahoney, W.: Die encephalen Tumoren der Oblongata, Pons und des Mesencephalons, Arch. Psychiat. Nervenkr. 110:1-74, 1939.

Glaser, M. A.: Tumours of the pineal body, corpora quadrigemina and third ventricle, the inter-relationship of their syndromes and their surgical treatment, Brain 52:226-262, 1929.

Globus, J. H.: Tumors of quadrigeminate plate; clinico-anatomic study of 7 cases, Arch. Ophthal. 5:418-444, 1931.

Globus, J. H.: Pinealoma, Arch. Path. 31:533-568, 1941.

Hatcher, M. A., and Klintworth, G. K.: Sylvian aqueduct syndrome, Arch. Neurol. 15:215-222, 1966.

Horrax, G., and Bailey, P.: Tumors of the pineal body, Arch. Neurol. Psychiat. 13:423-467, 1925.

Horrax, G., and Bailey, P.: Pineal pathology: further studies, Arch. Neurol. Psychiat. 19:394-414, 1928.

Horrax, G., and Wyatt, J. C.: Ectopic pinealomas in chiasmal region, J. Neurosurg. 4:309-326, 1947.

Jaensch, P. A.: Doppelseitige Trochlearisparese als einzige Motilitätsstörung bei Zirbeldrüsentumor, Z. Augenheilk. 75:58, 1931.

Lillie, W. I.: Homonymous hemianopia; primary signs of tumors involving lateral part of the transverse fissure, Trans. Sect. Ophthal. Amer. Med. Ass., pp. 200-212, 1929.

Nashold, B. S., and Gills, J. P.: Ocular signs resulting from brain stimulation and lesions in the human, Arch. Ophthal. 77:609-618, 1967.

Netsky, M. G., and Strobos, R. R. J.: Neoplasms within the midbrain, Arch. Neurol. Psychiat. 68:116-129, 1952.

Palomar-Gollado, F., and Palomar, P. F.: Tumores cerebrales y signo de Argyll-Robertson, Arch. Soc. Oftal. Hispano-Amer. 20:470-477, 1960.

Pia, H. W.: Klinik, Differentialdiagnose und Behandlung der Vierhügelgeschwülste, Deutsch. Z. Nervenheilk. 172:12-32, 1954.

Posner, M., and Horrax, G.: Eye signs in pineal tumors, J. Neurosurg. 3:15-24, 1946.

Schultz-Zehden, W.: Augenveränderungen bei Tumoren der Vierhügelgegend, Z. Ges. Inn. Med. 6:8-12, 1951.

Simonelli, M.: Case of Parinaud's syndrome, Riv. Oftal. 1:110-123, 1946.

Smith, J. L., Zieper, I., Gay, A. I., and Cogan, D. C.: Nystagmus retractorius Arch. Ophthal. 62:864-867, 1959.

Thiébaut, F., Philippides, D., and Buchheit, F.: Clonus rétractorius ou nystagmus retractorius, symptome neurochirurgical, Rev. Otoneuroophtal. 39:72-82, 1967.

Tönnis, W.: Tumoren des Corpus striatum. Diagnostik der intrakraniellen Geschwülste. In Olivecrona, H., and Tönnis, W., editors: Handbuch der Neurochirurgie, Berlin, 1962, Julius Springer, vol. 4, p. 431.

Tönnis, W.: Geschwülste der Vierhügelgegend. Diagnostik der intrakraniellen Geschwülste. In Olivecrona, H., and Tönnis, W., editors: Handbuch der Neurochirurgie, Berlin, 1962, Julius Springer, vol. 4, p. 436.

Tönnis, W.: Geschwülste des Mittelhirns. Diagnostik der intrakraniellen Geschwülste. In Olivecrona, H., and Tönnis, W., editors: Handbuch der Neurochirurgie, Berlin, 1962, Julius Springer, vol. 4, p. 441.

Walsh, F. B., and Hoyt, W. F.: Pineal tumors. In Clinical neuro-ophthalmology, Baltimore, 1969, The Williams & Wilkins Co., p. 2240.

Walsh, F. B., and Hoyt, W. F.: Midbrain tumors. In Clinical neuro-ophthalmology, Baltimore, 1969, The Williams & Wilkins Co., p. 2215.

Walsh, F. B., and Hoyt, W. F.: Tumors of the basal ganglia. In Clinical neuro-ophthalmology, Baltimore, 1969, The Williams & Wilkins Co., p. 2222.

Weber, G.: Tumoren der Glandula pinealis und ektopische Pinealozytome, Schweiz. Arch. Neurol. Psychiat. 91:473-509, 1963.

Yanagida, N.: Ocular symptoms of tumor in the aqueduct of Sylvius (in Japanese), Acta Soc. Ophthal. Jap. 57:254-258, 1953.

Young, S., Wu, C. P., and Chen, C. S.: Pinealoma, Chinese Med. J. 68:261-262, 1950.

**Tumors of the cerebellum
(including von Hippel-Lindau disease)**

Arseni, C., Maretsis, M., and Vasilesco, A.: La maladie de von Hippel-Lindau, Rev. Otoneuroophtal. 39:174-182, 1967.

Bailey, P.: Syndrom der Kleinhirnhemisphäre. Die Hirngeschwülste, Stuttgart, 1951, Ferdinand Enke, p. 206.

Bailey, P.: Syndrom des Kleinhirnwurms. Die Hirngeschwülste, Stuttgart, 1951, Ferdinand Enke, p. 199.

Best, F.: Die Augenstörungen bei Kleinhirnerkrankungen. In Schieck, F., and Brückner, A., editors: Kurzes Handbuch der Ophthalmologie, Berlin, 1931, Julius Springer, vol. 6, p. 510.

Cogan, D. G.: Ocular dysmetria, flutterlike oscillations of the eyes and opsoclonus, Arch. Ophthal. 51:318-335, 1954.

Cogan, D. G.: Dissociated nystagmus with lesions in the posterior fossa, Arch. Ophthal. 70:361-368, 1963.

Craig, W. M., Wagener, H. P., and Kernohan, J. W.: Lindau-von Hippel disease, Trans. Amer. Neurol. Ass. 65:98-101, 1939.

Craig, W. M., Wagener, H. P., and Kernohan, J. W.: Lindau-von Hippel disease, Arch. Neurol. Psychiat. 46:36, 1941.

Cushing, H., and Walker, C. V.: Distortions of the visual fields in cases of brain tumor: binasal hemianopia, Arch. Ophthal. 41:559, 1912.

Danis, P.: Aspects ophthalmologiques des angiomatoses du système nerveux, Acta Neurol. Psychiat. Belg. 50:615-679, 1950.

Delmas-Marsalet, Lafon, and Faure: Vollständiges Chiasma-Syndrom, ausgelöst durch Ventrikelhydrocephalie bei Kleinhirntumor, J. Med. Bordeaux 118:661-664, 1941.

François, J., Hoffmann, G., and Jadoul, P.: Les astrocytomes kystiques du cervelet et leurs manifestations oculaires, Confin. Neurol. 21:307-325, 1961.

Hall, C. S.: Blood vessel tumors of brain with particular reference to Lindau's syndrome, J. Neurol. Psychopath. 15:305-312, 1935.

Heine, L.: Über Angiogliosis retinae mit Hirntumor, Graefe. Arch. Ophthal. 88: 1914.

von Hippel: Über eine sehr seltene Erkrankung der Netzhaut, Graefe. Arch. Ophthal. 59: 1904.

von Hippel: Die anatomische Grundlage der von mir beschriebenen seltenen Netzhauterkrankungen, Graefe. Arch. Ophthal. 95: 1918.

Holmes, G.: The cerebellum of man, Brain 62:1, 1939.

Huber, A.: Cerebrale und retinale Angiopathie, Ophthalmologica 117:265, 1949.

Kestenbaum, A.: Mechanismus des Nystagmus, Graefe. Arch. Ophthal. 105:799, 1921.

Krayenbühl, H.: Die Symptomatologie der Tumoren der hinteren Schädelgrube, insbesondere des Kleinhirns, Schweiz. Med. Wschr. 2:901-905, 1938.

Lewis, P. M.: Angiomatosis retinae, Arch. Ophthal. 30:250, 1943.

Lindau, A.: Studien über Kleinhirncysten. Bau, Pathogenese, Beziehungen zur Angiomatosis retinae, Acta Path. Scand. (supp. 1), p. 128, 1926.

Lindau, A.: Vascular tumors of brain, Proc. Roy. Soc. Med. 24:363, 1941.

Lotmar, F.: Zur Kenntnis der Lindauschen Krankheit, Schweiz. Arch. Neurol. Psychiat. 36:2, 1935.

McDonald, A. E.: Lindau's disease; report of 6 cases with surgical verification in 4 living patients, Arch. Ophthal. 23:564-576, 1940.

McGovern, F. H.: Angiomatosis retinae, Amer. J. Ophthal. 26:184, 1943.

MacNab, G. H.: Lindau's disease, Proc. Roy. Soc. Med. 34:324-325, 1941.

Magendie: Précis de Physiologie 3:1, 380, 1833.

de Martel, T., and Guillaume, J.: Les tumeurs de la loge cérébelleuse (fosse cérébrale postérieure), Paris, 1934, Gaston Doin & Cie.

Martin, P.: L'angiomatose rétinocérébelleuse et sa pathologie chirurgicale, Acta Neurol. Psychiat. Belg. 50:457-461, 1950.

Olson, A. K., and Pinley, C. C.: Cerebellar tumors and their clinical signs, Guthric Clin. Bull. 17:141, 1948.

Palin, A.: Von Hippel-Lindau's disease, Proc. Roy. Soc. Med. **32**:1618, 1939.

Rumbaur, W.: Angiomatosis retinae (Hippel-Lindau disease): report on a genealogic tree, Klin. Mbl. Augenheilk. **106**: 168, 1941.

Ryan, E. P.: Angiomatosis retinae, Arch. Ophthal. **23**:623, 1940.

Staz, L.: Angiomatosis retinae, Brit. J. Ophthal. **25**:167, 1941.

Stewart, T., and Holmes, G.: Symptomatology of cerebellar tumors; a study of forty cases, Brain **27**:522-591, 1904.

Tönnis, W.: Geschwülste des Kleinhirs. Diagnostik der intrakraniellen Geschwülste. In Olivecrona, H., and Tönnis, W., editors: Handbuch der Neurochirurgie, Berlin, 1962, Julius Springer, vol. 4, p. 444.

Traquair, H. M.: Hemangioma of the retina, Trans. Ophthal. Soc. U. K. **52**:311, 1932.

Wagener, H. P., and Cusick, P. L.: Chiasmal syndromes produced by lesions in posterior fossa, Arch. Ophthal. **18**:887-891, 1937.

Walsh, F. B., and Hoyt, W. F.: Cerebellar tumor. In Clinical neuro-ophthalmology, Baltimore, 1969, The Williams & Wilkins Co., p. 2206.

Weinberger, L., and Webster, J. E.: Visual field defects associated with cerebellar tumors, Arch. Ophthal. **25**:128-138, 1941.

Wertheimer, P., Lapras, C., Wertheimer, J., Thierry, A., and Dusquesnel, J.: A propos de quelques aspects ophthalmologiques des angiomes de la fosse cérébrale postérieure, Bull. Soc. Franc. Ophtal. **77**:290-296, 1964.

Tumors of the fourth ventricle

Alpers, D. J., and Yaskin, H. E.: The Bruns syndrome, J. Nerv. Ment. Dis. **100**:115-134, 1944.

Bailey, P.: Das Syndrom des 4. Ventrikels. Die Hirngeschwülste, Stuttgart, 1951, Ferdinand Enke, p. 183.

Best, F.: Erkrankungen im Bereiche des 4. Ventrikels. In Schieck, F., and Brückner, A., editor: Kurzes Handbuch der Ophthalmologie, Berlin, 1931, Julius Springer, vol. 6, p. 500.

Craig, W. M., and Kernohan, J. W.: Tumors of fourth ventricle, J.A.M.A. **111**:2370-2377, 1938.

Walsh, F. B., and Hoyt, W. F.: Tumors of the fourth ventricle. In Clinical neuro-ophthalmology, Baltimore, 1969, The Williams & Wilkins Co., p. 2256.

Tumors of the cerebellopontine angle

Ageeva-Majkova, O. G.: The clinical findings in neurinoma of the eighth nerve from the oto-neurological viewpoint (in Russian), Vestn. Otorinolaring. **12**:3-10, 1950.

Bailey, P.: Syndrom des Kleinhirnbrückenwinkels. Die Hirngeschwülste, Stuttgart, 1951, Ferdinand Enke, p. 59.

Best, F.: Die Augenstörungen bei Kleinhirnbrückenwinkelschwülsten. In Schieck, F., and Brückner, A., editors: Kurzes Handbuch der Ophthalmologie, Berlin, 1931, Julius Springer, vol. 6, p. 515.

Brunner, H.: Brain tumors and ear, Trans. Amer. Acad. Ophthal. **40**:59-78, 1935.

Cushing, H.: Tumors of the nervus acusticus and the syndrome of the cerebello-pontile angle, Philadelphia, 1917, W. B. Saunders Co.

Edwards, C. H., and Paterson, J. H.: A review of the symptoms and signs of acoustic neurofibromata, Brain **74**:144, 1951.

Globus, J. H.: Pontofacial angle tumors, with particular reference to involvement of the acoustic nerve, Laryngoscope **51**: 1119-1138, 1941.

Graf, K.: Die Kleinhirnbrückenwinkelgeschwülste, Basel, 1955, S. Karger AG.

Hitselberger, W. E.: Acoustic neuroma diagnosis, Eye Ear Nose Throat Monthly **29**:49-54, 1967.

Lundberg, T.: The symptomatology and diagnosis of acoustic tumours from an otological viewpoint (in Swedish), Nord. Med. **44**:1520-1523, 1950.

Lundberg, T.: Diagnostic problems concerning acoustic tumours: a study of 300 verified cases and the Békésy audiogram in the differential diagnosis, Acta Otolaryng. (suppl) **99**:110, 1952.

Salgado Benavides, E.: Tumors of cerebellopontile angle, Arch. Soc. Oftal. Hispano-Amer. **4**:75, 1944.

Tönnis, W.: Geschwülste des Kleinhirnbrückenwinkels. Diagnostik der intrakraniellen Geschwülste. In Olivecrona, H., and Tönnis, W., editors: Handbuch der Neurochirurgie, Berlin, 1962, Julius Springer, vol. 4, p. 459.

Tönnis, W.: Neurinom des N. acusticus. Diagnostik der intrakraniellen Geschwülste. In Olivecrona, H., and Tönnis, W., editors: Handbuch der Neurochirurgie, Berlin, 1962, Julius Springer, vol. 4, p. 60.

Walsh, F. B., and Hoyt, W. F.: Cerebellopontine angle tumors. In Clinical neuro-ophthalmology, Baltimore, 1969, The Williams & Wilkins Co., p. 2229.

Winston, J.: Revision of cerebellopontile angle lesion syndrome, with analysis of vestibular findings in 34 cases of verified tumor of cerebellopontile angle, Arch. Otolaryng. **32**:877-886, 1940.

**Tumors of the pons and
the medulla oblongata**

Allerand, C. D.: Paroxysmal skew deviation in association with a brain stem glioma, report of an unusual case, Neurology **12**:520-523, 1962.

Bailey, P.: Syndrom des Hirnstammes. Die Hirngeschwülste, Stuttgart, 1951, Ferdinand Enke, p. 275.

Best, F.: Die Ponserkrankungen. In Schieck, F., and Brückner, A., editors: Kurzes Lehrbuch der Ophthalmologie, Berlin, 1931, Julius Springer, vol. 6, p. 494.

Brock, S., and Needles, W.: Tumors of the brain stem: a clinical study of 5 cases with autopsy findings, J. Nerv. Ment. Dis. **72**:531-534, 1930.

Buckley, R.: Pontile gliomas: a pathologic study and classification of 25 cases, Arch. Path. **9**:779, 1930.

Dandy, W. E.: In Lewis, Dean, editor: Lewis' practice of surgery, Hagerstown, 1936, W. F. Prior Co., Inc., vol. 12, pp. 534-556.

Daroff, R. B., and Waldman, A. L.: Ocular bobbing, J. Neurol. Neurosurg. Psychiat. **28**:375, 1965.

Fisher, C. M.: Ocular bobbing, Arch. Neurol. **11**:543-546, 1964.

Foerster, O.: Die encephalen Tumoren der Oblongata, des Pons und des Mesencephalons, Z. Neurol. Psychiat. **168**:482-518, 1940.

Foerster, O., Gagel, O., and Mahoney, W.: Die encephalen Tumoren des verlängerten Markes, der Brücke und des Mittelhirns, Arch. Psychiat. **110**:1-74, 1939.

Haire, C. C., and Wolf, A.: Intramedullary tumors of the brain stem, Arch. Neurol. Psychiat. **32**:1230-1252, 1934.

Horrax, G., and Buckley, R.: A clinical study of the differentiation of certain pontile tumors from acoustic tumors, Arch. Neurol. Psychiat. **24**:1217-1230, 1950.

Kaufmann, J.: Tumeurs pontines et bulbopontines, Schweiz. Arch. Neurol. Psychiat. **64**:197, 1949.

Mackensen, G.: Differentialdiagnose Kleinhirnbrückwinkel—pontiner Tumor, Ber. Deutsch. Ophth. Ges. **67**:430-431, 1965.

Sagebiel, J.: Intrapontine tumors; clinicopathologic study, Ohio Med. J. **33**:760-768, 1937.

Santha, K.: Zur Symptomatologie der Ponstumoren, klinisch-anatomischer Beitrag zur Kenntnis der pontinen Blicklähmung, Arch. Psychiat. **103**:539-551, 1935.

Simon, K. A., and Gay, A. J.: Optokinetic responses in brain stem lesions, Arch. Ophthal. **71**:303-307, 1964.

Smith, J., David, N. J., and Klintworth, G.: Skew deviation, Neurology **14**:96-105, 1964.

Tönnis, W.: Tumoren von Pons und Oblongata. Diagnostik der intrakraniellen Geschwülste. In Olivecrona, H., and Tönnis, W., editors: Handbuch der Neurochirurgie, Berlin, 1962, Julius Springer, vol. 4, p. 442.

Walsh, F. B.: Tumors of the brain stem. In Clinical, neuro-ophthalmology, Baltimore 1947, The Williams & Wilkins Co., p. 2215.

Walsh, F. B., and Hoyt, W. F.: Gliomas of the pons. In Clinical neuro-ophthalmology, Baltimore, 1969, The Williams & Wilkins Co., p. 2216.

CHAPTER FOUR *Relationship between type of tumor and ocular symptoms*

Bailey, P.: Metastatische Tumoren. Die Hirngeschwülste, Stuttgart, 1951, Ferdinand Enke, p. 303.

Bonkalo, S.: Die Bedeutung der Hirngeschwulstart und des Geschwulstsitzes für die Entstehung der Hirnschwellung, Deutsch. Z. Nervenheilk. **149**:243-253, 1939.

Dandy, W. E.: Metastatic tumors. In Lewis, Dean, editor: Lewis' practice of surgery, Hagerstown, Md., 1944, W. F. Prior Co., Inc., vol. 12, p. 669.

Gärtner, J.: Statische Untersuchungen an 654 intrakraniellen raumfordernden Prozessen: ein Beitrag zur Biologie der Hirngeschwülste, Zbl. Neurochir. **15**:333-351, 1955.

Klingler, M., and Condrau, G.: Lokalisatorisch irreführende Gesichtsfeldsymptome bei Hirntumoren, Ophthalmologica **120**:270, 1950.

Livingston, K. E., Horrax, G., and Sachs, E.: Metastatic brain tumours, Surg. Clin. N. Amer. **28**:805, 1948.

Mahoney, W.: Die Epidermoide des Zentralnervensystems, Z. Ges. Neurol. Psychiat. **155**:416-471, 1936.

Minkowski, M.: Über metastatische Hirngeschwülste, Schweiz. Arch. Neurol. Psychiat. **47**:9, 1941.

Orban, T., and Födö, V.: Rindenblindheit infolge beiderseitiger Metastase im Okzipitalpol, Klin. Mbl. Augenheilk. 148:700-704, 1966.

Pass, E. K.: Zur Klinik der zerebralen Karzinommetastasen, Nervenarzt, p. 385, 1939.

Rados, A.: Occurrence of glioma of retina and brain in collateral lines in same family: genetics of glioma, Arch. Ophthal. 35:1, 1946.

Reymond, A.: Classification anatomo-pathologique des tumeurs cérébrales, Ophthalmologica 125:204, 1953.

Scheid, W.: Zur Klinik der intrakraniellen Karzinommetastasen, Allg. Z. Psychiat. 113: 66-82, 1939.

Tönnis, W.: Augensymptome bei 3033 Hirngeschwülsten, 59 Zusammenk., Deutsch. Ophthal. Ges., Heidelberg, 1955.

Tönnis, W.: Erscheinungsbild und Verlaufsform der verschiedenen Geschwulstarten. Diagnostik der intrakraniellen Geschwülste. In Olivecrona, H., and Tönnis, W., editors: Handbuch der Neurochirurgie, Berlin, 1962, Julius Springer, vol. 4, p. 2.

Walsh, F. B., and Hoyt, W. F.: Hematogenous metastases to the brain. In Clinical neuro-ophthalmology, Baltimore, 1969, The Williams & Wilkins Co., p. 2328.

Weber, G.: Zur Diagnose und Prognose intrakranieller Tumoren, Ophthalmologica 125: 231, 1953.

CHAPTER FIVE *Pseudotumor cerebri, subdural hematomas, brain abscesses, and aneurysms*

Pseudotumor cerebri

Albrecht, H.: Stauungspapille bei nicht tumorösen Hirnprozessen. Sammlung zwangloser Abhandlungen aus dem Gebiete der Psychiatrie und Neurologie, Jena, 1964, VEB Gustav Fischer Verlag, Heft 26.

Bailey, P.: Contribution to the histopathology of pseudotumor cerebri, Arch. Neurol. Psychiat. 4:401-416, 1920.

Frazier, C. H.: Cerebral pseudotumors, Arch. Neurol. Psychiat. 24:1117-1132, 1930.

Krayenbühl, H.: Die Bedeutung der Angiographie für die Diagnose der cerebralen Thrombophlebitis, Acta Neurochir. Supp. 3:198-201, 1955.

Marr, W. G., and Chambers, J. W.: Occlusion of the cerebral dural sinuses by tumor simulating pseudotumor cerebri, Amer. J. Ophthal. 61:45-49, 1966.

Nonne, M.: Ueber Fälle von Symptomenkomplex "Tumor cerebri" mit Ausgang in Heilung (Pseudo-Tumor cerebri), Deutsch. Z. Nervenheilk. 27:169, 1904.

Stochdorph, O.: Zur Frage der chronischen Hirnschwellung, Arch. Psychiat. Nervenkr. 181:101, 1949.

Walsh, F. B.: Papilledema associated with increased intracranial pressure in Addison's disease, Arch. Ophthal. 47:86, 1952.

Walsh, F. B., and Hoyt, W. F.: Pseudotumor cerebri, In Clinical neuro-ophthalmology, Baltimore, 1969, The Williams & Wilkins Co., p. 593.

Walsh, F. C., Clark, D. B., Thompson, R. S., and Nicholsen, D. H.: Oral contraceptives and neuro-ophthalmologic interest, Arch. Ophthal. 74:628-640, 1965.

Subdural hematomas

Bailey, P.: Subdurales Hämatom. Die Hirngeschwülste, Stuttgart, 1951, Ferdinand Enke, p. 354.

Bradley, K. C.: Extra-dural haemorrhage, Aust. New Zeal. J. Surg. 21:241-260, 1952.

Fischer-Brügge, K.: Das Klivuskanten-Syndrom, Acta Neurochir. 2:36, 1951.

Furlow, L. T.: Chronic subdural hematoma, Arch. Surg. 32:688-708, 1936.

Govan, C. D., and Walsh, F. B.: Symptomatology of subdural haematoma in infants and in adults: comparative study, with particular reference to the ocular signs; an observation concerning pathogenesis of subdural haematoma, Arch. Ophthal. 37:701-715, 1947.

Guthkelch, A. N.: Extradural haemorrhage as a cause of cortical blindness, J. Neurosurg. 6:180-182, 1949.

Hanke, H.: Das subdurale Hämatom, Berlin, 1939, Julius Springer, p. 180.

Heyser, J., and Weber, G.: Die epiduralen Hämatome, Schweiz. Med. Wschr. 94: 2-7, 46-52, 1964.

Kennedy, F., and Wortis, H. Surg. Clin. N. Amer. 63:732, 1936.

Klingler, M.: Das Schädel-Hirntrauma, Stuttgart, 1961, Georg Thieme.

Krayenbühl, H., and Noto, G.: Das Intrakranielle subdurale Hämatom, Bern, 1949, Hans Huber.

MacDonald, E. A.: Ocular lesions caused by intracranial hemorrhage, Trans. Amer. Ophthal. Soc. 29:418, 1931.

Maltby, G. L.: Visual field changes and subdural hematomas, Surg. Gynec. Obstet. 74: 496-498, 1942.

McKissock, W., Richardson, A., and Bloom, W. H.: Subdural haematoma, a review of 389 cases, Lancet 1:1365-1369, 1960.

Metzger, O., and Philippides, D.: Étude d'un syndrome de Parinaud au cours de l'évolution d'un hématome sous-dural chronique, Rev. Otoneuroophtal. 20:377-378, 1948.

Poppe, J. L., and Strain, R. E.: Chronic subdural hematomas, Surg. Clin. N. Amer. 32:791-799, 1952.

de Quervain, F.: Die starre Pupillenerweiterung in der Diagnostik der Schädel- und Hirntraumen, Schweiz. Med. Wschr. 1:75, 1935.

Schörcher, F.: Über die Ursache der einseitigen Pupillenerweiterung beim epi- und subduralen Hämatom, Deutsch. Z. Chir. 248:420, 1937.

Sunderland, S., and Bradley, K. C.: Disturbances of oculomotor function accompanying extradural haemorrhage, J. Neurol. Neurosurg. Psychiat. 16:35-46, 1953.

Voris, H. C.: Diagnosis and treatment of subdural hematomas, Surgery 10:447-456, 1941.

Walsh, F. B., and Hoyt, W. F.: Subdural hematoma. In Clinical neuro-ophthalmology, Baltimore, 1969, The Williams & Wilkins Co., p. 2413.

Weber, G.: Über subdurale Empyeme, Schweiz. Med. Wschr. 80:51, 1349, 1950.

Weber, G., Heyser, J., Rosenmund, H., and Duckert, F.: Subdurale Hämatome, Schweiz. Med. Wschr. 94:541-548, 1964.

Welte, E.: Zur formalen Genese der traumatischen Mydriasis: Oculomotoriusschädigung durch einseitiges Vorquellen des Uncus hippocampi, Zbl. Neurochir. 8:217, 1943.

Woodhall, B., Devine, J. W., and Hart, D.: Homolateral dilatation of the pupil, homolateral paresis and bilateral muscular rigidity in the diagnosis of extradural hemorrhage, Surg. Gynec. Obstet. 72:391, 1941.

Brain abscesses

Bailey, P.: Hirnabscess. Die Hirngeschwülste, Stuttgart, 1951, Ferdinand Enke, p. 349.

Bucy, P. E., and Weaver, T. A.: Jr.: Paralysis of conjugate lateral movement of the eyes in association with cerebellar abscess, Arch. Surg. 42:839-849, 1941.

Colemann, C. C.: Brain abscess: a review of twenty-eight cases with comment on ophthalmologic observations, J.A.M.A. 95:568, 1940.

Cowan, A.: Ophthalmic symptoms in brain abscess, Ann. Surg. 101:56-63, 1935.

Evans, W.: The pathology and etiology of brain abscess, Lancet 1:1231-1235, 1931.

Grant, F. C.: Brain abscess; collective review, Int. Abstr. Surg. 72:118-138, 1941.

Gros, C., and Cazaban, R.: Stase papillaire contro-latérale et papillite homolatérale: complication d'um abcès du cerveau d'origine fronto-sinusienne, Montpellier Med. 37:134-137, 1950.

Krayenbühl, H., and Weber, G.: Zur Behandlung und Diagnose akuter Hirnabszesse und cerebraler Thrombophlebitiden, Acta Neurochir. 2:281, 1952.

Lillie, W. I.: The clinical significance of choked disks produced by abscess of the brain, Surg. Gynec. Obstet. 47:405-406, 1928.

Lundberg, N.: Diagnostik und Behandlung der Hirnabscesse: Klinische Übersicht, Acta Otolaryng. 29:36-55, 1941.

Portmann, G., Pouyanne, H., Leman, P. M., and Mesnage, S. J.: Troubles oto-neuro-ophtalmologiques par abcès du cerveau de diagnostic étiologique difficile, Rev. Otoneuroophtal. 24:138-141, 1952.

Porto, G.: Papilledema in encephalic abscesses. Paper read before the Association of Medicine of the Penido Burnier Institute, 1941.

Raimondo, N.: I sintomi oculari degli acessi endocranici. Studio su 100 casi, Minerva Oftal. 2:114-117, 1960.

Simonetta, B.: Osservazioni sulla terapia degli ascessi encefalici, Riv. Otoneurooftal. 26:1-29, 1951.

Weber, G.: Hirnabszesse im Rahmen infektiöser intrakranieller Erkrankungen, Stuttgart, 1956, Georg Thieme.

Weber, G.: Der Hirnabszess, Stuttgart, 1957, Georg Thieme.

Intracranial aneurysms (including subarachnoidal hemorrhage)

Abrahamson, L. A., and Bell, L. B.: Carotid cavernous fistula syndrome, Amer. J. Ophthal. 39:521-526, 1955.

Alpers, B. J., and Schlezinger, N. S.: Aneurysm of the posterior communicating artery, Arch. Ophthal. 42:353-364, 1949.

Amyot, R.: Céphalée orbito-frontale et ophtalmoplégie par anévrisme carotidien intracrânien, Union Med. Canada 79:756-761, 1950.

Bailey, P.: Aneurysma. In Die Hirngeschwülste, Stuttgart, 1951, Ferdinand Enke, p. 354.

Ballantyne, A. J.: The ocular manifestations of spontaneous subarachnoid hemorrhage, Brit. J. Ophthal. 27:383, 1943.

Bergouignan, Pouyanne, Arne, and Leman: Ophtalmoplégies par anévrismes artériels de la base du crâne, Rev. Otoneuroophtal. 25:54-55, 1953.

Bonnet, P.: Les anévrismes artériels intra-crâniens, Paris, 1955, Masson & Cie.

Cardarello, G.: I sintomi oculari nell'emorragia subaracnoidea, Rass. Ital. Ottal. 16: 50-63, 1947.

Cardarello, G.: Ulteriore contributo allo studio dei sintomi oculari nelle emorragie subaracnoidee, Rass. Ital. Ottal. 18:182-188, 1949.

Chavanne, H., and Devic, M.: Valeur sémiologique des hémorragies rétiniennes au cours de processus anévrismaux intracrâniens, Bull. Soc. Franc. Ophtal. 63:39-49, 1950.

Cushing, H. W., and Bailey, P.: Tumors arising from the blood vessels of the brain, London, 1928, Baillière, Tindall & Cox.

Dandy, W. E.: Carotid-cavernous aneurysms (pulsating exophthalmus), Zbl. Neurochir. 2:77, 1937.

Dandy, W. E.: Arteriovenous aneurysms of the brain, Arch. Surg. 17:190-243, 1938.

Dandy, W. E.: Intracranial arterial aneurysm, Ithaca, N. Y., 1944, Comstock Publishing Co.

Dandy, W. E.: Carotid-cavernous aneurysms (pulsating exophthalmos). In Lewis, Dean, editor: Lewis' practice of surgery, Hagerstown, Md., 1936, W. F. Prior Co., Inc., vol. 3, pp. 319-323.

Dandy, W. E., and Follis, R. H.: On the pathology of carotid-cavernous aneurysms (pulsating exophthalmos), Amer. J. Ophthal. 24:365-385, 1941.

Drews, L. C., and Minckler, J.: Massive bilateral preretinal type of hemorrhage associated with subarachnoidal hemorrhage of brain (with case report and pathologic findings), Amer. J. Ophthal. 27:1-15, 1944.

Duke-Elder, S.: Textbook of ophthalmology, St. Louis, 1949, The C. V. Mosby Co., vol. 4, p. 3535.

Ectros, L.: Contribution à l'étude des anévrismes intracrâniens de la carotide et de ses branches, Rev. Otoneuroophtal. 21:259-285, 1949.

Enoksson, P., and Brynke, H.: Visual field defects in arteriovenous aneurysms of the brain, Acta Ophthal. 36:586-600, 1958.

Feld, M., and Taptas, J.: L'ophtalmoplégie dans les anévrismes artériels intracrâniens, Rev. Otoneuroophtal. 20:244-249, 1948.

Frankel, K.: Relation of migraine to cerebral aneurysm, Arch. Neurol. Psychiat. 63:195-204, 1950.

Gerlach, J., Spuler, H., and Viehweger, G.: Ueber das Aneurysma der Orbita, Klin. Mbl. Augenheilk. 140:344-356, 1962.

Globus, J. H., and Schwab, J. M.: Intracranial aneurysms: their origin and clinical behavior in a series of verified cases, J. Mt. Sinai Hosp. 8:547-578, 1942.

Griffith, J. Q., Jeffers, W., and Fry, W. E.: Papilledema associated with subarachnoidal haemorrhage: experimental and clinical study, Arch. Intern. Med. 61:880, 1938.

Hamby, W. B.: Intracranial aneurysms: a general survey, Practitioner 162:313-320, 1949.

Hanafee, W., and Jannetta, P. J.: Aneurysm as a cause of stroke, Amer. J. Roentgen. 98:647-652, 1966.

Heppner, F., and Lechner, H.: Diagnostische Probleme bei einem gleichzeitigen Vorkommen von Hirnaneurysma und Hirntumor, Zbl. Neurochir. 13:269-275, 1953.

Holmes, J. M.: The ocular symptoms of intracranial aneurysm, Trans. Ophthal. Soc. U. K. 74:549, 1954.

Huber, A.: Cerebrale und retinale Angiopathie, Ophthalmologica 117:265, 1949.

Huber, A.: Les anévrismes de l'orbite et du sinus caverneux, Bull. Soc. Franc. Ophtal. 64:347-355, 1951.

Huber, A.: Arteriography and phlebography in the diagnosis of orbital affections, Bull. N. Y. Acad. Med. 44:409-430, 1968.

Ikui, H., Inomata, H., and Hayashi, J.: The pathogenesis of optic nerve sheath hemorrhage, J. Ophthal. 11:67-78, 1967.

Insausti, E., and Matera, R. F.: Sintomas oftalmologicos de los aneurismas intracraneanos, Arch. Oftal. B. Air. 28:105-127, 1953.

Irish, C. W.: Aneurysms of the cerebral vessels: with a study of 32 cases found at 12,503 consecutive necropsies, Ann Arbor, Mich., 1940, Edwards Brothers, Inc.

Jaeger, R.: Aneurysm of the intracranial carotid artery: syndrome of frontal headache with oculomotor nerve paralysis, J.A.M.A. 142:304, 1950.

Jamieson, K. G.: Aneurysms of the vertebrobasilar system: surgical intervention in 19 cases, J. Neurosurg. 21:781-797, 1964.

Jefferson, G.: Compression of the chiasma, optic nerves and optic tracts by intracranial aneurysms, Brain 60:444, 1937.

Jefferson, G.: On the saccular aneurysms of the internal carotid artery in the cavernous sinus, Brit. J. Surg. 26:267, 1938.

Jefferson, G.: Isolated oculomotor palsy caused by intracranial aneurysm, Proc. Roy. Soc. Med. 40:419-432, 1947.

Jefferson, G.: Chiasmal lesions produced by intracranial aneurysms, Arch. Neurol. Psychiat. 72:11, 1954.

Kaeser, E. H.: Das arteriovenöse Aneurysma im Sinus cavernosus, Ophthalmologica 126:257, 1953.

Klingler, M.: Compression des nerfs et du chiasma optique par des anévrismes, Confin. Neurol. 11:261-270, 1951.

Krayenbühl, H.: Das Hirnaneurysma, Schweiz. Arch. Neurol. Psychiat. **47**:155, 1941.

Krayenbühl, H., and Richter, H.: Die cerebrale Angiographie, Stuttgart, 1952, Georg Thieme.

Krayenbühl, H., and Yasargil, M. G.: Die vasculären Erkrankungen im Gebiet der Arteria vertebralis und Arteria basialis, eine anatomische und pathologische, klinische und neuroradiologische Studie, Stuttgart, 1957, Georg Thieme, p. 170.

Krayenbühl, H., and Yasargil, M. G.: Das Hirnaneurysma, Basel, 1958, J. R. Geigy.

Krayenbühl, H., and Yasargil, M. G.: Die Zerebrale Angiographie, Stuttgart, 1965, Georg Thieme.

Ley, A.: Compression of the optic nerve by a fusiform aneurysm of the carotid artery, J. Neurol. Neurosurg. Psychiat. **13**:75-86, 1950.

Lutz, A.: Ueber einige weitere Fälle von binasaler Hemianopsie, Graefe. Arch. Ophthal. **125**:103-124, 1930-1931.

MacDonald, A. E.: Ocular lesions caused by intracranial haemorrhage, Trans. Amer. Ophthal. Soc. **29**:418, 1931.

MacDonald, A. E., and Korb, M.: Intracranial aneurysms (reference bibliography of 1125 cases), Arch. Neurol. Psychiat. **42**:298-328, 1939.

Manschot, W. A.: Subarachnoid hemorrhage; intraocular symptoms and their pathogenesis, Amer. J. Ophthal. **38**:501, 1954.

Manschot, W. A., and Hampe, J. F.: The origin of ocular symptoms in spontaneous subarachnoid haemorrhage, Proceedings of the Sixteenth International Congress on Ophthalmology, London, 1950, vol. 1, pp. 356-368.

Martin, J. D., and Mabon, F. R.: Pulsating exophthalmos: review of all reported cases, J.A.M.A. **121**:330-335, 1943.

Mason, T. H., Swain, G. M., and Osheroff, H. R.: Bilateral carotid cavernous fistula, J. Neurosurg. **11**:323-326, 1954.

Meadows, S. P.: Aneurysms of the internal carotid artery, Trans. Ophthal. Soc. U. K. **69**:137-155, 1949.

Meves, H.: Zur Genese der Augenhintergrundshämorrhagien bei subarachnoidalen Blutungen, Klin. Mbl. Augenheilk. **106**:339-342, 1941.

Milletti, M.: Gli aneurismi dei vasi cerebrali. Diagnostica clinica a terapia chirurgica, Arch. Neurochir. **1**:433-656, 1952.

Moniz, E.: Die cerebrale Arteriographie und Phlebographie. Ergänzungsband zum Handbuch der Neurologie II, Berlin, 1940, Julius Springer.

Oberle, A.: Über irreführende Symptome bei Aneurysmen der basalen Hirnarterien, Dissertation, Zürich, 1951.

Olivecrona, H., and Ladenheim, J.: Congenital arteriovenous aneurysms of the carotid and vertebral arterial systems, Berlin, 1957, Julius Springer.

Ott, T., and Cuendet, J. F.: Hémorragies sous-arachnoïdienne et rétinienne par rupture d'anévrisme artério-veineux intracrânien, Confin. Neurol. **13**:3, 170-174, 1953.

Paton, L.: Ocular symptoms in subarachnoid haemorrhage, Trans. Ophthal. Soc. U. K. **44**:110-126, 1924.

Patrikios, J. S.: Ophthalmoplégie et Migraine, Arch. Neurol. Psychiat. **63**:902-917, 1950.

Penholz, H., and Mildbraed, I.: Über die Augensymptome intrakranieller arterieller und arteriovenöser Aneurysmen und ihre Behandlung, Klin. Mbl. Augenheilk. **124**:179, 1954.

Peter, R., Van'Sek, J., and Króo, M.: Eye symptomatology in intracranial. aneurysms (in Czech.), Čsl. Ofthal. Prague **9**:72-84, 1953.

Pfingst, A. O., and Spurling, R. G.: Intracranial aneurysms: their role in the production of ocular palsies, Arch. Ophthal. **2**:391-398, 1929.

Pool, L., and Potts, D. G.: Aneurysms and arteriovenous anomalies of the brain: diagnosis and treatment, New York, 1965, Harper & Row Publishers, Inc.

Rhonheimer, C.: Zur Symptomatologie der sellären Aneurysmen; ein Beitrag zur Differentialdiagnose der Chiasmasyndrome, Klin. Mbl. Augenheilk. **134**:1-34, 1959.

Richardson, J. C., and Hyland, H. H.: Intracranial aneurysms: a clinical and pathological study of subarachnoid and intracerebral hemorrhage caused by berry aneurysms, Medicine **20**:1-83, 1941.

Riddoch, G., and Goulden, C.: On the relationship between subarachnoid and intraocular haemorrhage, Brit. J. Ophthal. **9**:209-223, 1925.

Sands, I. J.: Diagnosis and management of subarachnoid hemorrhage, Arch. Neurol. Psychiat. **46**:973, 1941.

Schiefer, W., and Marguth, F.: Intraselläre Aneurysmen, Acta. Neurochir. **4**:344-354, 1956.

Sjöqvist, O.: Über intrakranielle Aneurysmen der A. carotis und deren Beziehungen zur ophthalmoplegischen Migräne, Nervenarzt **9**:233-241, 1936.

Smith, D. C., Kearns, T. P., and Sayre, G. P.: Preretinal and optic nerve sheath hemorrhage: pathologic and experimental aspects in subarachnoid hemorrhage, Trans. Amer. Acad. Ophthal. Otolaryng. **61**:201-211, 1967.

Sugar, H. S., and Meyer, S. J.: Pulsating exophthalmos, Arch. Ophthal. **23**:1288-1321, 1940.

Thiébaut, F., and Matavulj, N.: Troubles oculaires dans quelques cas d'angiome artério-veineux du cerveau, Bull. Soc. Franc. Ophtal. 74:759-770, 1961.

Thomas, M. H., and Petrohelos, M. A.: Diagnostic significance of retinal artery pressure in internal carotid involvement, Amer. J. Ophthal. 36:335, 1953.

Unger, H. H., and Umbach, W.: Kongenitales okulozerebrales Rankenangiom, Klin. Mbl. Augenheilk. 148:672-682, 1966.

Wagener, H. P., and Foster, R. F.: Ruptured intracranial aneurysm with hemorrhages into retina and vitreous, Mayo Clin. Proc. 10:225-229, 1935.

Walsh, F. B.: Diagnosis, localization and treatment of intracranial saccular aneurysms, Arch. Neurol. Psychiat. 44:671, 1940.

Walsh, F. B.: Visual field defects due to aneurysms at the circle of Willis, Arch. Ophthal. 71:15, 1964.

Walsh, F. B., and Hedges, T. R., Jr.: Optic nerve sheath hemorrhage, Trans. Amer. Acad. Ophthal. Otolaryng. 55:29-48, 1950; Amer. J. Ophthal. 34:509-527, 1951.

Walsh, F. B., and Hoyt, W. F.: Arteriovenous aneurysm. In Clinical neuro-ophthalmology, Baltimore, 1969, The Williams & Wilkins Co., p. 1690.

Walsh, F. B., and Hoyt, W. F.: Aneurysms of the cerebral arteries. In Clinical neuro-ophthalmology, Baltimore, 1969, The Williams & Wilkins Co., p. 1737.

Walsh, F. B., and Hoyt, W. F.: Carotid-cavernous fistula. In Clinical neuro-ophthalmology, Baltimore, 1969, The Williams & Wilkins Co., p. 1714.

Walsh, F. B., and Hoyt, W. F.: Subarachnoid and intracerebral hemorrhage. In Clinical neuro-ophthalmology, Baltimore, 1969, The Williams & Wilkins Co., p. 1782.

Walsh, F. B., and King, A. B.: Ocular signs of intracranial saccular aneurysms, Arch. Ophthal. 27:1, 1942.

Walsh, F. B., and King, A. B.: Ocular signs of intracranial saccular aneurysms: experimental work on collateral circulation through the ophthalmic artery, Arch. Ophthal. 27:1-33, 1942.

Weber, G.: Zur Diagnose und Behandlung der arterio-venösen Aneurysmen im Bereich der Grosshirnhemisphären, Schweiz. med. Wschr. 78:629-634, 1948.

Weber, G.: Das intrazerebrale Hämatom, Schweiz. Arch. Neurol. Psychiat. 91:510-552, 1963.

White, J. C., and Ballantine, H. T., Jr.: Intrasellar aneurysm simulating hypophyseal tumors, J. Neurosurg. 18:34-50, 1961.

Wyburn-Mason, R.: Arteriovenous aneurysm of mid-brain and retina, facial naevi and mental changes, Brain 66:163-203, 1943.

Yasargil, G. M.: Die Röntgendiagnostik des Exophthalmus unilateralis, Basel, 1957, S. Karger AG.

Ziedses des Plantes, B. G.: Subtraktion, Stuttgart, 1961, Georg Thieme.

Zielinski, H. W.: Augenhintergrundsveränderungen bei intracraniellen Aneurysmen und arteriovenösen Angiomen, Ber. Deutsch. Ophth. Ges. 59:48-52, 1955.

Index

Boldface page numbers indicate the main discussions of the particular subject.